**Nonwoven Fabrics**

*Edited by*
*Wilhelm Albrecht*
*Hilmar Fuchs*
*Walter Kittelmann*

*Books also of interest:*


Urban, D., Takamura, K.

**Polymer Dispersions and Their Industrial Applications**

2002, ISBN 3-527-30286-7


Gierenz, G., Karmann, W.

**Adhesives and Adhesive Tapes**

2001, ISBN 3-527-30110-0


Fakirov, S.

**Handbook of Thermoplastic Polyesters**
Homopolymers, Copolymers, Blends and Composites

2001, ISBN 3-527-30113-5


Wilkes, E. S.

**Industrial Polymers Handbook**
Products, Processes, Applications
4 Volumes

2000, ISBN 3-527-30260-3

# Nonwoven Fabrics

*Edited by*
*Wilhelm Albrecht*
*Hilmar Fuchs*
*Walter Kittelmann*

Editors

**Prof. Dr. Wilhelm Albrecht**
Dr.-Tigges-Weg 39
42115 Wuppertal
Germany

**Prof. Dr.-Ing. Hilmar Fuchs**
Sächsisches Textilforschungsinstitut e.V.
Annaberger Straße 240
09125 Chemnitz
Germany

**Dr.-Ing. Walter Kittelmann**
Sächsisches Textilforschungsinstitut e.V.
Annaberger Straße 240
09125 Chemnitz
Germany

■ This book was carefully produced. Nevertheless, editors, authors and publisher do not warrant the information contained therein to be free of errors. Readers are advised to keep in mind that statements, data, illustrations, procedural details or other items may inadvertently be inaccurate.

**Library of Congress Card No.:** applied for

**British Library Cataloguing-in-Publication Data**
A catalogue record for this book is available from the British Library.

**Bibliographic information published by Die Deutsche Bibliothek**
Die Deutsche Bibliothek lists this publication in the Deutsche Nationalbibliografie; detailed bibliographic data is available in the Internet at <http://dnb.ddb.de>

© 2003 WILEY-VCH Verlag GmbH & Co. KGaA, Weinheim

Printed in the Federal Republic of Germany
Printed on acid-free paper

**Composition**  K+V Fotosatz GmbH, Beerfelden
**Printing**  Strauss Offsetdruck GmbH, Mörlenbach
**Bookbinding**  Litges & Dopf Buchbinderei GmbH, Heppenheim

**ISBN**  3-527-30406-1

# Foreword

When in 1981 the world's first authentic and complete handbook on Nonwovens was published, the authors Albrecht and Lünenschloß already noted that these initially relatively simple substitution products had become an independent and technically sophisticated industry in its own right. Nonwovens owed their growth to an unusual multiplicity of raw materials and process options.

Since than 20 years have passed. Sales, distribution and diversity of an originally young and modest industry, whose focal points were clearly in Europe and USA, have multiplied. Experts expect a worldwide production of approx. 3.3 mio. tons at a market value of approx. US$ 14.6 billion in the year 2000. This means more than 5% of conventional textile production will already be represented by Nonwovens. In numerous market segments, Nonwovens already play a leading role. In certain areas they have assumed genuinely novel functions – for example in textiles for personal and medical care.

Without doubt, the Nonwovens industry has also suffered during the course of its 50-years maturing process. Several markets are not longer growing or do not allow economically acceptable returns on investment any more. In a number of regions and market segments undiscriminating investments and the availability of turn-key technology have done severe harm to the industry.

For the qualified and responsible producer, however, the Nonwovens industry continues to offer endless new challenges and opportunities. Not many other fields of endeavour offer such creative diversity of raw materials and processes as well as a limitless variety of finishing and application possibilities in order to fulfil customer demands with tailor-made solutions. In this respect innovative Nonwovens producers can have confidence in a successful future.

In this spirit I welcome the new up-dated and extended version of the Nonwovens handbook and wish both, the authors and the publishers, the success they deserve. We all shall stand to benefit.

H. N. Dahlström

# Preface

Twenty years ago, the reference book "Nonwovens" was kindly received by all concerned with textile manufacturing. In that book, more than 20 authors described in detail the raw materials, their processing into a wide range of nonwoven products, the characteristics of the products and the testing techniques then in use. "Nonwovens" was much asked for in industry, education and, with regard to new products, in R&D. Meanwhile, the quantity of nonwovens made worldwide has grown, the range of goods based on nonwovens is much wider, the technical equipment, the raw materials as well as the auxiliaries used have been further developed. Therefore, the idea did not come as a surprise to revise the book. This meant to find a team of authors fully conversant with the current state and the quantitative and qualitative developments going on in a field of industry which is – as hardly any other industry – run on a worldwide scale. A sophisticated project like this called for specialist co-ordination, which was provided by Sächsisches Textilforschungsinstitut in Chemnitz, a research institute preferably dealing with questions of nonwoven production and innovation in the field of nonwovens. This institute works closely together with companies that make or process nonwovens as well as with the suppliers of fibres, the manufacturers of the relevant equipment and the producers of auxiliaries, which has been very helpful.

Today, the nonwoven-producing industry is best characterized as an industry that has accomplished a rise in product quality which the user can see and feel. Its range of products has become ever larger. This has been achieved by creative work and successful co-operation with the suppliers of raw and auxiliary materials as well as the manufacturers of equipment. Based on this co-operation, there are good prospects for novel products coming. Future developments will, in the widest sense of the word, continue to focus on best-possible functionality and lowest-possible consumption of resources. To this end, it will be important all concerned work together even more closely. The editing team give their views of ways to go and aims to reach in the future in the last Chapter of this book headlined "outlook", thus outlining the potential which is still waiting to be exploited. This book is supposed to contribute to developing the nonwoven-producing industry.

We have been lucky one of the editing team has been in a position to do all the work in detail required to prepare this reference book. This meant spending much time talking to the authors of the single Chapters and co-ordinating them.

More help was provided by Wiley VCH Publishers, who will, except for the German edition, publish the book in English and Chinese, too. We are grateful to Dr. Böck for comprehensive advice. Thanks to her efforts, the book includes advertisement which will help the nonwoven-producing industry to deliver best quality and to develop new products. Our thanks go to all authors and those who have contributed in whatever way.

Today, nonwovens are part of what is known as the world of textiles. Due to their tailor-made characteristics, they are highly suitable to meet a wide diversity of requirements. Thus, nonwovens are more than products which are up-to-date. They give evidence that it is possible to master the challenge of the future.

We hope the reader can make good use of this book.

Wilhelm Albrecht
Hilmar Fuchs
Walter Kittelmann

# Contents

# List of editors and authors

## Editors

Prof. Dr. Wilhelm Albrecht
Dr. Tigges-Weg 39
42115 Wuppertal
Germany

Prof. Dr.-Ing. Hilmar Fuchs
Sächsisches Textilforschungsinstitut e.V.
P.O. Box 1325
09072 Chemnitz
Germany

Dr.-Ing. Walter Kittelmann
H.-Greif-Straße 35
01217 Dresden
Germany

## Authors

Dr.-Ing. Klaus Afflerbach
Voith Paper GmbH & Co. KG
Ravensburg
Betriebsstätte Düren
Veldener Straße 52
53349 Düren
Germany

Dipl.-Ing. Hans-Claus Assent
Albert-Ludwig-Grimm-Straße 18
69469 Weinheim
Germany

Lutz Bergmann
P.O. Box 2189
La Grange, GA 30241-2189
USA

Dipl.-Ing. Siegfried Bernhardt
Spinnbau GmbH
P.O. Box 71 03 60
28763 Bremen
Germany

Dipl.-Ing. Uta Bernstein
Sächsisches Textilforschungsinstitut e.V.
P.O. Box 1325
09072 Chemnitz
Germany

Dr. Walther Best
Thomas Joseph Heimbach GmbH & Co.
Gut Nazareth
52353 Düren
Germany

Dipl.-Ing. Bettina Bieber
Sächsisches Textilforschungsinstitut e.V.
P.O. Box 1325
09072 Chemnitz
Germany

Dipl.-Ing. Dieter Blechschmidt
Sächsisches Textilforschungsinstitut e.V.
P.O. Box 1325
09072 Chemnitz
Germany

Dr.-Ing. Peter Böttcher
Sächsisches Textilforschungsinstitut e.V.
P.O. Box 1325
09072 Chemnitz
Germany

Dipl.-Ing. Margot Brodtka
Sächsisches Textilforschungsinstitut e.V.
P.O. Box 1325
09072 Chemnitz
Germany

H. Norbert Dahlström
Freudenberg & Co.
P.O. Box 10 03 63
69465 Weinheim
Germany

Dipl.-Ing. Johann Philipp Dilo
OSKAR DILO Maschinenfabrik KG
P.O. Box 1551
69405 Eberbach
Germany

Dr.-Ing. Waldemar Dohrn
BGB Stockhausen GmbH & Co. KG
P.O. Box 10 04 52
47705 Krefeld
Germany

Dr.-Ing. Peter Ehrler
Bergstraße 19
01069 Dresden
Germany

Dr. Ir. J. J. Frijlink
Berkenweg 18
3941 JB Doorn
The Netherlands

Dr. Rainer Gebhardt
Sächsisches Textilforschungsinstitut e.V.
P.O. Box 1325
09072 Chemnitz
Germany

Dipl.-Ing. Bernd Gulich
Sächsisches Textilforschungsinstitut e.V.
P.O. Box 1325
09072 Chemnitz
Germany

Dr.-Ing. Victor P. Gupta
OSKAR DILO Maschinenfabrik KG
P.O. Box 1551
69405 Eberbach
Germany

Dr. Rainer Gutmann
Institut für Chemiefasern der DITF
Körschtalstraße 26
75770 Denkendorf
Germany

Dipl.-Phys. Jürgen Haase
Sächsisches Textilforschungsinstitut e.V.
P.O. Box 1325
09072 Chemnitz
Germany

Dr. Jan Hoborn
SCA Mölniycke Clinical Products AB
Bakstensgatan 5
40503 Göteborg
Sweden

Prof. Ji Guobiao
China National Textile Council
12 East Chang An St.
100742 Beijing
China

Prof. Dr.-Ing. Radko Krčma †

Dr.-Ing. Peter Kunath
OSKAR DILO Maschinenfabrik KG
P.O. Box 1551
69405 Eberbach
Germany

Dipl.-Ing. Ferdinand Leifeld
Von-Behring-Straße 34
47906 Kempen
Germany

Dipl.-Ing. Catrin Lewicki
Sächsisches Textilforschungsinstitut e.V.
P.O. Box 1325
09072 Chemnitz
Germany

Prof. Dr.-Ing. Klaus Lieberenz
Oberonstraße 8 d
01259 Dresden
Germany

Dr.-Ing. Matthias Mägel
Sächsisches Textilforschungsinstitut e.V.
P.O. Box 1325
09072 Chemnitz
Germany

Dipl.-Ing. Sabine Martini
Kirchplatz 24
47918 Tönisvorst
Germany

Guy Massenaux
26, rue V. Greyson
1050 Brussels
Belgium

Dipl.-Chem. Wolfgang Möschler
Rosentor 18
09126 Chemnitz
Germany

Dr.-Ing. Ullrich Münstermann
FLEISSNER GmbH & Co.
Maschinenfabrik
Wolfsgartenstraße 6
63329 Egelsbach
Germany

Dipl.-Ing. Ansgar Paschen
Institut für Textiltechnik
der RWTH Aachen
Eilfschornsteinstraße 18
52062 Aachen
Germany

Dr. Thomas Pfüller
Sächsisches Textilforschungsinstitut e.V.
P.O. Box 1325
09072 Chemnitz
Germany

Helmuth Pill
PILL Naßvliestechnik
Heilbronner Straße 274
72760 Reutlingen
Germany

Dipl.-Ing. Norbert Ritter
FLEISSNER GmbH & Co.
Maschinenfabrik
Wolfsgartenstraße 6
63329 Egelsbach
Germany

Prof. Dr. Hartmut Rödel
Institut für Textil- und
Bekleidungstechnik der TU Dresden
01062 Dresden
Germany

Dipl.-Ing. Manfred Sauer-Kunze
GEA Delbag – Lufttechnik GmbH
Südstraße 48
44625 Herne
Germany

Dipl.-Ing. Manfred Schäffler
AUTEFA Maschinenfabrik GmbH
Röntgenstraße 1–5
86316 Friedberg
Germany

Dipl.-Chem. Wolfgang Schilde
Sächsisches Textilforschungsinstitut e.V.
P.O. Box 1325
09072 Chemnitz
Germany

Dipl.-Ing. Elke Schmalz
Sächsisches Textilforschungsinstitut e.V.
P.O. Box 1325
09072 Chemnitz
Germany

Dipl.-Ing. Gunter Schmidt
Finkenweg 22
79312 Emmendingen
Germany

Dipl.-Ing. Jochen Schreiber
Sächsisches Textilforschungsinstitut e.V.
P.O. Box 1325
09072 Chemnitz
Germany

J. M. Slovaček
Colbond Nonwovens
73 Westervoortsedijk
6800 TC Arnheim
The Netherlands

Dr. Jürgen Spindler
EMS-CHEMIE AG
7013 Domat-Ems
Switzerland

Wolfgang Stein
Strassberger Straße 6
04288 Leipzig
Germany

Prof. Dr. Michael Stoll
Forschungsinstitut für Leder und
Kunstledertechnologie gGmbH
P.O. Box 11 44
09581 Freiberg
Germany

Karl-Heinz Stukenbrock
Panoramaweg 25
41334 Nettetal
Germany

Dipl.-Ing. Alfred Watzl
FLEISSNER GmbH & Co.
Maschinenfabrik
Wolfsgartenstraße 6
63329 Egelsbach
Germany

Dipl.-Ing. Alexander Wegner
KARL MAYER Malimo
Textilmaschinenfabrik GmbH
P.O. Box 713
09007 Chemnitz
Germany

Dr. Jochen Wirsching
FREUDENBERG
Haushaltprodukte KG
Corporate Technical Centre
Zwischen Dämmen, Bau 148
69465 Weinheim
Germany

Prof. Dr.-Ing. Burkhard Wulfhorst
Rolandstraße 37 A
52223 Stollberg
Germany

Dipl.-Ing. Walter Zäh
KARL MAYER
Textilmaschinenfabrik GmbH
Brühl 25
63179 Obertshausen
Germany

## This book was translated by:

| Chapter/Section | Translator |
| --- | --- |
| 0 | Guy Massenaux |
| 1.1 to 1.3 | Irene Mühldorf (Team Coordinator) |
| 1.2 | Ji Guobiao |
| 1.4 | Harald Linke |
| 2.1 | Harald Linke |
| 2.2, 2.3 | Rainer Gutmann |
| 2.4 | Sabine Martini |
| 2.5 | Waldemar Dohrn |
| 3.1, 3.2 | George Noot |
| 3.3 | Jürgen Spindler |
| 4.1.1 | Annemaria Köhler |
| 4.1.2 | George Noot |
| 4.1.3 | Annemaria Köhler |
| 4.1.4 | Oldřich Jirsák |
| 4.2 | Harald Linke |
| 5 | George Noot |
| 6 | Andrea Wunder |
| 6.1 | Andrea Wunder |
| 6.2 | Harald Linke |
| 6.3, 6.4 | Marie-Luise Moldenhauer |
| 6.5 | George Noot |
| 6.6 | Harald Linke |
| 7, 8 | Alison Sanders |
| 9 | Hartmut Rödel |
| 10 | Alison Sanders |

| Chapter/Section | Translator |
| --- | --- |
| 11 | Jan Hoborn |
| 12 | Jochen Wirsching |
| 13.1 to 13.4 | George Noot |
| 13.5 | Patricia van der Horst |
| 14.1, 14.2 | Alison Sanders |
| 14.3 | Michael Stoll |
| 15.1 | Alison Sanders |
| 15.2 | Elke Schmalz, Lutz Bergmann |
| 15.3.1 | Jacqueline Borstnik Billingsley |
| 15.3.2 | J. J. Frijlink |
| 15.4 | Harald Linke |
| 15.5 | Jacqueline Borstnik Billingsley |
| 15.6 | Jacqueline Borstnik Billingsley |
| 16 | Harald Linke |
| 17, 18 | Jacqueline Borstnik Billingsley |
| 19 | Marie-Luise Moldenhauer |
| 20 | Harald Linke |

# 0
# Introduction to nonwovens

G. MASSENAUX

Whilst the first production of a "nonwoven fabric" in Europe goes back to the thirties, the existence of a recognizable nonwovens industry in Europe can be dated to the mid-sixties.

This is also when the terms nonwovens or *Vliesstoffe* got used, preferably to others, in a small circle of manufacturers and converters.

Since then, the manufacture of nonwovens has expanded rapidly and the use of such products has penetrated many aspects of industry and of private life. Nonwovens are found in hygiene and health care, in rooting and civil engineering, household and automotive, in cleaning, filtration, clothing, food wrap and packaging, to name only a few end-uses. Confusion or ignorance about nonwovens remains large though. The present book comes out therefore at the right time to give a comprehensive view of what is to be understood by nonwovens, their manufacturing process, applications and possibilities. Presenting in a coherent way the present state of the art of nonwovens manufacturing and end-uses will be an invaluable help to all those within the industry or outside of it who deal with nonwovens or might get the opportunity to do so.

It is to be hoped that this book will also have a seminal influence in attracting young talents to this growing industry, where so much still is needed in order to further develop machinery, raw materials and properties of nonwovens to their best use.

## 0.1
## Definition of nonwovens

The term used to designate the products generally known as nonwovens, was coined in most languages in opposition to woven fabrics, which implicitly were taken as a reference. A nonwoven was something that was not woven.

Even the German name *"Vliesstoffe"* wasn't clear either as it could be confused with ceramical material and in any case remained ambiguous in its unusual spelling. Only specialists know that nonwovens are unique engineered fabrics which offer cost effective solutions as e.g. in hygiene convenience items, or as battery separators, or filters, or geotextiles, etc.

There is a formal definition of Nonwoven: ISO 9092[1] which bas been adopted by CEN (EN 29092) and consequently by DIN or AFNOR or any standardization office in the EU. Various legal or regulatory implications derive from it[2].

As a main characteristic the CEN definition indicates that a nonwoven is a fabric made of fibres, that is consolidated in different ways. Nonwoven fabrics are made out of fibres, without any restriction, but not necessarily from fibres. These can be very short fibres of a few millimetres length as in the wetlaid process; these can be "ordinary" fibres, as used in the traditional textile industry, or then very long filaments etc. Properties and characteristics of a nonwoven fabric depend for a large part from the type of fibre it is ultimately made of. These fibres can be natural or man-made, organic or inorganic; the characteristic of a fibre being that it is longer than its thickness, or diameter. Such fibres can also be produced continuously in connection with the nonwoven process itself and then cut to length, or then extruded directly e.g. from polymer granules into a filament and then fibrous structure.

To make good measure the ISO definition also excludes various types of fabrics to which, voluntarily or not, one might compare nonwovens. Nonwovens are not paper and indeed, when made out of very short, cellulose fibres, they essentially differ from paper because there aren't any, or hardly, hydrogen bonds linking such fibres together[3].

Nonwovens, as indicated by their English or French name, are neither woven fabrics, nor such other textiles as knitted fabrics. Behind these statements lies a fundamental characteristic of nonwoven: contrary to woven or knitted fabrics, fibres that ultimately make up the nonwoven fabric need not to go through the preparatory/transitory stage of yarn spinning in order to be transformed into a web of a certain pattern.

Some will remark that other textile fabrics were created in the past besides the weaving and knitting process, e.g. felting (which is also yarnless) or more recently stitchbonding. For this reason as well – especially in the early days – some have tended to literally classify as nonwovens all textile fabrics that are outside the weaving/knitting domain. Matters have settled since then and the reflexions at

1) A manufactured sheet, web or batt of directionally or randomly orientated fibres, bonded by friction, and/or cohesion and/or adhesion, excluding paper and products which are woven, knitted, tufted, stitch-bonded incorporating binding yarns or filaments, or felted by wet-milling, whether or not additionally needled. The fibres may be of natural or man-made origin. They may be staple or continuous filaments or be formed *in situ*. (This definition is completed by various notes.)

2) The CEN nonwovens definition is adopted by EDANA, the European nonwovens industry association. INDA, the North American Association has a slightly different, wider definition which has the merit of apparent simplicity: a sheet, web or batt of natural and or man-made fibres or filaments excluding paper, that have not been converted into yarns and that are bonded together by any of several means (such means are then listed).

3) There remain marginal cases with respect to paper or other fabrics which the ISO/CEN definition tries to deal with in its notes but we won't bother the reader with it. Like in nature there are some areas which can be contested between sea and terra firma and where the final accepted limit is somewhat arbitrary; or like in the plant/animal realm where the final distinction depends from the criteria that are finally adopted...

ISO and CEN helped clarify this. As far as textiles go, nonwovens are only part of a category of fabrics that exist besides weaving and knitting.

Nonwovens though go also beyond the limits of textiles. Fibres they ultimately are made of can be very short "unspinnable" ones like in the paper industry; the fibrous web can also originate from foils and other plastics. Nonwovens therefore share for a part manufacturing characteristics and properties with the paper industry or the chemicals/plastics industry to finally make a world of their own.

Nonwovens do not depend on the interlacing of yarn for internal cohesion. Intrinsically they have neither an organized geometrical structure. They are essentially the result of the relationship between one single fibre and another. This provides nonwoven fabrics with characteristics of their own, with new or better properties (absorption, filtration) and therefore opens them up to other applications.

## 0.2
## Nonwoven manufacturing processes

There are three main routes to web forming:

– the drylaid system with carding or airlaying as a way to form the web;
– the wetlaid system;
– the polymer-based system, which includes spunlaying (spunbonding) or specialized technologies like meltblown, or flashspun fabrics etc.

The lack of sufficient frictional forces however bas to be compensated for by the bonding of the fibres, which provides web strength. Consolidation of the web after its formation is the second step in the nonwoven manufacturing process.

This consolidation for a large part sets the final characteristics of the fabric and therefore, if possible, ought to be chosen with the end application in mind. Such consolidation can be done by use of chemical means (chemical bonding) like binders. These can be applied uniformly by impregnating, coating or spraying or intermittently, as in print bonding. The consolidation can also be reached by thermal means (cohesion bonding), like the partial fusion of the constituting fibres or filaments. Such fusion can be achieved e.g. by calendering or through-air blowing or by ultra-sonic impact.

Finally, consolidation can be achieved by mechanical means (frictional bonding), like needling, stitching, water-jet entangling or a combination of these various means.

Customers needs can be further met by modifying or adding to the existing properties of the fabric through finishing. A variety of chemical substances can be employed before or after bonding or various mechanical processes can be applied to the nonwoven in the final stage of the manufacturing process.

The choice of the raw material and the final constituting fibrous element, the depositing of the fibres as a fibrous material of a varying density, the choice of consolidating and finishing means, all this creates a series of parameters which can be played with in order to reach the required properties. This confirms what was indicated earlier that nonwovens are engineered fabrics par excellence. When ingredients, web formation and consolidation are chosen in order to best meet the characteristics needed at the end application, then for sure, we have a winner.

## 0.3
### Nonwoven properties and applications, including environmental considerations

Nonwovens are in fact products in their own right with their own characteristics and performances, but also weaknesses. They are around us and one uses them everyday, often without knowing it. Indeed they are frequently hidden from view. Nonwovens can be made absorbent, breathable, drapeable, flame resistant, heat sealable, light, lint-free, mouldable, soft, stable, stiff, tear resistant, water repellent, if needed. Obviously though, not all the properties mentioned can be combined in a single nonwoven, particularly those that are contradictory.

Their applications are multifold. Examples of their uses can be listed as follows:

- Personal care and hygiene as in baby diapers, feminine hygiene products, adult incontinence items, dry and wet pads, but also nursing pads or nasal strips.
- Healthcare, like operation drapes, gowns and packs, face masks, dressings and swabs, osteomy bag liners, etc.
- Clothing: interlinings, insulation and protection clothing, industrial workwear, chemical defence suits, shoe components, etc.
- Home: wipes and dusters, tea and coffee bags, fabric softeners, food wraps, filters, bed and table linen, etc.
- Automotive: boot liners, shelf trim, oil and cabin air filters, moulded bonnet liners, heat shields, airbags, tapes, decorative fabrics, etc.
- Construction: roofing and tile underlay, thermal and noise insulation, house wrap, underslating, drainage, etc.
- Geotextiles: asphalt overlay, soil stabilization, drainage, sedimentation and erosion control, etc.
- Filtration: air and gas, Hevac, Hepa, Ulpa filters
- Industrial: cable insulation, abrasives, reinforced plastics, battery separators, satellite dishes, artificial leather, air conditioning, coating.
- Agriculture, home furnishing, leisure and travel, school and office etc.

The origins of nonwovens are not glamorous. In fact, they resulted from recycling fibrous waste or second quality fibres left over from industrial processes like weaving or leather processing. They also resulted from raw materials restrictions e.g. during and after the Second World War or later in the communist dominated countries in Central Europe. This humble and cost dominated origin of course lead to some technical and marketing mistakes; it is also largely responsible for two still lingering misconceptions about nonwovens: they are assumed to be (cheap) substitutes; many also associate them with disposable products and for that reason did consider nonwovens as cheap, low quality, items.

There is nothing wrong in being a substitute; on the contrary if properties are similar and the cost and price lower, then the benefits to the user are obvious; there is better value for money. At the beginning the price differentials of nonwovens with regard to the products they did substitute was sometimes such that even some lessening of the properties still made the service acceptable. However, more often than none nonwovens turned out to be a substitute, not only with a cost advantage but with more and more additional or improved benefits to the user: think of interlinings, wipes, operation gowns, various air filters, etc. It is still to be hoped that the number of items (plastic, textile, paper, etc.) which nonwovens as flexible sheet structures should be able to substitute isn't finite yet and that the comparative cost and efficiency advantages of nonwovens remain. And of course, beyond substitution a part of nonwovens potential lies in their own creativity and increasing sophistication.

Not all nonwovens end in disposable applications. A large part of production is for durable end-uses, like in interlinings, roofing, geotextile, automotive or floor covering applications etc. However, many nonwovens especially light-weight ones are indeed used as disposable products or incorporated into disposable items. In our view this is the ultimate sign of efficiency. Disposability is only possible for cost-efficient products that concentrate on the essential required characteristics and performances and provide them without unnecessary frills.

Most nonwovens, disposables or not, are high-tech, functional items, e.g. with ultra-high absorbency or retention for wipes, or with softness, strike-through and no wetback properties for those used into hygiene articles, with outstanding barrier characteristics for medical applications in the operation room, or better filtration possibilities because of their pores dimension and distribution, etc. They weren't manufactured with the aim of disposability but in order to fulfil other requirements. They mainly became disposable because of the sectors they are used in (hygiene, healthcare) and of their cost efficiency. And disposability very often creates an additional benefit to the users. As disposable items have never been used before, there is then a guarantee that they do possess all the properties required as opposed to reused laundered fabrics.

At this point of the presentation, a ward maybe ought to be said about the environmental impact of the nonwovens industry and the waste management of its products. Even with over a million tonnes produced in 2000, the European nonwovens industry still remains a comparatively small industry (e.g. European textile industry 5 632,000 tons in 1998, paper & board industry 90 million tons). The nonwoven manufacturing process itself is modern and straight forward, without presently obnoxious air or water emission, including for chemically-bonded fabrics, which make out presently about 10% of the total nonwoven production. As a modern industry its record is at least comparable, or better, than the paper or textile industries.

The apparent solid waste of the nonwoven industry in Europe can be estimated at 110–115,000 tons, of which 25% are raw materials. This quantity is minute in comparison to the total waste in Europe resulting from industry. (Manufacturing waste: 17% of 2,200 million tons/year – OECD/Eurostat 1999.) At least 50% of the nonwovens industry waste is recycled. A growing part of the remaining waste is turned to energy through incineration, or then sent to landfill, etc. This, however, is often dependent of the national circumstances and regulations, which vary sharply throughout Europe.

Nonwoven waste can also result from the disposal of used nonwoven products (i.e. post consumer waste). Such waste and quantities depend of the life-cycle of the nonwoven products themselves.

Of the 1,025,000 tons nonwovens produced in Europe in 2000, one can estimate that about 640,000 tons will appear as disposables or part of disposable items in the municipal waste stream of the year. Such amount makes about 0,30% of the total estimated municipal solid waste (MSW) (if the total waste collection figures remain the same). The rest of the production will be slowly distilled into the post-consumer waste stream as it comes to maturity.

There again waste quantities of nonwovens are not only comparatively low with regard to the paper or textile waste, but the nonwoven products themselves don't create more intrinsic problems than for paper or textiles. As such, nonwoven waste can be handled safely (as far as the nonwoven part is concerned) and all waste management solutions can be applied, at least in theory.

Finally, one should not overlook the many environmental benefits which result from nonwovens use, e.g. in air and oil filtration, oil absorption, protective work-wear, geotextiles, agriculture etc.

## 0.4
## Development of the nonwovens industry

Nonwovens developed into an industry in the three main industrialized regions of the world, the USA, Western Europe and Japan, each of them contributing to the technological development of the nonwovens industry and of course fuelling its growth by new applications.

The categories where nonwovens are used in these main regions are broadly similar, although there remain sharp differences in consumers' expectation and needs. Coverstock types vary between Japan, the USA or Europe; medical items have a higher penetration rate in the States than in Europe; house-wrap is mainly a U.S. end-use; nonwoven interlinings and geotextiles developed in Europe before spreading worldwide, etc. Whilst in Western Europe nonwoven production amounted to about 63,300 tons in 1972, it had more than doubled within five years, and in 2000 reached 1,025,000 tons, growing by more than 10% in weight over the last year. This was achieved with a total manpower in the order of 16,000 people in 2000 which shows how highly capitalistic this industry is.

This production in Western Europe is achieved by about 130 companies. Despite some mergers taking place, and various companies becoming global players, the nonwovens industry at large remains an industry of medium to small companies, or of non-autonomous divisions or departments of larger groups.

The production per group of countries presently is as follows:

**Table 0-1** Production of nonwovens in Western Europe (1983–2000)

*in 1,000 tons*

| 1983 | 1985 | 1987 | 1989 | 1991 | 1993 | 1995 | 1996 | 1997 | 1998 | 1999 | 2000 |
|------|------|------|------|------|------|------|------|------|------|------|------|
| 231.4 | 272.1 | 338.2 | 414.0 | 480.6 | 554.5 | 646.4 | 684.4 | 759.5 | 836.0 | 909.8 | 1,025.9 |

Source: EDANA – European Nonwovens and Disposables Association.
Copyright: EDANA 2001.

**Table 0-2** Nonwovens production by group of European countries

| Countries | 1995 | | 1997 | | 1999 | | 2000 | |
|-----------|------|------|------|------|------|------|------|------|
| | 1,000 tons | Million m² | 1,000 tons | Million m² | 1,000 tons | Million m² | 1,000 tons | Million m² |
| Scandinavia and Finland | 100.5 | 2,922.4 | 115.8 | 3,250.0 | 131.5 | 3,661.8 | 137.8 | 3,569.9 |
| U.K. and Ireland | 56.6 | 1,326.7 | 63.4 | 1,550.5 | 78.7 | 1,796.4 | 91.0 | 2,154.8 |
| France | 67.1 | 1,784.4 | 75.6 | 1,919.9 | 93.2 | 2,348.7 | 98.5 | 2,613.4 |
| Benelux | 90.9 | 1,650.0 | 97.5 | 1,575.0 | 92.6 | 1,461.6 | 99.9 | 1,405.7 |
| Germany | 179.9 | 5,046.0 | 198.4 | 5,953.4 | 224.6 | 6,370.9 | 257.2 | 6,824.4 |
| Italy | 103.8 | 3,148.7 | 156.9 | 4,513.8 | 219.4 | 5,970.1 | 251.6 | 6,935.7 |
| Others | 47.6 | 903.0 | 51.9 | 1,128.5 | 69.8 | 1,629.4 | 89.9 | 2,267.5 |
| Total | 646.4 | 16,781 | 759.5 | 19,891.1 | 909.8 | 23,238.9 | 1,025.9 | 25,771.4 |

Source: EDANA – European Nonwovens and Disposables Association.
Copyright: EDANA 2001.

As far as Europe is concerned, it should be noted that from the start national references were of a secondary importance to the nonwoven companies, as the then production was larger than their national markets and outlets had to be sought beyond their national limit. (It is to be reminded that the speed of nonwovens production was from the beginning a multiple of what would be achieved through weaving or knitting.) This has remained so, and there aren't close links between the country of production and a company's market.

Germany and Italy are practically at par as the most important nonwoven producing countries in Europe, both in tones and in square metres. Germany though remains the largest market. Comparatively overseas figures are estimated at:

Reliable statistics are missing relating to the turnover of the nonwovens industry (without converting operations). European estimates put it at 4,100 million EURO for Western Europe in 2000; it is valued at 3,000 million US dollars for North America (Source: INDA). In Europe, the production of nonwovens per manufacturing process developed as follows:

**Table 0-3**  Worldwide production of nonwovens

| Countries | 1997 (tons) | 2000 (tons) |
|---|---|---|
| Europe (Western) | 759,500 | 1,025,900 |
| Japan [1] | 296,700 | 314,100 |
| North America [1] | 875,000 | 967,000 |
| Others (estimates) | 350,000 | 550,000 |
| Total | 2,281,200 | 2,857,000 |

1) Contrary to Europe, Japan and USA data also include *most* needlepunched or stitchbonded fabrics.
 Sources: Europe and others: EDANA.
 Japan: MITI.
 North America: John Starr (1997), INDA.
 Copyright: EDANA 2001.

**Table 0-4**  Nonwovens production by manufacturing process in Europe (in 1,000 tons)

| Process | 1991 tons | 1991 % | 1995 tons | 1995 % | 1997 tons | 1997 % | 1999 tons | 1999 % | 2000 tons | 2000 % |
|---|---|---|---|---|---|---|---|---|---|---|
| Spunlaid [1] | 197.3 | 41.1 | 267.9 | 41.5 | 318.0 | 41.9 | 368.1 | 40.4 | 409.1 | 39.9 |
| Wetlaid | 46.9 | 9.7 | 51.0 | 7.9 | 55.2 | 7.3 | 62.7 | 6.9 | 63.1 | 6.1 |
| Drylaid [2] | 213.9 | 44.5 | 278.7 | 43.2 | 326.6 | 43.0 | 401.8 | 44.2 | 459.5 | 44.8 |
| Others [3] | 22.5 | 4.7 | 48.8 | 7.4 | 59.7 | 7.8 | 77.2 | 8.5 | 94.2 | 9.2 |
| Total | 480.6 | 100.0 | 646.4 | 100.0 | 759.5 | 100.0 | 909.8 | 100.0 | 1,025.9 | 100.0 |

Source: EDANA – European Nonwovens and Disposables Association.
Copyright: EDANA 2001.
1) Also includes other polymer-based processes e.g. meltblown, flashspun, orientated nets, perforated films, as well as composites of these fabrics (e.g. SMS…).
2) Basically groups thermal and chemical bonding, plus needling or stitching only and hydro-entangled webs. Does not include any "airlaid papers". The weight of adhesives, additives and similar chemicals has been taken into account (in addition to binders which always were included).
3) Now essentially represents short-fibre airlaid webs.

Table 0.4 shows a sharp development of polymer-based manufacturing processes in the nonwovens industry. The drylaid process has continued its progression and remains the main one. It is however very varied, especially as far as bonding is concerned. In recent years, we have seen a sharp increase of hydro-entangling, partly linked to the development of all sorts of wiping applications. Short fibres airlaid fabrics, which are the latest newcomer, are also progressing fast. Although input of the wetlaid sectors more than doubled in that same time span, this pales into insignificance when compared to the other processes. The use of nonwovens per large groups of applications was in the last 10 years as follows:

One can see that as far as Europe is concerned, deliveries from the European industry to the hygiene sector have almost trebled in 10 years (in tonnes). They

**Table 0-5** Deliveries of nonwovens per end-uses (in 1,000 tons)

| End-uses | 1991 | 1995 | 1997 | 1999 | 2000 |
|---|---|---|---|---|---|
| Hygiene | 131.4 | 210.7 | 252.2 | 324.9 | 341.4 |
| Medical/surgical [1] | 19.1 | 27.8 | 24.5 | 23.7 | 24.9 |
| Wipes (from 1998) for personal care [2] | 41.3 | 57.9 | 76.1 | 54.2 | 78.6 |
| Wipes – others | | | | 47.5 | 73.9 |
| Garment | 10.2 | 6.2 | 14.1 | 10.1 | 12.5 |
| Interlining | 24.8 | 28.5 | 28.0 | 23.8 | 22.4 |
| Shoe/leathergoods | 14.0 | 19.9 | 18.0 | 20.6 | 19.3 |
| Coating substrates | | n.a. | 7.6 | 11.3 | 14.5 |
| Upholstery/table linen/household | 46.0 | 39.1 | 29.8 | 51.1 | 59.3 |
| Floor coverings | | 28.1 | 28.8 | 28.7 | 28.6 |
| Liquid filtration | 17.9 | 3.16 [3] | 22.0 [3] | 24.1 [3] | 28.2 [3] |
| Air & gas filtration | 8.0 | | 10.3 | 14.2 | 15.8 |
| Building/roofing | 89.0 | 60.6 | 99.6 | 115.5 | 134.4 |
| Civil engineering/underground | | 61.4 | 56.2 | 57.9 | 63.0 |
| Others [4] | 69.8 | 62.4 | 66.6 | 77.6 | 73.2 |
| Unidentified | | 11.2 | 18.8 | 13.7 | 21.1 |
| Total | 471.5 | 645.4 | 752.6 | 898.9 | 1,011.1 |

Source: EDANA – European Nonwovens and Disposables Association.
Copyright: EDANA 2001 .
* This refers to production/deliveries of nonwovens produced in Western Europe.
1) Excludes medical wipes.
2) Includes medical wipes.
3) Includes fabrics for tea and coffee bags.
4) Includes as well electric/electronic applications, abrasives, battery separators and agriculture.

now make 33.8% of total deliveries. This partly reflects the increased penetration of disposable diapers in the European markets reaching over 90–95% or more of potential markets. Other causes are increasing exports of nonwovens for hygiene purposes or the wider use of nonwovens for various applications within the baby care sector (e.g. coverstock, leg cuffs, tapes, acquisition/distribution layer, textile backsheet, …) or finally, the development of incontinence products and to a lesser degree of feminine protection items. Unfortunately, mainly due to the oligopolistic position of hygiene converters the nonwovens manufacturers haven't been able to reap sufficient profits in this sector.

Another sector, which in the last year took off dramatically, is the wipes sector, be it for personal, industrial or household applications.

Both sectors are relatively light-weight and therefore take up an even larger proportion of the nonwovens deliveries when expressed in $m^2$.

Finally the civil engineering (geotextiles) and roofing applications have trebled in 15 years time. On the other hand, the growth of nonwovens for medical purposes has not held its promises; levels of penetration don't compare to those in the U.S. The use of nonwoven interlining, although now dominant in the apparel industry, hasn't progressed as hoped in Europe, because of the difficulties of the European apparel sector.

**Table 0-6** Fibres consumption in the nonwovens industry (in 1,000 tons)

| Fibres/polymers | 1991 | 1995 | 1997 | 1999 | 2000 |
|---|---|---|---|---|---|
| Rayon viscose | 53.5 | 56.0 | 67.3 | 78.1 | 92.5 |
| Polyester | 90.8 | 118.7 | 150.7 | 189.8 | 228.5 |
| Polyamide | 13.6 | 13.5 | 12.5 | 12.6 | 12.9 |
| Polypropylene | 209.2 | 311.3 | 369.9 | 442.3 | 491.4 |
| Multi-components (1) | 16.9 | 32.6 | 29.1 | 35.2 | 34.2 |
| Other man-made fibres | 35.2 | 38.8 | 44.7 | 51.8 | 59.7 |
| Wood pulp | 43.9 | 71.4 | 85.3 | 108.0 | 114.0 |
| Natural fibres | 16.7 | 16.4 | 13.4 | 14.5 | 15.6 |
| Mineral fibres | 4.2 | 5.9 | 6.0 | 6.5 | 6.7 |
| Other materials | | 0.6 | 2.6 | 5.4 | 5.1 |
| Total | 484.0 | 665.2 | 781.5 | 944.2 | 1,060.6 |

Source: EDANA – European Nonwovens and Disposables Association
Copyright: EDANA 2001
The data relate to fibres used (including wood fibres) and to polymer granules turned into filaments during the manufacturing process (e.g. spunlaid) and for a smaller part to plastic (perforated) films. Fibres used include regenerated fibres.
1) In 1997, the definition of bicomponent and multi-component fibres has been made more accurate.

The increase in fibre production reflects the trends in the production of nonwovens. In the last 10 years, the use of polypropylene has more than doubled and is the main fibre or filament used in the whole nonwoven industry. The use of other fibres, which in the eighties seemed to regress, has picked up again as new technologies, or new applications develop and spread. On the other hand the use of wood pulp has increased sizeably and isn't at all limited to the wetlaid process. Other, mainly specialized, manmade fibres also made inroads. The use of natural fibres like cotton or wool in nonwovens remains minimal.

In 2000, polypropylene cut fibres made 37.9% of the total polypropylene used in the nonwovens industry, whilst polyester cut fibres made 61.9% of the polyester fibres used. Viscose was exclusively used as a cut fibre.

To round up this positive image one should add that the balance of external trade of the European nonwovens industry is positive. Despite their imperfection, statistics indicate that (Western) Europe exports more nonwovens than it imports, the surplus being in 2000 in the order of 83,000 tons and still growing. A deficit exists though vis-à-vis the United States where it is very much influenced by the trade policy of dominant U.S. companies with subsidiaries in Europe. Countries like the Czech Republic and Slovenia, and above all Israel, are now nett importers of nonwovens into the European Union.

**0.5**
**Future perspectives**

Questions however are being raised about the future of the nonwovens industry in Europe. Obviously growth isn't a problem as far as the rest of the industrializing world is concerned: compared to needs, the production and use of nonwovens, e.g. for hygiene or civil engineering application, has barely started there. But, isn't the nonwovens industry maturing in Europe? Are there novel speciality applications opening up that will relay the dominant hygiene (baby-care) end-use? Will it not undergo a fate similar to the European textile or apparel industry, which, for a large part, gave in to lower price imports from emerging markets?

We don't have any crystal ball but here are a few elements of answer. Technology hasn't stood still and large reserves of technological development for nonwovens no doubt still exist, especially within the European manufacturing equipment industry. The capital costs of entry have been brought down though to relatively low levels and newcomers do find it relatively easy to come in.

However the availability of production equipment is only part of the answer. Maybe more than in comparable industries, qualified manpower and know-how that exists in companies remains of paramount importance (and can be patented). Hence it is also a must that proper R & D departments are maintained within companies, or in sufficiently large research institutes, so as to exploit further possibilities which otherwise would remain ignored or neglected.

The nonwovens industry has induced transformation and discoveries not only in the equipment sector but also at raw materials and converting levels. As an example among many, the fibres for the nonwovens industry aren't identical to fibres for weaving or papermaking. Without a steady supply, at the right price, of fibres and polymers and an appropriate research and development of fibres specifically for the nonwovens industry a question mark could arise as to the future development of a dynamic nonwoven industry in Europe. But, on the other hand, the sheer size reached by the European nonwovens industry makes it more rewarding to develop raw materials for it, and this effect snowballs into better products.

Beyond efforts in reducing costs – e.g. in manufacturing waste – and increasing production speeds, there are still many reserves regarding the incorporated high-tech properties of nonwoven webs for which the European industry can keep its advantage: progresses in the regularity and uniformity of the web, at increasing speeds and diminishing fabric weight are only part of the answer in the race to keep the present advantages. The combination of multiple technologies into sophisticated composites is another one as long as the constraints of recyclability – as opposed to energy recovery – don't become excessive. The industry in Europe has also to position itself both vis-à-vis the world and its present or potential customers. Mergers are one of the attempted solutions, which can alter bargaining powers (and therefore price) and create opportunities. Such organizational changes though, shouldn't be detrimental to flexibility and response to the signals of NEW markets. Indeed sustained growth needs entry into new markets and requires also new products appropriate to the demands from such markets. It cannot be said that the markets for nonwovens in Western Europe have all been found and taken and have become a close and finite universe. (Also Eastern Europe and the Mediterranean basin are next door.) Nowadays nonwoven fabrics are quite different in properties, appearance, and costs from a decade or more ago. Society's expectations and products have also changed. Beyond incontinence which some see beaconing on the horizon as a future large outlet, new opportunities could be tapped (even if emerging markets aren't immediately very sophisticated or large), providing there is a will and a pioneering spirit ready for it. Therefore, in my view, increased nonwoven promotion and marketing investment by the industry are necessary in Europe in order to further boost off the second stage of the nonwovens rocket.

# Part I
# Raw materials for the production of nonwovens

Nonwovens are textile fabrics consisting of separated fibres which are arranged properly by means of enduse-oriented technologies. In order to guarantee serviceability of the finished product, they are bonded. For this reason the choice of fibres and possibly bonding materials is of special importance: This relates to fibre raw materials and fibre dimensions. As a rule they have a greater share in creating the specialities of the nonwovens than this is the case in textile fabrics made of yarns. The bonding agents can also have an impact on the quality of the nonwovens.

# 1
# Fibrous material
W. Albrecht

Virtually all kinds of fibres can be used to produce nonwoven bonded fabrics.

The choice of fibre depends on
– the required profile of the fabric and
– the cost effectiveness

To produce nonwoven bonded fabrics
– chemical fibres of both cellulosic and synthetic origin as well as
– natural fibres and
– inorganic fibres
are mainly used

Because such a wide range of fabrics is either being developed or is already in production, it is impossible to name and describe all fabrics and fibres. The most important details will be provided below and the relevant literature will be cited. For more details the interested reader would be advised to consult additional reference material and to assess the importance of e.g. chemical fibres experimentally.

## 1.1
## Natural fibres

### 1.1.1
### Vegetable fibres

The most important constituent of vegetable fibres is cellulose, which is hydrophilic and hygroscopic. Apart from cellulose, vegetable fibres also consist of several other substances which affect their properties.

Cotton is the most important vegetable fibre used to produce nonwoven bonded fabrics. Table 1-1 shows the development of cotton production.

**Cotton (Gossypium)**
All varieties of cotton belong to the mallow family. To grow properly, the plants need moisture as well as dry heat alternately at the right times. Cotton is an an-

**Table 1-1** Development of cotton production world-wide from 1981 to 2000

| Year | Quantity (tons) | Area (km²) | Yield/area (kg/hectare) |
|------|-----------------|-----------------|-------------------------|
| 1981 | 14,995,000 | 330,690 | 471 |
| 1986 | 15,264,000 | 292,010 | 523 |
| 1991 | 20,805,000 | 349,390 | 595 |
| 1998 | 19,548,000 | 337,670 | 579 |
| 2000 | ~ 20,000,000 | ~ 335,000 | ~ 595 |

nual plant and grows to a height of approximately 1 m to 2 m. It grows fruit the size of walnuts which contain seeds covered with cotton fibre. The ripe fruit shells burst open and the cotton swells out in thick white flocks. The crop is usually harvested by machine, so that the cotton fibres are more likely to become contaminated than if harvested by hand. After harvesting, the seeds are removed with cotton gins and the cotton is packed into bales. The short fibres (linters) are removed by means of specialized machines and are used to produce a wide range of products, including the raw material for the production of cupro and acetate fibres. In fact, linters are also used in the production of nonwoven bonded fabrics.

Raw cotton contains:
– cellulose (80% to 90%)
– water (6% to 8%)
– waxes and fats (0.5% to 1.0%)
– proteins (0% to 1.5%)
– hemicelluloses and pectins (4% to 6%)
– ash (1% to 1.8%)

The quality of cotton and hence the grading depend on the following qualities:
– fibre length (10 mm to 50 mm)
– linear density (1.0 dtex to 2.8 dtex)
– colour
– purity (trash and dust)
– tensile strength (25 cN/tex to 50 cN/tex)
– elongation (7% to 10%)

The cotton fibres have to be scoured in an alkaline solution and/or have to be bleached to obtain the proper qualities and purity standard required for various purposes. To develop their typical fine sheen, the fibres have to be mercerised hot or cold. One of the most important characteristics of wet cotton is that it is some 10% stronger than dry cotton. Its good mechanical properties and serviceability are due to its structure.

As shown in Fig. 1-1 a and b, cottons shape and structure make it suitable for use for the production of nonwoven bonded fabric: cotton has a ribbon-shaped cross-sectional form, a spiral twist, a hollow structure, a high wet strength for a high module and it is hygroscopic.

**Table 1-2** World production of cotton by countries and regions in 1997 and 1998

| Country/region | Quantity (tons) | Area (km²) | Yield/area (kg/hectare) |
|---|---|---|---|
| EU | 465,000 | 4,980 | 934 |
| Formermly USSR | 1,587,000 | 25,510 | 622 |
| – Usbekistan | 1,150,000 | 15,050 | 764 |
| China | 4,400,000 | 46,000 | 957 |
| Asia (without China) | 6,156,000 | 138,650 | 444 |
| – India | 2,450,000 | 88,060 | 278 |
| – Israel | 53,000 | 290 | 1,828 |
| USA | 4,100,000 | 53,760 | 763 |
| America (without USA) | 1,040,000 | 23,840 | 445 |
| Africa | 1,797,000 | 44,790 | 401 |
| – Egypt | 350,000 | 3,730 | 937 |
| Eastern Europe | 3,000 | 140 | 231 |

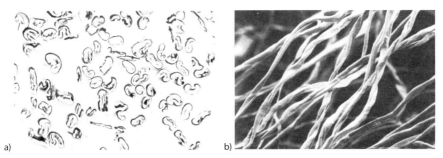

a)        b)

**Fig. 1-1** a) Cross section, b) longitudinal section of cotton fibres

Thus its use for the production of nonwoven bonded fabrics met with a fair degree of success in the early years. Its success, however, faded quickly because of the impurities which affected the production and even the quality of the finished product. This problem could not be solved, for it is impossible to remove all impurities during the production process or counteract their effect. This technical problem caused the noted decline of cotton usage in the production of nonwoven bonded fabrics.

### Jute (Corchorus)

Basically, two types of jute are grown to produce bast-fibre; with Bengal jute generally being preferred because of its pliability. The strands of fibro-vascular tissue from the inner cortex are prepared and turned into fibrous material in a special process. The quality of raw jute depends on the quality of the soil, the climate and roasting, and the method used to separate the bast from the cortex after it has been removed from the 3 m to 5 m long stems. To soften the vegetable glue (gliadin) in the ribbon of bast, the jute needs to be batched with softening oils and crushed repeatedly to allow further processing. The long bast ribbons are cut into pieces of 25 cm to 35 cm length with strong carders, second breakers, or special machines and are turned into mats.

Chemically, jute is a highly lignified fibre, which consists of:
– 60% cellulose
– 26% hemicellulose
– 11% lignin
– 1% proteins
– 1% waxes and fats
– 1% ash

with the substances cellulose and bastose forming a compound (lignocellulose, bastose), whose properties differ from those of other bast fibres. Jute is important for special usages of nonwoven bonded fabric. As it is quite inexpensive and has good physical properties it is predominantly used

– as the basic material for floor coverings
– as the base or intermediate layer in tufted floor coverings
– in filling pieces as, for example, in upholstery.

After the basic material has been subjected to months or years of wear, as it is the case in floor coverings, the surface deforms in the direction of tread, which must be considered when the floor is laid. The reason for this deformation is the rigidity of the individual fibres, which prevents them from altering their shape under pressure, so they slide off one another in course of time. Furthermore, it must be noted that jute may rot.

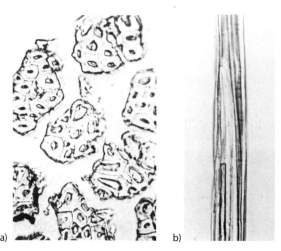

a) b)

**Fig. 1-2**   a) Cross section, b) longitudinal section of jute

### Flax (Linum usitatissimum)

Flax, an annual, is harvested shortly before the seed grows ripe for the extraction of fibre. The harvest comes to approximately 4,000 kg/ha, which yields 600 kg to 1,000 kg of raw flax are gained. The fibres embedded in the parenchyma of the stem in a high concentration are freed by retting. Then the flax is washed, dried and broken to loosen the brittle wood from the bast and to separate the fibres from each other. The wooden parts are removed by means of scutches (rotary crushing machines). Finally the fibres are combed by means of hackles. Flax typically has a high tensile strength and low elongation and crimp. It is also used for nonwoven bonded fabrics, mostly for the fabrication of filling pieces.

### Manila hemp (Musa textilis)

Manila hemp is one of the mock or skereuchym bast fibres. It is derived by drying and beating the mock stems, which are in fact rolled leaf bast. The fibres are yellow to brown in colour, about 5 mm to 8 mm long and very firm, light and shiny. They have a very high wet strength and rot resistance (Fig. 1-3).

Manila hemp is used to produce tea bags and manila paper on adapted machines. The firmness of the fibres and their pectin content give these special papers their unique qualities.

### Coconut fibre (Cocos nucifera)

Coconut fibre is obtained from unripe coconut fruit. The coconut is steeped in hot sea water, and subsequently the fibres are removed from the shell by combing and crushing. The raw fibres are between 15 cm and 35 cm long and between 50 μm and 300 μm in diameter. Coconut fibre is used to produce matting as well as coarse filling material and upholstery.

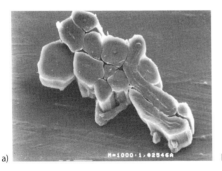 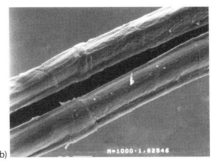

a) b)

**Fig. 1-3** a) Cross section, b) longitudinal section of hemp (photographs ACORDIS, microlaboratory)

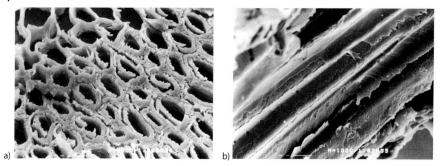

**Fig. 1-4** a) Cross section, b) longitudinal section of coconut fibre
(photographs ACORDIS, microlaboratory)

## 1.1.2
## Animal fibres

### Sheep's wool (Ovis aries)

Of all the animal wool and hair, only sheep's wool is of any importance for the
production of nonwoven bonded fabrics. As its price is high, it is used mainly in
the form of reclaimed wool or cuttings. The variations in quality and the impurities in reclaimed wool as well as the chemical and physical properties determined
by its provenance impose restrictions on its use.

The longitudinal section (Fig. 1-5 b) clearly shows the imprecate structure of
wool. This structure is less marked in reclaimed wool, but for filling material,
wadding and base layers it is still sufficient to guarantee a firm fabric. In chemical terms, wool is a suitably stiff and permanently crimped bicomponent fibre.
The distinct variations in thickness are in most cases favourable to produce nonwovens. Like the traditional woolen felts, nonwoven bonded fabrics made of wool
feature a relatively good shape stability, are high-bulking, and also good insulators
because of the air trapped between the fibres.

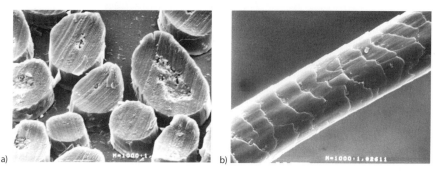

**Fig. 1-5** a) Cross section, b) longitudinal section of wool
(photographs ACORDIS, microlaboratory)

**Table 1-3** World production of wool and production by countries and regions

| Year | Quantity (1,000 tons) | Country/region | Quantity in 1996/97 (1,000 tons) | Contribution to total production in % |
|------|------|------|------|------|
| 1981 | 1,616 | Australia/New Zealand | 642 | 44.0 |
| 1986 | 1,789 | China | 150 | 10.3 |
| 1991 | 1,734 | Eastern Europe | 119 | 8.1 |
| 1996 | 1,456 | Western Europe | 108 | 7.4 |
| | | Uruguay | 60 | 4.1 |
| | . | Argentina | 41 | 2.8 |
| | | South Africa | 35 | 2.4 |
| 2000 | ~1,400 | | | |

### Silk (Bombyx mori)

Silk fibre, composed of a fibroin core and a sericin casing, is relatively rigid because the sericin causes the filaments to adhere to one another. Since silk is expensive and scarce, it cannot play any sizeable role in the nonwovens industry. The remarkable tensile strength and fineness of silk fibres makes it suitable for the manufacture of special expensive types of papers.

## 1.2
## Chemical fibres

Table 0-6 shows the importance of chemical fibres in the production of nonwoven bonded fabrics. The study of these fibres which follows focuses particularly on properties which are relevant to the production of web, the compacting into and the use of nonwoven bonded fabric. All other properties are described at length in numerous other publications. The significance of chemical fibres in the manufac-

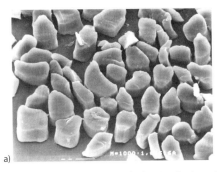
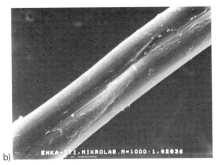

a) b)

**Fig. 1-6** a) Cross section, b) longitudinal section of silk (photographs ACORDIS, microlaboratory)

ture of nonwoven bonded fabrics has increased rapidly since the end of the 1950s. The evolution of chemical fibres was the decisive factor for the systematic development of nonwoven bonded fabrics for many new purposes.

## 1.2.1
## Chemical fibres made from natural polymers

### 1.2.1.1 Cellulosic chemical fibres

Cellulosic chemical fibres can be used alone or mixed with other fibres to make nonwoven bonded fabrics. Generally, there are two methods to produce such fibres:

- the viscose process: regeneration of cellulose fibres from solutions of derivatives (e.g. viscose, modal fibres)
- the solvent process: regeneration of cellulose fibres from solutions of cellulose (e.g. cuprammonium, NMMO)

As fibres produced by the copper oxide-ammonia process have not as yet gained any importance in the manufacture of nonwoven bonded fabrics, the following section mainly deals with fibres produced by the viscose process. It must be noted, however, that some properties of fibres made by the copper oxide-ammonia process can be useful for the production and the application of nonwoven bonded fabrics. The Asahi Company in Japan, for example, seizes upon the conglutination tendency of cuprammonium fibres in the regeneration of the cellulose for making cellulosic spunbonded material without bonding agents.

### 1.2.1.2 Viscose staple fibres

By varying the conditions in the viscose process, e.g. the composition of the viscose and/or of the precipitating bath, or by using different methods of drawing the newly spun rayons, a number of basic fibres for spinning can be produced. Table 1-4 shows these basic fibres and their characteristic properties. Some of these figures are, however, more relevant to firms producing and processing yarn than to the nonwoven bonded fabrics sector, since the stability and elongation properties can hardly come to bear in nonwoven bonded fabrics. But the thickness of the fibre, the relative wet strength, the water retention value, the wet module and the suitability for mercerising may be of importance, directly or indirectly, to the production and the use of nonwoven bonded fabrics. The thickness (titre) of the fibres used determines a number of properties of the nonwoven bonded fabric that is to be produced:

- Distribution of the fibres in web formation: Provided that the mass per unit area is the same, the substance surface (covering power) is greater using fine fibres, but it may be more difficult to produce the even surface which is necessary for the frequently desired low mass per unit area.

**Table 1-4** The most important viscose staple fibres and their properties

| Properties | | Viscose staple fibres – basic types – | | | Modal fibres | |
|---|---|---|---|---|---|---|
| | | normal type | highly crimped type | high wet strength | polynosic | high wet modulus |
| Titre | dtex | 1.3–100 | 2.4–25 | 1.4–7.8 | 1.7–4.2 | 1.7–3.0 |
| Maximum tensile load in dry state | cN/tex | 27–7.5 | 24–18 | 36–28 | 45–32 | 45–36 |
| Maximum tensile load extension in dry state | % | 16–30 | 20–30 | 21–28 | 8–14 | 14–18 |
| Relative wet strength | % | 60–65 | 60–65 | 65–80 | 72–65 | 75–65 |
| Water retention | % | 90–115 | 90–115 | 65–80 | 65–75 | 65–75 |
| Suitable for mercerising | | no | no | no | good | with reservations |

- Surface of the fibre in the web or in the nonwoven bonded fabric: It can be relevant to bonding as well as to the use of the nonwoven bonded fabric (e.g. filter) and to the choice of the mass per unit area.

- Stiffness of the nonwoven bonded fabric: Although it depends more on the mass per unit area and on the type of bonding agent, it also depends on the thickness of the fibre.

Fibres with thicknesses of 1.0 dtex to 5.0 dtex, especially those between 1.7 dtex and 3.3 dtex, have shown good results in large-scale production. But both finer and coarser fibres are used for special purposes as well.

The maximum tensile load and the maximum tensile load extension are familiar characteristics of fibres. Whereas these properties are of great significance for yarn production and processing they are much less relevant for the production and the use of nonwoven bonded fabrics because they can hardly have an effect in the web itself and it is only the combination of fibre and bonding agent which results in the effective stability and elongation properties.

The cross tenacity and the hollow volume of the fibres are important for the wearability particularly of nonwoven bonded fabrics made with chemical bonding agents. These two properties also make clear why viscose fibres with high wet strength have become so important especially for interlinings: they show by far the greatest cross tenacity and hollow volume of all fibres featured in Table 1-4. The void volume allows the fibre to take up the substance displaced under flexural strain more easily.

The wet strength in the web or in the nonwoven bonded fabric can be relevant to both production and use, since this property may have effects on the generally continuous manufacture of nonwoven bonded fabrics in the wet state or during chemical bonding. Certain difficulties which used to occur during manufacturing have largely been eliminated through modification of the production process.

The water retention capacity affects the choice of production in different ways. First of all, it is of great importance in the wet production of nonwoven bonded fabrics, because a high percentage (~120%) or even very high percentage (~300%) leads to a more even suspension than a lower one (~90%). This is because the swollen fibres behave like a hose filled with liquid and thus do not form water coils. In this state, longer fibres can be used during the wet process and suspensions will last longer even if there is a low fibre-to-water ratio.

The high water retention capacity inevitably presents a disadvantage for drying. Nonwoven bonded fabrics made of such fibres are suited for use in various medical fields, particularly as a high water retention capacity in general goes hand in hand with a quick absorption of liquids. However, even whilst moisture is being absorbed, the high water retention capacity may cause a separation of liquids, if water is more quickly absorbed than larger and less mobile molecules, which may even settle on the surface. Furthermore, when drying, the swollen fibres predominantly contract cross-directionally, but also in the direction of the axis of the fibre. If the bonding system then has a lowered shrinkage rate, the bonding agent no longer completely surrounds the fibres and micropores will develop between the fibre and the bonding agent. Fig. 1-7 shows the water retention capacity of various cellulose fibres and may help with the choice of the right fibres and aid when determining what manufacturing process to use.

Although the dimensional stability and the stiffness of nonwoven bonded fabrics are contingent rather on the method of bonding than on the fibres used, the wet modulus has a certain impact on the properties. If small forces affect the fibre in the wet state, they can cause a relatively significant deformation, so the bonding system has to be even more loadable, which can be a disadvantage for the textile property of nonwoven bonded fabrics. The concepts "initial module" and "reference load" are explained in Fig. 1-8.

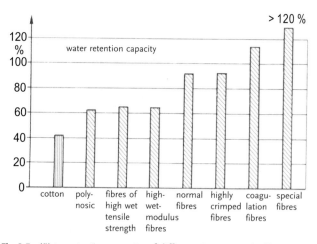

**Fig. 1-7** Water retention capacity of different viscose staple fibres compared to cotton

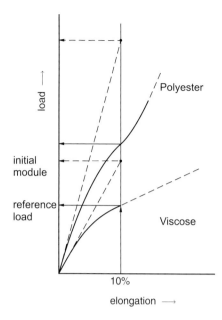

**Fig. 1-8**  The load extension graph for polyester and viscose fibres to determine the initial module and the reference load

When producing nonwoven bonded fabrics – especially when short fibres are used – the fineness ratio plays an important role. It is even more important when producing nonwoven bonded fabrics either with the wet method or with the aerodynamic process.

The following formula is used to express the fineness ratio:

$$\text{fineness ratio} = \frac{100 \cdot \text{length of a fibre}}{\sqrt{\text{fibre titre}}}$$

The fibre length is given in mm and the fibre titre in dtex.

That the importance of the ratio of the fibre length should not be underestimated is shown by an example: The fineness ratio of a 1.6 dtex fibre with a length of 6 mm is 474, whereas the ratio of a 10 mm fibre is already 791. Thus, higher fineness ratios can lead to problems in processing. However, these problems are often accepted to get the longest possible fibre length and to give especially nonwoven bonded fabrics made with the wet process the desired textile properties. This example shows very clearly how important it is to take into consideration not only the nominal ratings but also the individual effective ratings.

The reaction to mercerising is a criterion which has been adopted from conventional textile practice and which has to be adjusted to nonwoven bonded fabrics. It may be necessary to carry out alkaline processes while manufacturing nonwoven bonded fabrics. Similar treatment of nonwoven bonded fabrics is also possible after finishing. Once the mercerising properties have been established, we can predict how the fibres of nonwoven bonded fabrics will react in more or less concentrated alkaline media. Other important fibre properties – not stated in Table 1-4 – are the shape of the cross-section and the surface condition.

Fig. 1-9 shows the cross-sectional shape and the fibre structures of standard, high tenacity and high wet tensile strength viscose staple fibres. What is most striking is that the shape of the cross-section changes from lobate to rounded. The differences between the fibre types become evident in the processing of non-woven bonded fabrics when various degrees of opening-readiness and draft properties can be observed. Furthermore, it is striking that the fibre structure varies throughout the cross-section. As an example, the more or less thick fibre sheet and the full-sheet structure of the wet crease resistant type can be cited. Additionally, the structures determine the rate of the water retention capacity illustrated in Fig. 1-7. This means that in practice the standard and the high wet strength viscose fibres react differently to liquids: the high wet strength fibres absorb liquids more quickly but at a lower retention rate.

This means that, as the substance of the high wet strength viscose fibres has a denser structure, the fibres have to have microvoids. These microvoids are also the reason for the high bending strength of this fibre type.

The higher substance density throughout the cross-section of the fibre together with the low module has further practical consequences, too. For instance, needling is made more difficult. Experts know how to avoid needle smash by choosing the right fibre: higher titres or the addition of standard viscose staple fibres guarantees the desired needling properties. In practice, the differences in structure between these two fibre types have the following effects on processing non-woven bonded fabrics:

- Standard type: easy to process, has the standard properties expected of non-woven bonded fabrics, low in price, readily available.

- High wet strength type: easy to process if appropriate care is taken, better finishing, better exploitation of the substance, better flexibility and consequently a noticeable improvement of wearability, restricted availability.

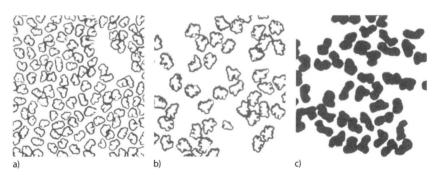

a)   b)   c)

**Fig. 1-9**  Cross-sections of different viscose staple fibres: a) normal viscose, b) high tenacity viscose, c) high wet strength viscose

Fig. 1-10 shows cross-sections and longitudinal sections of highly crimped viscose staple fibres. The cross-section of highly crimped fibres is the same as that of normal fibres, but their skin is irregular, causing tension throughout the cross-section and in the end leading to crimping. Crimping is strong when the fibres are dry; however, it becomes less stable when fibres become damp and swell. In this way crimp disappears as soon as the water content reaches approximately 20%.

Certain kinds of paper, such as those used for cigarettes and vacuum cleaner bags, are made from polynosic fibres. Fibrillation (shown in Fig. 1-11) occurs rapidly when "milled" fibres are wet, it is important in the production of these papers and gives them their special properties. When combined with cellulose, such "milled" fibres increase stability considerably and facilitate adjustment for a specific porosity of fabrics. This also applies when fibres are wet. Lyocell fibres may be used for similar purposes.

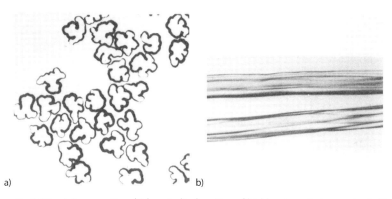

a)                                                                      b)

**Fig. 1-10**   a) Cross-section, b) longitudinal section of highly crimped viscose staple fibres

**Fig. 1-11**   Wet fibrillated polynosic fibres (lyocell fibres have similar properties)

Fig. 1-12 shows different crimp properties in viscose fibres. Crimping can vary considerably, as already explained in the section about the treatment of the fibre structure. Crimping is between 80 to 100 turns per 100 mm for normal dry staple fibres, while crimping is set at 120 to 140 turns for highly crimped fibres. It is not very difficult to obtain even higher crimping figures. However, problems arise in the production of nonwovens, since fibres get caught up in one another. It is important to note that the crimping frequency and amplitude figures apply only to dry fibres. This reservation is not valid for synthetic staple fibres.

The causes of crimping in chemical fibres are presented in Table 1-5 below. It shows that the processes adopted for cellulosic man-made fibres can also be used for synthetic chemical fibres, often with better results. It also shows that selecting the proper fibre makes it possible to achieve specific volume effects during the production of nonwovens.

Viscose filaments: Viscose filaments are produced with the same structural properties with which viscose staple fibres may be produced. However, their use in the field of nonwovens is not very significant.

There are two ways to produce such fabrics:

- Production immediately after spinning the filament. Spinnerets can be shaped in a specific way for this purpose. The production of such filaments requires additional modifications to ensure that the filaments guided in water or aqueous solutions are also mixed horizontally to give nonwovens the required transverse strength.

- Reeled, non-twisted, still damp or dry filaments can be drawn off and turned into nonwovens through air or liquids.

In practice, no market has been found so far for nonwoven bonded fabrics made from viscose filaments. The cost of the raw materials is rather high and the adjustment of the required transversal strength is relatively difficult for such nonwoven bonded fabrics. To a certain extent man-made filament yarns have been used successfully as fibre web to strengthen nonwoven bonded fabrics.

Cellulosic fibres for nonwoven bonded fabrics: Since the bonding of nonwoven bonded fabrics is a problematic production stage with downstream effects, the obvious thing to do was to try to develop suitable cellulosic bonding fibres. At first, ribbon-like fibres, used as glittering fibres in yarns, were processed on a trial basis.

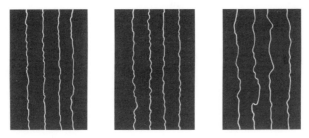

**Fig. 1-12**  Crimping of different viscose staple fibres

**Table 1-5**  Crimping in fibres, yarns and fabrics

| Causes of crimping | Use | Permanency |
|---|---|---|
| Mechanical distortion due to e.g.<br>a) pushing<br>b) gear treatment | Seldom in cellulosic man-made fibres, mainly in synthetic fibres | Cellulosic man-made fibres: poor<br>Synthetic staple fibres: satisfactory to good |
| Mechanical distortion followed by setting due to:<br>a) pushing/setting<br>b) twisting/setting | Synthetic fibres | Very good |
| Tension in fibres due to: skin/core structure or some other bicomponent structure | High-crimped cellulosic and synthetic chemical fibres | Cellulosic man-made fibres: dry: satisfactory<br>wet: poor<br>Synthetic fibres: very good |
| Mechanical longitudinal distortion due to:<br>a) filament breaking<br>b) mixing with normal fibres<br>c) shrinking in yarns or fabrics | Tapes of synthetic fibres mixed with shrinking natural or chemical fibres | Very good |
| Mixing of high-shrinking and normal-shrinking fibres with following shrink treatment | All kinds of high-shrinking synthetic fibres mixed with low-shrinking fibres | Very good |

Fig. 1-13 shows cross-sections and longitudinal sections of such fibres. The ratio of thickness to width – often called axial ratio – is approximately 1:12 in the depicted type of fibres. It can still be increased up to approximately 1:40. The bonding effect of such ribbon-like fibres was evident, but not as strong as expected. For this reason hollow fibres were produced, as shown in Fig. 1-14. In non-bonded woven fabrics of the wet process type, they distinctively increased firm-

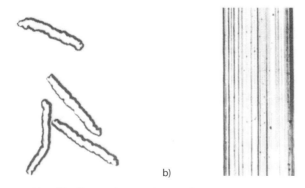

a)                                          b)

**Fig. 1-13**  Ribbon-like fibres with an axial ratio of approx. 1:12

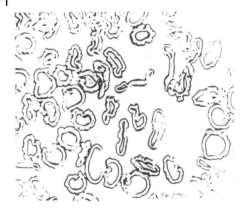

**Fig. 1-14** Cross-sections of hollow fibres

ness, without causing paperiness, as do many bonding agents. In forming nonwoven bonded fabrics, hollow fibres fault and enfold the fibres supposed to be bonded. Chemical bonds – so-called hydrogen bonds – such as those developing in the sheet forming of paper between the pulp fibres, do not occur when ribbon-like fibres or hollow fibres are used. But instead the water evaporating from the hollow space in the course of the drying process – as shown in Fig. 1-15 can "burst" the fibre and form a "double bonding area".

**Fig. 1-15** Bonding of nonwoven bonded fabrics with hollow fibres

Another method to produce cellulosic bonding fibres was found in the USA. BAR (Bonding Avisco Rayon) fibres consisted of a partly dissolved cellulose, were easily workable in the delivered state and on the wet fleece folding machine dissolved into a gel-like bonding agent that solidified the fibres, which were insoluble under these circumstances (Fig. 1-16). Nonwoven bonded fabrics produced in this way were particularly soft to the touch and had textile-like characteristics.

Lenzing company supplies another special fibre for the wet process. If a grafting process is used as long as the fibre is swollen, a very high water retention value is obtained and a special surface property is attained. A dispersibility of an un-

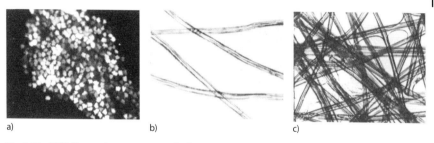

a)                                b)                                c)

**Fig. 1-16** BAR fibres: a) cross-section, b) longitudinal section, c) state after bonding

precedented magnitude is created. The suspensions are clot-free and are quite stable. This method also makes it possible to use longer fibres that make the finished product more textile-like (see water retention value). An additional method, the polybondic method, makes wet-bonded fabrics from fibres longer than 20 mm. For this process incompletely regenerated viscose staple fibres, which are more swellable than others, are used. A higher dispersibility and a more textile character of the product is obtained, because no chemical bonding is necessary.

Highly swellable cellulose fibres: Apart from the already briefly described method of producing fibres with a high water retention value, there are still other possibilities to produce fibres, which are swellable within broad limits and preferably contain cellulose. On the market, these fibres are often called "super slurpers". To produce them, alkalised celluloses are etherified and crosslinked. Different etherification agents – ethylene oxide, chloricetic acid, methyl chloride and others – and different crosslinking agents, like mono- and multifunctional compounds, can be used. These highly swellable fibres, which are rendered insoluble by crosslinking in water and many other liquids, can depending on their structure possibly absorb several thousand per cent of their weight in water relatively quickly and also retain it relatively well. This effect makes these fibres especially suited for the use in sanitary articles and special technical fields. Such fibres are characterised by their extremely high water retention value (1,000% to 3,000%), the faster and higher absorption of the surrounding moisture (2 to 3 times that of cotton), the fact that they can be reused to absorb moisture and the ability to absorb water from saline solutions. These properties are highly dependent on the degree of crosslinking. Additionally, the production of nonwoven bonded fabrics and the manufacturing of the finished articles influence their serviceable properties because their properties are so distinct.

### 1.2.1.3 **Summary**

Cellulosic chemical fibres of all lengths and degrees of refinement and with clearly different properties, are at the disposal of the industry of nonwoven bonded fabrics. They are all characterised by the ability to absorb a fairly high amount of moisture. That recommends their use wherever this property is useful for the production of nonwoven bonded fabrics and/or the use of nonwoven bonded fabrics is even a precondition. The use of cellulosic fibres confirms time and again how advantageous it is for the production and the use of nonwoven bonded fabrics that the fibres are free from impurities and are easy to handle at all stages of processing.

## 1.2.2
### Man-made fibres from synthetic polymers

The field of nonwoven bonded fabrics has become so broad that it includes nearly all kinds of existing fibres to some extent. However, specific fibre types have become predominant in certain areas within this field, a fact which will be explained in the sections below.

### 1.2.2.1 Polyamide fibres

Synthetic man-made account for the largest part of the raw material used in manufacturing nonwoven bonded fabrics. In this group of synthetic nonwoven bonded fabrics, polyamide fibres are the not only the oldest ones used in production, they also increase the serviceability of the product. This improved quality is of importance for various purposes, e.g.:

– where nonwoven bonded fabrics are subjected to frequent folding, as in the case of paper reinforced with synthetic fibres
– where exceptional resistance to abrasion is required, as is the case with needled floor coverings

The two main types of fibre are polyamide 6, usually known as Perlon, and polyamide 6.6, which is generally called Nylon to distinguish it from Perlon. The number or numbers after the word 'polyamide' indicate how many carbon atoms there are in each molecule making up the polyamide. The fact that there is only one number in one instance and two in the other shows that polyamide 6 contains only one basic module and polyamide 6.6 contains two, with six carbon atoms in each molecule. The figure also draws attention to the fact that the basic modules differ in size (e.g., polyamide 6.10 or polyamide 11 = Rilsan). Thus the number does not always have to be 6. A matter of course is that the properties of the polyamides change along with the different basic modules. Changes in water absorption capacity are important for the field of nonwoven bonded fabrics. Compared to the standard – polyamide 6 – it rises as the number of carbon atoms decreases and declines as the number increases. The other properties do not change in principle.

The bonds that link the basic molecules are the same for all polyamides. Macromolecules of this structure are referred to as polyamides because great numbers of molecules have to be present in order to form the macromolecules in the fibre.

$$-\begin{bmatrix} \overset{O}{\underset{\parallel}{-C}} & - & \overset{H}{\underset{|}{N-}} \end{bmatrix}$$

Polyamide 6 is made from ε-caprolactam, and polyamide 6.6 from hexamethyl-diamine and adipic acid. For fibre production, the resulting polyamide has to have the capacity to be spun into filaments, i.e.

- it must have the capacity to be melted without decomposing and to be forced through a jet

- the molten mass must be such that the filaments that are still ductile when formed do not break during cooling. Certain conditions must be met, one of them being a minimum prescribed length for the macromolecule

Fig. 1-17 shows the method of manufacturing melt-spun filaments as used for polyamide filaments.

The molten mass is forced through the holes in the spinneret by pressure pumps and metering pumps, after which it is pulled off in the form of filaments. They cool rapidly in the (air) blasting chamber and are then either baled or wound onto bobbins at a constant speed. The macromolecules are still randomly distributed in the filaments, which is why they are stretched so that molecules are more longitudinally oriented. Once they have this orientation, the filaments take on their characteristic physical properties and can be cut to the lengths needed to make the fibres. Then the filaments of staples are prepared to ensure that they retain their processing properties. This is how all of the fibres listed in Fig. 1-18 are produced.

The most important values for the physical properties of normal spun polyamide fibres are listed in Table 1-6, which covers various fibre thicknesses, degrees of lustre and cross-section forms. The term 'normal' is of great significance for nonwoven bonded fabrics, because:

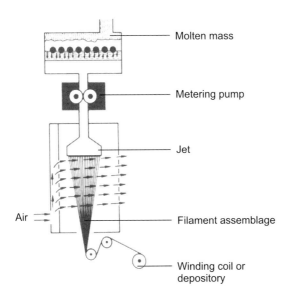

Molten mass

Metering pump

Jet

Air

Filament assemblage

Winding coil or depository

**Fig. 1-17** Method of production of man-made fibres by melt spinning

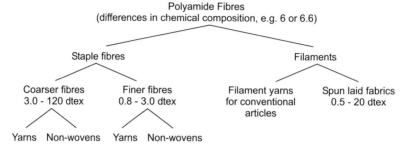

**Fig. 1-18**   Simplified subdivision of the polyamide fibres in production

**Table 1-6**   Typical values for normal polyamide fibre properties

| Polyamide fibres | Maximum tensile strength<br><br>(cN/tex) | Maximum elongation<br><br>(%) | Relative wet strength<br><br>(%) | Water retention value (WRV)<br>(%) | Water content at 20°C and 65% rel. humidity<br>(%) |
|---|---|---|---|---|---|
| 1.6 dtex/40 mm bright | 50–60 | 45–55 | 80–90 | 10–15 | 4 |
| 3.0 dtex/40 mm semi-dull | 45–55 | 50–60 | 80–90 | 10–15 | 4 |
| 17 dtex/80 mm semi-dull | 40–50 | 65–75 | 80–90 | 10–15 | 4 |
| 22 dtex/80 mm semi-dull | 40–50 | 55–65 | 80–90 | 10–15 | 4 |
| 35 dtex/100 mm bright prof | 30–40 | 70–80 | 80–90 | 10–15 | 4 |

- copolyamide fibres can also be used (see the section on synthetic bonding fibres)

- the filaments used in spun laids are produced under different conditions than the textile or technical man-mades with regard to their production conditions (see Section 4.2.1)

- very strong fibres are not used at all in nonwoven bonded fabrics, whereas they are used in tarpaulins, conveyor belts and tyres, but

- fibres which react differently when dyed can be used together, for example in needled floor coverings

The term 'prof' used in reference to Type 35 dtex, 100 mm bright (Table 1-6) stands for 'profiled', indicating that the fibre does not have the usual round cross-

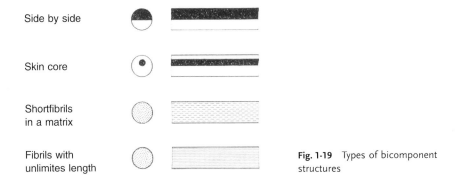

Side by side

Skin core

Shortfibrils
in a matrix

Fibrils with
unlimites length

**Fig. 1-19** Types of bicomponent structures

section. In this case, it is three-sided. Further information on cross-sections is given in the section on polyesters.

Other variations in melt-spun fibres are the result of bicomponent spinning, i.e., the combination of two more or less different raw materials. Such types are logically called bicomponent fibres. Fig. 1-19 shows a schematic diagram of bicomponent fibre structures.

These types can be varied even more widely by changing their form and the relative fibre proportions of their raw materials, thus producing different external effects. In this way, for example, side-by-side fibres can be made with considerable variations in their crimping or curling effect (frequency, amplitude, volume and permanency).

The core in skin/core fibres is not always to be found exactly at the centre or indeed in the same position over the whole length of the fibre. This likewise creates a tension in the fibre, which results in crimping. Far more important, however, is the fact that bicomponent fibre spinning technology enables polymers with different properties to be spun together, thereby producing fibres with a polyester core and polyamide skin, for example. In such a case, the core guarantees the dimension stability of the fibre and the skin ensures that the fibre will dye easily and well. The properties of bicomponent fibres are governed by:

– the two raw materials
– the relative quantities of the two components
– their arrangement within the fibre
– the thickness of the fibre

Fig. 1-20 gives a schematic representation of how bicomponent fibres are produced and explains how it is in fact possible to mix polymers in such a way as to make fibres other than the bicomponent fibres described.

If, for example, the melting points of the polymers being mixed are different, the component which melts at a higher temperature will solidify in the still-molten mass of the other polymer, which will set at a lower temperature. When the fibres are drawn, elongated inclusions are formed (Fig. 1-21 a). These may even be endless, i.e., they are present in the matrix as very fine filaments. Such fibres are exception-

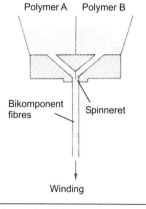

Polymer A    Polymer B

Bikomponent
fibres        Spinneret

Winding

Examples of fibre cross-sections

1. Side by side       2. Core/skin

**Fig. 1-20**  Bicomponent fibre production process

ally stiff and display special physical and chemical properties. Especially in the case of nonwovens, bicomponent fibres open up new fields of application in which particularly large fibre surfaces in relation to their mass per unit area are required. Although relatively fine fibres can be manufactured for this purpose, they are practically impossible to handle at thicknesses of less than approx. 0.5 dtex. For this reason, spinning begins with bicomponent fibres with a total denier that supports the reliable production of fabrics, followed by the splitting of bicomponent fibres into their individual components through chemical or physical processes.

Ultrafine fibres are made by removing the matrix which holds the individual fibres together (Fig. 1-21b). These ultrafine fibres are similar in form to skin fibrils and can therefore be used in the manufacture of artificial poromeric leathers.

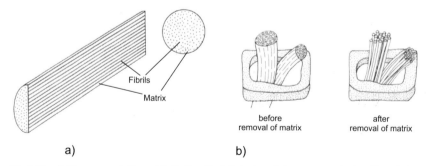

Fibrils

Matrix

before
removal of matrix

after
removal of matrix

a)                                    b)

**Fig. 1-21**  a) Bicomponent fibre with fibrils of indefinite length in a matrix,  b) bundles of ultrafine fibres before and after removal of the matrix

Other man-made fibres can also be produced by this method, which works most easily with melt-spun fibres.

Another way of modifying polyamide fibres is to vary the number of amino end groups in the macromolecule. If the two molecules 'A' and 'B' (see below) do not combine to form the amide group, the 'free ends' must be saturated in some way.

$$
\begin{array}{ccc}
\underset{\displaystyle A}{\overset{\displaystyle \begin{array}{ccc} H & & O\;\;\; H \\ \diagdown & & \diagup\;\;\; \diagdown \\ N\!\sim\!\sim\!\sim C & + & N\!\sim\!\sim\!\sim \\ \diagup & & \diagup \\ & HO - H & \end{array}}{}} 
& 
\underset{\displaystyle B}{\overset{\displaystyle \begin{array}{cc} O\;\;\; & H \\ \diagup\;\; & \diagdown \\ \sim\!\sim\!\sim C & N\!\sim\!\sim\!\sim \\ \diagdown\;\; & \diagup \\ OH\;\; & H \end{array}}{}}
\end{array}
$$

or

$$
\underset{\displaystyle C}{\overset{\displaystyle \begin{array}{cc} O\;\;\;\;\;\; & H \\ \diagdown\;\;\;\; & \diagup \\ C\!\sim\!\sim\!\sim & N \\ \diagup\;\;\;\; & \diagdown \\ HO\;\;\;\;\; & H \end{array}}{}}
$$

This can be done with water, for example, resulting in a molecule with an amino group at one end and a carboxyl at the other, as in C below.

The amino group is reactive, and it is well known for binding acidic dyes and other such textile auxiliaries. In the case of polyamide 6.6, which is formed from hexamethyldiamine (with one of these amino end groups at either end) and adipic acid, some of the molecules must have one or even two amino end groups. The number of amino end groups depends on the length of the molecule; but since in practice minimum and maximum molecule lengths are required for the manufacture of fibres, the direct addition of substances with amino end groups will also affect the reaction of fibres with specific dyestuffs (chemicals). The affinity to acidic dyes can thus be increased (deep type), reduced (light type), and even blocked (nontype). This makes differential dyeing possible, i.e., the use of such fibres in combination with others which have a 'normal' affinity to dyestuffs.

## 1.2.2.2 Polyester fibres

As its name indicates, this type of fibre consists of macromolecules of esters, which are chemicals made of acid and alcohol. If many of these basic molecules are joined, they will form polyesters. This description of the structure indicates that polyesters may develop from completely different acids and alcohols. Although this absolutely holds true, the term 'polyester fibre' generally comprises only fibres which consist of terephthalic acid and glycol. Fig. 1-22 illustrates the manufacture of polyester from different components. It also shows that it is not necessarily the acid that has to serve as basis, but that a simple, easily manageable ester, i.e. dimethylterephthalate (DMT), may be taken as well. A monomer, i.e. terephthalic acid diglycolester, develops by transesterification with the help of

**a) from DMT and glycol**

dimethylterephthalate + 2 glycol ⟶ diglycolterephthalate + 2 methyl alcohol
    DMT                                    DGT


n-diglycolterephthalates manufacture polyester by glycol splitting
    DGT                                                      PET


**b) from TPA and glycol**

terephthalic acid + glycol manufacture polyester by dehydration
    TPA                                        PET

**Fig. 1-22**   Manufacture of polyester from a) DMT and glycol, b) TPA and glycol

glycol, which is a bivalent alcohol used for forming fibres. During this reaction the methyl alcohol needed for developing DMT is freed again and recovered.

By condensation (chemical condensation is defined as the fusion of molecules by dehydration or evaporation of alcohol; the term polymerisation means that molecules merge completely, as e.g. when polyamide 6 is formed) of terephthalic acid diglycolester, polymer develops, which is usually poured out in ribbons, cooled, chopped to small chips (granulate), remelted after drying and pressed through nozzles to produce filaments. The still liquid formed polymer may also be directly spun, if it seems appropriate. The newly spun filaments, which are still highly elastic, are wound or put into big containers and then stretched when hot. They may, however, also be directly stretched, i.e. without intermediate storage. When manufacturing spun fibres, the stretched filaments are towed, crimped, cut to the desired length and baled. In manufacturing nonwovens, spun polyester fibres are more important than filaments (see Table 1-7). Furthermore, spun-laids made of polyester filaments meet a ready market and are being developed further.

**Table 1-7**   Selection of available spun polyester fibres

| Spun polyester fibres | Type | Thickness in dtex |
|---|---|---|
| Normal fibres | Fine fibre type | 1.0–2.4 |
|  | Wool type | 2.4–5.0 |
|  | Filling type | 3.3–22 |
|  | Carpet type | 6.7–17 |
| Low-pilling fibres |  | 1.7–4.4 |
| High-shrinkage fibres (classical and linear) |  | 1.7 |
| Fibres with abnormal dye affinity (low-pilling at the same time) |  | 4.4–17 |
| Bicomponent fibres |  | 3.0–17 |
| Bonding fibres |  | 1.7 and coarser |
| Flame-retardant types |  | 1.7–4.4 |

The lengths of cut are adapted to the respective manufacturing procedure. Fibres are available in different degrees of lustre and cross-sectional forms.

As shown in Table 1-7, polyester fibres may be modified in many different ways. That is why they have been made to suit a wide range of different purposes and uses. It is, however, difficult to give the basic properties of polyester fibres without indicating the name of the respective type of fibre. Thus it should only be mentioned that, in general, polyester fibres

– may be used for a great number of purposes because they
– may be relatively easily adapted to their intended use
– while retaining all their serviceable qualities,
– especially their dimensional stability as well as their high light and weather resistance.

In addition to these generalised but characteristic features, Table 1-8 contains figures for some properties of the most important types of polyester fibres.

In the stretching process, the molecules come to lie parallel to one another, which causes inner tension in the fibres. Because of this tension, the fibres may shrink again when heated, if they are not held tight at both ends. Therefore the stretched, unfixed fibres are called unstabilised. If the molecules now lying more in parallel to one another are able to attain optimum alignment as a result of heating, the fibres will not shrink any further in subsequent processing. Such fibres are referred to as stabilised or fixed.

Fig. 1-23 shows that tension-free stabilisation also influences the tensile properties of fibres. These properties, however, remain unaffected, if the stabilising process is executed under tension. The tensile strength graph for the stabilised type in Fig. 1-23 is supposed to show unstabilised fibres that were able to shrink freely when subjected to heating. Shrinkage on boiling is up to 10% in unstabilised fibres, while it is from about 2% to less than 6% in stabilised fibres depending on the conditions of stabilisation (temperature, time and tension). Unstabilised fibres

**Table 1-8** Important properties of the basic types of polyester fibre

| Property | | Fine fibre unstabilised | Fine fibre stabilised | Wool type normal | Wool type low-pilling |
|---|---|---|---|---|---|
| | dtex/mm | 1.7/40 | 1.7/40 | 3.3/60 | 3.3/60 |
| Maximum tensile load | cN/tex | 55–60 | 55–60 | 45–55 | 36–42 |
| Maximum tensile load extension | % | 24–30 | 22–28 | 40–50 | 35–45 |
| Wet strength (relative) | % | 100 | 100 | 100 | 100 |
| Shrinkage on boiling | % | 4–7 | 1–4 | 1–3 | 2–3 |
| High temperature shrinkage | % | 6–9 | 2–5 | 2–5 | 3–5 |
| Hot air shrinkage | % | 16–18 | 5–7 | 9–12 | 10–12 |
| No. of abrasion revs | | 4,000–6,000 | 3,000–4,000 | 3,500–4,500 | 2,000–2,500 |

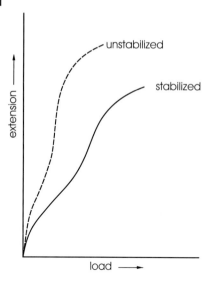

**Fig. 1-23** Load-extension graph of unstabilised and stabilised PET fibres

are often used for the manufacture of nonwoven bonded fabrics, as they will shrink a little further in finishing. Thus the final product will be appropriately close with a little more volume. After such treatment the fibres, which now will not shrink any further, have also got the desired dimensional stability.

Shrinkage in polyester fibres may be directly influenced by varying the physical conditions (how the fibres are stretched) or chemical conditions (use of copoly-esters) during manufacture. Such polyester fibres are called polyester high shrinkage (HS) fibres. In Fig. 1-24, which illustrates their characteristics, shrinkage is completely different in each of the two HS types depending on temperature.

The physically modified fibres begin to shrink at 50 °C to 60 °C, with maximum shrinkage at 100 °C. Furthermore, the dotted line between points A and B shows that further shrinkage is blocked when heating is interrupted and continued later on, i.e. shrinkage will come to a halt at point A if the temperature is increased to

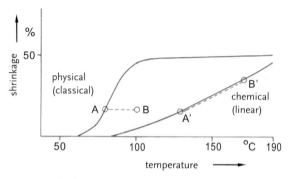

**Fig. 1-24** Shrinkage characteristics of different PET HS fibres

e.g. 100 °C. In chemically modified fibres, the conditions are totally different. The copolymers added to the polyesters at first block shrinkage up to about 100 °C. Then, however, they even promote shrinkage and, in addition, prevent it from being blocked. This is indicated by the dotted line from A′ to B′. When using this type of fibre, it is thus possible to achieve a certain degree of shrinkage by changing the temperature accordingly.

In textile practice, the available shrink power is even more important than the measurable shrinkage of individual fibres in a tensionless state. Above all, the shrink power is the factor which makes it possible to surmount the resistance to distortion of the added fibres which shrink little or not at all, thus allowing these fibres to be crimped so that the yarn or fabric will be of the desired volume. Table 1-9 gives figures for shrinkage and shrinkage power of various types of Diolen spun fibre. Both properties are very important for the manufacture of nonwovens.

The cross-sectional form of melt-spun fibres may be altered fairly simply by using nozzles with appropriately shaped holes. Normally the nozzle holes are round, which makes it possible to obtain almost round fibres, as shown in Fig. 1-25.

The photograph of the slightly bent fibre below, however, reveals two further aspects which are essential for the manufacture of nonwovens. The smoothness of the surface will certainly not improve the adhesive power of thin bonding films, and the folds on the inside of the curve will, in addition, soon loosen any not firmly adhering bonding agent.

Fig. 1-26 shows various cross-sectional structures of melt-spun fibres: three-lobal, round and hollow. Apart from that, other shapes such as four-, five-, six- and eight-lobal as well as ribbon shaped are used quite frequently. Solid fibres with their different cross-sectional shapes fulfil two interesting demands that the end product makes on them. Firstly, they are chosen for a certain visual effect. Especially the outer surfaces and the degree of transparency determine the amount of light reflected and absorbed and thus the appearance of the textiles manufactured from these fibres. Secondly, the stiffness of fibres depends on the cross-sectional shape, which influences the texture and volume of fibres as well as the integration of the individual fibre into the whole fabric. These properties are also taken

**Table 1-9** Shrinkage properties of various types of Diolen spun fibre

| | | *Hot air 3′ 190 °C* | | *Boiling water 10′* | |
| --- | --- | --- | --- | --- | --- |
| | *dtex/mm* | *Shrinkage (%)* | *Shrinkage power (cN/dtex)* | *Shrinkage (%)* | *Shrinkage power (cN/dtex)* |
| Diolen 12 | 1.7/40 | 5 | – | 2 | – |
| Diolen 11 | 1.7/40 | 12 | – | 8 | – |
| Diolen 21 | 3.3/60 | 6 | 0.025 | 2 | – |
| Diolen 31 | 3.3/60 | 50 | 0.045 | 45 | 0.065 |
| Diolen 33 | 3.3/60 | 50 | 0.055 | 6 | 0.060 |

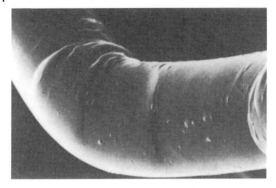

**Fig. 1-25**  Scanning electron microscope photograph of a PET fibre

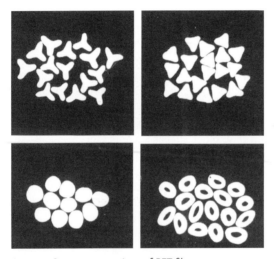

**Fig. 1-26**  Some cross-sections of PET fibres

advantage of in needled floor coverings, where the use of profiled fibres helps to create a subtle sheen and makes the material pleasant to walk on.

As hollow fibres provide much volume and increased stiffness compared to solid fibres, the end product will be lighter in weight. Depending on the size of the tube, the density of the fibres is proportionally lower than the density of the substance ($1.38 \, \text{g/cm}^3$). The tube extends all throughout the fibre, as is shown in the longitudinal view of Fig. 1-27. The diameter of the tube can be modified; it can amount to up to 20% of the cross-sectional area. Because of varying conditions during manufacture, the tubes are rarely perfectly round. When equal quantities of dye are absorbed, hollow fibres appear to be slightly darker than round solid fibres.

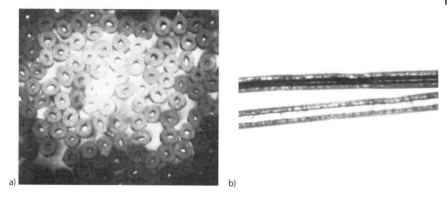

a)                                          b)

**Fig. 1-27**   Hollow polyester fibres: a) cross-sectional view, b) longitudinal view

Fibres resistant to pilling, or to be more accurate, low-pilling types are another type of polyester fibre which is of some importance also for manufacturing non-wovens. Pilling in textiles means that individual fibres which are fairly well integrated into the whole fabric are pushed outwards when they are subjected to frequent bending, abrasion or strain, and subsequently form a loop with their middle parts. If this effect continues, the loop connects with adjacent fibre ends, and together they develop pills or little knots (see Fig. 1-28).

Since especially in nonwoven bonded fabrics the individual fibres cannot be firmly integrated by twisting, as in the case of yarns, the problem should be solved by the right choice of fibres or by good bonding in the relatively few cases where pilling is detrimental. Table 1-10 lists different types of polyester fibres to show how much man-made fibres industry can influence pilling. When compil-

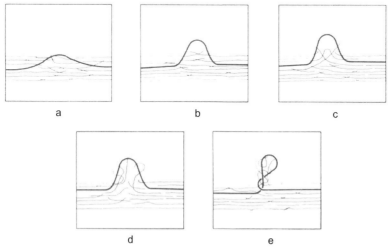

a                    b                    c

d                    e

**Fig. 1-28**   Diagram of how pilling is caused

**Table 1-10** Characteristic test figures for four different polyester fibres

| Fibre types with round cross-sections | | Normal | Hollow | Pill-resistant | Super pill-resistant |
|---|---|---|---|---|---|
| **Fibre characteristics initial state** | | | | | |
| Thickness | dtex | 3.3 | 3.3 | 3.3 | 3.0 |
| Maximum tensile load | cN/tex | 50 | 45 | 40 | 30–33 |
| Maximum tensile load – extension | % | 35 | 40 | 45 | 32–37 |
| Relative loop resistance | % | 95 | 90 | 90 | 80 |
| Bending resistance | turns | 150,000 | 150,000 | 50,000 | 900–1,300 |
| **After four hours of high-temperature dyeing** | | | | | |
| Thickness | dtex | 3.6 | 3.6 | 3.6 | 3.0 |
| Maximum tensile load | cN/tex | 45 | 40 | 35 | 22 |
| Maximum tensile load – extension | % | 35 | 40 | 40 | 25 |
| Relative loop resistance | % | 85 | 80 | 80 | 70 |
| Bending resistance | turns | 70,000 | 120,000 | 20,000 | 1,000 |

ing this list, a clear distinction was made between the initial state of the fibres and the state in which the fibres are in a textile ready for use. The four hour high-temperature dyeing, which is part of testing, should comprise all possible finishing processes, including the chemical bonding in nonwoven bonded fabrics.

Table 1-10 clearly shows that mechanical and technological properties of fibres can be altered by modifying the cross-sectional shape and by adding copolymers (super pilling resistant). It is important to reduce flexing resistance so as to decrease pilling.

When pills are about to be formed, the fibres that have been pushed or pulled out should preferably break off, a circumstance which depends on maximum tensile load and elongation behaviour. – Evidence gained from experience shows that in the finished product, super lowpilling fibres should have a thickness/maximum tensile load ratio of 30 cN/tex as well as a maximum load extension of 35%. Fig. 1-29 illustrates the effects that can actually be achieved in similarly constructed plain woven materials made exclusively of one type of fibre.

The curves of the graph show that pills are formed when fabric is exposed to abrasion, which is also the case for wool. The number of pills developed per unit area, however, varies considerably depending on the type of fibre. If abrasion continues after the maximum number of pills per unit area has been reached, pills will break off. Therefore the total number of pills will decrease, despite the formation of new ones. If normal polyester fibres are used, still many pills will remain visible on the fabric, whereas in the other three types of fibre, pills will break off almost entirely or even entirely. Finally, also those polyester fibres should be mentioned that take dyes in a different way. The ability of such fibres to take dyes is improved and thus they will absorb the dye more quickly and more intensely. In

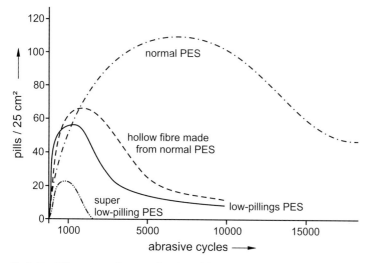

**Fig. 1-29** Pilling curves for normal and modified polyester fibres

addition, types of fibre that take basic dyes as do polyacrylonitrile fibres are also important. For that purpose acidic groups are integrated into the fibre substance. The remaining polyester substance of the same fibre can, however, also absorb dispersion dyes just like normal polyester. This explains why no true differential dyeing is possible when polyester fibres with an affinity to normal and basic dye-stuffs are mixed.

Polycarbonate fibres are a special kind of polyester fibre. Polycarbonate is the polyester of carbonic acid and is made from bisphenol A and phosgene. The fibres can be spun either dry or wet, and their density is about 1.2 g/cm$^3$. As polycarbonate is very heat-resistant and non-flammable, it is used in other fields, too, e.g. in air and gas filtration. Other important features are good electrical insulation as well as low water absorption.

**Polyolefine fibres**

In contrast to paraffins, olefins are unsaturated hydrocarbons. Therefore it is relatively easy to put them together to molecular chains. Although there is a great number of olefins, a fibre expert only thinks of polyethylene or polypropylene when talking about polyolefins. Both products are well-known as raw materials for the production of wrapping films and sheetings, containers and mouldings. The basic material is either ethylene or propylene, which occur as by-products of oil distillation or a special cracking process. Polymerisation is carried out by a high or low-pressure method using special catalysts, whereas spinning and making into fibres is done by melt spinning as for polyamide and polyester fibres. Polyolefins can also be made of films which are produced by melt spinning, as well. That is why fibre experts are trying hard to turn cast or blown-extruded films within a continuous process into fibre form, to put the fixed widths of the produced film fibres together to a yarn and to wind it immediately afterwards. The first stage of this method can also be used to produce nonwoven bonded mats. The raw materials that are most frequently employed for this process nowadays are low-pressure polyethylene and polypropylene. Increasingly, polymers which have been polymerized with the help of metallocene catalysts are processed. One of the remarkable features of these polymers is the greater uniformity of their macromolecules, which simplifies the processing. The respective raw material is melted in an extruder and forced through

– slot nozzles (see Fig. 1-30) or
– ring nozzles (see Fig. 1-31)

to form either plain films or different kinds depending on whether profiled nozzle lips or nozzle rings are used.

The shape of the nozzle determines whether

– flat films or
– tubular-blown films, which are cut at a later stage, are produced.

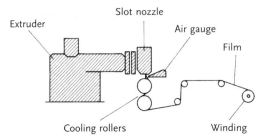

**Fig. 1-30**   Flat sheet extrusion with cooling rollers

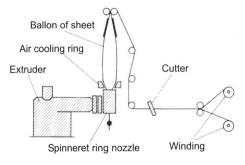

**Fig. 1-31**   Blow (extrusion film)

Further stages may vary, depending on the way the film has been made:
– cutting the films into narrow stripes (flat threads)
– splicing, controlled or uncontrolled, of the distended films
– splitting of the films that have been profiled during the processing
– fibrillation of multicomponent films

In order to produce cut flat fibres the films can be

– cut first and the resulting flat fibres then stretched monoaxially,
   or the films keeping
– constant width
– stretched first and then cut

Splicing of the stretched films can be done

– mechanically, when dealing with highly stretched films made of polypropylene
   which have a high melt-flow index by brushing, rubbing, using air jets or ultrasound, all of which lead to an uncontrollable and practically unrepeatable splitting up of the films longways;
– controlled mechanically by using a roller covered with fine needles and rotating
   in the direction of the film feed;
– uncontrolled, partly chemical, partly mechanical. In this case an additive is put
   into the raw material of the films. This additive, if statistically distributed inside
   the film, produces irregularities that simplify mechanical splicing at a later stage.

The splitting of profiled films leads to regular network-like fabrics which can be either
used as they are, bunched together into "filament yarns" or processed into webs.

In case that two or more polymers are
– mixed before extrusion or
– put on top of one another in layers,
they are called multicomponent films.

When these multicomponents are fibrillated, the result naturally depends on the
structure of the film. As for the mixed polymer films, their fibrillation tendency
determines the type of polymer used, the proportions of the mixture and the distribution. It is worth testing the fibrillation tendency of the respective film. The
easiest way to do that is to pass the film at constant tension over a small roller
covered with needles. In this way, a split factor can be calculated. The higher it is,
the easier it will be to achieve fibrillation in the film. Fig. 1-32 shows a basic example of the interdependence of split factor and proportion of mass.
   How strongly the fibrillation tendency depends on the polymer mixture is additionally influenced by the stretch factor of these films. Three examples in Fig. 1-33 show
these relationships.
   When fibrillating films consisting of two or more layers, one has to assume
that the individual components have a different shrinkage behaviour, which will
affect both the processing and especially the crimping of fibrous structures. When
producing multicomponent films by means of lamellar extrusion, also known as

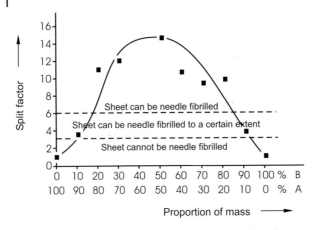

**Fig. 1-32** Dependence of the fibrillation tendency on the polymer mixture

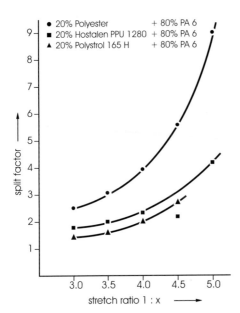

**Fig. 1-33** Dependence of the fibrillation tendency on the second polymer and the stretch ratio

"the Rasmussen principle", two incompatible polymers are extruded through the special nozzle shown in Fig. 1-34. Cutting produces ribbons with thin layers, which split when stretched.

Fig. 1-35 shows how the components can possibly be arranged and when a third component, a binder, is added, sandwiches are produced. Their multilayered structure makes the sandwiches self-crimping.

In practice, fibre networks, as i.e. shown in Fig. 1-36, always develop when the film is turned into fibres, unless special cutters are used. The structure of these fibre networks makes them well suited for use as open webs.

## raw materials

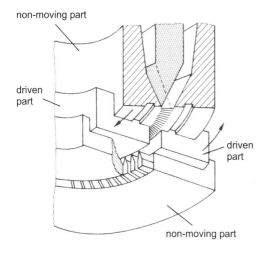

non-moving part

driven part

driven part

non-moving part

**Fig. 1-34** The Rasmussen principle of nozzle construction

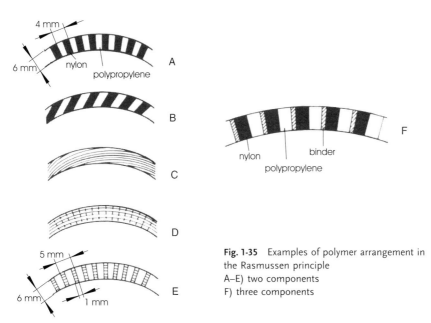

4 mm

6 mm

nylon

polypropylene

A

B

C

D

5 mm

6 mm

1 mm

E

nylon

polypropylene

binder

F

**Fig. 1-35** Examples of polymer arrangement in the Rasmussen principle
A–E) two components
F) three components

The possibilities extrusion offers are listed in Table 1-11. They are still rather theoretical because the methods for the production of web from film are still scarcely used on a commercial scale. However, wherever their physical properties render them useful, the fine networks are processed. This technology is used more frequently for a simplified (shortened) yarn-making process today. In prac-

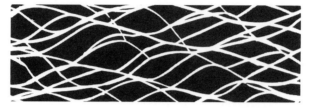

**Fig. 1-36**   Fibre network

**Table 1-11**   Various methods of film extrusion for web production

| | *Type of polymer* | *Web formation* | | *Web bonding* | *Nonwoven finishing* |
|---|---|---|---|---|---|
| Process parameter | type of polymer | extrusion fibre thickness crimp fibre cross-section | laying web organisation | thermal chemical mechanical | dyeing printing finishing |
| Styling possibilities | physical- and physico-chemical properties | | volume handle appearance | handle appearance mechanical properties | volume handle appearance special properties |

tice, the individual fibre thickness ranges from 3 to 10 dtex. One can also produce finer fibres of i.e. 1 to 3 dtex.

The characteristics of the most important polyolefine fibres are compiled in Table 1-12. They also apply to the film fibres outlined above. The most interesting figures are given in Table 1-12, the low melting points and the values for the density. In this context, it must be noted that shrinkage occurs before melting, which appears when work is done at temperatures close to the already low melting point of the polyolefine fibres.

Other characteristics of the fibres important for nonwoven bonded fabrics are that

- the customary polyolefine fibres cannot be dyed with the usual procedure. That is why they are dyed before spinning. Chemically modified fibres are also available, which, though they are dyeable, are expensive as well

- the light resistance of commonly used polyolefines is not sufficient. However, this circumstance can be ameliorated fairly easily and at rather low cost. In most cases the spun-dyed fibres are light-resistant enough for the purpose in question

**Table 1-12** Characteristic properties of the most important polyolefine fibres

|  |  | *Polyethylene* | *Polypropylene* |
|---|---|---|---|
| Maximum tensile strength dry | cN/tex | 50–72 | 40–94 |
| Relative wet strength | % | 95–100 | 100 |
| Maximum elongation | % | 35–20 | 22–15 |
| Shrinkage (95 °C, H20) | % | 5–10 | 0–5 |
| Melting point | °C | 100–120 | 164–170 |
| Density | g/cm$^3$ | 0.92–0.96 | 0.90–0.91 |

- if stretched and set under pressure even in a cold state, the fibres "float" irreversibly

- thus, their recovery is insufficient for many purposes. The circumstance that there is irreversible creep can sometimes be offset by choosing a specific type of construction for the fabrics made from these fibres, e.g. by increasing density to an appropriate high level in the raw state for needled floor coverings

- the fibres do not absorb any moisture, which makes them fit for outdoor use or use on water

As polyolefine fibres are relatively inexpensive compared to other synthetic fibres, the producer of nonwoven bonded fabrics cannot ignore them. There are in fact special fields of application, where the mentioned fibre properties are not a liability but can even be employed in a useful way. It is always advisable to consider mixing polyolefine fibres with other fibres, e.g. as intermediate layers. Something that can lead to complications, though, is the use of polyolefins as a simple substitute for other fibres, without taking into account their special properties.

**Polyacrylonitrile fibres**
The raw material polyacrylonitrile results from the polymerisation of acrylonitrile, also known as vinylcyanide. Polyacrylonitrile is made into fibres by wet or more often dry spinning; it deviates from the previously discussed synthetic chemical fibres which are produced in the melt spinning process. This is necessary, as polyacrylonitrile cannot be melted without being decomposed.

As Fig. 1-37 shows in a diagram, the white polyacrylonitrile powder dissolved in dimethyl formamide (DMF) or in similar solvents is wet-spun in a water bath containing only few additives.

During the dry spinning process the spinning solution is forced through nozzle holes into a vertical 5 m to 6 m high chimney, where hot air flows towards the filaments, causing most of the solvent to evaporate. Most wet- or dry-spun polyacrylonitrile filaments are then made into spun fibre (Fig. 1-38). Pure polyacrylonitrile cannot be dyed using traditional methods. Therefore approximately 10% of a fibre forming chemical (copolymer) is included by polymerisation in the polyacrylonitrile which is to be spun, a chemical that ensures good and easy dyeing with basic

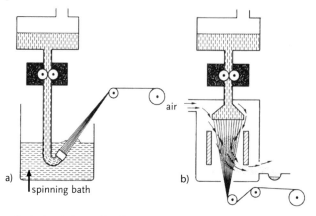

a)

b)

air

spinning bath

**Fig. 1-37** a) Wet and b) dry spinning process for the manufacture of polyacrylonitrile fibres

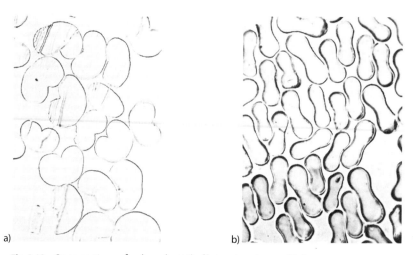

a)

b)

**Fig. 1-38** Cross-sections of polyacrylonitrile fibres: a) wet-spun, b) dry-spun

dyestuffs. The maximum amount of copolymer allowed is 15% if the name poly-acrylonitrile is still to be applied. As explained in the section about bicomponent fibres, this additive also accounts for the good crimp properties of the polyacrylo-nitrile fibres. These fibres are nearly exclusively processed into spun fibre and give yarn and fabric a remarkable volume.

The characteristic properties of polyacrylonitrile fibres are listed in Table 1-13, where no differentiation of interest for manufacturers of nonwoven bonded fab-rics between wet- and dry-spun fibres has to be made.

**Table 1-13** Characteristic properties of polyacrylonitrile fibres

| | | |
|---|---|---|
| Maximum tensile strength dry | cN/tex | 28–56 |
| Relative wet strength | % | 80–95 |
| Elongation | % | 40–15 |
| Shrinkage at 95 °C H$_2$O | % | 0–5 [1] |
| Melting point | °C | not determinable: decomposes |
| Density | g/cm$^3$ | 1.12–1.19 |

1) Except for high-shrinkage fibre.

Further properties not shown in Table 1-13 can be of interest for the producer of nonwoven bonded fabrics:

– low absorption of humidity less than 2% under normal climatic conditions
– the resistance of pure polyacrylonitrile to light and weather and
– the good resistance to chemicals

Furthermore it should be mentioned that the usual kind of polyacrylonitrile fibre
– shows signs of discoloration at temperatures above 140° C and
– is quite easily flammable in loosely bonded fabric, such as cotton and viscose; that is why modacrylic fibres, which normally contain 50% polyvinylchloride (PVC), are used for certain purposes. According to the definition, they must contain at least 35% polyacrylonitrile. The PVC, which is usually added, alters the properties of the fibres depending on its share. Thus the density increases to 1.30–1.42 g/cm$^3$. The flame-retardant effect must not be compromised on the account of the admixture of PVC. The Limiting Oxygen Index (LOI) indicates the smallest amount of oxygen in the air necessary to keep the substance burning. The LOI index for 50%/50% modacrylic fibres is 27% compared to 18% for polyacrylonitrile fibres. Modacrylic fibres are well-suited to being used in mixtures. It has to be mentioned, however, that during thermal decomposition of modacrylic fibres, hydrochloric acid is given off, which is poisonous and causes metal corrosion.

## Polyvinyl alcohol fibre (PVAL)
J. GUOBIAO

When it comes up to PVAL, aftertreated polyvinyl alcohol fibres (PVA) on formaldehyde, called VINYLON in Japan, are usually meant. Most of them consist of staple fibres.

Early in 1924, W. O. Herrmann and W. Haehnel produced PVA on polyvinyl chloride acetate (PVCA) on alcoholysis and made fibres by dry spinning. In 1930, Wacker Company (Germany) manufactured this fibre, named SYNTHOFIL, mainly for surgical threads.

In 1939, Sakurada successfully turned PVA into water-insoluble fibre by aldolization. Industrialization of this fibre was achieved in 1950. Since then, the production has been steadily growing up till other synthetic fibres got on the market.

Today, countries producing PVA fibres are mainly Japan, China and Korea. Because of poor dyeing property and low elasticity, the application of PVA fibre in clothing has withered while in technical textiles, agriculture, fishery and medical use the markets still exist. Now the world output of PVA is more than 100,000 tons per year (about 40,000 tons in Japan, 60,000 tons in China).

**Production of PVA**

Vinyl alcohol (VA) is not able to exist at free state, it will turn into aldehyde:

$$CH_2 = CH \rightarrow H_3C - C = O$$
$$\quad\quad\; | \quad\quad\quad\quad |$$
$$\quad\quad OH \quad\quad\quad H$$

Thus, it is not possible to obtain PVA directly from ethylene. Usually, vinyl acetate is first polymerized to polyvinyl acetate and then PVA is produced on alcoholysis or hydrolysis.

**Synthesis of vinyl acetate**

There are two methods: the acetylene process and the ethylene process.

**(1) Acetylene process**

Acetylene and acetic acid are converted during the vapour phase at about 200 °C, normal pressure and zinc acetate on carbon as catalyst.

$$HC \equiv CH + CH_3COOH \rightarrow CH_2 = CH$$
$$\quad\quad\quad\quad\quad\quad\quad\quad\quad\quad\quad |$$
$$\quad\quad\quad\quad\quad\quad\quad\quad\quad OCOCH_3$$

This conversion is carried out both as carbide acetylene and natural gas acetylene method.

**a) Carbide acetylene method**

The starting component is calcium carbide. It reacts with water into acetylene:
This method consumes lots of electric energy and produces large amounts of carbide slag.

**b) Natural gas acetylene method**

The starting component of natural gas is methane ($CH_4$). It is burnt at 1,300 ~ 1,500 °C with non-sufficient oxygen whereby acetylene is formed.
This method is preferred in practice.

## (2) Ethylene process

Ethylene and acetic acid are converted to vinyl acetate monomer at temperature $>100\,^{\circ}\mathrm{C}$ when Pd/Au catalyst and $NaCH_3COO$, as auxiliary catalyst, are present.

$$CH_2 = CH_2 + CH_3COOH + \frac{1}{2}O_2 \rightarrow H_2C = CH + H_2O$$
$$\underset{OCOCH_3}{\big|}$$

### Polymerization of vinyl acetate

For producing PVA fibre, polyvinyl acetate is usually made by polymerization using methyl alcohol as a solvent.

$$n\, H_2C = CH_2 \rightarrow [-CH_2 - CH-]_n \qquad + 892\ KJ/mol$$
$$\underset{OCOCH_3}{\big|} \qquad\quad \underset{OCOCH_3}{\big|}$$

### Alcoholysis of PVCA

Polyvinyl acetate is converted to PVA by methyl alcohol and caustic for fibre forming.

### Properties of PVA

*Physical properties* (see *Table 1-14*)

### Chemical properties

PVA molecules contain a large amount of hydroxyl groups, so that its chemical property resembles cellulose in many aspects.

Table 1-14  Physical properties

| Appearance | Creamy particles, flakes or powder |
|---|---|
| Apparent density (g/cm³)<br>– low alkali process<br>– high alkali process | <br>0.45<br>0.27 |
| Glass state temperature (°C) | 75 ~ 85 |
| Thermal stability | at 100 °C slow colour changing<br>at 150 °C fast colour changing<br>above 200 °C decomposition |
| Melting temperature (°C) (not plasticized)<br>– totally alcoholysed<br>– partly alcoholysed | <br>230<br>180 ~ 190 |
| Light stability | good |
| Burning ability | like paper |

a) Esterification: PVA can react with many kinds of acids, acid anhydrides, acyl chlorides to form correspondent esters of PVA.

PVA can form intramolecular or intermolecular compounds with several metallic salts. For example: PVA reacts with titanous sulphate to form following structure.

$$-CH_2 - CH - CH_2 - CH - CH_2 -$$

with the $O$ atoms bonded to $Ti$: $O - Ti - O$, and $Ti$ double-bonded to $O$.

b) Ethers of PVA lowers the strength, specific gravity, softening point and hydrophilic property.

c) PVA can be acetalized when acidic catalyst is present.

### Quality indexes and specification of PVA

PVA used for fibre forming must meet several quality indexes, such as degree of polymerization, Ac-content, swelling value, purity, density etc.

### Preparation of spinning solution

The preparation of spinning solution are effected by various steps: water scrubbing and dehydrating lead to the starting component PVA that is inserted into water to 14 ~ 18%, mixed, solved, filtered and defoamed.

### Spinning of PVA solution

There are two methods for PVA spinning: wet and dry process. Wet for staple fibres and dry for filaments for specific application.

In the wet spinning process several salt bathes can be used for coagulation. Sodium sulphate water solution (400–420 g/l, 45 °C) is the preferred one.

With dry spinning, filaments are produced which show a high uniformity, high strength, low extension, high modulus, good dyeing behaviour and have a silk-like characteristic curve. The concentration of PVA-solution is usually 30–40%, the energy consumption is high and the spinning speed relatively low.

### After treatment of PVA fibres

Due to drafting of the spinning fibres, the macromolecules are arranged with a more or less degree of orientation and crystallization and, thus, the final fibre properties are fixed. Drawing takes place at different sectors in the fibre producing process.

Thermal aftertreatment to increase dimension stability, improve mechanical properties and hot water resistance is necessary for fibres to withstand the following acetalization.

Acetalization serves for blocking hydroxyl groups of PVA macromolecules. Commonly used aldehyde in industrial production is formaldehyde. Acetalized PVA fibre has good heat resistance, softening point in water is not until to $110\sim 115\,^{\circ}$C.

$$-CH_2 - CH - CH_{2-} + HCHO \rightarrow -CH_2 - CH - CH_2 - CH - CH_{2-} + H_2$$
$$\qquad\qquad | \qquad\qquad\qquad\qquad\qquad | \qquad\qquad\qquad |$$
$$\qquad\quad OH \qquad\qquad\qquad\qquad\quad O - CH_2 - O$$

Acetalization mainly takes place at the hydroxyl groups that are not involved in crystallization of fibre macromolecules. With increasing degree of acetalization heat resistance of fibre is improved. The degree of acetalization is usually between 20–30% and occurs at about $70\,^{\circ}$C within 10 to 30 minutes.

**Behaviour of PVA fibre**

Hygroscopicity of PVA fibre is the best among synthetic fibres. Blended with cotton at 50/50, the strength is 60% higher than pure cotton yarns, abrasion resistance exceeds pure cotton yarns by 5 times (Table 1-15).

Heat resistance in comparison with other fibres appears to be good (see Section 6.3). Weatherability of PVA fibres in comparison with other fibres turns out to be good to very good (Table 1-16). Chemical resistance of PVA fibres is comparable good to very good (Table 1-17).

As remarkable particularities for application of PVA fibres, the following characteristics should be taken into consideration: poor dyeing behaviour, not in bright

**Table 1-15** PVA-properties in comparison with other fibres

|  | *Strength (dN/tex)* | *Elongation (%)* |
| --- | --- | --- |
| PVA fibres | $\sim 65$ | $\sim 21$ |
| PA 6 | $\sim 50$ | $\sim 27$ |
| Polyester | $\sim 50$ | $\sim 23$ |
| Viscose | $\sim 28$ | $\sim 20$ |
| Cotton | $\sim 35$ | $\sim 8$ |
| Silk | $\sim 50$ | $\sim 13$ |

**Table 1-16** Weatherability of different fibres

|  | *Strength retain after one years' exposure in sun (%)* |
| --- | --- |
| PVA filament (dry process) | 70 |
| PVA (regular) | 34 |
| Polyester | 37 |
| PA 6 | 15 |
| Polypropylene | 11 (after six months) |
| Cotton | 30 |

**Table 1-17** Chemical resistance of different fibres

| Testing conditions | Residual strength (%) | | | | |
|---|---|---|---|---|---|
| | PVA | PA 6 | Polyester | Cotton | Viscose |
| 10 % $H_2SO_4$ 20°C, 10 h | 100 | 54 | 100 | 55 | 55 |
| 10 % NaOH 20°C, 10 h | 100 | 80 | 95 | 69 | 19 |

**Table 1-18** General behaviour of PVA fibre

| Behaviour indexes | | Staple fibre | | Filament | |
|---|---|---|---|---|---|
| | | regular | high tenacity | regular | high tenacity |
| Breaking force (dN/tex) | dry | 4.1–4.4 | 6–8.8 | 2.6–3.5 | 5.3–8.4 |
| | wet | 2.8–4.6 | 4.7–7.5 | 1.9–2.8 | 4.4–7.5 |
| Breaking elongation (%) | dry | 12–26 | 9–17 | 17–22 | 8–22 |
| | wet | 13–27 | 10–18 | 17–25 | 8–26 |
| Loop strength (dN/tex) | | 2.6–4.8 | 4.4–5.1 | 4–5.3 | 6.2–11.5 |
| Knot strength (dN/tex) | | 2.1–3.5 | 4–4.6 | 1.9–2.7 | 2.2–4.4 |
| Elastic recovery after 3% elongation (%) | | 70–85 | 72–85 | 70–90 | 70–90 |
| Density (g/cm$^3$) | | 1.28–1.30 | | | |
| Thermal property | | softening point (dry heat) is 215–220°C, no melting point, flammable, it forms grey or black irregular lumps | | | |
| Sunlight resistance | | good | | | |
| Acid resistance | | not affected by 10 % chloric acid, but swell and decompose in concentrated chloric acid, sulfuric acid and nitric acid | | | |
| Alkali resistance | | strength is almost the same in 50 % custic soda and concentrated ammonia | | | |
| Other chemical resistance | | good | | | |
| Solvent resistance | | not dissolvable in common organic solvents. Swell or dissolve in hot pyridine, phenol, cresol and formaldehyde | | | |
| Abrasion resistance | | good | | | |
| Insect and fungus resistance | | good | | | |
| Dyeing behaviour | | dyeable by direct dyes, sulfurs, azo dyes, vat dyes and acid dyes etc. But the dye uptake is comparable lower, and colour is not very bright | | | |

colour, bad elasticity and crease resistance, poor hot water resistance and at wet state, it will significantly shrink and deform.

General behaviour of PVA fibre includes in Table 1-18.

### High performance and water-soluble PVA fibre

*Wet spun high modulus PVA fibre*

At the end of the sixties, Kurary developed a kind of PVA fibre, called FWB-fibre, which is spun by wet method and boron-added process. For production of high modulus PVA-fibres it is necessary to increase the degree of crystallization and orientation. The resulting properties compared to those of other fibres are shown in Table 1-19.

**Table 1-19** Comparison of (PVA-) FWB fibre with other synthetic fibres

|  |  | PVA FWB fibre | PET | PA 6.6 | HT-CV |
|---|---|---|---|---|---|
| Standard | Tenacity (cN/tex) | 9.3 | 7.5 | 7.5 | 6.0 |
|  | Elongation (%) | 5.7 | 15.7 | 20.6 | 11.0 |
|  | Initial modulus (cN/tex) | 224.0 | 95.0 | 41.0 | 101.0 |
| High temperature (~ 120 °C) | Tenacity (cN/tex) | 8.3 | 5.0 | 4.9 | 5.0 |
|  | Elongation (%) | 6.9 | 18.5 | 22.5 | 9.0 |
|  | Initial modulus (cN/tex) | 120.0 | 41.0 | 11.0 | 84.0 |

FWB fibre is mainly applied for cord or cement substitute of asbestos. The tenacity of this fibre is up to 2.65 GPa, modulus is about 50 GPa.

*Gel spinning method to make high strength and high modulus PVA fibre*

The basis process of gel spinning is as follows: PVA solution of 2–15% (wt.) is first prepared, then extruded into gas or liquid medium to form gel state as-spun fibre after cooling. Then solvent is removed off as-spun fibre by extraction. The influence of high degree of crystallization and drawing to the fibre properties shows Table 1-20.

**Table 1-20** The effects of concentration of PVA solution and drawing on mechanical properties

| C (%) | Drawing | Tenacity (cN/tex) | Modulus (cN/tex) | Elongation (%) |
|---|---|---|---|---|
| 13 | 4.5 | 19.5 | 79.6 | 219.0 |
|  | 22.0 | 83.1 | 86.5 | 15.8 |
| 10 | 4.5 | 30.6 | 14.8 | 91.5 |
|  | 22.0 | 98.9 | 112.0 | 9.6 |

The usual solvent for preparing the PVA spinning solution is ethylene glycol or propanol. As extracting agent are used methylalcohol, ethylalcohol, ethylether etc. Drawing temperature is 200 to 250 °C.

*Water-soluble PVA fibre*

Water-soluble PVA fibre is the only fibre of all synthetic fibres which dissolves in water. Both dry spinning and wet spinning methods can be used for production. Dry spinning method is comparatively simple, especially for fibre that dissolves in water at normal temperature.

To get fibres that dissolve in water of different temperature, following ways of producing process are used.

(1) Changing conditions of fibre forming, drawing and thermal treatment
(2) Reducing the degree of polymerization and alcoholysis of PVA (DP $\sim 1000$)
(3) Changing the chain structure of PVA by graft modification (e.g. with propenol) or oxidative degradation.

## Applications of PVA fibre

Since the eighties, PVA fibres have been rarely used in clothing but more frequently in technical textiles, such as building enhanced material, tarpaulin, industrial sewing thread, braiding flexible pipe, tire cord, transmission belt, agricultural anti-coldness yarn.

Water-soluble PVA fibre is used as fine yarn in textile industry, adhesive fibre to increase strength and flexibility in paper making industry, embroidery base cloth in textile arts and as surgical threads.

## Polytetrafluorethylene fibres (PTFE)

PTFE fibres are extremely resistant to heat and chemicals; they have the lowest known friction coefficient, are nearly inflammable, ultra-violet light-proof, and weather-proof. They also keep their pliability even at very low temperatures. Only the macromolecular structure provides the variety of such unusual properties.

PTFE fibres consist of a linear, unbranched, and not interwoven chain of carbon atoms with two fluorine atoms each. PTFE is mainly used for technical purposes. The wax-like feel, the impossibility of dying, and the high price of the raw materials forestall its use for textiles. The most important area of use is braided cords for packings in glands of pumps, fans, tube tanks and valves.

Another very important area of use of PTFE fibres is the filtration of aggressive gases and liquids. To this end carded webs made of PTFE spun fibres are needled onto a PTFE filamentyarn woven. Reinforced needled webs of this kind can also solve other filtration problems requiring e.g. resistance of the filter medium to chemicals and heat. Apart from a long service life with continuous operation the easy removal of the filter cake from the filter area be an advantage.

**Special synthetic fibres – bonding fibres** (see also Section 3.3)
It was only natural that the industry of man-made fibres foster the development
of nonwoven bonded fabrics by producing systematically constructed bonding fibres. Bonding fibres are understood as fibres which are able to firmly bind other
fibres onto themselves or to each other because they have a different solvent or
melting behaviour. For this end special, homogeneous fibres and also bicomponent fibres are the obvious choice.

Bonding fibres can be divided into three groups:
– soluble fibres, such as polyvinylalcohol fibres and alginate fibres,
– melting fibres, such as copolyamide, bicomponent, mixed polymeric, and thermoplastic fibres with lower melting points than the fibres that are to be
  bonded,
– adhesive fibres like undrawn polyesters.

*Soluble bonding fibres*

Special polyvinylalcohol fibres (PVA) are probably the oldest known bonding fibres. Their behaviour in water at different temperatures is shown in Fig. 1-39.
The original fibre, which was well suited to being made into a web, swells in hot
water and after gelling dissolves step by step as the temperature is increased. At
the right temperature an effective bonding effect with and between other fibres
can be achieved. Polyvinylalcohol fibres can easily be modified more or less
strongly. For example three types, which dissolve at 60 °C, 70 °C, or 80 °C respectively, are available (see also Section PVA fibres).

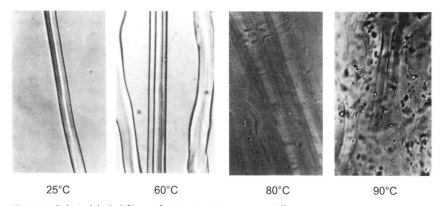

| 25°C | 60°C | 80°C | 90°C |

**Fig. 1-39** Polyvinylalcohol fibres after treatment in water at different temperatures

*Melting bonding fibres*

Various copolyamide fibres are known at present. Fig. 1-40 shows the behaviour of
CoPA K 115 in water at 25 °C and 100 °C. The copolyamide fibre CoPA K 140 be-
haves as shown in Fig. 1-41.

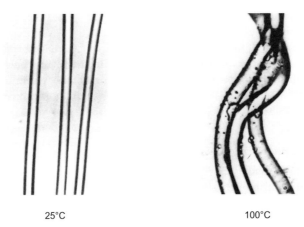

25°C                                    100°C

**Fig. 1-40**  Copolyamide fibres CoPA K 115 after treatment in water at different temperatures

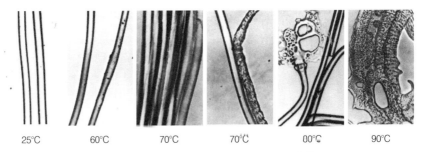

25°C          60°C          70°C          70°C          80°C          90°C

**Fig. 1-41**  Copolyamide fibres CoPA K 140 after treatment in water at different temperatures

Before softening and melting on, which already starts at a little over 70 °C, the fibres shrink with only insignificant shrink strength. But this bonding effect takes place at a considerably lower temperature than it does in CoPA K 115. The lowering of the melting point in the water/copolyamide system backs up this process. Before reaching the bonding temperature the copolyamide fibres shrink by 45% to 50% (Fig. 1-42). Therefore a nonwoven bonded fabric consisting of only 50% CoPA K 115 can shrink by up to 35%, depending on its mass and construction.

Fig. 1-42 shows the reaction of the two bonding fibres. The figures for shrinkage, however, do not indicate the shrink strength, so it must be emphasized that the shrink strength in the CoPA K 140 type is practically insignificant, whereas it is a principal feature of the CoPA K 115 type.

Fig. 1-42 shows also that the copolyamide fibres' in the binary water/fibre system react differently to a shock-like increase of temperature and to a slower temperature increase. Furthermore the figure shows when the bonding effect begins in hot air. The fibres' different reactions under these three conditions mentioned show that it is essential to find the desired bonding effect by experimenting. The low melting points impede the production of cut fibres in both types. There is a

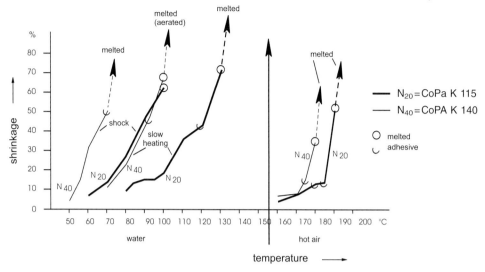

**Fig. 1-42** Reaction of the copolyamide fibres CoPA K 115 and CoPA K 140 in wet or dry medium depending on the temperature
Shock shrinkage = rapid immersion in hot water (length of time 10 min)
Slow heating = immersion in cold water, slow heating (length if time 1 h)
Hot air = length of time 3 min (new material was taken for each point measured)

risk of melting at the fibre ends, but suitable opening agents can be used to separate these again.

*Bicomponent fibres*

Skin/core fibres consisting of two polymers, with a coating polymer which melts at a lower temperature than the core polymer, are suitable as bonding fibres. Various types of fibres are available:

• Heterofil fibres consisting of a polyamide 6 skin and a polyamide 6.6 core. Perfect bonding can be achieved by heating the whole web or smaller areas, throughout the web, to temperatures well above the melting point of polyamide 6, with the polyamide 6.6 core remaining intact. Fibres with a PP skin and a PET core have similar properties maybe even more distinct.

*Mixed polymer fibres*

The following types belong to this group:

• MP fibres, which have been available for years, consist of 85% PVC and 15% polyvinyl acetate. They soften at a temperature between 70 °C and 80 °C and melt at 159 °C. As the bonded fabric is hard to the touch, the use of MP fibres is restricted.

- Efpakal L90 from Japan consists of 50% PVC and 50% polyvinyl alcohol. The polyvinyl alcohol part dissolves in water at 90 °C whereas the PVC part softens, which makes the dissolving and softening reaction of this fibre quite interesting.

*Adhesive fibres*

Unstretched PET fibres belong to this group. When they are warmed up to the glass-transition temperature, which is ~80 °C when fibres are dry and plainly lower when they are wet, the fibre turns soft and sticky for the time of the transition period. Fig. 1-43 shows how the surface of the fibre changes after treatment in water at 100 °C. When such sticky fibres now come into contact with other fibres or if they are pressed against them, for example during calendering, strong irreversible bonding develops. The bond cannot be undone even if the fibres are heated again. Moreover it is notable that the bonding points are almost punctiform, which is positive for the feel and flexibility of the nonwoven bonded fabric.

During the production and the use of such unstretched PET fibres it is necessary to keep them unstretched since even a slight stretch has an adverse effect on the desired result. The binding capacity of the fabric decreases rapidly as crystallization and the orientation of the polyester molecules increase. The few examples of available bonding fibres show which methods are used to produce such fibres. But they also show that it is possible to have special fibres tailor-made for interesting purposes. The only problem is that the cost is high because demand is relatively low.

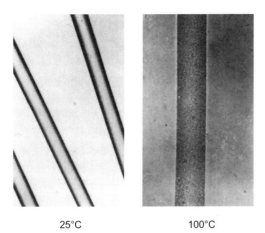

**Fig. 1-43** Unstretched PET fibres after treatment in water at various temperatures

25°C          100°C

*Monocrystalline fibres (whiskers)*

Whiskers are extremely fine, monocrystalline fibres that are produced under carefully controlled conditions. Their diameter is small (between 1 μm and 30 μm), but nowadays it is also possible to produce thicker fibres. This development has positive effects on the fibre length. So it is possible to produce whiskers that are longer than 20 mm. The tensile strength of these mono crystalline fibres is incredibly high. The bond is almost as strong as an atomic bond. In composite material whiskers increase the fabrics' strength as soon as the fabric contains approximately 5% of whiskers. Their high tensile strength and modulus makes them even suitable as reinforcement of metals. The high price for whiskers prevents them from being used broadly. The types of whiskers available are:

Aluminium oxide (sapphire) ($Al_2O_3$), graphite (C), silicon carbide (SiC), silicon nitride ($Si_3N_4$). During production and especially during processing, safety regulations must be strictly observed.

*Polycarbonate fibres – Polyester of carbonic acid*

The fibre is made from bisphenol A and phosgene, and is either wet- or dry-spun.

Polycarbonate fibres
– have a density of 1.2 g/cm$^3$
– are heat resistant up to 300 °C
– are non-flammable
– have excellent insulating capacity
– have low water absorption (see Section "PET fibres").

*Melamine fibres*

Melamine fibres are extremely suitable for the production of heat resistant and inflammable textiles. BASOFIL by BASF with its mesh structure has the characteristic features of such condensation resins such as hardness, chemical resistance and the above-mentioned outstanding fire resistance. The characteristics are:

| | | |
|---|---|---|
| fibre fineness | dtex | 2.2 |
| density | g/cm$^3$ | 1.4 |
| tensile strength | cN/dtex | 2–4 |
| breaking elongation | % | 15–25 |
| LOI | | 32 |
| dyeability | | dyeable |
| shrinkage at 200 °C | % | 2 |
| permanent temperature resistance | °C | approx. 200 |
| drain-off reaction | | no draining off |
| melting point | | no melting point |

The melamine fibre BASOFIL may be processed to textiles using the usual procedures. For the production of finer yarns and applications requiring more stability, the addition of stronger fibres like PES, aramides, highly strong polyethylene or others has proved worthwhile. The required temperature and burning reaction are to be taken into account, too. The use of BASOFIL fibres, pure and in adjusted mixtures, proved highly suitable for nonwoven bonded fabrics used in hot environments and at high temperatures and as a fire barrier.

The main areas of application are:
– protective textiles exhibiting special properties
– heat and noise insulating materials
– nonwoven bonded fabrics for special filters
– textiles for fire barriers

Synthetic fibres produced by unconventional methods also belong to the group of special fibres, which includes drawing of fibre collectives: basic materials of various kinds are covered with thermoplastic polymers, and the heated polymer is subsequently drawn off to form fibres. Depending on the composition of the polymer and the production conditions, a pile consisting of fibres of various fineness and length still anchored in their polymer base forms on the surface. The characteristics of the fibres are determined by the polymer and the processing conditions. The most suitable polymers are those with a low melting point and a good fusibility, i.e. polyolefines.

Another method belonging to this group is fibre spraying from polymer solutions or molten polymers, as shown in Fig. 1-44. The indicated electrostatic field aids the process. The properties of the produced fibres vary widely, which need not necessarily be a disadvantage for some purposes, e.g. filters.

Fig. 1-44 shows that the sprayed polymer particles take on the shape of fibres after the spraying process and are then aligned by the electrostatic field. The fibres may then be carried out on a sieve or on a base material necessary for the process.

Fig. 1-45 shows another unconventional method: the polymer, usually polypropylene or polyethylene, is melted by extruders and then pressed through many small openings set in a row. At or below the spinneret, the molten polymer is hit by hot air, which stretches the filaments up to a diameter of 0.5 µm to 3.0 µm. The force of the streams of hot air and the fineness makes the filaments likely to break, so very fine filaments of various lengths are directly laid into a web on a conveyor belt.

Concluding, there are some ideas that have been published by the press but have not yet been developed to the production stage. One such concept is to pour monomers, which are capable of forming polymer compounds, together with catalysts and filling material onto a conveyor belt, to polymerise them and then to re-

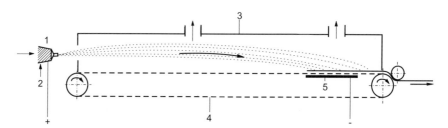

**Fig. 1-44**  Fibre spraying from polymer (solutions) and web formation in an electrostatic field:
1 spinneret, 2 air supply, 3 drying chamber, 4 feeder belt, 5 backplate electrode

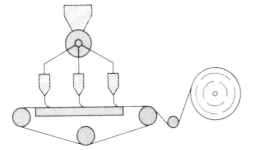

**Fig. 1-45** Schematic representation of the melt-blowing process

move the supplementary substances, if necessary. Such processes are conceivable and could be designed to produce specific features.

### Spun laid fabrics made of cellulose filaments

If the machinery is adapted, the different methods to produce chemical fibres can also be used for the direct manufacture of nonwovens. These nonwovens are also called "spun laid fabrics". Usually they are referred to as nonwovens produced by way of the melt spinning process. Asahi Chemical Industry Co. Ltd. in Japan, however, produces spun laid fabrics consisting of 100% cellulose filaments, using the "copper method", a wet spinning process. It is sold under the name "Bemliese". As is usual in the "copper method", in order to produce Bemliese, the cleaned cotton linters are dissolved in cuprammonium. Then the solution is deairated, filtered and spun into fibres (Fig. 1-46).

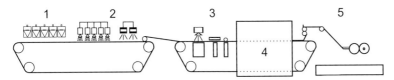

**Fig. 1-46** Production of Bemliese: 1 spinning, 2 refining, 3 water-jet process, 4 drying, 5 rolling-up

The spinning process is characterised by mild coagulation and a high stretching of the forming filaments. They are laid to form the nonwoven on a vibrating perforated belt, then washed and in a next step hardened by waterjets, and finally dried and wound. The copper method makes it possible to control the coagulation process precisely. Thus it is possible to reach the appropriate degree of "natural bonding" between the single, very thin cellulose filaments, which gives the nonwoven particular characteristics (Fig. 1-47), which can furthermore be obtained by the waterjet consolidation.

**Fig. 1-47**   Fleece structure of Bemliese (magnified approx. 1,000 times)

The manufacturing conditions and the characteristics of this kind of nonwovens can be summarised as follows:

**Manufacturing**

| | |
|---|---|
| Raw material | cleaned cotton linters |
| Solvent cupric | sulphate/ ammonia solution |
| Agents | none |
| Regeneration bath | 1. water |
| | 2. diluted sulphuric acid |
| Wash baths | water |
| Web formation | controlled deposit of filaments, formation of punctual bonds, waterjet consolidation |
| Manufacturing process | continuous |

**Characteristics of the nonwoven**

| | |
|---|---|
| Fibre fineness | adjustable between 0.9 and 1.7 dtex, very even |
| Fibre length | endless filaments |
| Avivages | none |
| Binder | none |

The resulting serviceable qualities are the following:
– practically pure cellulose
– lint-free
– non-toxic
– high and fast absorption of humidity
– high water retention
– antistatic
– uniform in density and thickness
– high-bulking and soft
– biodegradable

Because of these characteristics Bemliese is used in different fields:

Consumer goods:         cleaning clothes, cosmetic pads, tea bags etc.
Medical textiles:       sponges, gauze, towels, adhesive bandages etc.
Industrial textiles:    clean room cloths, cleaning cloths for high-tech products

Depending on their use, different types of nonwovens are offered (thickness of filaments, mass per unit area, solidity, purity, strengthening fibres etc.). In this way, the market's range of nonwovens is enriched by cellulose fibres.

## 1.2.3
## Modification of synthetic fibres

What has already been said about the manufacture of synthetic fibres that are suitable for nonwoven bonded fabrics proves that it is not only possible to use conventional spun fibres and filaments, but also to design special fibres for particular uses. Techniques for fibre production are usually special and independent from the raw material. See the summary in Table 1-21.

**Table 1-21**   Modifications in synthetic fibres

| *Additives* | *Effect* | *Modification taken in production process* | *Effect* |
|---|---|---|---|
| Delustering agent | Dye affinity, lustre | Amount of polymer per nozzle hole | Fibre thickness |
| Light-proofing agent | Light resistance | Cross-section of nozzle hole | Fibre cross-section |
| Barium sulphate | X-ray contrast effect | Extent of stretching | Strength, elongation behaviour |
| Dyestuffs | Colour, colourfastness | Temperature during stretching | Shrinkage a.o. |
| Optical brightener | Degree of whiteness, fluorescence | Cutting | Length of fibre |
| Flame-retardant agent Modifier | Flame-retardation Modification of physical and chemical properties | | |
| Two components | Bicomponent effect, crimp, physical and chemical properties | | |
| Special dyestuff-couplers | Colour, visual effects on tissue | | |
| Antistatic | Antistatic properties | | |

The first kind of substance to be discussed is the delustering agent. This is usually titanium dioxide, which is spun into the fibre in amounts varying from 0.04% to 4%. There are two modifications of titanium dioxide which are available: anatase and rutile. The delustering effect is less pronounced with anatase than with rutile. Nevertheless it is used more often, because it is not as hard as rutile and so causes less abrasive damage to machinery and guide parts during fibre manufacturing and processing. Rutile, with its so-called rounded corners, does not offer a satisfactory solution either. – In theory, barium sulphate and silicates can also be used as delustering agents. They perform reasonably well, but affect the colour of the crude fibre, shifting it from white into yellow, which is not desirable.

Suitable *light-proofing agents* are bivalent manganese compounds and phosphates. They are deposited on the surface of titanium dioxide. Thereby the negative effect of titanium dioxide is eliminated and the fibre substance is protected. Exposing titanium dioxide to ultra-violet rays in a moist environment may cause decomposition. But in the production process lightproofing agents can also be added to the fibre raw materials themselves before spinning. Quite often the impact of light-proofing agents is considered completely independently from agents used to stabilise temperatures. This is certainly not warranted, since light is only a form of energy after all. Of course, a wavelength can be different from those of other forms of energy. It is often found that the light-proofing effect decreases with time. This is because the usually small amounts of light-proofing agents, 100 ppm to 500 ppm, get washed out. Therefore, to improve lightfastness, the use of fibres made of light-resistant raw materials is recommended. Given in order of increasing light-resistance, the conventional synthetic fibres are polyamide → polyester → pure polyacrylonitrile. In cases where resistance to light is an important factor, it is advisable to use bright fibres.

*Barium sulphate*, in concentrations of about 40% and upwards in fibres, shows up perfectly on x-ray plates, e.g. of people. Even parts of the body with many bones give impeccable contrasts, provided the x-ray density is sufficient. Therefore such fibres are used increasingly in the medical sector. In comparable textiles, yarns are usually worked into the fabric.

*Dyestuffs* in concentrations ranging from less than 1% to 15% are applied in the manufacture of so-called "spun-dyed" or "dyed in the mass" fibres. Generally, inorganic and organic pigment dyes are used. During wet spinning these dyestuffs have to disperse (e.g. in viscose) and must not changed when the cellulose is regenerated. Also they must not dirty the regenerating bath. The dyes have to withstand temperatures of about 200 °C to over 300 °C encountered in melt-spinning and dry-spinning. Moreover, coloration must not be impaired when the fibres are stretched. Spun-dyed fibres are usually perfectly colourfast and have proved their worth especially in furnishing fabrics (e.g. also in needled floor coverings) and automotive textiles.

For practical use, *brighteners* can be regarded as dyestuffs. They can be applied to the fibres in baths – e.g. with the avivage – or added to the mass during the production of fibre, a method which is used more commonly. They are "light converters". They partly convert short-wave rays which are rich in energy into longer, visible rays. The resulting fluorescence also influences the colour of the fibres before and after dyeing and the visual result of the fabric made from them. According to the current regulations, textiles used in the medical sector must not contain visual brighteners, neither on, nor in the fibres. In this context it is important to be aware of the fact that in Germany a difference is made between surgical and cosmetic cotton wool. This means that the use of brightened fibres for cosmetic cotton wool is permitted.

Analogously to the agents that have already been discussed, *flame-retardant agents* can also be applied to the surface of the fibres. Flame-retardant agents are special products which contain phosphorus, sometimes also nitrogen and usually halogens, and are applied by means of conventional or improved techniques. Such products can also be bonded to the basic fibre if the process is suitable. This makes the textile very durable. Furthermore, chemicals of similar composition can be spun into the fibres. In this case the effect is much greater than when the chemicals are applied to the surface. However, in most cases much higher concentrations are needed, which raises the cost of the fibres considerably and affects their physical properties. Besides, the production of all types of fibres becomes more complicated. Basically, highly "flameproof" synthetic fibres can be produced that way. But special raw materials can also be used, e.g. aromatic polyamide (aramide), metalliferous chelate, PVC-mixed polymers such as PVA/PVC, mod-acrylics, and also inorganic substances such as carbon. But all these fibres involve problems of varying magnitude with them as far as price, melting point, dye affinity, the original colour of the fibre or physical textile properties are concerned. If such fibres are heated, they may give off corrosive or even poisonous gases and large amounts of smoke. The fibre has to be chosen to suit the specific use. This requires a profile of the exact requirements of the respective finished textile product and a comparison with the performance profiles of the various fibres. It is also important to consider how the synthetic fibres react during melt. On the one hand the melting can be dangerous, on the other hand it removes great amounts of energy from the source of fire and so makes such textiles less flammable. The same problem arises when textiles, e.g. a lady's dressing gown lined with polyester-fill, catch fire while they are worn. The melted fibre will certainly cause serious burns of parts of the body. However, if the gown is made from easily flammable fibres such as cotton or viscose, it is highly probable that the fire will spread quickly throughout the textile. This brief description of how synthetic fibres are melt emphasises how important it is to fully consider the consequences of a fibre's characteristics when compiling the performance profiles for textiles.

*Modifiers* are added to the chemical fibre raw material in relatively small amounts – from under one percent to a few percent – but still they have a relatively strong effect. For example, they can influence the physical characteristics of viscose fibres enormously, but are hard to detect in the finished fibres. They have opened up new possibilities for fibre production, and their use has led to improvements of quality and new fields of use.

Without explaining in greater detail how modifiers work, it should be pointed out that, for example, in the viscose process they slow down the regeneration of the cellulose from the viscose with help of the regenerating bath (also called spin-bath). This makes it possible to carry out the stretching in such a way as to obtain a better order of macromolecules. This again leads to the macromolecules – which are already quite long themselves (the average degree of polymerisation of the cellulose is raised from about 300 under normal conditions to 400 or 500) and which are used in such cases – display more suitable physical characteristics in the fibre. Such fibres typically have increased dry-, wet- and crosswise tenacity.

As a *second component*, many modifiers – apart from the already mentioned ones – can be added to the different spinning solutions or melted mass. Mostly, they remain foreign bodies in the fibre substance itself and more or less distinctly change the physical and chemical characteristics of the fibre (melting point, crimp, shrinkage, dye affinity, tenacity etc.). For example, as already described in the section about polyamide fibres, increasing amounts of polyamide 6.6-conden-sate can be added to polyamide 6. When this mixture is remelted, passed through jets and then – as in the production of synthetic fibres – drawn off and at the same time quickly cooled down, the polyamide 6.6-condensate will solidify first and hinder the formation of the usual orientation of the polyamide 6-polymeride, also during the following stretching process. The result will be fibres with a high crimp, shrinkable at a certain temperature and a lower melting point (Fig. 1-41). But a second component can also be used for totally different reasons. In such a case, under certain production conditions, the second component links the mole-cules of the basic substance laterally. This linking increases the shape stability of the molecular compound and also changes the rest of the physical and chemical fibre characteristics, like the softening and melting behaviour, tenacity, stretching, shrinkage, dyeing, etc. The second component, however, can also have a support-ing function only, as you can see in Fig. 1-21. The surrounding matrix of a bicom-ponent fibre with very thin continuous fibrils of 0.1 to 0.01 dtex the surrounding matrix is dissolved away fully or partly. The thus uncovered fibril bunches, which can be produced only in this fashion, are similar to the fibril bunches of genuine leather. Therefore they can be made successfully into a poromeric synthetic leather with very attractive characteristics. This interesting way of producing very fine fibres can also be used for other purposes, but it is very labour-intensive.

Of course, *special dyestuff couplers* can be used as a second component. But the basic molecules can also be modified, e.g. in polyamide, as has already been out-lined. For example, couplers allow the number of amino-groups per unit of the substance to be varied. This modification influences dyeing when acidic dyestuffs are used and can be suited to special uses. All in all, the reactivity of the whole fi-bre substance can be influenced in a positive or negative way, or it can be modi-fied in a specific way.

*Antistatics* that are spun especially into polyamide fibres are usually organic ad-ditives which permanently give antistatic characteristics to the fibres and the tex-tile fabrics made from them (e.g. needled floor coverings). They lower the charge in a person to 2 kV to 3 kV at 25% to 30% relative humidity, i.e. to be below the

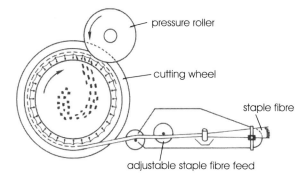

**Fig. 1-48** Schematic view of the staple fibre cutting technique

so-called "shock barrier". That means that the shocks that usually occur when a person touches conductive objects after stepping onto textile floor coverings no longer occur – or more correctly – are no longer felt. These antistatics have replaced the follow-up treatment of floor coverings. Another type of antistatic is steel fibres, which can also be incorporated into floor coverings. Therefore it is important to define the demands on the nonwoven bonded fabrics precisely and to select the fibre that suits those demands best.

After this short description of the manufacture of chemical fibres and their possible modifications, we will discuss only one more specific way of cutting filaments into short fibres. Conventional techniques, which are still employed today for spun fibre production, used special cutting devices with rotating blades cutting the chemical staple fibre into respective lengths. For the production of short cuts of, for example, 3 mm to 15 mm, for the wet-laid nonwovens cutting machine, guillotine-like cutting machines were used. Both of these techniques have now been replaced by the cutting technique shown in Fig. 1-48. The results of this technique have a much better quality for short cut processing, because only few fibres have overlengths. The somewhat greater variations in length above and below the length of the prescribed length are of hardly any importance.

As you can see in Fig. 1-48, the staple fibre is wound around a cutting wheel at delivery speed. The cutting wheel has blades that are placed at specific distances from one another as it comes from production. The pressure roller presses the staple fibre against the blades where it is cut. The cut fibres are moved towards the interior of the cutting wheel.

The methods to produce modified fibres "made to measure" that are summarised in Table 1-21 are especially interesting for the field of nonwovens, because they help with the choice of the proper fibre for a specific task. Thus nonwovens become fabrics of special purposes with advantageous characteristics. With these fabrics it is possible to produce fibres suited to specific demands, whereas with conventional textiles the range of applications is extremely limited.

## 1.3
## Other fibres made in industrial processes

### 1.3.1
### Glass fibres

As to the raw material glass used to make glass fibres or nonwovens of glass fibres, the following classification is known:

- A-glass:    With regard to its composition, it is close to window glass. In the Federal Republic of Germany it is mainly used in the manufacture of process equipment.
- C-glass:    This kind of glass shows better resistance to chemical impact.
- E-glass:    This kind of glass combines the characteristics of C-glass with very good insulation to electricity.
- AE-glass: Alkali resistant glass.

Generally, glass consists of quartz sand, soda, sodium sulphate, potash, feldspar and a number of refining and dying additives. The characteristics, with them the classification of the glass fibres to be made, are defined by the combination of raw materials and their proportions.

Textile glass fibres mostly show a circular cross-section and a fibre diameter of no more than 18 μm. In industry, fibres of 8–12 μm are usually processed, which means ~ 1.2–2.8 dtex.

For use in special-purpose filters, fibres of 1–3 μm may be used. Fig. 1-49 describes the manufacture of the glass melt.

The subsequent manufacture of glass fibres may be executed to the direct melting process. However, in most cases glass rods or balls are made first which then may undergo a variety of further processes.

**Nozzle-drawing.** As can be seen in Fig. 1-50, the glass fed in is melted in a heated melt tub at 1250–1400 °C. Then, it emerges at the bottom of the melt tub from

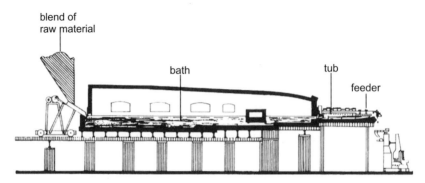

**Fig. 1-49** Manufacture of glass melt

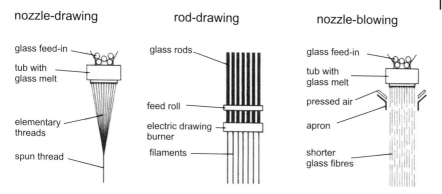

**Fig. 1-50**  Processes to make glass fibres

nozzle holes of 1–25 mm diameter and it is taken off and drawn. The filaments solidify and are finished and wound. One can find them in the shops as various kinds of "glass silk". To make them into webs, the filaments are cut to length (mostly, between 6 and 25 mm).

**Nozzle-blowing** (Fig. 1-50, right). The same as with nozzle-drawing, glass balls are melted in the tub. The melt emerging from the nozzle holes is then taken by pressed air, which draws the liquid glass so as to make fibres of 6–10 μm diameter. A fluttering effect is caused by the flow of pressed air, which results in fibres of lengths from 50 to 300 mm. A lubricant is put on and the fibres are laid down on a sieve drum which sucks them in. The dry web received is held together by the long fibres, the short ones lying in between them as a filling material. Then, the slivers of glass fibre material are cut.

**Rod-drawing** (Fig. 1-50, centre). By means of a burner, bundles of glass rods are melted at their bottom ends. This results in drops which, as they fall down, draw filaments after them. The filaments are taken by a rotating drum, a squeegee laying them down onto a perforated belt. Thus, a dry web is received which can be wound as glass fibre slivers. – Machine performance being limited by the number of glass rods fed in, the rotating drum may be combined with nozzle-drawing, which results in drum-drawing. This multiplies machine performance. The dry web is again laid down onto a perforated belt and solidified or, after winding it so as to receive slivers, cut for further processing on machines producing wetlaid nonwovens. Using and processing glass fibres is not without any problems. For example, fine pieces of broken fibres may disturb if the work place is not well prepared for the purpose. Using the nonwovens to manufacture glass-fibre reinforced plastics, it is important the surface of the plastic material is fully even. Ends of fibre looking out may be pulled out or loosened by outward stress (temperature, gases, liquids), which may influence material characteristics. In some cases, it is advisable to cover up such layers of glass fibre with suitable chemical fibres.

### 1.3.2
### Silicate fibres

Silica fibres and quartz fibres consist of silicon dioxide. Silica fibres are received by washing E-glass as long as only the silicon dioxide skeleton remains. Even if such fibres are not very strong, their resistance to heat is extremely high.

Quartz fibres are made by means of spinning quartz glass melts. They show a number of outstanding characteristics such as excellent resistance to heat or chemical substances and surprising tenacity as well as the high modulus. [1]

### 1.3.3
### Carbon fibres

In recent years, carbon fibres have found much attention. They are made by means of thermal degradation (pyrolysis) of viscose fibres or polyacrylonitrile fibres at temperatures up to $1000°$ C or even $1500°$C. The fibres contain between 95 and 98% carbon. Additional thermal treatment of the pyrolized polyacrylonitrile fibres at temperatures between 2000 and $3000°$C makes them into graphite, which shows an excellent grid-like structure, carbon contents amounting to $\sim 99\%$. The high modulus of carbon fibres makes them best suitable for application in composite materials used with high-performance functions, e.g. in the manufacture of aircraft or in space research. However, making carbon fibres is cost-intensive, which dampens sales. As to modulus, strength and elongation, the range of fibres available corresponds with the different applications.

### 1.3.4
### Boron fibres

Generally, boron fibres are made by means of boron vapour condensing on tungsten wires of $\sim 12\ \mu m$ diameter. The boron fibres received are $\sim 100\ \mu m$ in diameter. For carrier materials, glass, graphite, aluminium and molybdenum may be used alternatively. Preferably, boron fibres are used to make pre-impregnated ribbons, which contain 65–70% of fibre and 30–35% of epoxy resin or of phenolic resin or of polyimide resin.

### 1.3.5
### Metal fibres

Steel fibres of diameters from $\sim 75$ to $250\ \mu m$ are widely found today. Fibres of smaller diameters (down to $25\ \mu m$) are known of lengths as are usual with spun fibres. With corresponding composite materials, the high density of steel fibres reduces the strength to mass ratio. In addition, fibres of stainless steel of diameters

---

[1] Excellent resistance to heat is also found with ceramic fibres and fibres of stone. Together with glass fibres, they may be processed into mats showing good thermal and sound insulation capacity (see Section 15.1).

from 5 to 25 µm are also available. They are on the market as filament yarns, rovings, mats and wovens.

Except for steel, metal fibres of aluminium, magnesium, molybdenum and tungsten are made in the laboratory which, in single cases, are used for particular purposes.

## 1.4
## Reclaimed fibres
B. GULICH

Reclaimed fibres are textile fibrous materials as are found in the second or further production cycles. They are produced from used textiles and textile waste. Textile fibrous materials with their particular characteristics and their inherent functionality are best suited for repeated use. The collection of used textiles and their further use as textiles and, in addition, the re-use of the textile raw materials, has been known for many centuries. Up to the 17th century textile waste was exclusively used to make paper.

### 1.4.1
### Basics

With the development of processes to reclaim used textiles it became possible to recover longer fibres and spin them into yarns. At that time, a well-functioning system of textile circulation was mainly encouraged by the high raw material costs and limited availability of textile raw materials. Its core was the manufacture of reclaimed fibres.

Since the time synthetic fibres came up, primary fibrous materials have been available in sufficient quantities. Today, it is mainly for ecological reasons that textile recycling as a traditional branch of the textile industry continues to be of high interest. Sensible re-use, preferably in textile but also in non-textile applications, saves raw materials and contributes to avoid damage to the environment. However, recycling by tearing should preserve the functionality of the textile fibre and be economical. Used clothes and waste textile materials represent the classical source of raw material for the production of reclaimed fibres. In these days, they come from both private and household consumption as well as from the industry. For long, non-textile parts contained in the materials, dirt as well as material and structure variety made low-cost recovery difficult. Recently, effective equipment-related and technological solutions to separate non-textile parts have been found and transferred into practice. In Germany there are several complete systems to economically produce reclaimed fibres from textile waste. So far, german producers of reclaimed fibres mostly take their raw materials from production waste as found in the textile and garment-producing industries. Table 1-22, except for a steady rise in textile waste quantities, shows an enormous potential of raw material [31].

**Table 1-22** Quantity of textile waste available in Germany 1995 (according to [31])

| Type of waste | Quantity in tons |
|---|---|
| Production waste of textile industry | 65,000 |
| Production waste of the garment-prusing industry | 35,000 |
| Production waste of the chemical fibre industry | 70,000 |
| Waste textiles | 960,000 |
| Used industrial textiles | 250,000 |

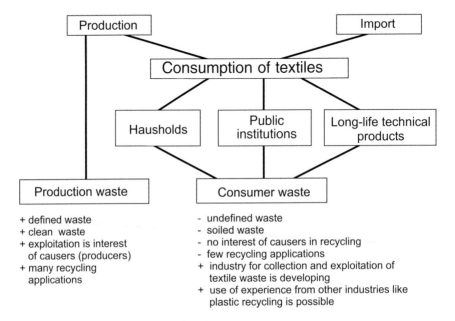

**Fig. 1-51**  Origin and characteristics of textile waste

Production waste can be clean-sorted and collected at reasonable cost. This is how, to a high degree, it can be distinguished by structure and composition (Fig. 1-51). Due to the use of non-textile materials (membranes, coating etc.) to make composites for technical applications, problems may occur as to processing.

1.4.2
**Making textile waste into reclaimed fibres**

1.4.2.1  **Pre-treatment**
The tearing process requires a number of steps to pre-treat the material. In this context, cutting the textile waste is particularly important as it largely influences the result of reclaiming. Pre-cutting is a prerequisite and the first step into a continuous process from the waste bale to the reclaimed fibre. Cutting machines for

textile waste work according to the principles of rotation cut and guillotine cut. Due to more accurate cutting and less inclination to lap-forming, numerous manufacturers prefer machines working according to the guillotine principle. To achieve even edge lengths, the material is usually fed by means of two cutting machines arranged to one another at a right angle. Semi-automatic cutting machines are especially economical, the textile waste being fed into them directly from the bale. The edge lengths of the pieces cut range, depending on the design of the material feed-in, from 40 mm to 150 mm. To achieve even product qualities and for logistic reasons, it is advisable to install boxes for the temporary storage of the cut material. In addition, it is helpful to moisten the materials with water or to grease them, which promotes the process of reclaiming and may contribute to easy processing in the steps to follow.

### 1.4.2.2 **Principle of reclaiming**

Reclaiming is the main processing step in textile recycling and it preserves the fibrous structure to a large degree. The way of how the textile structures are broken down is the key to higher or lower-quality fibres for new textile or non-textile products. This is how a reclaiming machine works (see Fig. 1-52): Roughly pre-cut material is supplied by a fast-rotating drum whose feed-in system both conveys and grips.

The feed-in system may consist of two rotating rollers or the combination of a rotating roller with a rigid trough. Today, the trough system is more frequent as it allows to react to the characteristics of the material fed in. On the surface of the drum, a large number of sword-shaped or round steel pins can be found. Together with the gripping system, they break down the textile structure. One partic-

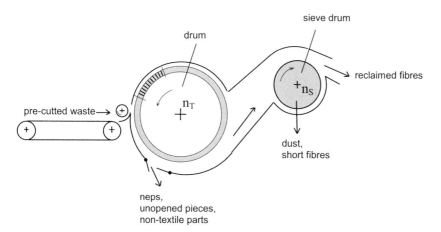

**Fig. 1-52** Structure in principle of a breaking-down unit (number of revolutions of the drum $n_T \gg$ number of revolutions of the sieve drum $n_S$)

ular design of tearing drum shows pickers to process used garments, another is equipped with sets of saw teeth to process threads and nonwovens. From the trajectory the material takes around the drum, unopened material known as neps, may be ejected (due to their higher mass). The material broken down is sucked in and cleaned by a sieve drum. Subsequently, the material (the felt created around the sieve drum) is conveyed to the next breaking-down unit to be fed in there. Usually, a suction unit is found at the last breaking-down unit which connects the machine with the baling press. Depending on the type and structure of the textile waste, several tearing cycles may be necessary to receive single fibres. To reclaim production waste, both single drum machines and system courses with 3 to 6 breaking-down units are usual. With systems of more than one unit, the number of pins on the drum and their fineness need to increase so as to better break down the material. Differences in the equipment of two subsequent drums should not be too large so damage to the material and machine overload can be held at a low degree. Working widths of systems up to six breaking-down units are usually between 1,000 and 2,000 mm. Depending on the material, manufacturers state throughput performances up to 1000 kg per hour and per meter of working width. Single and two-drum machines are mainly suitable to process a closely limited assortment of constant product properties and of small quantities. Such machines are frequently found with in-plant circulation systems, particularly in the field of nonwoven manufacturing. Working widths between 500 and 1,000 mm permit a throughput from approximately 100 to 450 kg per hour.

The economical manufacture of reclaimed fibres from textile waste requires comparatively high technical and technological expenditure. One to two coarse pickers applied before the reclaiming process itself and the repeated material passage via different separating mechanisms (to separate non-textile parts) are characteristic of such systems.

### 1.4.2.3 Subsequent treatment

Generally, subsequent treatment focuses on further quality improvement so the reclaimed fibre, as compared to primary fibrous materials, is better suitable for application. Process steps as freeing from dust, blending (with primary fibres) and short fibre removal result in more economical material preparation, higher efficiency (e.g. by optimal cleaning cycles) and smoother processing (fewer material-caused deadlocks) with the user of reclaimed fibres. Reclaimed fibres are transported in bales. Frequently, the manufacture of reclaimed fibres is directly combined with fibre processing.

### 1.4.3
### Reclaimed fibre quality

Reclaimed fibres are quite different from primary fibres. The damage they suffer during their manufacture is characterized by a wide spectrum of fibre lengths with a high share in short fibres as well as a proportion of unopened pieces of thread and fabric. Factors of influence on the properties of reclaimed fibres result from the textile waste, its pre-treatment and the execution of the tearing process itself. Reclaimed fibres are mostly available as blends. Their accurate composition regarding the contained types of polymer can hardly be described. Usually, specifications refer to the polymer predominant in the blend [32]. To make them into nonwovens, proportions as high as possible of broken-down fibres of lengths sufficient for the appropriate web-forming process are necessary. Pieces of yarn and or thread still contained in them contribute directly to matrix formation or are further broken down during the carding process. Short fibres and dusts, the same as remaining pieces of fabric, disturb processing. Their share should be kept as small as possible, which may be achieved by means of an optimal material-related tearing technology. At present, 55–25% of reclaimed fibres at least 10 mm long can be further processed using textile technologies. (Exception: reclaimed fibres from easy-to-break textile structures like less twisted yarns or nonwovens slightly to medium-compacted). On the market, reclaimed fibres are in the lower price segment, which is continuing to shrink due to prices of primary fibrous materials of both synthetic and natural types rapidly falling.

### 1.4.3.1 Description of reclaimed fibre quality

The quality of reclaimed fibres is hard to describe using conventional textile-physical methods and comparing primary fibrous materials. Reasons why existing measuring techniques and instruments available are not very helpful to describe reclaimed fibre quality are the inhomogeneity of the fibre blends used, the high share of short fibres and the proportions of non-fibrous textile materials still remaining in them (neps, unopened pieces). This requires deviations from the test conditions determined in the DIN standards for primary fibre test methods and, partially, increased inspection expenditure [33]. Speaking of reclaimed fibre quality today, special attention is drawn to grade of opening, average fibre length and staple diagram. The grade of opening represents the material spectrum available in the result of the tearing process and is determined as follows:

- Quantitative determination of the proportion of fibres, threads, neps and unopened fabric remaining by means of manual separation (with reference to a sample of 3 g or 10 g of reclaimed fibre blend)
- Specification of the mass proportions in percent

Fibre and thread shares can be looked upon in summary if the subsequent treatment of the material is intended by carding (to nonwovens). Mechanical methods to separate material components (e.g. Trashtester/Zellweger Uster) may, due to further material break-down, cause false results [33, 34].

Fibre length and distribution is, apart from the grade of opening, the most important quality parameter with regard to reclaimed fibres and of particular importance for the selection of the processing technology. The determination of fibre length can be carried out by means of manual single fibre measuring techniques (Johannsen/Zweigle), semi-automatic measuring systems (AFIS L-module, Almeter, HVI) or by image processing, which is currently under development [35]. In the result, average fibre length is present as a statistic average value. The distribution of the fibre lengths is represented as a histogram (absolute frequency) and as a curve (sum frequency). The fibre finenesses contained in blends of reclaimed fibres of different distributions depend on the raw material. During the reclaiming process, they are not influenced in any significant way. With reclaimed fibres and given circular cross-sections, fibre fineness can be determined by means of the AFIS-D-module. It is of interest with selected applications (e.g. geotextiles or nonwovens used to attenuate noise). From the distribution of fibre fineness, conclusions are possible as to the main components contained in the reclaimed fibre blend in question. Fibre tensile strength and fibre elongation can be measured according to DIN EN ISO 5079, however, they are of very little interest when describing the quality of reclaimed fibres. For both manufacturers of reclaimed fibres and processing companies it is essential to know what materials the fibre blends are composed of. To this end, measuring systems are being developed which use NIR spectroscopy to describe material composition with sufficient accuracy.

### 1.4.3.2 Ways clear to technologically modify the characteristics of reclaimed fibres in the manufacturing process

For long, the technology of tearing textile waste has not changed very much. Today, the wide variety of fibrous materials and material structures requires product-related approach to the tearing process. Technical and technological measures designed to influence fibre length and grade of opening may improve the material parameters of the final product. A new method already transferred into tearing practice is controlling the power input (which depends on the structure of the waste in question). For this purpose, the number of drum revolutions is varied [36]. Thus, apart from improved quality, the reclaiming process can be carried out more economically by saving energy. Table 1-23 is to clarify the effects achievable by varied numbers of revolutions of the drum (reclaimed fibre production from a fabric of worsted/synthetic material).

Further possibilities to improve reclaimed fibre quality are seen in novel process design and new ways of processing that avoid damage to the fibres. The idea

**Table 1-23** Effects to be achieved by optimised numbers of revolutions of the drum – comparison between standard and optimal version by the example of worsted/blended fabrics

| Version | Passage revolutions of the drum (rpm) | | | Average fibre length (mm) | Grade of material opening (%) | Energy consumption (kW) |
|---|---|---|---|---|---|---|
| | 1 | 2 | 3 | | | |
| Standard | 1000 | 1000 | 1000 | 8.4 | 95.6 | 45.0 |
| Optimal | 1250 | 1000 | 750 | 11.9 | 97.7 | 42.9 |

is to protect single fibres that have been extracted from the textile structure from later damage in the course of processing (multi-drum breaking-down systems [37]). This is known as good-fibre separation, which utilizes the trajectories occurring at the circumference of the drum that result from the differences in particle weight. At present, several ways are being examined to filter these material proportions from the process.

## 1.4.4
## Reclaimed fibre application

Taking into consideration all essential commercial aspects, reclaimed fibres can today be used in the modern spinning processes (adaptation to reclaimed fibres necessary), e.g. open-end rotor spinning, **covered spinning process** PARAFIL and the friction spinning process DREF. By far more important are reclaimed fibres in the manufacture of nonwovens [38]. The different web-forming processes mean different requirements to be met by reclaimed fibre quality, as is shown in Table 1-24.

Examples of nonwovens from or with reclaimed fibres are automotive textiles and building textiles with the main functions insulation and covering, agrotextiles and geotextiles (erosion protection). Further examples are versions of nonwovens for the upholstery and mattress-producing industries as well as textile secondary backs for floor covering [38].

The application of reclaimed fibres is based on a performance profile sufficient for the use in question and on reasonable prices. When manufacturing technical textiles for functional purposes, ways are open into processing secondary fibres more inexpensive than primary ones, in particular, with uses not allowing to process primary fibres. Reclaimed fibres from wool in laminated nonwovens, aramid fibres used in garments for cut protection or nonwovens from micro-fibres serving insulation or polishing purposes are well-known examples. Product-related reclaimed fibre characteristics are achievable by appropriate materials and corresponding reclaiming technologies. Selected examples are given in Table 1-25. In addition, there will remain a large range of products where, due to the function wanted, primary fibrous materials need to be used [39].

**Table 1-24** Minimum requirements to be met by reclaimed fibre quality as depending on the web-forming process [39]

| Web forming-principle | Average of fibre length (mm) | Quality of reclaimed fibre | |
|---|---|---|---|
| | | Percentage of short fibres | Grade of opening |
| Aerodynamic | 5 to 30 | low to medium | medium |
| Mechanical | 15 to 50 | low | high |
| Hydrodynamical | to 5 | complete | high |

**Table 1-25** Examples of requirements to be met by reclaimed fibres for use in technical textiles

| Application/product | Main requirement to be met | | | | |
| --- | --- | --- | --- | --- | --- |
| | *Polymer* | *Fineness* | *Strength* | *Length* | *Colour* |
| Reinforcing fibres | × | | × | × | |
| Covered yarns for technical textiles | × | × | | × | × |
| Nonwovens for erosion protection | × | | | | × |
| Geotextiles | × | × | × | × | |

## References to Chapter 1

[1] Bayer AG (1973) X400–Perlon-Spinnfaser für Vliesstoffe, text praxis internat 28: 493

[2] Salomon M, Hagebaum HJ, Wandel M (1974) Synthesefasern für die Herstellung von Vliesstoffen auf nassem Wege, Chemiefasern Text Ind 24/76:639–642

[3] Wild U (1974) Gekräuselter Synthesekurzschnitt, Wochenblatt der Papierfabrikation No 15

[4] Ehrler P, Janitza J (1973) Gekräuselter Faser-Kurzschnitt und voluminöse naß gelegte Vliese, Melliand Textilber 54: 466–470

[5] Welfers E (1974) Moderne Herstellungsverfahren von Chemiefasern, Melliand Textilber 55:313–317, 410–412, 507–509, 584–590

[6] Albrecht W, Knappe PE (1969) Über die Einsatzmöglichkeiten von Diolen-Hochschrumpf-fasern, Chemiefasern 19: 440–449

[7] Kratzsch E (1972) Chemiefasern der zweiten Generation, Chemiefasern Text Ind 22: 781–785

[8] Strunk K (1971) Herstellung und Verarbeitung von Glasfaservliesen, Referat anläßlich der Tagung des RKW

[9] Hansmann J (1979) Glas und Glasfaservliese, Text Prax 25: 396–399, 476–478

[10] Okamoto M (1979) Feinsttitrige Synthesefasern und ihre Anwendungsgebiete, Chemiefasern Text Ind 29/81: 30–34, 175–178

[11] Albrecht W (1976) Modifikationen von Fasereigenschaften in der Faserherstellung, Textilveredlung 11: 90–99

[12] Albrecht W (1977) Entwicklungstendenzen in der Chemiefaserherstellung, Melliand Textilber 58: 437–440, 528–532

[13] Albrecht W (1977) Eigenschaften von Polyesterfasern und ihr Nutzen für Bekleidungstextilien, Chemiefasern Text Ind 27/79: 883–890

[14] Jörder H (1979) Die Bedeutung der Chemiefasern für die nicht gewebten textilen Flächengebilde (Textilverbundstoffe), Chemiefasern 20: 764–767

[15] Egbers G (1974) Vliesstoffe der zweiten Generation, Angew Makromol Chem 40/41: 219

[16] Lauppe W (1976) Der Einfluß des Rohstoffes Fasern und seiner Eigenschaften auf Herstellung und Verwendbarkeit von Vliesstoffen, Melliand Textilber 57: 290–300

[17] Gürtler HC, Dietrich H (1976) Anforderungen an Synthesefasern bei der Herstellung und Anwendung von Nadelfilz, vorzugsweise für die Trocken- und Naßfiltration, Melliand Textilber 57: 301–305

[18] Hüter J (1971) Verfilzbare Chemiefasern, Chemiefasern 21: 1060

[19] Ott K (1996) A new synthetic fiber based on melamin resin, INDEX 96, Nonwovens Congress, Raw Materials Session EDANA Geneva/CH

[20] Watson D (1996) Metal fibers in nonwovens, TANDEC 6th Annual Conf, Knoxville, USA 18.–20. 11. 1996

[21] Cowen Ph (1998) New Cellulosic Fibres for Nonwovens, Int Nonwovens Symp EDANA 04./05.06. 1998

[22] Cheng CYD, Permentier JWC, Kue GC, Richeson (1997) Processing characteristics of Metallocene-bases, Polypropylene TANDEC 1997, The University of Tennessee, Knoxville

[23] Eichinger D, Lotz C (1996) Lenzing Lyocell-potential for technical textiles, Lenzinger Ber 75: 69–72

[24] Fust G (1998) Eine neue Fasergeneration für die Interliningbranche, 2. Intern Interlining Symp Flims/Waldhaus/CH

[25] Bobeth W et al. (1993) Textile Faserstoffe, Beschaffenheit und Eigenschafen, Springer-Verlag Berlin Heidelberg

[26] Faserstofftabellen nach P-A Koch, Neuauflage Inst f Textiltechnologie der RWTH Aachen, Chem Fibers Intern, Editorial Department Frankfurt/M Mainzer Landstr. 251

[27] Yamane C, Mori M, Saitoh M, Okajima K (1996), Polym J 28, 12: 1039–1047

[28] Sato J, Matsuse T, Saitoh M (1998) Seńi Gakkaishi 54, 8: 93–100

[29] Okamura H (1998) 37 Intern Manmade-Fibres Conf Dornbirn

[30] Nishiyama K (1997) Cell Manmade-Fibres Summit Singapore 4

[31] Dönnebrink H (1996) Textile Produktionsabfälle nach Prozeßstufen, Forschungsstelle für allgemeine und textile Marktwirtschaft an der Universität Münster, Arbeitspapier No. 22

[32] Eisele D (1996) Reißfasergut – Merkmale – Zusammenhänge, Melliand Textilber 77, 4:199–202

[33] Mägel M, Bieber B (1993) Erste Untersuchungsergebnisse zur Bestimmung ausgewählter textilphysikalischer Parameter von Reißfasern, Kolloquium Reißfaser '93, Sächsisches Textilforschungsinstitut e.V. Chemnitz

[34] Elsasser N, Maetschke O, Wulfhorst B (1998) Recyclingprozesse und Reißfaseraufkommen in Deutschland, Melliand Textilber 79, 10: 768–771

[35] Fischer H, Rettig D, Harig H (1999) Einsatz der Bildverarbeitung zur Messung der Längenverteilung von Reißfasern, Melliand Textilber 80, 5: 358–360

[36] Gulich B, Schäffler M (1995) Einfluß der Trommeldrehzahl auf die Reißfaserqualität, Kolloquium Reißfaser '95, Sächsisches Textilforschungsinstitut e.V. Chemnitz

[37] Fuchs H, Gulich B (1998) Prozeßsicherheit und Qualitätsverbesserung von Reißfasern für höherwertige Produkte, Melliand Textilber 79, 5: 366–369

[38] Watzl A (1992) Vom Textilabfall zum Nonwovenprodukt – Nutzen durch Recycling, Melliand Textilber 73, 5: 397–401 73, 6: 487–495, 7: 561–563

[39] Böttcher P, Gulich B, Schilde W (1995) Reißfasern in Technischen Textilien – Grenzen und Möglichkeiten, Techtextil-Symp Frankfurt/M

# 2
# Other raw materials

## 2.1
## Cellulose (Pulp)
W. ALBRECHT

So far, cellulose has been the most often used raw material to make paper. In the last decades, it has become more and more important with regard to nonwovens, too. It has found a wide field of applications, preferably, together with super-absorbent powders (SAP), for hygienic purposes. This is why the final products based on it are designed to meet a number of particular requirements, such as take-in and transport of moisture, absorbency as well as low dust emission. Buckeye, an important supplier, gives these specifications:

- water contents in %      6
- mass per unit area in g/m$^2$      709
- density in g/cm$^3$      0.57
- whiteness (ISO) in %      86
- viscosity (0.5%, CED) cP      22
- methylene chloride extract in %      0.02
- length of fibre (Kajaani FS-200)      2.7–2.8
- classification of length of fibre cumulative in %
  - 8-mesh      4.0
  - 14-mesh      43.4
  - 30-mesh      68.6
  - 100-mesh      96.7

Taking into consideration these cellulose characteristics are largely wanted with hygienic articles expected to be kind to the human skin, it is most essential they are constant throughout the full lifetime of the products in question. In 1997, about 3.2 million tons were processed worldwide, even higher quantities can be expected in the future.

Generally, cellulose is made from wood by means of chemical opening-up and bleaching processes. Another sensible approach is to make it from one-year plants. The base substance of the short fibres is cellulose. As the specifications

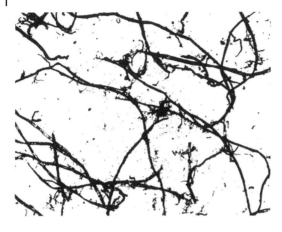

**Fig. 2-1**   Long staple pulp

above show, this kind of cellulose contains as much as no additional substances. Fig. 2-1 describes cellulose of this kind.

## 2.2
## Granules
R. GUTMANN

Apart from natural fibres, today an ever increasing part in production of nonwovens is covered by man-made fibres. As a starting product for industrial manufacture of the most important man-made staple fibres and filament yarns used for production of nonwovens, basically all of those thermo-plastic polymers are used which are applied in other textile production processes, too (knitting, weaving, warp knitting). With respect to quantity, polyolefines, polyesters, and polyamides are the most important raw materials. The economic significance of each polymer thereby is mirrored by its respective production figures. According to the latest information these figures add up globally for all polymers applied for nonwovens production to two million tons a year, alone 60% thereof are processed in Western Europe and North America [1]. The figures of the respective portions for the different polymers as well as their absolute amounts are taken from a CIRFS report and are indicated in Table 2-1.

These products are supplied to those different industrial branches which apply nonwovens for processing to final products like hygiene and technical nonwovens, textiles for apparel, home and furniture applications, as well as carpets and automotive products. With raw material, finally, is applied for production of a specific end product, is determined on the hand by its polymer and processing properties and on the other hand by its foreseen end use.

Predominantly, polymers use for fibre or nonwovens production are applied in form of granules, however, instead of the term 'granules' very often the terms 'chips', 'scraps', or 'flakes' are in use. How these small pieces will be shaped is de-

**Table 2-1** End use of non-spun fibres in the year 1995

|  | *[%]* | *[1000 t]* |
| --- | --- | --- |
| Polypropylene | 34.5 | 425 |
| Polyester | 32.5 | 400 |
| Cellulosics | 17.5 | 215 |
| Cotton | 6 | 75 |
| Polyamide | 5 | 60 |
| Acrylics | 3.5 | 40 |
| Wool | 1 | 15 |

termined by the kind of art they were produced, most frequently strand-like granules with a circular or rectangle shaped cross-section, spherical or lentil shaped granules, and, more seldom, even irregularly shaped granules are commercially available. As wide as variations of shape are those with respect to product's dimensions, however, length of the edges or diameters mainly are within a range between 1–2 mm and 6–7 mm. Thus, polyesters and polyamides, which are synthesized according to a polycondensation reaction and subsequently are delivered from the reactor as a strand, are made to pieces of appropriate length by passing a cutter [2, 3]. In contrast to these polymers the polyolefines like polypropylene or polyethylene, especially when produced according to up-to-date polymerization processes, exhibit a spherical form, which is due to the applied catalyst systems. As an example for irregularly shaped starting material may serve recycled products polymers from all kinds of polymers. They can be manufactured from consumer goods by passing those waste products through a shredding machine and an appropriate grinder where they are broken to arbitrary pieces. Another possibility, which can be applied to manufacture irregularly shaped granules without passing through the melt stage of a polymer, makes use of a compacting system followed by an agglomeration and granulation device, this combination will, however, only allow processing of especially formed waste [4]. Hence, the shape of all commercially available recycled polymers is achieved by pure chance and is merely depending from the equipment used by the supplier. This is the reason why these products have only gained importance for processing on small extruders with screws having a low thread height, which on the other hand might cause some feeding problems in case of there are some big granules.

The physical-chemical properties of the raw materials are determined by the chemical components which make up the polymer structure as well as by the processing conditions applied at polymer synthesis. The interplay of chemical and physical polymer functions is also responsible for those important parameters of industrial engineering processes like minimum processing temperature, thermal stress in the molten state or flow behaviour of the melt. Based on these parameters not only the melt spinning device is planned and designed but also the maximum achievable productivity will be limited.

2.2.1
## General discussion of physical properties [5]

The first and most important feature of all thermoplastic polymers, which are most exclusively used for production of a fibre-like structure, is the existence of a melting point, this holds e.g. in contrast to duroplastic polymers. Each polymers exhibits a characteristic melting point which besides other criteria may serve as a means to identify an unknown polymer sample. These temperatures together with other relevant additional thermo-physical data are indicated in Table 2-2 for the most important starting materials of man-made fibre production.

The fact that there can be observed a melting point is due to the existence of a crystalline structure. For melting a crystalline structure needs to absorb a specific amount of thermal energy – the so-called heat of fusion – while melting is charac-terized by a phase transition from the solid to the liquid state. With the reverse transition from the liquid to the solid state the same amount of energy is set free being now termed as heat of crystallization. That is what happens at fibre forma-tion, too, i.e. generating a crystalline structure at solidification from the cooling melt. These processes of melting and crystallization, in principle, can be passed as often as will and hence makes this kind of polymers of interest with respect to re-use, alike metallic materials.

The portion of crystalline structures generated at fibre formation, their size and crystallization rate are decisively influenced by two parameters: chemical structure of the polymer and processing conditions of the melt. By these parameters in ad-dition arrangement as well as morphology of crystalline domains are influenced, which may consist of crystallites or spherulites, in case of bigger sized entities. Opposite to the crystalline units there exists a matrix of less ordered structure, which often is merely called amorphous phase and which is characterized by an arbitrary arrangement of the polymer chain molecules without formation of any higher ordered regions. All of the physical properties of the polymer and espe-cially the mechanical properties of the later on produced yarn depend on this so-called 'Two-Phase-Structure' which can be considered of forming a network con-

**Table 2-2** Physical properties [1] of the most important man-made polymers

| | Glass transition temperature [°C] | Melting point [°C] | Moisture take-up [%] | Specific heat $[Jg^{-1}K^{-1}]$ |
|---|---|---|---|---|
| Viscose | | (175–205) | 9–11 | 1.3–1.5 |
| PET | 80–110 | 250–260 | 0.2–0.5 | 1.1–1.4 |
| PA6 | 80–85 | 215–220 | 3.5–4.5 | 1.5–2.0 |
| PA6.6 | 90–95 | 255–260 | 3.5–4.5 | |
| PAC | 30–75 | >250 | 1–2 | 1.2–1.5 |
| Modacryl | 85–95 | 130–170 | 0.5–4 | unknown |
| PP | −10 | 160–175 | 0 | 1.6–2.0 |
| PE | −35 | 125–135 | 0 | 1.4–2.0 |

**1)** Taken from Denkendorfer Fasertabelle

sisting of those two structural units [6]. While the crystalline domains are thought to make up the centres of network, the 'holes' in between are formed by the less ordered sections which are responsible for the plastic behaviour and the extensibility due to the polymer chain mobility on the molecular level. By means of X-ray investigation the highly ordered crystalline structure, i.e. the crystal lattice, can be characterized [7]. Within the crystal lattice the polymer chains possess a specific arrangement at a fixed, invariable gap. What concerns the distance between the polymer chains, it depends from the polymer chain geometry which for its part is determined by the chemical structure. Via their functional groups the chain molecules are able to establish higher or lower forces of interaction which are responsible for the macroscopic properties of the polymers that are, finally, measured and which are among others also responsible for the observed characteristic melting points. The crystallite size normally is less than one ten thousandth of a millimetre and a result of crystallization rate on the one hand as well as cooling rate and freezing temperature of molecular chain motion on the other hand. Freezing temperature of molecular chain motion, also called glass transition temperature (see Table 2-2) is defined by that temperature at which mobility of polymer chain segment in the less ordered regions is stopped. That, however, does mean that above this temperature those parts of a chain molecule which are not fixed into a crystalline structure are able to perform molecular movements of entire chain segments due to their intrinsic thermal energy. By this kind of mobility, which also affects collisions with other chain molecules and chain segments, the respective non-crystalline domains become re-structured. In this way different chain segments get in touch with each other and formation of new nuclei or growth onto existing crystallites are able to take place. In contrast to this thermally induced crystallization process, crystallization is supported by additional mechanical orientation of the polymer chains, too. This e.g. is brought about at fibre formation in the melt spinning process by the draw-down forces necessary for winding. In this case one talks about crystallization induced by spinning stress [8]. Which significance crystallinity and crystallization behaviour do have on processing of polymeric materials can be seen on the example from polyester. Before drying and extrusion this polymer needs to be crystallized slowly at temperatures lower than drying [9]. By doing this, the heat of crystallization, which always is set free at heating amorphous polyester, can be controlled and, thus, it is prevented that too high temperatures, as normally applied with drying conditions, will initiate too fast crystallization and hence would lead to a sticking together of the granules and subsequently to a blockage in conveying the granules from the dryer to the feed pocket of the extruder. The amount of thermal energy set free at crystallization depends on the molar heat of crystallization and the increase of crystallinity which can be reached.

Along with macroscopic form and microscopic structure there is another relevant physical variable which must not be left out of consideration, that is polymer chain length. As already indicated molecule size or molecular weight, which is an appropriate term generally used in chemistry, does have a decisive influence on properties and processing behaviour, especially on flow characteristics of the different polymers. Thus, viscosity exhibits to be used as a measure for determination of molecular weight from flow behaviour. For that two methods are available which are based on measurement of the polymer melt either, or measurement of a polymer solution prepared by means of a suitable solvent. Besides these there are some additional methods, e.g. light scattering, osmometry size exclusion chromatography, ultra centrifugation, and others which are suited, too, in order to characterize molecular size [10]. In practical use for specification of the most important parameters, as e.g. indicated on a product's data sheet, for solvent based systems polymer producers often prefer to use different formats of their own or in case of olefins refer to the so-called melt flow rate (MFR). The latter stands for the amount of melt forced through a specified die by a defined load within ten minutes. The higher the MFR-value the less viscous is the melt and the lower is the molecular weight of the polymer. Applying the same settings for temperature and load it is possible to directly compare melt flow rates of different polymers from different suppliers by weighing the melt underneath the die. In case of different solution viscosity formats are used, a comparison of polymer specifications coming from different suppliers only is possible if except from applying the same conditions for measurement in addition the same solvents and the same condition for dissolving have been used. Under these precondition one can get from a series of solution viscosity results by extrapolation towards infinite dilution the so-called limiting viscosity number (intrinsic viscosity) which allows comparison of different granules from different suppliers. Determination of intrinsic viscosity (I.V.), however, is quite time consuming.

In the following section the properties of the most important polymers will be discussed starting with the polyolefins.

### 2.2.2
### Polyolefins [11–14]

As already mentioned the interaction between the polymer chains of the respective polymers are responsible for their physical properties. As a result from that all physical data as well as the corresponding differences in processing depend from the height of these forces of interaction. Thus, due to their weak interactions based on Van-der-Waals forces, all polyolefins exhibit very low melting points. In order to achieve a sufficient tenacity with the yarn it is necessary to process extremely long polymer chains which are able to establish those forces of interaction between the polymer chains rather by means of a high number of points of interaction than by the strength of interaction. In this respect polyolefins exhibit very low interactions, that is the reason why highest molecular weights are required in order to be applied in textile manufacturing. More favourable with regard to possible interactions are those polyolefins which bear a regularly repeating side group attached to the linear polymer backbone. This already hold for isotactic polypropylene, polyisoprene, polystyrene, or bi-cyclic polyolefins. Due to the more or less bulky side groups the polymer chain becomes less mobile, i.e. in the molten state the polymer takes more energy to flow or to be sheared and as a result builds up a higher melt pressure. In practice at processing this can be counteracted by increasing processing temperature somewhat further above the melting point as it is usually done compared with processing of polyesters or polyamides. As a consequence, this comparably higher processing temperature in relation to the melting point causes an increased thermal degradation which in practice – associated, however, in connection with the above mentioned drawbacks – can be cured by applying a higher molecular weight in the starting material or by addition of a thermo-stabiliser. As aside from their lower thermal stability polyolefins also exhibit a reduced light stability in the latest past ever more effective stabilizing system have been developed by the auxiliary producers. These products can be added to the polymeric raw materials at extrusion, e.g. in form of a masterbatch or powder. With regard to best efficiency the individual stabilizers if used in common have to fit well chemically with each other. Here for example the following compounds are frequently added: alkyl-phenols, organic sulphur containing compounds, derivates from organic phosphonic acid or organic nickel salts, benzotriazols and sterically hindered amines (HALS-type stabilizers) [15]. The latter are especially applied in order to improve light stability.

With respect to stability and processibility it turned out that those recently available polyolefins with a narrow molecular weight distribution are somewhat more advantageous [16, 17]. As with the standard Ziegler-Natta catalysts used in polymerization of olefins only a broad molecular weight distribution can be reached, the high molecular weight portions severely reduce flow behaviour. Using the latterly available metallocene catalysts which are able to generate polymers with a narrow molecular weight distribution with these polymers one will find in addition a markedly improved flow behaviour. This allows processing at a lower temperature, at higher spinning speeds, or of yarns with a lower filament count. Thus, characterization of polyolefins by means of their flow behaviour is commonly used in practice and mainly is indicated as MFR-value (melt flow rate, in the past MFI: melt flow index) according to the German DIN 53735. The MFR-figure denotes the amount of melt given in grams which flows from a specified die within ten minutes when a defined temperature (e.g. 230 °C) and load (e.g. by a mass of 2.16 kg) are applied. A common range of melt flow rates related to those products available on the market and being mainly used for yarn production may be specified by a value between MFR 10 to MFR 40. With special CR-type polypropylenes – CR stands for controlled rheology – however, polymers with MFR-values of several hundreds are offered, too.

## 2.2.3
### Polyesters [18, 19]

As far as quantity is concerned polyethylene terephthalate (PET) is regarded to be the most important fibre forming polymer, markedly less attention is paid to polybutylene terephthalate (PBT). Very recently a third type of polyester industrially became available which might gain some importance in the future due to its elastic properties, that is polypropylene terephthalate (PPT) or more frequently termed as polytrimethylene terephthalate (PTT). What concerns chemistry, these products differ in chain lengths of their diols applied for synthesis. While with PET synthesis glycol together with terephthalic acid or dimethyl terephthalate are used, with the synthesis of PBT and PTT besides the same acid component 1,4-butanediol and 1,3-propanediol, respectively, are applied. In all cases the polycondensation reaction is catalyzed either by antimony or titanium compounds.

Especially within the field of the so-called functional textiles, which showed a strong increase in the last years, PET has found new applications in form of microfibre nonwoven aside from those applications up to now. As a result of their chemical structure which allows polar interactions between the ester groups two main differences become apparent in processing PET compared to polyolefins. Due to the comparably higher polymer melting point on the one hand higher processing temperatures – up to 300 °C – have to be applied (see Table 2-2). On the other hand it is possible to process polymers with comparably lower viscosity, i.e. lower molecular weight, or rather have to be applied for processing due to the higher forces of interaction which also act in the molten state. Polyester molecular weights normally used for fibre production range from about 15,000–25,000 g/mol and therefore make up only one tenth of molecular weight used with polyolefin fibres. An other decisive difference between polycondensation products – like polyester or polyamide – compared to polyolefins is the narrow molecular weight distribution found with the first. Molecular weight distribution is termed also as polymer non-uniformity U. It is defined as the ratio of mean molecular weight by mass divided by mean molecular weight by number minus 1. At the end of a polycondensation process normally for that ratio a value of about 2 is reached. Apart from viscosity and molecular weight derived from that the so-called numbers of ends can be used for characterization of polyesters. Mostly characterization is confined to the content of carboxylic ends (CEG) which is found for the commercial products to be in the range of 10–40 µval/g. In combination with the content of hydroxy ends (HEG) which frequently reaches values higher than 40 µval/g from the reciprocal value of their sum it is possible to assess a figure for the mean molecular weight by number. In addition to these parameter, on the product data sheets of the polyester producers there will be also indicated the content of diethylenegylcol which is for PET normally less than 1% as well as the amount of oligomers. This amount is different for the different types of polyester, in case of PET about 1.5–2%, for PBT about 1%, and for PTT more than 2.5% may be determined [20].

## 2.2.4
### Polyamides [21, 22]

Looking upon the broad area of application which the most important polyamides have gained in fibre production up to now, one will hardly find anything comparable within any other class of polymers. Except from application in production of nonwovens, this area extends from manufacturing panties, top clothes and underwear up to furnishing fabrics, carpets and technical fabrics as well as artificial suede or flock fibres. Regarding the types of raw material processed within this class of polymers, Nylon 6 or polyamide 6 (PA6) and Nylon 66 or polyamide 6.6 (PA6.6) are economically the most important. Because of their respective chemical structure PA6 is a member of the poly-$\omega$-amino acid series while PA6.6 is part of a series of polyamides which are synthesized from $a,\omega$-diamines and $a,\omega$-dicarboxylic acids. From this results a certain difference in physical properties which can be seen most clearly with their difference in melting points, thus, the diamino-diacid polymers exhibit melting points about 40 °C higher than poly-$\omega$-amino acids with the same number of carbons in the chain segments. Based on the condition of PA6 synthesis which is an equilibrium reaction, a high amount of about 8% monomer ($\varepsilon$-caprolactam) remains in the polymer and has to be washed out before further processing [23, 24]. In spite of this additional processing step and, although, physical properties are not quite favourable, PA6 has gained within the poly-$\omega$-amino acid series the highest economic significance. But this only because of the fact that $\varepsilon$-caprolactam can be produced very easily by applying industrial techniques while the necessary purity is obtained, too.

What concerns the starting components for PA6.6 synthesis: 1,6-diamino hexane and adipic acid, they are also available on industrial scale with sufficient purity. However, not only with synthesis but also with extrusion of polyamides appropriate demands on purity have to be fulfilled, these especially refer to exclusion of oxygen and moisture. While oxygen from the air already starts at about 70 °C to degrade the polymer by irreversible yellowing, moisture will influence the balance of the equilibrium in the polymerization reaction [25]. Thus, at a moisture content higher than 0.1% increasing hydrolysis will reduce molecular weight. For processing at very low moisture content in the polymer one will observe post-condensation reaction, hence, an increase of molecular weight will take place as long as by condensation reaction that amount of water has been set free, which corresponds to the water content of the equilibrium state at processing temperature. As an increase in molecular weight is accompanied by an undesirable increase in melt viscosity, there are added chemical compounds for regulation of chain length to the starting compositions of PA6 and PA6.6 synthesis. These most frequently comprise aliphatic amines or diamines and mono-carboxylic acids or dicarboxylic acids. In this way polymers for fibre production are generated possessing molecular weights in the range of about 25,000–45,000 g/mol. In addition additives which control chain length also influence amino and carboxylic ends and therefore are important for end use properties, too. For example dyeability of the polymers may be changed, as very often especially the acid-base interaction between an acid dye and the basic amine is used for dyeing polyamides. From the discussion above it becomes clear that amino ends are a quite variable parameter. But with respect to the dyeing properties it can be stated that after extrusion to fibre-like products the content of amino ends (AEG) must not be less than 40–50 μval/g.

**2.3**
**Powders**
R. GUTMANN

In addition pure synthetic raw materials which are used as granules in the area of nonwovens production, there are also some polymers applicable which are processed as powders. Among these there are fibre forming polymeric compounds which can be processed to fibres directly, and there are low molecular weight organic and inorganic compounds. The latter substances which are used as additives together with the fibre forming polymer either take over the task to stabilize the product or act as a modifier with respect to visual appearance of the fibres produced from these polymers. In the following some of the most important products are going to be introduced, starting with the group of polymers.

2.3.1
**Polymer powders**

2.3.1.1 **Polyacrylonitrile** [26]
The infusuble polymer which degrades before melting mainly is synthesized starting from acrylonitrile by means of a precipitation polymerization taking place in water as a solvent and being initiated by a redox system. After processing and drying a powdery product is gained, that directly can be dissolved in an appropriate solvent suitable for fibre production according to a dry or wet spinning process. The molecular weights normally achieved at industrial production range from 80,000–180,000 g/mol. With respect to apparel applications 100% pure polyacrylonitrile hardly is used, but is mainly applied with technical products. Most of the polymers based on acrylonitrile consist of two or more comonomers, with acrylonitrile making up more than 85%. Beside that acrylic acid, methacrylic acid, vinyl esters as well as sulphonates containing double bonds may be incorporated in the polymer backbone in order to realize specific physical properties or to improve dyeability. If there is only a reduced content of 50–85% acrylonitrile present in the polymer then it is termed as a modacryl type polymer. This kind of polymers preferably is modified with halogen containing monomers which improve processibility and flame retardancy. While polyacrylonitrile fibres are solvent-spun most frequently from solution based either on dimethyl formamide, diethyl formamide or dimethyl sulphoxide, with modacryl type polymers the much cheaper solvent acetone can be used. In order to get a high quality polymer solution in any case it is necessary have polymer powders with a particle diameter in the range of 0.01–0.1 mm. Before dissolving, the polymer requires to be very carefully dispersed preferably at low temperature, then the temperature in the stirring vessel is raised up to 100 °C and a clear, slightly yellowish solution is formed. Too high temperatures and long times of stirring will increasing yellowing. With respect to get a homogeneous white fibre, yellowing must be avoided and hence is counteracted by working under a protective atmosphere, by addition of complex forming agents in order to mask traces of iron or by the use of reductive agents.

### 2.3.1.2 Further copolymers

Comparable to polyacrylonitrile polyvinyl chloride [27] has found some minor applications in form of a copolymer which can be solvent-spun to fibres. Thus, a copolymer containing 85% vinyl chloride and 15% vinyl acetate has gained some importance in the nonwovens area as it can be used as a binding fibre in thermobonding due to its low softening temperature of about 70 °C.

From this functionality as a binding fibre for thermobonding of nonwovens some further copolymers derive their continuous demand. Among others there have to be regarded the low melting copolyesters and copolyamides which are manufactured as powders and as granules, as well. Depending on the selection of the monomer component it is possible to adjust melting and softening points within a wide range of temperatures what allows to offer a special designed binding fibre for different kinds of application.

### 2.3.2
### Additives [28]

The group of low molecular weight compounds being important not only for production of nonwovens but for production of fibre-like products in general, can be subdivided with regard to their respective activity into auxiliaries possessing a stabilizing function, on the one hand, and pigments, on the other hand. Apart from the possibility to insert suitable, low molecular weight compounds into the polymer backbone possibly as comonomers directly at synthesis, most of these additives are mixed to the polymers just before manufacturing of the granules or before processing the latter into fibres. In the first case dosing is finished at the producer of the granules, in the second case addition takes place only at extrusion of the granules by that the fibre producer is able to adjust the amount of additive individually. As the percentage of the applied additives, normally, is quite low there might result an inhomogeneous distribution of the components added to the polymer matrix, especially if the additives are applied as a powder and their solubility in the molten polymer is low or they tend to form agglomerates [29]. With respect to industrial processing of such systems there exist some technical solutions which are leading to good results and hence are applied for production. The easiest way is for example to 'powder' the additive onto the polymer granules. That can be done in a rotating vessel wherein each granule particle is covered with the additive and in this form is fed to the extruder. Passing the extruder the system is transferred into the molten state and now it is possible to homogenize the components of the system further by means of an extruder screw bearing a dynamic mixer on its top or by static mixers installed in the melt pipe leading to spin packs. If case of additives are used which are compatible with or even soluble in the polymer a sufficient distribution can be achieved. Processing of incompatible additives or pigments, however, will be a problem. Here a twin-screw extruder can be used which because of its specific shear and mixing effect makes sure that a homogeneous distribution is achieved. If this alternative is not available a different route can be followed which comprises production of a masterbatch. A masterbatch is a blend made of a polymer with an additive in a very high concentration. It is also produced on an extruder with mixing facilities, but, at first, there will be extruded a strand which is cut to pieces. The granules produced in this way then will be used together with an appropriate smaller amount of polymer in order to reach the desired additive concentration. When this mixture is melt spun to fibres there is no danger anymore of getting an inhomogeneous distribution as the components have been mixed twice in the molten state. It must be mentioned, however, that there might take place a marked degradation because of the thermal stress acting twice at processing and, hence, one should be aware that mechanical and optical properties might be reduced. Which route, finally, will be followed at industrial production, depends not only from the kind of additive going to be incorporated, but is mainly determined by the technical equipment available at the respective producer.

### 2.3.3
### Stabilizers [15]

As all thermoplastic polymers have to be heated above their respective melting point at processing, thermal as well as thermal oxidative degradation of the polymer chains can not be avoided. Particularly affected in this way is the polypropylene which is processed at 100 °C about its melting point and which is due to its high number of tertiary carbon atoms along the polymer backbone very easily attacked in form of a radical degradation reaction. In order to prevent this kind of radical initiated degradation and a subsequently possible oxidative chain scission, respectively, e.g. sterically hindered phenols or fatty acid thioester are added which are able to take over the radicals and thus protect the polymer.

Regarding the later on application and the end use of textile products from man-made fibres it is essential to get a good thermo-oxidative stabilization next to a very effective flame retardancy [30, 31], especially against the background that in some countries very restrictive regulations are applied concerning flammability of synthetic materials for domestic or apparel use. This holds for nearly all polymers, except those containing a high halogen concentration. That is why polymers are stabilized by addition of various phosphorus acid esters, antimony oxide or alumina and as already mentioned by addition of organic or inorganic compounds with a high percentage of halogen. The effect of the above mentioned stabilizers to decrease flammability in general is based upon their activity to reduce formation of inflammable gases which are generated along with thermal degradation of a polymer.

As important as stabilization of man-made fibre polymers against influences of heat at processing and in use or against flammability is stabilization in order to reduce degradation by UV-light. Here again polymer degradation is brought about by a radical reaction which is initiated by the UV-portion of the sunlight. By the energy content of this shortwave radiation which is in the range of the bonding energies of organic compounds a homolytic scission of the chemical bonds in the polymer chain can be affected, similar to the thermally initiated reactions. Thus, in order to counteract UV-light degradation an existing thermo-stabilizer may already contribute to light stabilization [32]. Especially with stabilization of polymers against UV-light – and here of polypropylenes in detail [33] – benzophenones, benzotriazoles, sterically hindered amines (HALS-types) or nickel chelate complexes are applied. In any way it would be, however, particularly effective if interaction of UV-radiation with the polymer would be prevented as early as possible. That can be reached at best by the so called UV-absorbers as e.g. benzophenones or benzotriazoles which absorb disintegrative UV-radiation and convert it to heat. The most effective way to achieve that is by addition of carbon black, as far as the resulting black colour doesn't matter.

## 2.3.4
## Pigments [34]

The addition of fine powdered organic or inorganic pigments with a powder grain diameter of less than 1 μm is one of the most important methods to get coloured products at processing of fibre forming polymers. Some of the remarkable features of such coloured products are their depth of colour as well as their fastness of colour. Especially with respect the polyolefins up to now pigmentation is the most important technical possibility to get coloured fibre materials with a high fastness of colour. The one and decisive criterion for application of a pigment, however, is its stability to heat. That means, there must not take place any degradation of the pigments at processing together with the molten polymer – of course, this holds also with the later on end use for the influences due to light, environment, chemicals (for cleaning), etc. – and for their part the pigments may not catalyze degradation of the polymer. Generally inorganic pigments have an advantage due to their temperature of degradation which is high above their respective melting points, while the latter again are much higher than the melting points of the polymers.

Looking, at first, to the most relevant inorganic pigments, there titanium dioxide has gained highest importance especially in form of its anatase modification. This is due to the fact that this pigment can be added to all polymers in order to reduce the lustre of man-made fibre products. Thus, the incident light no longer can pass the polymer unimpeded, but is scattered at the pigments. An other positive effect related to the addition of titanium dioxide to fibre polymers is an intensification of the depth of colour at dyeing compared to an appropriate dyeing carried out on fibre material without addition of titanium dioxide. In this context titanium dioxide not only is referred to as a white pigment but also as a matting agent. This function can be fulfilled by barium sulphate, zinc sulphate or zinc oxide, too, but their commercial importance remained very low.

The number of inorganic pigments being available for colorants is fairly limited. That is among others related to the fact that these pigments consist of chemical compounds containing heavy metals like lead, cadmium, mercury, etc. which are nowadays for health and ecological reasons even less desirable. If, however, a high fastness to light is demanded these pigments are still the state-of-the-art in case of yellow (vanadates, molybdates, chromates) or red colours (iron oxides) have to be realized. As mentioned, grey and black colours are produced by addition of different quantities of carbon black.

For pigmentation of polyolefins and especially of polypropylene there has been developed another system in the past which is able to deliver products which exhibit deep colour shades as well as high fastness to light. This is achieved at processing of polypropylene by addition of nickel salts which are able to fix appropriate disperse dyes via the metal atom in form of a chelate complex. Regarding the use of nickel compounds there hold, however, the same reservations as already discussed with the heavy metals. That is why it is reasonable to assume that the use of this system will become ever less important.

Nevertheless, in order to allow melt dyeing, and here especially of polyolefins, today a series of organic pigments are available which fulfil the above mentioned demands with respect to stability and fastness. While these pigments are added separately or as a mixture it is possible to realize the whole range of colour shades. Thus, perylene pigments are used to produce red shades, azo and diazo pigments are used to produce orange and yellow shades, and the series of phthalocyanine pigments is applied in order to get green and blue shades.

At the end of this chapter one property has to be mentioned which may be responsible for some problems observed with the production of pigmented polymers. This property is related to the crystallinity of the pigments resulting their high hardness, and holds especially for the inorganic pigments close to the fibre sheath protruding from the surface. Caused by these particles abrasion is affected on those parts of the processing and production machines the particles get in touch with. In addition problems at production may often result from the spinneret area by formation of agglomerates, deposits and products of degradation from the applied pigments.

## 2.4
## Absorbent polymers
S. MARTINI

Absorbent Polymers are water-insoluble, cross-linked polymers which absorb large quantities of aqueous liquids by forming a hydrogel. The gel-like mass responds in an elastic manner to external mechanical pressure. The liquid is retained even under pressure. Owing to their characteristic absorption properties, these polymers are used worldwide in the hygiene industry (baby diapers, adult care articles and hospital products, sanitary napkins) [35].

Even though **S**uperabsorbent **P**olymers (SAP) do not belong directly to the group of primary raw materials for the production of nonwovens, they are inseparably linked to the nonwovens industry through a number of technical innovations in hygiene applications. EDANA (European Disposables and Nonwovens Association) stated for the year 2000 that the main end-use for the Western European nonwovens industry is the hygiene market with 341,000 tons (34% of total production) or 14,5 billion m$^2$ (57%) [36].

## 2.4.1
### Absorption mechanism

In contrast to other liquid-binding raw materials (cellulose fibres, foams, etc), cross-linked, partly neutralised polyacrylates do not only absorb large quantities of liquid, but they also store these permanently, even under pressure. The mechanism of absorption of SAP is shown in Fig. 2-2 [37, 38].

The carboxylate groups in the polymer are strongly solvated in contact with aqueous liquids. There is an accumulation of similarly charged groups along the polymer chains which repel each other electrostatically. This process opens the polymer clusters, resulting in the transformation of the absorbent polymer into a hydrogel. In consequence of the cross-linking, the polymer chains remain firmly connected to one another at some points so that the liquid absorption only results in swelling. The hydrogel does not liquefy even if it consists of 99% water [39–41].

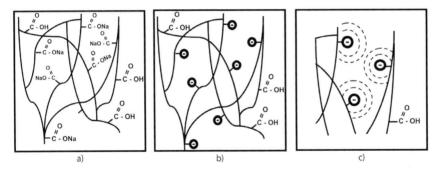

**Fig. 2-2** a) SAP in a dry state, b) SAP in contact with aqueous liquids, c) SAP after swelling and liquid absorption

## 2.4.2
### Production process

Patent literature shows a broad spectrum of possible variations in manufacturing methods of granular superabsorbents. The usual grain sizes are within a range of 150–800 μm. The two most suitable processes are described below.

### 2.4.2.1  Suspension polymerization
In the inverse suspension polymerization, partly neutralized acrylic acid is dispersed. This dispersion is supported by suspension aids in a hydrophobic, organic solvent. The polymerization is initiated by radical initiators. The crosslinking reac-

tion is carried out through copolymerization of a polyfunctional crosslinker which is added to the monomer solution and/or by the reaction of suitable crosslinking agents with functional groups of the polymer. This leads to the production of small porous droplets. After the polymerization is finished, these droplets are dried. [42] provides an overview of the technical details.

### 2.4.2.2 Solution polymerization

The more dominant manufacturing process for the production of superabsorbent granules is the radical solution polymerization, consisting of the following steps (Fig. 2-3): In this method monomeric acrylic acid is converted by partial neutralization to sodium acrylate (55–75%) before polymerization. The addition of a crosslinking system then brings about the formation of a three-dimensional network (Fig. 2-4). Important process parameters are the cooling of the partly neutralized monomer solution, the removal of oxygen by gassing with nitrogen, the exact addition of the initiators (redox systems, UV), as well as polymerization and temperature control [42].

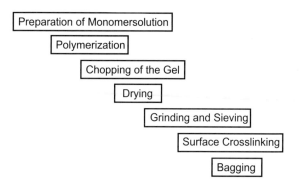

**Fig. 2-3** Solution polymerization

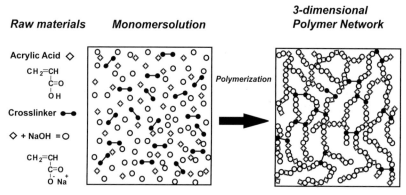

**Fig. 2-4** Polymerization and gel formation [43]

After the polymerization, the more-or-less viscid gel (depending on the concentration of crosslinker and the degree of neutralization) is treated in various mixing and kneading processes. Belt dryers with temperature and speed control dry the granules at a constant layer thickness to the desired residual moisture. Depending on the product specification the dry gel is reduced in size by grinding and sieving.

### 2.4.2.3 Surface crosslinking

The surface crosslinking production step comprises the application of reactive substances to the base polymer. They form an additional network on the surface of the granules. This method, first developed by Tsubakimoto [44] and Dahmen [45], conducted a significant improvement of the SAP performance. It was the most important precondition for the development of ultra thin baby diapers with a low proportion of fluff and a high ratio of superabsorbents (Fig. 2-5).

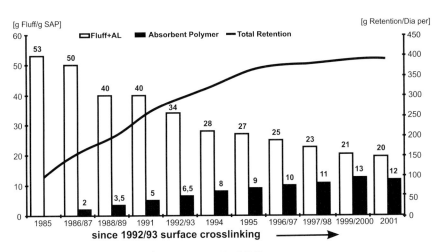

**Fig. 2-5** Baby Diaper Core Development (size: Maxi) [43]

### 2.4.2.4 In-situ polymerization

Experimental and theoretical work by Houben, Herrmann und Dahmen [46] has provided significant progress in the development of "in-situ" polymerization of absorbent polymers. Kaltenecker [47] has shown that three-dimensional networks can be produced directly on nonwovens (e.g. PP or viscose) by both chemically initiated copolymerization and graft polymerization induced by electron beam. Whether these innovative production processes will be used on an industrial scale remains to be seen.

2.4.3
**Test methods**

The performance profiles of absorbent polymers are obtained using different test methods. Distinction is made between physical data and performance of the polymer itself according to Edana [48], properties of absorbent polymers in a fluff-matrix and the performance of a hygiene article (e.g. baby diaper) [49].

2.4.3.1
**Characteristic data of absorbent polymer**

The maximum retention, also termed CRC (Centrifuge Retention Capacity) defines the maximum possible liquid retention of the hydrogel after centrifugation. The test liquid used for all methods is physiological sodium solution (0.9% NaCl) in order to classify the different SAP types in conditions as realistic as possible. The absorption against pressure (AAP) measures the time-related swelling of the polymer at 0.3 or 0.7 or 0.9 psi gravimetric load. Here, the swelling capacity of the superabsorbents under the weight of a wearer (e.g. infant) is simulated. This test clearly shows the difference between base polymers and surface crosslinked polymers (see Table 2-3).

The physical properties of the absorbent polymers include flowability, bulk density and particle size distribution [48].

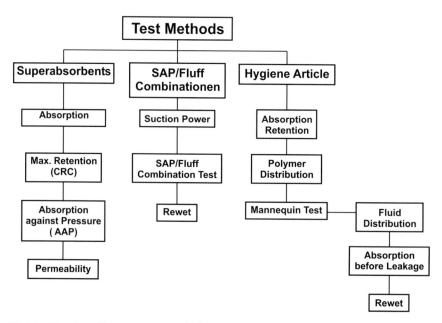

**Fig. 2-6** Overview of important test methods

**Table 2-3** Significant performance data of base polymers (A) in comparison to surface cross-linked polymers (B) [50]

|  | A | B |
|---|---|---|
| Maximum Retention – CRC | 32 g/g | 30 g/g |
| Absorption under 20 g/cm$^2$ pressure – AAP 0.3 psi | 10 g/g | 29 g/g |
| Absorption under 50 g/cm$^2$ pressure – AAP 0.7 psi | 6 g/g | 23 g/g |

2.4.4
### Field of application

Graham [42], Rohe [51] and consultant companies such as Chem Systems [52] publish worldwide market figures on production capacities of absorbent polymers as well as estimates of demand and fields of application. In 2001, the world wide production capacity reaches about 1.2 million tons [53]. The actual quantity marketed is approximately 85% of the total capacity. More than 95% of absorbent polymers produced are used in hygiene articles. The increasing number of publications and patent specifications show the growing importance of ultra thin (volume-reduced) hygiene articles during the last decade. The typical application of absorbent polymers is described by Brandt [53] and Kellenberger [54]. Modern prefabricated thermobonded absorbent cores essentially consist of fluff, PE/PP bicomponent fibres and superabsorbents. These liquid-storing three-dimensional materials (airlaid/composite) are described by Knowlson [55] and Herrmann [56] (Fig. 2-7).

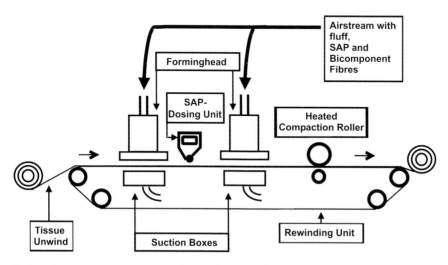

**Fig. 2-7** Airlaid process for the production of thermobonded absorbent cores [56]

**Table 2-4** Significant performance data of SAP fibres in comparison to SAP granules [50]

| | SAP Fibres | | SAP Granulates |
|---|---|---|---|
| | Camelot® 1241 | Oasis® Type 101 | Favor® SXM 4750 |
| Maximum Retention – CRC | 31 g/g | 26 g/g | 30 g/g |
| Absorption under 20 g/cm² pressure – AAP 0.3 psi | 23 g/g | 14 g/g | 29 g/g |
| Absorption under 50 g/cm² pressure – AAP 0.7 psi | 10 g/g | 6 g/g | 23 g/g |

Such composite materials are increasingly being used as absorbent cores for ultra thin high-quality hygiene articles. Another use of superabsorbents is the direct fixation on fibres. Table 2-4 lists a performance comparison of these SAP fibres with commercially available granular superabsorbents. Allan [57] gives details of further different types of fibre and publishes comparative test results.

### 2.4.5
### Summary

Not all possibilities of superabsorbent polymers have been exhausted by any means. They form an ideal complement to nonwovens whose practical uses can be considerably extended by tailor-made combinations with absorbent polymers. In addition to that further product innovations and new applications can be expected due to the current dynamic developments resulting from SAP research and development (see Part V).

### 2.5
### Spin finishes
W. DOHRN

### 2.5.1
### General

#### 2.5.1.1  Definitions
In the field of man-made fibres, the specialist term "spin finishes" has a wide range of meanings today. As there are no clear definitions, there is an explanation of terms below in order to make the following information more comprehensible.

Each man-made fibre receives its first spin or primary finish shortly after being spun. This may, for example in the case of filaments or even staple fibres from compact plant, be the product with which the fibre is supplied to the customer. In the case of staple fibres it is a help in the further processing along the fibre path, where it is intended to make stretching or crimping easier. The staple fibre then receives its final spin finish along or at the end of the fibre path, as shown in Fig. 2-8.

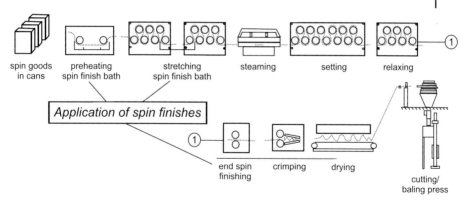

spin goods in cans | preheating spin finish bath | stretching spin finish bath | steaming | setting | relaxing

Application of spin finishes

end spin finishing | crimping | drying | cutting/ baling press

**Fig. 2-8** Possibilities of application of spin finishes along a staple fibre path

The term "secondary spin finish" is also used for the application of texturizing oils on to textured filaments, but also frequently for finishing or softening in staple fibre yarn production.

In the nonwovens sector there are a number of terms which in reality only signify the process of applying the spin finish before the combing or carding process. Here, the terms spin finishes, finishing, afteroiling, lubricating, needling auxiliaries, combing and carding auxiliaries or oils are used.

It is certainly meaningful to make only one subdivision in man-made fibre production into filament spin finishes and staple fibre spin finishes.

In the first processing step in the field of filaments the exact technological categories such as texturizing, coning or warping oil should be used, whereas in the field of staple fibres the word finish should be used for the textile spinning and nonwovens sector. This provides a better delineation of the term finishing as the final stage of wet finishing.

### 2.5.1.2 The requirements placed on spin finishes

With the developments in techniques and technology in the manufacture of man-made fibres and their processing, new and greater demands were placed on the properties of spin finishes. The main reason for these demands was the rapidly increasing speeds of the processing machines and the associated friction, abrasion and electrostatic behaviour. The avoidance of spray and aerosol formation became a necessity. These requirements are shown in Table 2-5, using a staple fibre spin finish as an example.

Currently the required development of technical and technological properties is placed in a new light owing to the requirements placed on spin finishing agents in terms of environmental protection. These requirements represent a difficult challenge to the manufacturers of spin finishes and will certainly lead to the development of new products.

**Table 2-5** Requirements placed on a staple fibre spin finish

- Optimum sliding and friction properties (fibre/fibre, fibre/metal, fibre/ceramics), static and dynamic
- Optimum antistatic properties
- Positive effect on the crimping behaviour
- High stability of the spin finish film
- Good wetting behaviour
- No aerosol formation
- No yellowing
- No foaming
- No tendency to conglutinate
- No waste gases
- Sufficient thermostability
- Long storage life

**Table 2-6** Environmental requirements placed on spin finishes

- Biodegradability
- Low emissions – thermostability
- Skin tolerance
- No oral toxicity
- No fish toxicity
- No alga toxicity
- No bacterial toxicity
- No heavy metal or halogen content

Table 2-6 shows a list of these new requirements, which apply both to the filaments and the staple fibres. Modern spin finishing agents should always form a film on the fibre surface which is stable and elastic in terms of external influences.

### 2.5.1.3 Compositions of spin finishes

Today spin finishes consist of various components which are intended to create a system with various properties. This system comprises the following components: bonding agents, gliding agents, antistatic agents, emulsifiers, anti-splash and thread cohesion agents, bactericides, wetting and moisture retaining agents, corrosion inhibitors, etc., whereby gliding agents, antistatic agents and emulsifiers make up approximately 90% of the system [58].

Whilst in the past mineral oils of various degrees of purity, natural fats and oils played a dominant role in the range of spin finishes, these are today supplemented by silicones, ester oils, phosphoric ester and polyalkylene glycol ether in their simplest forms, the ethylene and propylene oxides (EO/PO). However, environmental protection requirements as shown in Table 2-6 cannot be fully met by these products.

**Table 2-7** Ecological and human toxicological properties of carbonic esters, using a texturising oil as an example

| Property | Result | Test method |
|---|---|---|
| Biodegradability | over 90% | OECD 302 B, 303 A |
| Emission | 20 g C/kg product | calculated |
| Water hazard class | 1 | own classification |
| Heavy metal content | 0 | |
| Adsorbent org. halogens | 0 | |
| Acute fish toxicity | not toxic | OECD 203 |
| Acute Daphnia toxicity | low | OECD 202 |
| Chronic alga toxicity | low | OECD 201 |
| Chronic bacterial toxicity | not toxic | DIN L8 28412 |
| Orientating acute oral toxicity | not toxic | OECD 401 |
| Acute skin irritation | none | OECD 404 |
| Sensitisation of the skin | none | OECD 406 |
| Eye/mucous membrane irritation | low | OECD 405 |

**Table 2-8** Comparison of the thermostability of texturising oils

| Product | Carbon emission in g C/kg of product |
|---|---|
| Mineral Oil | 350 |
| Ester Oil | 150 |
| Carbonic polyester | 20 |

Test conditions: 190 °C, 90 s

However, new opportunities are offered by polyether/ester compounds connected by a carbonyl group. In literature and in practice these products appear under the heading carbonic polyesters [59]. These are self-wetting thermostable products which are readily soluble in water and have a good greasing effect. Table 2-7 shows the very good ecological and human-toxicological properties of these products [60]. Table 2-8 contains a comparison of the thermostability with a classical mineral oil and an ester oil.

## 2.5.2
## The application of spin finishes

### 2.5.2.1 Man-made fibre production
Spin finishes are applied in the form of neat-oil or from a watery medium. Solvents are rarely used. The production of the corresponding emulsions requires a good deal of time and energy, such as many hours of stirring and high temperatures. Pastes have to be melted down beforehand. Large plant used in the manufacture of man-made fibres contain their own preparation units. Often a stock

preparation is produced at first which is then diluted down to the desired application concentration and pumped into the storage containers of the spinning machine. This then usually becomes a cyclic procedure. Here too, carbonic esters make handling considerably easier owing to their solubility in water.

There are then various technical possibilities for the application of the spin finish. Rollers or disks transport the spin finish from a bath. Filaments or tows contact the disk and take over the solution or emulsion. The precondition for good spin finishing is good wetting of the disk. Difficulties may arise in the case of a change to the spin finish between incompatible substances, as disks made of sintered corundum cannot be cleaned easily, for example. It is possible to vary this system by changing the speed and direction of the disk. In order to improve evenness it is possible to use two discs on one side or one on either side.

One modern technique is to apply exact quantities using metering pumps (finger system).

In staple fibre production the tow is often led through one or several baths. Here such a bath can also serve the purpose of preheating the material before stretching.

The method of spraying is not used so often, as this is associated with greater unevenness of the application. In some staple fibre plant, however, antistatic agent is sprayed on to the fibres after the dryer.

### 2.5.2.2  Processing

Texturizing, warping and stretch-warping oils are almost always applied using rollers. Modern schooling and twisting machines work with various metering devices.

Finishes are almost always applied using spraying devices both in the manufacture of staple fibre yarns and in the nonwovens sector. Spraying is carried out in the pneumatic fibre stream, in mixing chambers or also on the pick-ups or outputs of fibre openers. Through the spraying an additional drying process is avoided. Here too, work can be carried out with undiluted products or from a watery medium. In order to achieve greater evenness of the application of the finishing agents on the fibre some firms work with a wet application, e.g. combed tops in a backwashing machine. However this procedure requires subsequent drying of the treated material.

### 2.5.3
### Test methods

### 2.5.3.1  Tests on the spin finishing agent

Nowadays spin finishes can be, and indeed are, subjected to a large number of various tests. These range from simple visual assessment to sophisticated analytical methods [61, 62]. Here it is intended to deal only with methods which enable the manufacture of nonwovens to test a finishing agent as to its suitability and quality and above all its constancy of type. In most cases parameters are involved which are stated by the producer in his technical leaflets or which are laid down in the quality agreements. Visual assessments often provide the operator with an initial but sufficient indication of the constancy of the product. Changes to the colour or opacity, as well as any cloudiness, should be a reason to carry out further tests, to ask the producer for further information or to look for causes within the operator's own company (e.g. non-adherence to storage conditions). The visual assessment may also be an initial evaluation of the compatibility of various products.

The viscosity (determined according to ASTM-D 1824-66) is important for the thickness of the oil pick-up and the friction properties of the fibre. Higher viscosities mean an increase in the coefficients of friction for fibre/fibre and fibre/metal [61]. The pH value (determined according to DIN 19268) is determined from a watery solution and provides an indication for the application conditions.

The refraction (determined according to DIN 53491) makes it possible to distinguish and define products, as well as determine their concentration in the watery initial bath.

The TO point (determined according to DIN 53917 for non-ionogenic surfactants) directly indicates in the form of a turbidity index a temperature barrier which, if exceeded, results in a marked change in characteristics.

The iodine colour value (determined according to DIN 6162) is well suited to determining the constancy of a product as it shows changes in the manufacturing conditions.

The water content (after Karl Fischer DIN 51777) is important for making up the recipe, as this determines the effective substance of a product.

The density (DIN 51757), acid number (DIN 53402), phosphorus, nitrogen or ash content are further parameters which can be used in defining the properties of a spin finish.

It is certainly never necessary to use all these methods simultaneously (e.g. in a quality agreement or incoming goods inspection), but they should rather be selected specifically according to the relevant processing technique and the field of application.

Tests on the behaviour in the environment are almost always carried out by the producer of the spin finishes or other suitable institutes owing to the complexity of the task and the sophisticated equipment required. In this field a standardization of the methods in the direction of the OECD norms is becoming evident.

### 2.5.3.2 Tests on spin finished fibre material

Here too there are a number of tests which can, for example, play a large role in the development of a new spin finish [61], whereby methods for the filament field differ markedly from tests on staple fibres [58].

In both cases the most significant aspect is the testing of the oil pick-up of the product, or rather its active ingredient, on the fibre. The fibre manufacturer is always interested in methods which produce quick and reliable results, if possible on-line [62, 63]. Modern spectroscopic measurements form a basis for this. However, they do have the disadvantage that a large number of calibrations are required, and can only be used if the composition of the spin finish is known exactly. For this reason the classical methods of solvent extraction are of greater importance for the textile industry. The precondition for achieving reproducible results is adherence to exact experimental conditions such as the acclimatization of the fibre material before the test or the selection of the correct solvent. Table 2-9 lists corresponding recommendations. This selection is very important, as an unsuitable solvent can cause considerable errors. For example, polar solvents cause oligomers to be dissolved out of PET fibres, resulting in the pick-up value becoming too high [64].

A classical test method for determination of the pick-up is Soxhlet extraction according to DIN 12602. This test can be automated through the use of Soxtherm or Soxtec equipment [65].

In a similar way, the testing of the electrostatic behaviour of the fibre is extremely important in the case of filaments and staple fibres. Here too there exist various test methods and principles. In practice, however, testing of the electrostatic behaviour in accordance with DIN 54345 is sufficient.

Especially in staple fibre processing the fibre/fibre adhesion is of particular interest. It is the significant precondition for the quality of a card web or a drawing-frame sliver and also influences the strength of a nonwoven fabric.

The simplest method to do so is to test the fibre cohesion with a suitable strength testing device [66]. With a numerical value or by recording an adhesion-sliding curve it is possible to determine the static friction behaviour. It is possible to achieve very good results by carrying out a test on the rotor ring device. Here, using the parameters bandwidth and opening performance it is possible to obtain good results relating to the designation of dynamic fibre/fibre and fibre/metal friction. After the corresponding data has been collected and the relevant experience gained, these results make it possible to estimate the adhesion and processability [67]. For both tests it is necessary to prepare the samples using a sample carding machine. Purely subjective examinations by the operator of the bulk and handle of a fibre [68] should not be underestimated. These do, however, require a good deal of experience.

**Table 2-9** Recommendations for extracting agents

| Fibre type | Viscose | Polyester | Polyamide | Polyacrylic | Polypropylene |
|---|---|---|---|---|---|
| Extracting agent | Ethanol or methanol | Petroleum ether/i-propanol (1:1) or methanol | Petroleum ether 40–60 °C | Methanol | Methanol (with restrictions) |

The extraction method is not suitable for slightly stretched PP fibres.

2.5.4
**Spin finishes on nonwovens**

2.5.4.1 **General**

Nonwovens have taken on an important position in the textile world in the last few years [69, 70]. This applies not only in terms of quantities, but also to the increasing number and type of applications. For the man-made fibre manufacturer this means that he not only has to develop new fibres, but also has to use new spin finishes for this field. It is necessary to take into account such trends as fibre attenuation, faster-running processing machines, ecological and toxicological requirements and new, sophisticated fields of application. Spin finishes and other finishes therefore become more important and have to be adapted to these requirements. They not only make up a decisive proportion of the manufacture of nonwovens, but also directly affect the properties of the end product.

There is also a clear trend developing in which spin finishes are becoming more important in comparison to other finishes. This means that the spin finish should be so well suited to processing and final application that the use of other finishes should only be necessary in exceptional cases. Spin finishes should not be considered "miracle products", however. They are a component of the system of fibre production – processing – field of application. Their effectiveness is often associated with compromise and may also be covered up by other influences.

2.5.4.2 **Nonwoven fabric manufacture and spin finishes**

The various methods of nonwoven fabric manufacture place various requirements on spin finishes.

In the case of nonwoven formation by a carding or crimping process with subsequent needle punching, the spin finish should satisfy the following main requirements [71]:

- lubrication of the fibre surface, i.e. the attainment of an optimum gliding/friction ratio
- sufficient fibre/fibre cohesion
- protection against electrostatic charge

These three points must always be seen within the framework of working conditions:

- type and state of the processing machines and their working speeds
- processing climate (temperature and humidity fluctuations markedly change the antistatic behaviour of a spin finish)
- fibre properties such as fibre type, diameter, cross-section, crimping, staple length, surface changes through delustering or spin dyeing
- storage conditions and storage time of the fibres before processing (danger of evaporation, degradation or absorption of the spin finish)

The requirements and working conditions result in a very high number of influential factors which seriously affect the manufacture and properties of a nonwoven fabric.

When nonwoven formation is considered more closely in a carding process, it quickly becomes clear that the effects of a spin finish are always associated with compromise [71]. The fibres must have enough sliding properties in order that there is not too much resistance to opening or thinning, i.e. the fibre/fibre friction must not be too high. On the other hand, if there is too little fibre/fibre friction, there will be too little cohesion, with the result that the strength of the nonwoven will be insufficient. It should not be forgotten, however, that the cohesion effect of a spin finish is far less than that of the fibre crimping. The fibre/metal friction must be sufficient for processing on the carder, but should not have a negative effect on the needling process in nonwoven punching. This is the reason for the differences between a spin finish for spinning fibre yarns and one for nonwovens. The latter should be lower in terms of fibre/metal friction, and higher in terms of fibre/fibre friction so that higher needle service times and nonwoven strengths can be achieved.

In cases of aerodynamic nonwoven formation, slight separation of the fibres is desirable, i.e. low fibre/fibre friction and a high antistatic effect of the spin finish [72].

In hydrotechnical nonwoven formation it is necessary to ensure very good dispersibility of the (mainly) short fibres.

Also the thermal or chemical bonding of nonwovens presupposes suitable spin finishes.

### 2.5.4.3 End product and spin finish

The number of fields of application for nonwovens today is vast and is being constantly extended. In contrast to end products made of spinning fibre yarns and filaments, end products based on nonwovens are affected by spin finishes to a greater extent. The reasons for this are as follows:

In the nonwoven sector, only a small amount of washing or rinsing is carried out. Textile dyeing processes are also not so common, but in comparison the proportion of spin dyed fibres is higher. The spin finish therefore often remains on the fabric until the final product and influences its quality and properties. Of course this means that is possible to influence certain properties of the end product specifically with the spin finish. Here are some examples to document the relationship between end product and spin finish: When used in hygiene articles and for medical purposes, not only the toxicological properties are significant, but also direct applicational values such as wetting-back and moisture penetration. The EDANA regulations provide very good information in this respect.

A diaper requires both hydrophobically and hydrophilically prepared fibres in order to meet these requirements.

In motor vehicle construction – one of the largest customer for nonwoven fabrics – a spin finish must make a contribution towards a low fogging value and a low level of unpleasant smells [73, 74]. In the manufacture of moulded parts, a high degree of thermal resistance of the spin finish is an advantage in order to avoid smoke and smells.

For geotextiles, ecotoxicological properties which do not impair the quality of the groundwater are of extreme importance.

Thermal and chemical resistance are important in the filter sector and for the quality of coated nonwovens or those with a special finish.

In the food sector, the suitability of a spin finish is assessed together with the substrate.

## 2.5.5
### Future prospects

In future the nonwovens sector will not only grow in terms of quantities, but will also conquer many new fields of application. It is certain that new manufacturing processes will be developed. In addition, there will be the familiar trends towards increasing production speeds, fibre attenuation and mixed processing [70, 75].

The development of special fibres will also continue. Spin finishes will have to follow these tendencies, always taking into account the aspects of processability and applicational suitability. Developing the correct products is only possible with good co-operation between all partners such as fibre manufacturers, spin finish producers and nonwoven fabric manufacturers. From case to case it will certainly also be necessary to involve mechanical engineering specialists.

## References to Chapter 2

[1] Koslowski HJ (1998) Nonwovens – a fast expanding business, Chem Fibers Intern 48: 360
[2] Mülhaupt R, Rieger B (1996) Übergangsmetall-Katalysatoren für die Olefinpolymerisation: 2. Herstellung von stereoregularen Poly(1-olefinen), Chimica 50: 10–19
[3] Steinmetz B, Tesche B, Przybyla C, Zechlin J, Fink G (1997) Polypropylene growth on silica-supported metallocene catalysts: A microscopic study to explain kinetic behavior especially in early polymerization stages, Acta Polymer 48: 392–399
[4] Jungbauer A (1994) Recycling von Kunststoffen, Vogel Verlag, Würzburg
[5] Brandrup J, Immergut EH (1989) Polymer Handbook, 3rd Ed, John Wiley & Sons, New York
[6] Hearle JWS (1982) Polymers and their Properties, Ellis Horwood Ltd, Chichester
[7] Alexander LE (1969) X-Ray Diffraction Methods in Polymer Science, Wiley-Interscience, New York
[8] Shimizu J, Okui N, Kikutani T, in: Ziabicki A, Kawai H (1985) High-Speeed Fiber Spinning, Chap. 15, John Wiley & Sons, New York
[9] Wichert H (1964) Kontinuierliche Trockner dringen weiter vor, Chem Ind 16, 9: 581
[10] Kuhn R (1981) Molekülmassen und Molekülmassenverteilung, in: von Falkai B, Synthesefasern, Chap 9.3, Verlag Chemie, Weinheim
[11] Ahmed M (1982) Textile Science and Technology Vol 5: Polypropylene Fibers – Science and Technology, Elsevier Scientific Publishing Comp, Amsterdam

[12]  Boor J (1979) Ziegler-Natta Catalysts and Polymerization, Academic Press, New York

[13]  Kissin YV (1995) Isotactic Polymerization of Olefins, Springer-Verlag, Berlin

[14]  Fink G, Mülhaupt R, Brintzinger HH (1995) Ziegler Catalysts, Springer-Verlag, Berlin

[15]  Allan NS (1983) Degradation and Stabilization of Polyolefins, Applied Science Publishers, London

[16]  Gleixner G, Vollmar A (1998) Fibers of metallocene polyolefins, Chem Fibers Intern 48: 393–394

[17]  Blechschmidt D, Fuchs H, Vollmar A, Siemon M (1996) Metallocene-catalysed polypropene for spunbond applications, Chem Fibers Intern 46: 332–336

[18]  Tetzlaff G, Dahmen M, Wulfhorst B (1993) Polyesterfasern, Faserstoff-Tabellen nach P-A Koch, Deutscher Fachverlag GmbH, Frankfurt

[19]  Ludewig H (1975) Polyesterfasern, Chemie und Technologie Akademie-Verlag, Berlin

[20]  Zeitler H (1985) Cyclische Oligomere in Polyester, Melliand Textilber: 132–138

[21]  Zaremba S, Steffens M, Wulfhorst B, Hirt P, Riggert KH (1997) Polyamidfasern, Faserstoff-Tabellen nach P-A Koch, Deutscher Fachverlag GmbH, Frankfurt

[22]  Klare H, Fritzsche E, Gröbe V (1963) Synthetische Fasern aus Polyamiden, Akademie-Verlag, Berlin

[23]  Rane E (1997) Karl Fischer – active in PA6 from its origins, Chemical Fibers Int 47: 465–468

[24]  Heintze A, Wolff K, Gries Th (1997) Polyamide 6 – Technologies for polymerization and spinning of textile filaments, Chem Fibers Intern 47: 469–470

[25]  Gutmann R, Herlinger H (1994) Thermische Belastbarkeit von Garnen und Textilien aus PA6, Chemiefasern Text Ind 44/96: 752–757

[26]  Frushour BG, Knorr RS (1968) Acrylic Fibers, in: Lewin M, Pearce EM, Handbook of Fiber Science and Technology, Vol IV: Fiber Chemistry, Marcel Dekker Inc, New York

[27]  Gord L (1968) Polyvinyl Chloride Fibers, in: Mark HF, Atlas SM, Cernia E, Man-Made Fibers, Science and Technology, Vol. 3, Interscience Publishers, New York

[28]  Gächter R, Müller H (1984) Plastic Additives Handbook, Hanser Verlag, München

[29]  Dietze R, Schultheis K, Kaufmann S (1998) Determination of size distribution of titanium dioxide agglomerates, Chem Fibers Intern 18: 414–415

[30]  Gröbe A, Herlinger H, Metzger W (1978) Herstellung und Untersuchung schwerent flammbarer Polyester- und Polyamidfasern durch Einspinnen von Additiven, Chemiefasern Text Ind 28/80: 231–235

[31]  Herlinger H, Einsele U, Püntener A, Meyer P, Metzger W, Gröbe A, Veeser K (1980) Prinzipien und Grenzen der Möglichkeiten zur Entwicklung schwerentflammbarer Fasern, Lenzinger Ber Folge 48, 3: 65–76

[32]  Herlinger H, Küster B, Essig H (1989) Untersuchungen zum thermo- und photooxidativen Abbau von Polyamidtextilien in Gegenwart und Abwesenheit von Stickoxid, text praxis internat 44: 655–663

[33]  Todesco RV, Diemunsch R, Franz T (1993) Neue Erkenntnisse bei der Stabilisierung von PP-Fasern für die Anwendung im Automobil, Techn Textilien/Technical Textiles 36: T197–T199

[34]  Krässig HA, Lenz J, Mark HF (1984) Fiber Technology, From Film to Fiber, Marcel Dekker, New York

[35]  Werner G (1984) Verwendung von hochsaugaktiven Polymeren in technischen Vliesstoffen, AVR-Allgem Vliesstoffrep: 178–182

[36]  EDANA statistik. www.edana.org

[37]  Kerres B (1996) Superabsorber für wässrige Flüssigkeiten, Textilveredlung 31: 238–241

[38]  Buchholz F (1998) Modern Superabsorbent Polymer Technology, Wiley-VCH, Weinheim: 6–14

[39]  Tanaka T (1981) Gele, Spektrum der Wissenschaft, 3: 79–93

[40]  Thiel J, Maurer G, Prausnitz J (1995) Hydrogele: Verwendungsmöglichkeiten und thermodynamische Eigenschaften, Chem Ing Techn (67): 1567–1583

[41] Elias, H-G (1992) Makromoleküle, Vol 2: Technologie, Hüthig & Wepf Verlag, Heidelberg: 735–740

[42] Graham A, Wilson L (1998) Commercial Processes for the Manufacture of Superabsorbent Polymers in *Modern Superabsorbent Polymer Technology*, Wiley-VCH, Weinheim: 69–114

[43] Martini S (1998) Superabsorber und ihre Anwendungen, Melliand Textilber: 717–718

[44] Tsubakimoto T, Shimomura T, Irie Y (1987) U.S. Patent 4666983

[45] Dahmen K, Mertens R (1991) Deutsche Patentschr 4020780

[46] Houben H, Herrmann E, Dahmen K (1994) Deutsche Patentschr 4420088 C2

[47] Kaltenecker O (1996) Superabsorbierende Beschichtung auf Vliesstoffen – Herstellung und Eigenschaften, Dissert, Univers Stuttgart Fakult Chemie

[48] edana Recommended Test: 400–470

[49] Meyer S, Werner G (1996) Flüssigkeitsspeichernde Polymere – Innovationsmotor für moderne Kinderwindeln, Intern Textilbull 42, 1: 32–38

[50] Geschäftsbereich Superabsorber, Anwendungstechnik, Stockhausen GmbH & Co KG, Krefeld (1998)

[51] Rohe D (1996) Das Wasser geht ins Netz, Chem Ind: 12–15

[52] Chem Systems (1996) Developments in Superabsorbent Polymer Technology, New York

[53] Brandt K, Goldman S, Inglin T (1988) U.S. Pat 4654039

[54] Kellenberger S (1989/1991) EP 0339461 A1 und EP 0443627 B1

[55] Knowlson R (1995) Efficient use of Superabsorbents in Composite Structures, AVR-Allgem Vliesstoffrep: 39–44

[56] Herrmann E (1997) Pre-manufactured Airlaid Composites containing Superabsorbents, edana Nordic Nonwovens Symp, Göteborg

[57] Allan D (1998) Other Superabsorbent Polymer Forms and Types in *Modern Superabsorbent Polymer Technology*, Wiley-VCH, Weinheim: 223–245

[58] Kleber R (1984) Tenside bei der Herstellung von Stapelfasern Chemiefasern Text Ind 6: 412–415

[59] Winck K (1992) Ökologische und toxikologische Aspekte bei der Entwicklung umweltfreundlicher Faserpräparationsmittel, Chemiefasern Text Ind 11: 893–895

[60] Dohrn W, Winck K (1996) Neues Präparationssystem für PES – Filamente im Wirkwarenbereich, Melliand Textilber 11: 770–774

[61] Gutmann R et al. (1992) Spinnpräparationen – Zusammenhänge zwischen Struktur und Wirksamkeit, Chemiefasern Text Ind 11: 886–893

[62] Rothermel CE (1988) Messung der Präparationsauflagen mit Infrarot-Analysator, Chemiefasern Text 6: 568–570

[63] Rossa R (1992) Online-Messung von Spinnpräparationen im Spinnprozeß, Chemiefasern Text 11: 896–897

[64] TEGEWA (1998) Auflagenbestimmung von Faserpräparationen, 2. Entw 08.05.98

[65] Stockhausen Prüfanweisung A00TC109, Bestimmung der Präparations- bzw. Avivageauflage an Stapelfasern und Filamenten durch Extraktion mit dem Gerhard Soxtherm 2000 automatic

[66] Stockhausen Prüfanweisung A00TC106, Bestimmung der Bandhaftung von Stapelfasern

[67] Stockhausen Prüfanweisung A00TC107, Bestimmung der Rotorringwerte von Stapelfasern

[68] Bobeth W (1993) Textile Faserstoffe – Beschaffenheit und Eigenschaften, Springer-Verlag Berlin-Heidelberg: 114

[69] Dahlström N (1993) Weltweite wirtschaftliche und technische Einflüsse auf die Vliesstoffindustrie Chemiefasern Text Ind 3: 156–164

[70] Albrecht W (1988) Vliesstoffe: ihre Bedeutung und Entwicklungstrends, Chemiefasern Text Ind 7/8: 672–676

[71] Billica HR (1977) Funktionen der Faserpräparation bei der Verarbeitung von Chemiespinnfasern, Chemiefasern Text Ind 4: 322–328

[72] Albrecht W (1987) Chemiefasern für die Vliesstoffindustrie: Standard- und Spezialtypen, Eigenschaften und Einsatzmöglichkeiten, Chemiefasern Text Ind 8: 682–687

[73] Hardt P, Weihrauch T (1994) Prognosen zum Fogging- und Abluftverhalten von Textil-hilfsmitteln, text praxis internat 1/2: 79–82

[74] Eisele D (1992) Nadel-/Polvliesbeläge (NVB/PVP) für den Automobilbau, text praxis 8: 723–728

[75] Sames G (1998) Application of cellulosic fibres in nonwovens, Chem Fibers Inter 6: 203–206

# 3
# Binders
P. EHRLER

## 3.1
## Introduction

Using binders is one of the basic methods for consolidating webs, for producing nonwoven fabrics. They bind the fibres in a web positively with one another, adhesively as a rule. Generally speaking, a nonwoven attains its maximum strength at minimum bending rigidity when all the fibre intersecting points in the staple fibre web are bonded positively and in punctiform manner by binders. An additional binder component, located between the intersecting points, has a stiffening effect.

Although positive bonding restricts the relative movement of the fibres, it stabilizes their position in the case of open web constructions. Thus this method of bonding has a decisive effect on the basic characteristics and specific end-use properties of bonded fibre fabrics: compared to frictionally consolidated webs, for example spunlace products, positively consolidated nonwovens tend to exhibit a larger amount of elastic deformation in the case of bending, tensile and compression (bulking) stresses. The 'recovery margin' of the nonwoven is improved vis-à-vis such stresses which are met with in daily usage. In addition the tendency to surface roughening is reduced and the resistance to washing/cleaning stresses is improved.

The concept of binders includes a very wide range of polymer products which can be subdivided into two large groups:

- binder fluids
- binder fibres

The term 'binder fluid' serves as an overall concept for the synonyms latex, adhesive, dispersion, synthetic dispersion, polymer dispersion and emulsion polymer. This concept emphasizes the practical closeness to binder fibres: both forms serve the same purpose: the consolidation of webs.

Binder fibres and binder fluids can be compared with one another from different angles, whereby a detailed examination shows considerable differences: 'Punctiform' bonding combines softness with stability and therefore corresponds to the current image of a 'soft nonwoven with high-end use strength'. The form of bonding achieved with binder fibres approaches this ideal image because the substance which effects bonding necessarily concentrates on the points of intersection because of the high melt viscosity. To guide a binder fluid exclusively to these intersection points, a controlled surface tension dependent on the binder fluid and the fibre finish is needed. Nevertheless it is almost impossible to prevent 'sail-shaped' binder skins at the points of intersection. It is therefore simpler to satisfy the requirement 'maximum strength at minimum stiffness' with binder fibres than with binder fluids.

Binder fibres and binder fluids also differ considerably from one another as regards the contribution of adhesion to the strength of nonwovens. Reactive groups which improve the adhesion between the solidified/crosslinked binder droplets and the fibres are added to the binder fluids. Binder fibre developments are not aimed at greatly improving the adhesion, but at optimizing the melt viscosity at a specific melting point. Despite the disregard of the adhesive contribution, however, there is no proof that binder fibres give rise to a lower nonwoven fabric strength than binder fluids.

Moreover, binder fibres have not won a greater share of the market at the expense of binder fluids just because of these differences. The trend which can be detected since the eighties is based on the clear disadvantages of binder fluids during production:

– slow process rate because of drying and crosslinking
– additional stages in the process
– high energy consumption because of drying
– poor overall environmental compatibility

Typical binder fractions on nonwovens come to 10–40% by weight, a high proportion owing to the effectiveness of binders. Because of this binders are commercially of interest.

Binder fluids are not specific to any web. They are used with the same or similar formula in different parts of industry, for example in textile finishing, the paper industry, printing technology and the plastics industry.

In addition to binder fibres, bonding powder can also be used for consolidating webs, preferably those based on polyamide, polyethylene or polyester [5] (see also Section 2.3).

## 3.2
## Binder fluids

Meanwhile, because of the rapidly rising importance of binder fibres, the end has been forecast for binder fluids. This prognosis has not been fulfilled, mainly for three reasons [1–5]:

– the formulae of binder fluids can be customized and thus adapted to product requirements

– the relatively low cost of many binder fluids
– and the possibility of subsequently producing different nonwovens from the
  same web, which facilitates planning

In the meantime the wide range of available binder fluid classes exceeds the
range of web fibre types. The shrinking range of fibres contrasts with a constantly
expanding range of binder fluid classes and types – a development that can be observed in the case of numerous technical textiles.

## 3.2.1
## Chemical structure, construction principle [1–6]

### 3.2.1.1 Monomers

The binder fluids are produced from monomers by emulsion polymerization
(Fig. 3-1). The wide range of monomers and comonomers used for this has been
analyzed in detail by Morris and Mlynar [5], with the vinyl monomer $CH_2=CH–R$
as the basic monomer and "R" representing different comonomers:

- ethylene
- styrene
- vinyl acetate
- vinyl chloride
- acrylic esters: ethyl acrylate, butyl acrylate
- acrylonitrile
- methyl methacrylate

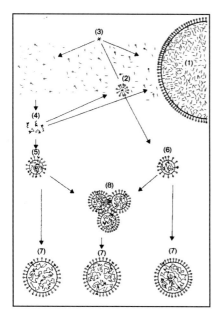

**Fig. 3-1** An idea of the particle formation in
emulsion polymerization (polymer)
"/" monomer molecule
"-o" emulsifier molecule
(1) monomer droplet
(2) monomer-containing emulsifier micelle
(3) water-soluble radical
(4) oligomer radicals at the primary agglom-
    eration stage
(5), (6) latex particles
(7) polymerized latex particles
(8) latex particles at secondary agglomeration
    stage

**Table 3-1** The most important binder fluid classes

| Vinyl-basis [5] | Further classes |
|---|---|
| Acrylate | Elastomers |
| Styrene acrylates | Polyurethane |
| Vinyl acetate | Silicon elastomer |
| Vinyl acrylate | Natural rubber |
| Ethylene vinyl acetate | Thermosetting plastics (crosslinked) |
| Styrene butadiene | – Phenolic resin |
| Polyvinyl chloride | – Melamine resin |
| Ethylene vinyl chloride | – Urea resin |
| Vinyl alcohol | – Formaldehyde resin |
| Butadiene acrylate | – Epoxy resin |
| | – Alkyd resin |
| | – Polyester resin |

A comparable procedure is found elsewhere (Devry [2]), for example in the combination of polyethylene with vinyl acetate, vinyl chloride, styrene or acrylonitrile.

The wide choice of co-monomers permits correct polymer constructions, which results in a wide range of binder fluid classes (Table 3-1). Their price is essentially determined by the monomer costs [4].

The co-monomers considerably influence the properties of the emulsion polymer (the binder fluid), where a few basic properties act as a guide:

- **Glass transition temperature** (Table 3-2). The lower the glass transition temperature, the softer the resultant polymer.
  This basic property is particularly important.

- **Hydrophilicity/hydrophobicity.** To ensure high nonwoven wetfastness, it is essential to use hydrophobic co-monomers, e.g. styrene. Hydrophilic co-monomers such as acrylonitrile can be considered if high resistance to solvents is required.

**Table 3-2** Comonomer/glass transition temperature ranking

| Monomer | Glass transition temperature ($°C$) |
|---|---|
| Ethylene | −125 |
| Butadiene | −78 |
| Butyl acrylate | −52 |
| Ethyl acrylate | −22 |
| Vinyl acetate | +30 |
| Vinyl chloride | +80 |
| Methyl methacrylate | +105 |
| Styrene | +105 |
| Acrylonitrile | +130 |

- **Molecular weight.** The extensibility of the binder depends on the glass transition temperature and the molecular weight. Compared to vinyl chloride or methyl methacrylate, butadiene ensures better elastic behaviour, for example.

### 3.2.1.2 Functional groups, crosslinking agents

Functional groups are incorporated in addition to the co-monomers. They also influence the properties of the polymer, and hence of the nonwoven: the mechanical properties, solvent resistance, adhesion and viscosity. The following are well-known functional groups:

- acrylic acid
- methacrylic acid
- acrylamide
- n-methylol acrylamide

The last-named, n-methylol acrylamide, acts as a crosslinking agent: it is self-crosslinking at high temperatures. In combination with other substances (e.g. ammonium nitrate, melamine formaldehyde) the crosslinking temperature can be reduced considerably. These crosslinking agents came in for much criticism during the discussions on formaldehyde, with the result that the search has been on for alternative crosslinking systems in recent years (cf. Section 3.2.4).

Binder fluids containing crosslinking agents are preferred in the case of high water and solvent resistance requirements because the crosslinked binders contribute to this resistance.

In addition to these crosslinked polymers, thermoplastic polymers which have the same chemical basis and do not contain crosslinking agents can also be used.

### 3.2.2
### Binder fluids and their processing

The polymer emulsion, which forms the basis of the binder fluid, has to be stabilized before use. Here use is made of surface-active agents, the ionic properties of which (anionic or nonionic) also affect the properties of the resultant binder fluid. The molecular weight and particle size also have an effect. Such as basic component may contain approximately 5% by weight auxiliaries in total.

From the polymer emulsion a binding liquid (Table 3-4) is achieved by adding a variety of agents (Table 3-3), the liquid being tailor-made for particular processing parameters and application-oriented requirements [1, 7].

In practice, it is important – analogues with textile finishing – to avoid the binding liquid migrating while drying, which would disturb. The liquid is thermally sensitive, which is the result of a low coagulation temperature (40 °C to 50 °C; if necessary, even ambient temperature: for example LEC-acrylate of Rohm and Haas). This advantage being helpful with migration-free drying, there is, unfortunately, danger such a binding agent coagulates while being stored, which will also reduce its adhesive capacity. Due to this dilemma, it is common practice to escape

**Table 3-3** Typical supplementing materials in binder fluids

Thickeners such as polyvinyl alcohol or cellulose ether
pH regulators
Antifoaming agents
Wetting agents
Antibacterial agents
Plasticizers
Substances having a repellent effect, e.g. perfluoro polymers, silicon elastomers
Salts for flame-resistant finishing
Optical brighteners

**Table 3-4** Characteristic features of binder fluids [1, 2, 4, 11]

**Dispersion**
Particle size
Viscosity
Solids content
Ionic character
pH value
Surface tension
Colloidal stability
Compatibility with oleophobic substances
Pigment can be added

**Film**
Minimum film forming temperature
Glass transition temperature
Coagulation point
Curing temperature (with crosslinkable binder fluids)
Tackiness of film
Bending rigidity of film

**Processability**
Foam height (in the case of an undesirable tendency to foam)
Stirrability
Foamability
Foam stability (with useful foam)

migration using binding agents containing maximum solid substance or foam binding agents.

Stepanek [8], for instance, summarizes the advantages of foam: more even distribution of binding agent, in particular with unstable foam; higher concentration of binding agent, faster drying; more porous binding film with stable foam and smaller binding elements.

Fibre/binder adhesion, i.e. binder adhesiveness, has lost its thematic prominence because more efficient binder fluids have been developed and binder-compatible fibre lubricants have been formulated. Nonetheless adhesion improvement

is an enduring theme in the case of multi-use nonwovens, i.e. easy-care interlinings. Against this background Sigurdson [9] has reported on the possibility of improving nonwoven fibre/binder adhesion using plasma pre-treatment (low pressure, i.e. a technical vacuum).

3.2.3
**Binder fluids and the properties of nonwoven fabrics**

The properties of binder fluids can be adapted to the requisite nonwoven properties
within wide limits while conforming to the principle of tailor-made binder fluids. So
only a few nonwoven properties (stability, wet strength, hydrophilicity/hydrophobi-
city) influence the decision as to which binder fluid class should be used for a particular application.

Of course, many complex nonwovens properties have to be taken into account when
selecting the binder fluid (Table 3-5), whereby a distinction has to be drawn between
processing properties and the properties which depend on the end-use (Table 3-6).

**Table 3-5** Significant binder-dependent features of nonwovens [1, 2, 4, 11]

**General remarks**
Adhesion: to specific surfaces
Blocking tendency
Heat sealability
Capable of ultrasonic welding
Temperature-dependent yellowing tendency

**Selected use-specific features**
Hydrophilic property
Hydrophobic property
Oleophobic property
Fluid transport
Recovery after compression
Fogging tendency
Resistance to alkalis
Launderability
Drycleaning resistance
Heat resistance
Resistance to sterilizing
Oxidation resistance

**Table 3-6** Characteristic uses of nonwovens containing binders

| | |
|---|---|
| Cleaning cloths | Plasticizer-free packaging [29] |
| Hand towels | Carriers for coating |
| Roofing | Car interiors |
| Medical products | Nonwoven carpets |
| Interlinings | Bedding |
| Filters | Upholstery fabrics |
| Nappies | Clothing |

Binder fluid criteria such as plasticity, hardness levels, etc. no longer play a decisive role in the selection, as there are now many ways of modifying the end-product. Basic properties and characteristic features [1–3, 5–8, 10–13, 15–18]:

- Acrylates [22, 24]
- can be used universally
- high lightfastness, heat resistance and resistance to oxidization, high resistance to washing and dry cleaning
- high wet strength and bulk resistance
- wide scope for varying the bending strength, the adhesion and the hydrophobicity

- Vinyl acetate [4, 6]
- yellowing tendency, hydrophilic
- adequate dry strength
- wide scope for modifying the properties, no adequate softness as a pure product
- low-cost binder fluid

- Vinyl acrylate [5]
- more hydrophobic than vinyl acetate
- occupies a position between acrylates and polyvinyl acetates as regards the properties and price

- Ethylene vinyl acetate [3, 5, 25]
- high wet strength, favourable liquid absorption
- soft to medium hard
- costs less acrylates

- Acrylonitrile butadiene (NBR) [1]
- limited heat resistance and fastness to light and ultraviolet radiation; limited resistance to solvents
- favourable softness and 'resilience'; elastomer type, i.e. high extensibility in the case of a high elastic component, favourable abrasion resistance
- nonwovens can be very easily split and ground because of the low thermal plasticity of the NBR.

- Styrene butadiene (SBR) [1, 13]
- excellent hydrophobicity in the crosslinked state, can be vulcanized, limited resistance to heat, light and ultraviolet radiation
- the styrene-/butadiene ratio affects vital properties, including elastic recovery and softness
- low-cost binder fluid

- Vinyl chloride [26]
- on its own vinyl chloride is too hard for consolidating nonwovens; co-monomers are needed for improving its properties
- has thermoplastic properties; hence its use as a 'heat-sealable' and a high frequency weldable binder
- limited flame-resistant properties
- yellowing tendency

- Ethylene vinyl chloride [3, 26]
- limited flame-resistant properties, yellowing tendency
- softer than vinyl chloride products
- good adhesion to synthetic fibres, high acid resistance

- Polyvinyl alcohol [1, 12]
- properties depend on the molecular weight and level of hydrolysis (fraction of non-saponified vinyl acetate)
- too hard on its own
- softer as a result of using polyethylene glycol; high dry strength when used in combination with vinyl acetate
- adjustable water solubility, high resistance to oil/fat and organic solvents
- limited biodegradability

- Polyurethane [1, 10, 25]
- the usual aqueous dispersions are based on polyester-polyurethane; polyurethane solutions have largely vanished from the market
- polyurethane is characterized by a special, positive range of properties: high resistance to hydrolysis, good adhesion and lightfastness, adjustable softness and 'resilience'
- is of particular interest because of the current attempts to produce 'elastic' nonwovens

### 3.2.4
### Focal points of development [3, 5, 7, 14–21, 25–27]

In the seventies, and also sometimes in the eighties, developments were chiefly aimed at producing binders for softer nonwovens with improved adhesion and better washfastness [1], as well as maximizing the solids content of the binder fluids, without endangering the stability of the dispersion and the proccessability. These problems have been largely resolved.

The developments over the past decade related to other problems:

- improved environmental compatibility
- reduced levels of toxicity
- better binder stability
- reduced flammability

Significant emphasis was and is being placed on work to improve environmental compatibility and to increase safety by reducing toxicity levels.

The modifications to binder fluid formulations necessary to achieve this are largely similar, so that they will be treated together from now on.

Current requirements:

– no volatile residual monomers
– no volatile adhesives
– no compounds likely to cause fogging (volatile additives in car interiors)
– no formaldehyde emissions
– no phenols
– no preservatives
– no heavy metals
– complete biodegradability in water
– no noxious compounds
– good degradability in landfills

These problems cannot be solved in isolation, i.e. solely for the given polymer system. The various additives to a binder fluid (for example, pH regulators, flame retardants and fillers) must be included in the optimization.

Confalone et al. [20] discuss ecologically safe, self-crosslinking ethylene vinyl acetate binder fluids. A whole family of environmentally safe acrylates is presented by Schumacher et al. [14, 28]. The products are suitable for degradable viscose rayon nonwovens and also for recyclable nonwovens and are, in addition, almost formaldehyde-free (less than or equal to 2 ppm).

Formaldehyde reduction was one focal point of developments because binder fluids contain at least three sources of formaldehyde: crosslinking agents, emulsifiers and biocides [17]. Walton [19] introduced ethylene vinyl acetate binders containing less than 15 ppm formaldehyde.

A reduced 'free formaldehyde content' was achieved for melamine formaldehyde resin by Kajander [16] by adjusting the formaldehyde and nitrogen fractions in a way which is different from previous relationships.

Young [21] sees in the use of epoxy resin (based on bisphenol A) the chance of reducing the free formaldehyde content of various binder fluids.

Schumacher et al. [14] emphasize the analytical problems when determining low concentrations of free formaldehyde: only by integrating an additional HPLC separation stage in the analysis is it possible to exclude artefacts.

The strict environmental requirements are also valid for the binder fluids with flame-resistant properties [27], whereby production and application are equally affected. No dioxins may be released during burning, for example.

Koltisko [26] refers to the favourable flame-retardant properties of vinyl copolymers.

The great demand for baby napkins and incontinence pads encouraged attempts to develop binder fluids for biodegradable nonwovens; Schumacher et al. [14, 28] pointed out that, although no really biodegradable bonded-fibre nonwovens could be developed in the foreseeable future, low-formaldehyde formulations would result in extensive compostability, which has been confirmed by burial, pilot compostability and controlled compostability tests.

The demand for increased stability refers to different features. Donno [15], for example, introduces acrylate binder fluids, the heat resistance of which is considerably above the conventional limit (approximately 160 °C). The yellowing tendency should be appreciably reduced.

## 3.3
## Adhesive fibres
J. SPINDLER

The use of adhesive fibres is one of the most elegant and convenient methods to consolidate nonwoven structures. Adhesive fibres are fibres, which are able to provide adhesive bonds to other fibres due to their solubility or their fusible character.

Adhesive fibres can be subdivided in two categories:
• Soluble fibres are fibrous products, which become sticky under the influence of a solvent, like polyvinyl alcohol- (PVA) or alginic fibres.
• Hotmelt adhesive fibres are fibres made from hot melt adhesives or general thermoplastic polymers, which exhibit a lower softening point compared to the matrix fibres.

It has to be mentioned that these polymers can also be applied as powders onto or into a nonwoven. A good example is the powder-bonding process, where a powdered hotmelt adhesive is scattered into the nonwoven. The use of powdered binders requires additional equipment and the fusing has to be initiated immediately after applying the adhesive.

Adhesive fibres are easy to apply by blending without additional process steps or equipment. This requires that the adhesive fibres are easy to blend with other fibres and that the mixture can be manufactured into a nonwoven structure. Therefore hotmelt adhesive fibres have to be easy to open.

### 3.3.1
### Soluble fibres

PVA-fibres are most probably the oldest type of adhesive fibres. If these fibres are treated with water they first swell and afterwards dissolve under the influence of elevated temperature. Selection of the correct temperature is very important for an effective bonding. The dissolving temperature can be varied within limits by modification of the polymer. A typical application for PVA-fibres (see PVA fibres) are wet-laid nonwoven [30].

3.3.2
**Hotmelt adhesive fibres**

Different aspects can classify hotmelt adhesive fibres. On the one hand, according to the appearance and the chemical structure, and on the other hand by the way of application. Table 3-7 shows without claim of completeness examples for commercial hotmelt adhesive fibres.

3.3.2.1 **Appearance**
Hotmelt adhesive fibres are available either as monocomponent fibres, which are made from 100% of a hot melt adhesive polymer, or as bicomponent fibres. In case of bicomponent fibres, the fibre sheath consists of a polymer which has a lower melting point compared to the core component (Fig. 3-2).

3.3.2.2 **Chemical structure**
In principle, all thermoplastic polymers can be used for the manufacture of hotmelt adhesive fibres. The heat resistance of the "to-be-bonded" fibres of the nonwoven construction, however, limits the selection. Hotmelt adhesive fibres are primarily based on polyolefins, polyesters or polyamides, but the use of fibres made of polyvinyl chloride/vinyl acetate copolymers is also known.

**Polyolefins**
Polyethylene (PE)- and polypropylene (PP)- fibres are suitable as bonding fibres because of their low melting point. However, the high degree of crystallinity, the very high melt viscosity and the relatively low affinity to non-olefin fibres restricts their application possibilities.

**Fig. 3-2**   Cross-section of bicomponent hotmelt adhesive fibres

**Table 3-7** Selection of commercial available bonding fibres

| Name | Manufacturer | Type of polymer | Melting point resp. $T_g$ of the adhesive component | Special properties | Applications |
|---|---|---|---|---|---|
| ES-C | FiberVisions | PP/PE | 125 °C | BICO | hygiene nonwovens |
| ES-E | FiberVisions | PP/PE | 125 °C | BICO, helical crimped | hygiene nonwovens |
| Grilamid HP 1200 | EMS-GRILTECH | Polyamide 12 | 178 °C | | composites |
| Grilon KA 115 | EMS-GRILTECH | Copolyamide | 115 °C | activation by steam | paper machine clothing |
| Grilon KA 140 | EMS-GRILTECH | Copolyamide | 140 °C | | |
| Grilon BA 115 | EMS-GRILTECH | PA 6/Copolyamide | 115 °C | BICO | paper machine clothing |
| Grilon BA 140 | EMS-GRILTECH | PA 6/Copolyamide | 140 °C | BICO | |
| Grilon BA 165 | EMS-GRILTECH | PA 6/Copolyamide | 165 °C | BICO | |
| Grilon BA 3100 | EMS-GRILTECH | PA 66/PA 6 | 220 °C | BICO | shoe lining |
| Grilene KE 150 | EMS-GRILTECH | Copolyester | 150 °C | | bulky fibre fill |
| Grilene KE 170 | EMS-GRILTECH | Copolyester | 170 °C | | bulky fibre fill |
| Fossfiber 565 | FOSS | PETP/PE | 125 °C | BICO | |
| Fossfiber 571 | FOSS | PETP/Copolyester | $T_g$ 70 °C | BICO | |
| Fossfiber 410 | FOSS | Copolyester | $T_g$ 78 °C | | |
| Fossfiber 566 | FOSS | PETP/Copolyester | $T_g$ 78 °C | BICO | bulky fibre fill |
| Fossfiber 531 | FOSS | PETP/Copolyester | 160 °C | BICO | filtration, molded parts |
| Fossfiber KB | FOSS | PETP | $T_g$ 80 °C | amorphous | bulky fibre fill, molded parts |
| Cellbond J58 | KoSa | PETP/Copolyester | 155 °C | BICO | filtration, molded parts |
| Tergal T 190 | Tergal Fibre | PETP | $T_g$ 80 °C | amorphous | nonwovens for upholstery |
| ELK | Teijin | PBTP/Copolyester | 170 °C | BICO, elastomeric binder, helical crimp | |

**Table 3-7** (continued)

| Name | Manufacturer | Type of polymer | Melting point resp. $T_g$ of the adhesive component | Special properties | Applications |
|------|--------------|-----------------|-----------------------------------------------------|--------------------|--------------|
| Trevira 254 | TREVIRA | PETP/Copolyester | $T_g$ 70°C | BICO | bulky fibre fill |
| Trevira 255/256 | TREVIRA | PETP/PE | 125°C | BICO | hygiene nonwovens |
| Trevira 259 | TREVIRA | PETP | $T_g$ 80°C | amorphous | filtration, molded parts |
| MELTY 4080 | Unitika | PETP/Copolyester | $T_g$ 70°C | BICO | bulky fibre fill |
| MELTY 7080 | Unitika | PETP/Copolyester | 160°C | BICO | bulky fibre fill, molded parts |
| Wellbond 1439 | Wellman | PETP/Copolyester | 160°C | BICO | bulky fibre fill, molded parts |
| Wellbond 1440 | Wellman | PETP/Copolyester | $T_g$ 70°C | BICO | bulky fibre fill |
| Type T-201 | FIT | PETP/Copolyester | 110°C | BICO | |
| Type T-202 | FIT | PETP/Copolyester | 185°C | BICO | |
| Type T-207 | FIT | PETP/Copolyester | 130°C | BICO | |
| Type T-215 | FIT | PETP/Copolyester | 145°C | BICO | molded parts |

These fibres should be better called thermobonding fibres because pressure is necessary for bonding in most cases. Copolymerization or modification of the polymer with additives, however, can improve the bonding characteristics of polyolefins.

Olefin bonding fibres are well accepted for applications in the hygiene nonwoven market. Sheath/core bicomponent fibres [3] based on PE/PP like the ES fibre (FiberVisions) or fibres based on PE/polyethylene terephthalate (PET) (Trevira 255) are well known in the industry.

**Polyesters**
The modification of PET and polybutylene terephthalate (PBT) with comonomers like adipic acid or isophthalic acid results in polyesters with low softening points, which are suitable for the manufacturing of bonding fibres.

The partially crystalline fibres Grilene KE 150 and KE 170 (EMS-GRILTECH) are examples for monocomponent hotmelt adhesive fibres based on polyester. In addition, amorphous fibres with a softening point of about 80 °C (Fossfiber PETG Type 410) are available as bonding fibres.

Bicomponent fibres, like Wellbond 1440 (Wellman), MELTY 4080 (Unitika) and Trevira 254, have a PET core which is covered with an amorphous copolyester sheath having a glass transition temperature of about 70 °C. Their main use is bonding of fibrefill nonwoven applications.

Due to their better heat resistance, bicomponent fibres with a hotmelt adhesive sheath based upon partially crystalline copolyester, like Cellbond J58, FIT T-215 or Wellbond 1439, are suitable binder fibres for the manufacture of polyester automotive headliners as well as for nonwovens used as foam replacement [32]. The bonding fibre ELK from Teijin, with an elastomeric copolyetherester sheath, offers interesting properties. The sheath component with a melting point of 170 °C is not spun in a concentric position to the PBT core. Nonwovens made from these fibres offer excellent resilience properties [33].

The patent literature describes also bonding fibres made of aliphatic polyesters with softening points down to 60 °C [34].

Amorphous PET fibres, like Tergal T 190, Trevira 259 or Fossfiber PET type KB are special cases. These fibres become sticky when they are heated above their glass transition temperature. Before they start to recrystallize they can be bonded with pressure (e.g. by calendering) to themselves or onto another matrix. This process is irreversible because as soon as the PET starts to crystallize the tackiness disappears. Afterwards the created bond is stable even above the bonding temperature. Amorphous PET fibres are used for the manufacture of filters or in nonwoven forms for the production of molded parts [35].

**Polyamides**
Using comonomers like laurinlactam can easily lower the softening point of standard polyamides like polyamide 6 (PA 6) or polyamide 66 (PA 66).

Examples for copolyamide hotmelt adhesive fibres are Grilon KA 140 and KA 115 (EMS-GRILTECH) or the bicomponent fibre Grilon BA 140 with a copolyamide sheath melting at 140 °C and a PA 6 core.

Special types of bonding fibres made from homopolyamides are sheath/core bicomponent fibres consisting of a PA 66 core and a PA 6 sheath. They are mainly used for thermobonded shoe liners. PA 12 fibres can be used, for instance, as the matrix fibres for the production of automobile interior trim parts in combination with flax.

Adhesive fibres based upon copolyamide hotmelt impress by offering very high specific adhesive strength and good resistance to mechanical loading and solvents. Because of this reason they are used, for example, to improve the fibre anchorage in paper machine clothing [36].

A very interesting feature of copolyamide hotmelt adhesive fibres is the possibility to further reduce the softening temperature with moisture. Steam, as the medium for the fusing process is therefore used.

### 3.3.2.3 Mechanism of bonding

The functionality of bonding fibres is based upon the possibility to liquefy the fibres by a solvent or heat. The fibres, which have to be bonded together, are surrounded by the liquid polymer and anchored inside mechanically after cooling or drying. The local distribution of the adhesive in the web, the viscosity, the wetting properties and the mechanical strength of the adhesive polymer influence the resulting bond strength.

An important parameter for the processing of bonding fibres is the open time, which describes how long the adhesive polymer remains sticky after its activation.

The higher the open time, the longer it takes until the nonwoven can be handled after the bonding process. The higher the bonding fibres have to be heated above their glass transition temperature or melting point, the longer the open time will be. If the nonwoven batt has to be pressed into a molded part after the activation, a too short open time can be very disadvantageous. In case of partially crystalline polymers the open time depends on the rate of recrystallization.

The fibre count and the staple length can influence the distribution of the binder in the web. However, it is more important whether monocomponent or bicomponent fibres are used. In case of monocomponent hotmelt adhesive fibres, the adhesive polymer contracts to droplets as soon as it melts and forms relatively big bonding points with high bond strength (Fig. 3-3).

To the opposite, the core of the bicomponent hotmelt adhesive fibres prevents the adhesive from forming droplets. This results in a much bigger adhesive surface, allowing to create more bonding points. On the other hand there is a risk, that the fibres, which have to be bonded, are not completely surrounded by the hotmelt adhesive. This can cause lower bond strengths. If a sufficient concentration of bicomponent hotmelt adhesive fibres is used, the bicomponent fibres will create with themselves a very stable three-dimensional network (Fig. 3-4).

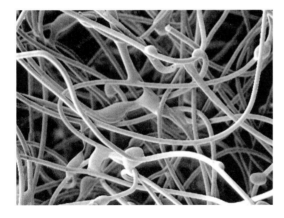

**Fig. 3-3** Nonwoven bonded with monocomponent hotmelt adhesive fibres (20% adhesive fibres)

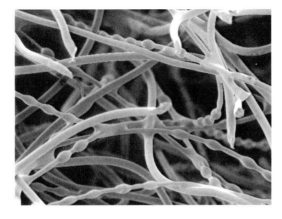

**Fig. 3-4** Nonwoven bonded with bicomponent hotmelt adhesive fibres (20% hotmelt adhesive fibres)

The lack of shrinkage of the bicomponent fibres can be responsible for undesired bonding or binder deposits on machine parts during the processing, because the adhesive polymer does not retract from the surface of the nonwoven.

Depending on the application, either the use of monocomponent or bicomponent fibres can be advantageous [37].

The lower the melt viscosity of the hotmelt adhesive during bonding process, the better the fibre is anchored in the adhesive.

The processing temperature, the molecular weight and the chemical structure of the adhesive influences the melt viscosity. In addition additives or moisture can impact the melt viscosity.

The level of the melt viscosity of the adhesive polymer is limited because the polymer has to be processed into fibres and the adhesive must have a sufficient cohesive strength.

In addition, an important difference exists between amorphous and partially crystalline hotmelt adhesive fibres. Amorphous hotmelt adhesive fibres become sticky above the glass transition temperature, but their viscosity is often extremely high. Therefore the bonding temperature has to be in most cases 60 to 80 °C higher than the glass transition temperature. Partially crystalline hotmelt adhesive fibres can, depending on the polymer used, already be activated 5 to 15 °C above the melting point.

### 3.3.2.4 Properties

Commercially available hotmelt adhesive fibres require processing temperatures between 110 and 230 °C. The selection of a hotmelt adhesive fibre type depends upon the available technical equipment, the type and materials of the nonwoven, which has to be bonded, and the requirements for the final product.

For the production of voluminous nonwovens only products with relatively low bonding temperature can be used, otherwise the hotmelt adhesive fibres inside the nonwoven would not be activated due to the insulation properties of the nonwoven.

Automotive interior parts made from nonwovens require heat resistance, recyclability and low fogging behaviour. Polyester bicomponent hotmelt adhesive fibres with partially crystalline adhesive components melting between 135 and 175 °C are performing very well in these applications.

Nonwovens used for apparel interlinings have to be resistant against laundering and dry cleaning, usually in combination with low activation temperatures. Partially crystalline polyester and polyamide hotmelt adhesive fibres meet these requirements best.

### References to Chapter 3

[1] Ehrler P (1982) Bindemittel, in Lünenschloß J, Albrecht W „Vliesstoffe", G Thieme Verlag, Stuttgart; New York

[2] Devry W (1992) The chemistry and formulation of latex nonwoven binders, Nonwoven Fabrics Forum Clemson USA

[3] Fischer K (1992) Binder und Binder-Additive – Variationsbreite unter Umweltgesichtspunkten, 7. Hofer Vliesstoff-Seminar

[4] Fischer K (1996) Polyvinylacetat-Dispersionen für die Textilveredlung, Melliand TB 7/8: 486–490

[5] Morris HC, Mlynar M (1995) Chemical binders and adhesives for nonwoven fabrics, INDA-TEC St. Petersburg USA

[6] Fischer K (1987) Polymer-Dispersionen für Technische Textilien, tex praxis internat 5: 518–522

[7] Pangrazi R (1992) Chemical binder application technology, Nonwoven Fabrics Formu Clemson USA

[8] Stepanek W (1989) Vliesstoffverfestigung mit verschäumten Acrylharzdispersionen, Taschenb f d Textilind: 213–220

[9] Sigurdson S (1996) Plasma treatment of polymers and nonwovens for improving their functional properties, INDEX 96

[10] Dieterich D (1983) Polyurethane – Neuere Entwicklungen, Kunststoff-Handbuch, Vol 7: Polyurethane, Verlag Hanser, München, Wien

[11] Loy W (1988) Technische Daten dispergierter Bindemittel und ihre Bedeutung für die Vliesverfestigung, Melliand Textilber 11: 836–839

[12] Goldstein JE, Koltisko BM (1992) Nonwoven applications for polyvinyl alcohol, INDA-TEC 92 Fort Lauterdale

[13] Williams MM, Rose KR (1992) Styrene butadiene latex polymers for nonwoven applications, INDA-TEC 92 Fort Lauterdale

[14] Schumacher K-H, Hummerich R, Kirsch H, Rupaner R, Wuestefeld R (1993) Eine neue Generation von formaldehydfreien Acrylatbindemitteln für Hygiene- und medizinische Vliesstoffe, INDEX 93 Genf

[15] Donno T (1994) New acrylics for nonwovens provide better color stability at high temperatures, Nonwovens Industry 25, 12: 93

[16] Kajander R (1994) Optimizing melamine bonded glass fiber nonwovens, INDA-TEC 94 Baltimore

[17] Schumacher KH, Kirsch HP (1994 u. 1996) Latest European developments in formaldehyde-free high performance binders, Nonwoven Industry 25 (1994), 12: 94–96. Medical Textiles 96 Bolton GB 17.–18.07.96

[18] Fischer K (1996) Polyester-Spunbond mit Polymer-Binder für Dachbahnen, Techn Textilien 39, 2: 60–65

[19] Walton JH (1996) Low formaldehyde, high performance nonwoven binders for airlaid pulp applications, INDA-TEC 96 Crystal City USA

[20] Confalone PA, Parsons JC, Nass DR (1997) New ethylene-vinyl acetate binders for nonwoven applications, INDA-TEC 97 Cambridge MA USA

[21] Young GC (1996) Waterborne epoxy resin systems for the use as binders in nonwovens and textiles, TAPPI Proc Charlotte USA

[22] BASF AG, Produkte für Vliesstoffe und Beschichtungen, Techn Mitt der BASF AG Ludwigshafen

[23] Bayer AG Leverkusen, Technische Informationen über „Impranil"

[24] Polymer-Latex GmbH, Acrylpolymer-Dispersionen, Chemie und Anwendung, Techn Inform der Polymer-Latex, 45764 Marl

[25] Fischer KK (1993) Polymer-Dispersionen – Einsatz als Bindemittel für Vliesstoffe unter Umweltgesichtspunkten, Textilveredlung 28, 7/8: 212–219

[26] Koltisko BM (1992) Vinyl copolymer materials, INDA-TEC 92 Fort Lauterdale

[27] Ulyatt J (1990) Nonwoven binder developments to help the manufacturer design safer nonwoven fabrics, INDEX 90 Genf

[28] Schumacher KH, Kirsch HP, Rupaner R (1996) New developments in acrylic binders for toxicologically safe and environmentally friendly nonwovens for medical applications, Medical Textiles 96 Bolton GB

[29] BedarfsgegVer (1992), Bedarfsgegenständeverordnung

[30] Ehrler P, Janitza J (1973) Text Anwendungstechn, Text Ind 23, 8: 746–751

[31] EP 0691 427 A1 (03.07.95) Horiuchi S (Prior. 04.07.94)

[32] Kmitta S (1995) Polyester-Faservlies – ein alternativer Polsterwerkstoff für PKW-Sitze? Vortr 34. Internat Chemiefasertag Dornbirn

[33] USP 5,298,321 (02.07.92) Isoda H. et al. (Prior. 05.07.91)

[34] EP 0 572 670 A1 (10.09.92) Mochizuki M.

[35] Wild U (1984) Thermische Vliesstoff-Verfestigung mit Copolyester-Schmelzklebefasern. Vortr INDEX 84

[36] EP 0 741 204 A2 (03.05.96) Gstrein H. (Prior. 04.05.95)

[37] Fust G (1998) Schmelzklebefasern und deren Anwendung, Vortr 37. Internat Chemiefasertag Dornbirn

# Part II
# Processes to manufacture nonwovens

As compared to textile fabrics of threads, nonwovens are mainly manufactured continuously to dry, wet or extrusion processes. With all these processes, the main steps are

- raw material supply
- web formation
- web-bonding
- nonwoven finishing

The ready-made bales of nonwoven are mostly finished in discontinuous processes. The process-related differences concern raw material supply and the ways into web formation. With fibre nonwovens or extrusion nonwovens, fibres or granulates can be supplied as raw materials, web formation being a dry process. With wet processes, a suspension is made of cellulose and/or short fibres, web formation being based on suspending (a modified paper-making process). Fig. 4-1 shows the possible manufacturing processes taking into consideration the base processes of nonwoven production.

The following is classified to the raw materials used (fibres, filaments, films) and to the web formation technologies available.

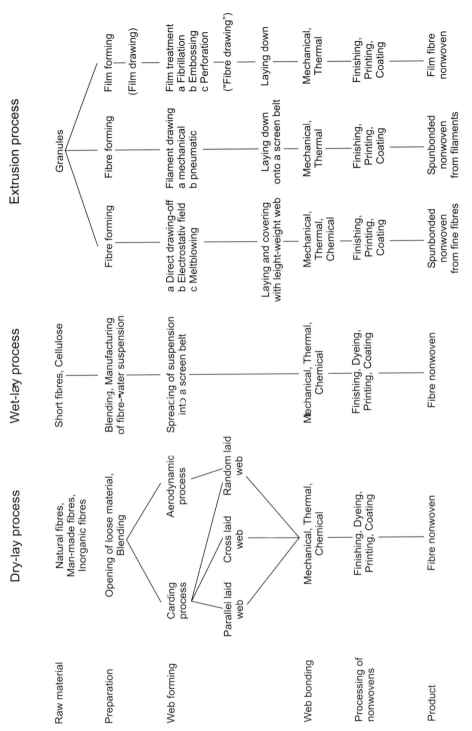

**Fig. 4-1** Manufacturing methods of nonwovens

# 4
# Dry-lay process

## 4.1
## Nonwoven fabrics

### 4.1.1
### Fibre preparation
F. Leifeld

The target of web production is to obtain a final product with specific features. Constant good quality, high production rates and low cost are the preconditions for an economic success. All three factors are influenced by the fibre material used and the machines and aids employed in the process.

For fibre preparation, which comprises the field from the bale up to the feeding device of the web forming machine, the machines and installations must been adapted to the fibre material to be processed and to the final product desired as early as during the planning stage. Everything that is stated here for web forming following the carding procedure is also valid for aerodynamic web forming in an analogous way. From the technological point of view, the essential process steps of fibre preparation are:

– bale opening
– dosing
– blending
– fine opening
– forming the feed web

These processes are carried out partly one after the other, partly simultaneously and mixed up in the various machines. Some jobs can only be done well if the process steps are repeated several times.

For example, this is true of opening. You have got to be aware here that the repeated process steps in the course of the process become more intensive. As for opening, the material is first coarsely opened and then increasingly finer.

## Opening

For the opening process, this procedure can be numerically demonstrated by calculating the theoretical tuft weight. Here it is initially assumed that an opening roll, no matter whether it is a spiked, needle or saw-tooth roll, will always take over the material to be opened from the clamping gap in such an ideal way that each spike or each needle will have the same amount of material. Then, the theoretical tuft weight per spike or tooth can easily be calculated on the basis of the material throughput in time, the working width of the roll, the speed of the roll and the number of spikes per surface unit of the roll. When sticking to certain, empirically found rules and by a successful gradation of successive opening rolls, the theoretical tuft weight is very close to the real tuft weight. In Fig. 4-2 an exemplary result of such a calculation for opening is shown. The tuft weights are plotted on a logarithmic scale. The individual points correspond to the successively used opening rolls. In this case, the first point refers to the rolls of the automatic bale opener BDT 020 and the last point to the result on the cylinder of the roller card.

Opening degree      Theoretical tuft weight (g/tuft)

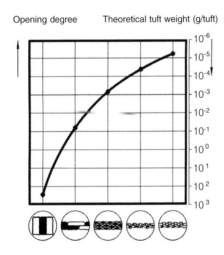

**Fig. 4-2** Graph of the theoretical tuft weight

## Dosing

By dosing, one understands on the one hand the observance of a constant and even material flow and on the other hand bringing exactly together different material components to a – in terms of weight – precisely adjusted blend, as it is for example carried out by means of weighing units. Here, the rule of the right gradation from coarse to fine in material flow direction particularly applies for the control of the whole material flow in order to obtain high evenness of the final product. Serious mistakes at the beginning of the process cannot be corrected later. Therefore, good levelling is essential even in preparation. The problem of exact adaptation of material flows for several machines in sequence can be solved by means of intermediate storing units.

## Blending

Several different process steps are mutually conditional. Blending of initially coarsely opened fibres is at first done only in a coarse manner. Later on, during the fine opening process, the fibres are mixed with each other in a very fine way without any further effort. For all process steps, the rules of gradation mentioned before must be kept. This leads to good results with a minimum effort and to little material stressing. For adding waste or reclaimed fibres, these must first be processed via recycling lines. There are new, performing installations with highest fibre yield particularly for cotton waste and linters. These installations are e.g. described in [1–10].

Fibre blending has a great significance for one-component as well as for multi-component blends. A good blending is an indispensable precondition for an even web quality for multi-component and for one-component blends. The blending constancy must be kept for the long-term as well as for the short-term range. It influences deviations of the final web for several kilometres, metres and even in the range of less than centimetres.

## Feeding

It is the feed web forming machine that must solve the problem to guarantee the mass distribution of the fibres in the web over the working width and the time flow. Here is the last and most important possibility to influence the long- and short-term deviations of the finished web prior to the web forming machine. Here the fine tuning takes place. A top quality can only be reached if already the preceding machines guarantee a chronologically even, continuous tuft flow for feeding the feed web forming machine. In the course of steadily increasing production rates and wider machines, the requirements made on the feed web forming machine are growing higher and higher.

## Machines

The bale opener BDT 020 (Fig. 4-3) represents a modern, fully automatic solution, even for automating the bale supply. As here many bales are worked off in the bale laydown, the blend is optimally influenced even at the very beginning, especially regarding long-term constancy. In practice, however, there are still hopper feeders in frequent use. Here the bale supply can be partly automated via a conveyor belt.

The following machines are mixers (for one-component installations) or pneumatically fed weighing feeders or weighing hopper feeders (for installations with several components). The latter can also be directly fed by hand via a supply belt, supplying layers or bales. The weighing feeder is followed by the mixers. Mixing chambers with a high volume are very common in the field of long-staple fibres. The large individual party, corresponding to the chamber's filling volume, is still in use here. But the trend towards continuously working, automatic installations with multi-chamber flow mixers is growing. This applies particularly to the short-staple fibre range.

**Fig. 4-3** Automatic bale opener
BLENDOMAT BDT 020

For high opening requirements multi-roll openers can be used as fine openers. The variants are one, two, three or four roll openers. In the long staple range, normally several one roll openers are used in sequence.

Hopper feeders with volumetric dosing of the feed web in the vibrating chute are used as feed web forming machines for feeding the roller card. By means of scales, partly in the form of belt weighers, an attempt is made to improve the evenness in the course of time. For high productions and high evenness requirements, however, the roller card feeding unit with two successively working chutes with pneumatic condensing has prevailed.

The most recent development of such a feeder (Fig. 4-4) in addition solves the problem of controlling the web profile over the working width and of the exact determination of the fed fibre mass over width and over time.

Starting with a storage unit in front of the roller card chute feed and via a controlled dosing opener, a continuous stream of tufts is fed by air into the upper chute of the feeder in exactly dosed quantities. From the upper chute, the controlled web feed is taken. Via tray – feed roll and opening roll the material is fed into the bottom

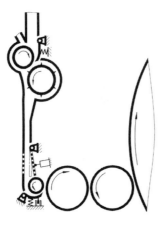

**Fig. 4-4** Tuft feeder Scanfeeder TSC

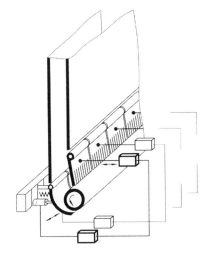

**Fig. 4-5**  Web profile control VPR

chute in the form of tufts and is pneumatically condensed and then transported. Via sectional trays that scan the thickness of the delivered web in several zones over the width, control values are determined and then transferred to servomotors (Fig. 4-5).

These servomotors change the distance of the walls in the web forming zone of the bottom chute in the corresponding, adjustable side zones. Thus the mass distribution over width and over time can be influenced by a levelling intervention. The thickness values can be converted into web weights. The delivery roll of the bottom trunk may at the same time be the feed roll of the roller card. Thus deviations and errors are avoided, which may appear in conventional transfer. Furthermore, the measured value of the total mass over the working width can serve as a basis to change the speed of the feed roll via controllers in such a way that the incoming fibre flow is once more levelled out. As a result, a new dimension in web evenness in MD and CD direction can be reached. This becomes even more significant with the increasingly wider machines (up to 5 m) and the growing production rates up to 400 kg per 1 m of working width and hour. This solution meets the requirement that the levelling devices must be the quicker and more exact, the nearer they are to the final product in material flow direction.

**Installations**
The choice of the processing machines and their combination to installations is determined by the raw material to be processed, the final product desired and economic aspects. Due to the great variety of different final products, there is also a great variety of installation concepts that cannot be treated here in detail. Fig. 4-6 illustrates selection criteria for the configuration of installations.

The most important question concerning machine selection is whether short- or long-staple fibres are to be processed. The difference between short- and long-staple fibres leads to two very different machine ranges for selection. It leads to differently designed machines. Longer fibres require larger distances between clamp-

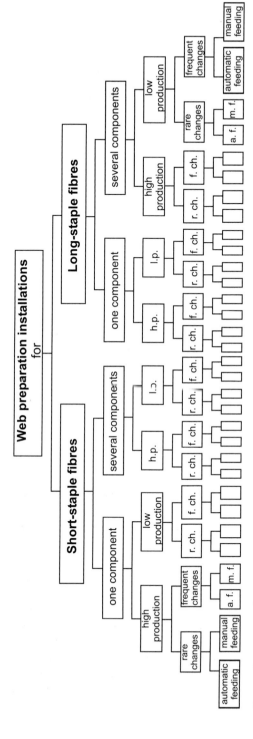

**Fig. 4-6** Decision tree for arranging web preparation installations

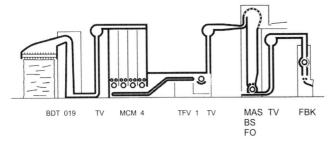

BDT 019     TV     MCM 4          TFV 1  TV          MAS TV     FBK
                                                      BS
                                                      FO

**Fig. 4-7**   Installation for processing one component

ing points, larger opening rolls, larger distances of spikes and longer spikes as well as more space in delivery zones and material storage units.

In practice, there are essentially two different types of machines that were developed in adaptation to the fibre length. Machines for short-staple fibre processing are designed for fibres in the range of 10 to 60 mm length. Long-staple fibre processing machines are designed for fibre lengths of 50 to 130 mm. For longer fibres, modifications of the long-staple machines are necessary. Thus a decision tree develops for short-staple and long-staple installations, the four subordinate levels of which reflect whether one or several components shall be processed, whether the production rate shall be high or low and whether there are rare or frequent changes. Finally, one must decide whether the material feeding shall be manually or automatically. Thus, in the left branches the fully-automatic installations with high production will be found, in the right branches the small installation with manual feed. Figs. 4-7 and 4-8 show examples of installation concepts. Detailed information about the state-of-the-art, about technological know-how as well as descriptions of machines and installations can be found in [1-10].

The abbreviations used in the figures have the following meanings:

| | | | |
|---|---|---|---|
| BDT 019 | bale work-off | BS | feedtrunk |
| TV | material transport | FO | universal opener |
| MCM | multi-mixer | FBK | tuft feeder Scanfeed |
| TFV | opener | BOWA | weighing bale opener |
| MASTV | material separator | FM | tuft blender |

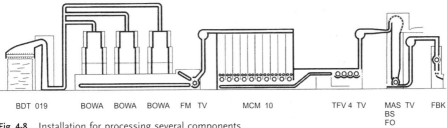

BDT 019        BOWA  BOWA  BOWA  FM  TV          MCM 10          TFV 4  TV     MAS TV     FBK
                                                                               BS
**Fig. 4-8**   Installation for processing several components                 FO

4.1.2

## Production of fibrous webs by carding

W. KITTELMANN, S. BERNHARDT

Fibre tufts are discretized to form single fibres, and the web is formed by a flat card (Latin carduus=thistle) or roller card (OHG Krampe=hook).

In the main roller cards are used for the production of fibre webs. The invention of a rotating card consisting of cylinders fitted with teasels goes back to patents dating from 1748 and formed the basis for Arkwright's carding machines. This textile technology for turning natural fibres into a web for yarn production, which has been known for more than 250 years, is still used today in web forming with the basic elements of carding – main cylinder/flat (flat card) or main cylinder/with worker and clearer rollers (roller card).

The function of the roller card can be defined as follows: the task of the roller card is to produce an orderly fibre layer from the tangled fibre mass resulting from the fibre opening or feed. This should result in individual fibres, i.e. in the disentangling of the fibre tufts and bundles. The discretization or opening should be carried out in such a way that a parallel layer of fibres or a tangled layer is produced. As a rule this occurs in the two-dimensional plane in the machine direction (MD) and across (CD) the machine direction. The individual fibres are subse-

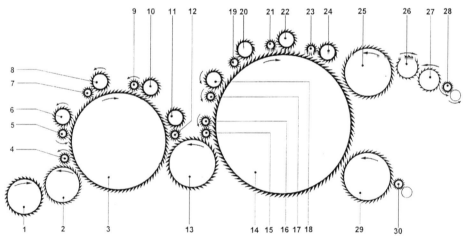

**Fig. 4-9** Principle of a universal roller card from Spinnbau GmbH Bremen:
*1* drawing in roller; *2* breast roller; *3* licker-in; *4* compacting roller; *5* worker at the licker-in; *6* clearer at the licker-in; *7* clearer at the licker-in; *8* worker at the licker-in; *9* clearer at the licker-in; *10* worker at the licker-in; *11* worker at the licker-in; *12* clearer at the licker-in; *13* transfer roller; *14* main cylinder; *15* worker at the main cylinder; *16* clearer at the main cylinder; *17* clearer at the main cylinder; *18* worker at the main cylinder; *19* clearer at the main cylinder; *20* worker at the main cylinder; *21* clearer at the main cylinder; *22* worker at the main cylinder; *23* worker at the main cylinder; *24* worker at the main cylinder; *25* upper doffer; *26* first upper stuffing roller; *27* second upper stuffing roller; *28* upper take-off roller; *29* lower doffer; *30* lower take-off roller

quently brought together to form a coherent, uniform web which is then transported away to a further processing stage.

The processing of ever-finer man-made, organic, inorganic and waste fibres at wide working widths and greater production rates has necessitated the development of new scientific and technical methods for application to carding and web forming.

The four functions of the roller card are:

– constant feed of fibre tufts per unit time with respect to length and width
– clearing and removal of foreign bodies, dirt and short fibres, for example
– opening the fibre tufts with minimum stressing of the fibres
– arrangement of the fibres in the web with stressed positioning of the fibres in the machine direction as a web of parallel fibres, or with unstressed positioning as a tangled fibre web

To carry out these functions a standard roller card [11] (Fig. 4-9) has the following units:

– a tuft feed unit comprising a feed shaft, feed roller and a drawing-in belt
– a trough or roller drawing-in mechanism, comprising a drawing-in roller, a drawing-in table (upper or lower) and a breast roller
– a main cylinder with worker and clearer rollers or stationary carding segments
– a doffing mechanism with take-off in the form of a roller or doffing comb

### 4.1.2.1 Roller carding theory

The basic quality feature of a web is its uniformity. We can proceed from the fact that a roller card produces no irregularities in its optimum setting. This means prerequisite for a regular fibre web is regular fibre feed.

### Feed

With feed (see also Section 4.1.1) a distinction is drawn between the discontinuous method, the weighing feeder, and the continuous method (bulk feed).

The weighing feeder feeds a specific quantity of fibres to the roller card in a defined and controlled manner, irrespective of the fibre fineness, fibre type, level of opening, moisture content, etc. The feed has periodic accumulations because of the balance movements. The mass variation in the running direction is determined by the number of balance movements per unit time and by the drawing-in rate in m/ min. With an increase in the drawing-in rate there is also an increase in the amplitude of mass variation at a constant number of balance movements. The product leaving the roller card is irregular because of the depositing intervals over the running length and width. The bulk feeder can be used for all fibres and ensures better feed uniformity than the balance feeder, with optimum compression of the raw material which can be carried out mechanically as well as with air. In the case of a mechanical bulk feeder the uniformity is determined by the hopper feeder content, the speed of the spiked feed lattice, the speed of the stripping and knock-off rollers and their settings, the height of the fibre column, the width of the vibrating lift shaft, the vibration frequency and the drawing-in speed. Phenomena which occur as a result of the differences in the opening level or fibre composition can be evened out to such an extent by weighing mechanisms that the feed rollers of the carding machine can be controlled in such a way that the production is constant per unit time.

$$P = m_{vo} \cdot v_E \cdot \frac{60}{1000} \tag{1}$$

P  = production rate [kg/h]
$m_{vo}$ = feed mass [g/m$^2$]
$v_E$  = feed rate [m/min]

The volumetric vibrating shaft feeder RS and the silver balance ME2 from Spinn-bau GmbH ensure a uniform tuft feed to the card of the order of CV = 1.0%. Combining gravimetric and volumetric feed by using the PMF combined feeder helps to improve the CV index by 0.8% [11].

**Drawing-in mechanism**
This ensures that the fibres are fed in for carding. There are different ways of feeding fibres to the carding process. It has been developed from processing of long-staple wool fibres as a six-four-or-two-roller drawing in mechanism. The purpose of the paired direction in order is to ensure straightforward feed to the lick-er-in/breast roller. These rollers may be located at the bottom – which is ideal for trash removal – but they may also be located at the top. The disadvantage of this system is that with a roller diameter of 120 mm, for example, the nip-point distance to the center of the next breast roller is 60 mm, which means that fibres less than 60 mm in length can migrate into the feed zone and may be fed to the breast roller in non-controlled manner, which leads to 'slubs' or thick spots.

So-called trough-type drawing-in mechanisms represent a further development, i.e. the classical form. These consist of one drawing-in roller about 200 mm in diameter, a feed table, and a breast roller are between 2 and 8 mm and feed the breast roller with a regular supply of very small fibre tufts. The disadvantage is that delicate fibres are subjected to greater stress because they are stretched over the table radius. To counteract this the table-type drawing-in mechanism can also operate at the top; of course, this necessitates an additional transfer roller.

The overhead table-type drawing-in mechanism represents the state-of-the-art development. Fig. 4-10 shows the overhead table-type drawing-in/taking-in mechanism from Spinnbau GmbH. With a drawing-in roller diameter of 412 mm and an overhead table, which can be adjusted both with respect to height and radially with respect to the fibres, both short and long fibres can be processed gently. These drawing-in mechanisms are also suitable for wider working widths of up to 5000 mm. The high-production card from OCTIR Nonwoven Machinery Division is also fitted with an overhead table-type drawing-in mechanism.

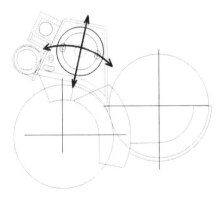

**Fig. 4-10** Section through an overhead trough-type (table-type) drawing-in mechanism from Spinnbau GmbH Bremen

**Carding**

Today carding is predominantly carried out using roller cards. Previously the rollers were fitted with flexible clothing, where the wire hooks are secured to a felt or rubber base and can be moved when subjected to stress. Rigid all-steel clothing is mostly used today however. With the advent of all-steel clothing that necessary evil, the cost of 'cleaning' the flexible clothing, has been decisively minimized and the card output has been increased. According to Damgaard [12], interlinked sawtooth wires with hardened tips have proved themselves. Fig. 4-11 features the most important clothing dimensions. In addition to the tooth dimensions the number of tips per unit area (generally still specified as per square inch) is important in carding. An optimum carding result is achieved when the number of points increases with increased opening of the fibre tufts right down to individual fibres. The fibre fineness also influences the number of points. Finer fibres require a greater number of points.

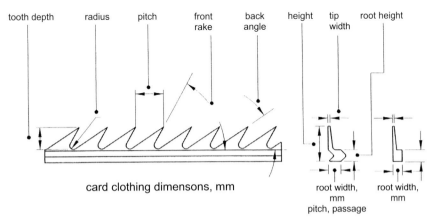

**Fig. 4-11** Important card clothing dimensions [12]

In addition to the choice of roller clothing, the number of carding points – worker and clearer pairs on the main drum, which is also known as the swift, the speeds of the workers and main drum, the fibre covering in g/m² on the main drum, the roller diameters of the workers and the main drum are important for carding.

Fig. 4-12 shows the most important combinations of the working units on a roller card. This features the co-operation of

– the main drum-workers and clearers (Fig. 4-12a) for opening, aligning and blending
– main drum-doffer (Fig. 4-12b) for transfer
– main drum-tangling roller-doffer-stuffing roller (condenser) (Fig. 4-12c) for taking over, entangling and stuffing/compressing the fibre web.

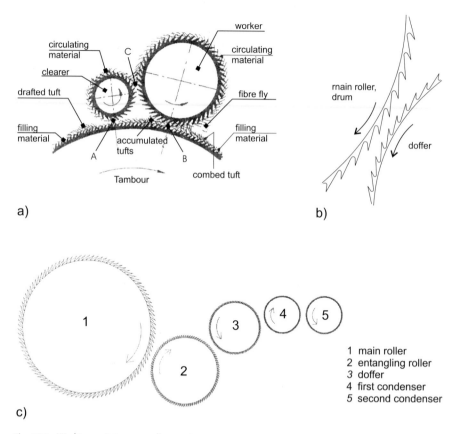

**Fig. 4-12**  Working points on a roller card:
a) Main cylinder – worker – clearer
b) Main cylinder – doffer
c) Main cylinder – entangling roller – doffer – condenser

The carding process between the main drum and the worker is determined by the speed difference between the two rollers and the forces acting on the fibres in the bundle. If the carding force is greater than the friction between the fibres, some of the fibres are removed from the worker and some remain on the main drum. This process takes place between the tips of the clothing and necessitates accurate setting of the clothing sections with respect to each other. The clearer or stripper removes the fibres from the worker and feeds them to the main drum again. This process is repeated several times and results in good fibre blending. The doffer removes the oriented fibres from the main drum. As Fig. 4-12c shows, the position of the fibres in the web can be varied as the fibres are transferred if the entangling roller rotates in the opposite direction to the main cylinder.

On transfer to the main cylinder the fibres come to mg/m². This amount is increased by the amount $m_W$ which is transferred from the clearer to the main cylinder. When the conditions are stable $(m+m_W)$ g/m² fibres are transferred from the main cylinder to the workers.

$$m = \frac{Q}{AB \cdot v_T \cdot t} \tag{2}$$

Q  = mass of fibres processed in t seconds [g]
AB = drum width [m]
$v_T$ = peripheral speed of main cylinder [m/min]

The mass $m_A$ of fibres reaching 1 m² of the worker surface depends on how much $m + m_W$ the worker transfers and on the speed ratios of the main cylinder $v_T$ and the worker $v_A$. The following relationship is valid:

$$m_A = (m + m_W) \cdot \frac{v_T}{v_A} \tag{3}$$

It is clear from equation (3) what influence the changes in the drum and worker speeds have on the fibre mass $m_A$ and the carding. An increase in the main cylinder speed also permits an increase in the worker speed, and thus an increased fibre throughput with good carding results. A reduction in the distance between the main cylinder and the worker increases the number of fibres $m_A$ on the worker. The fibre load on the main cylinder should be in the range 2–4 g/m² in the case of 1.1 dtex fibres. The transfer factor between the main cylinder and worker is of the order of approximately 15%. The mean dwell time of a fibre between the feed point to the main cylinder and the fibre removal by the doffer is of decisive importance for the carding and evening-out capacity of the card with regard to the feed irregularity and the blending of the fibres. The mean dwell time TFM, also known as the 'delay factor', is defined as the quotient of the separate fibre loads on the workers, clearers and main cylinder clothing, with the exception of the doffer clothing and the doffer production.

$$T_{Fm} = \frac{m_{Fa}}{P_A} \qquad (4)$$

$T_{Fm}$ = mean fibre delay factor [sec]
$m_{Fa}$ = overall fibre load on the clothing [g]
$P_A$ = doffer production rate [g/s]

Taking into consideration the technical specifications in Fig. 4-13, it is possible to calculate the delay factors (t) of the fibres at the working points and the relevant fibre quantities (m). In this case the fibre-coated roller curve (c) has to be taken into consideration.

Time on the worker with $c_A = 270°/360° = 0.75$
$$t_A = c_A \cdot 60/u_A = \frac{0.75 \cdot 60}{49} = 0.918 \text{ s}$$

Time on the clearer with $c_W = 234°/360° = 0.65$
$$t_w = c_w \cdot 60/u_w = \frac{0.65 \cdot 60}{530.5} = 0.074 \text{ s}$$

Time on the main cylinder
$$t_{Tr} = \frac{0.5 \cdot (d_A + d_W) \cdot 60}{d_{Tr} \cdot \pi \cdot u_W} = \frac{0.5 \cdot 0.38 \cdot 60}{1.2 \cdot \pi \cdot 530.5} = 0.006 \text{ s}$$

The time for fibre circulation about a working point ($t_G$) is obtained from

$$t_G = t_A + t_w + t_{Tr} = 0.918 + 0.074 + 0.006 \approx 0.998 \approx 1.0 \text{ s}$$

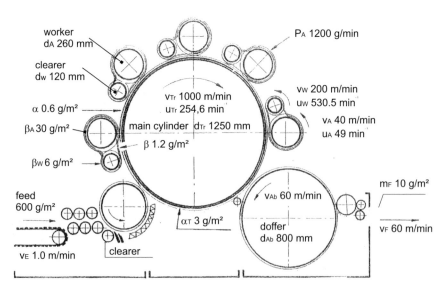

**Fig. 4-13** Standard roller card with technical/technological information

The quantities of fibres per m of working unit are as follows:

- for the workers $\quad m_A = \beta_A \cdot d_A \cdot \pi \cdot c_A = 30 \cdot 0.26 \cdot \pi \cdot 0.75 = 18.4\ g$

- for the clearers $\quad m_W = \beta_A \cdot \dfrac{v_A}{v_W} \cdot d_W \cdot \pi \cdot c_W = 30 \cdot \dfrac{40}{200} \cdot 0.12 \cdot 3.14 \cdot 0.65 = 1.47\ g$

- for the main cylinder $\quad m_{Tr} = \beta_W \cdot \dfrac{v_A}{v_{Tr}} \cdot \left( \dfrac{d_A + d_W}{2} \right) = 30 \cdot \dfrac{40}{1000} \cdot 0.19 = 0.228\ g$

Thus the number of fibres per working point

$$m = m_A + m_W + m_{Tr} = 18.4 + 1.47 + 0.228\ g \approx 20\ g$$

The total number of fibres on the main cylinder $M_{Tr}$ is obtained on the basis of the amounts ( between the drawing-in mechanism and the doffer and $a_{Tr}$ on the full circumference.

$$M_{Tr} = d_{Tr} \cdot \pi \cdot (a \cdot c_{Tr} + a_{Tr}) = 1.25 \cdot \pi \cdot (0.6 \cdot 0.75 + 3.0) = 13.5\ g$$
with $c_{Tr} = 270° / 360° = 0.75$

The total number of fibres in the carding machine $m_K$ is equal to the fibre mass at all n working points and that at the main cylinder.

$$m_K = (n \cdot m) + M_{Tr} = 5 \cdot 20 + 13.5 = 113.5\ g/m\ width$$

The quantity of fibres which is now taken off by the doffer and the quantity which remains on the main cylinder will then give the actual transfer factor. We can then determine how great the fibre load on the main cylinder actually is:

$$C_T = \frac{P}{AB \cdot v_T} \tag{7}$$

$C_T$ =load on main cylinder $[g/m^2]$
P $\;$ =production rate $[g/h]$
AB =working width $[m]$
$V_T$ =main cylinder speed $[m/min]$

For example, width $P=60{,}000\ g/h$, $AB=2\ m$ and $v_T=1{,}000\ m/min$, the result is $C_T=0.5\ g/m^2$.
With a transfer factor of 15% the actual value for $C_T$ is:

$$C_T = 0.5\ g/m^2 \cdot \frac{1}{0.15} = 3.3\ g/m^3 \tag{8}$$

This indicates that the greater the transfer factor the higher the card output; at the same time the evening-out capacity and the web quality decrease, and vice-versa. The maximum fibre load on the card clothing can thus be determined with respect to good fibre fed and carding quality – in particular with regard to nepping and fibre shortening. Fibres can be damaged during overcarding, for example. One very critical yardstick for evaluating the quality of carding is the number of neps in the web. The risk of neps being formed during is the greater, the finer and more slender the fibres. Tests with 1.7 and 40 mm viscose rayon fibres show that the neps are formed predominantly on the workers and consequently depend on the worker load. The number of neps increases with an increase in the production rate or a low transfer factor as a consequence of the higher main cylinder and worker loads. The reason for the excessively low transfer factor may be too low a main cylinder speed or doffer speed, unsuitable doffer clothing geometry, and too great a distance between the main cylinder and the doffer. The number of neps also increases too large a worker clothing top rake, as the fibre load is then too high. It also increases if the worker setting is too wide with respect to the main cylinder despite a low fibre load on the workers.

The number of carding points on the main cylinder depends on the geometrical relationships resulting from the main cylinder diameter and the arrangement of the transfer roller and the doffer. They determine the remanent surface which is divided into the number of worker/clearer pairs with the appropriate spaces. This also determines which fibre lengths can be usefully processed on the card. Nine work points are located on the basic card (Fig. 4-9). These working points result in optimum throughput and quality in the processing of fine fibres in the range 1.7 to 6.7 dtex. With only five working points there is a quality reduction of approximately 40% and a decrease in the production rate. If there are only four working points/work stations the quality decreases by more than 50%, which also means a corresponding fall in production rate. The smaller number of workstations/working points, the poorer the quality. The card output has to be altered appropriately for the same web quality.

The distance setting between the worker and the main cylinder considerably alters the fibre loading capacity of the worker. If the optimum setting of 0.15 mm is altered to 0.6 mm, the fibre pick-up changes by about 50%. This means that fibre throughput and the web quality deteriorate by 50%. At a constant fibre throughput and a constant spacing between the worker and the main cylinder, the number of fibres thereon varies with worker speed. A high main cylinder speed and a low worker speed lead to high stressing of the fibres. A high worker speed results in low fibre stressing. The advantage of a higher worker speed is that more fibres per unit time can be taken off the main cylinder. The prerequisite here is that the fibres must be suitable for processing. In addition a high worker speed results in considerably better fibre blending, better opening, and gentler fibre processing because the stresses are lower.

### Doffing

The quality and production requirements relating to nonwoven fabrics determine not only the fibre selection, but also the demands on the webs. These include the fibre positions in the web, i.e. whether they are parallel or entangled. Under certain circumstances a combination of the two is suitable. In addition to the fibre orientation in the web, the production rate in kg/h can also be influenced by the doffer. One of the factors which affects the selection of the doffer diameter is the contact surface, for example. This is obtained from the main cylinder and doffer diameters.

A large contact surface favours fibre parallelization.

In addition to the production rate, napping of the web is a further factor relating to selecting the doffer diameter. It is affected by the V-shape between the main cylinder and the doffer below the point of contact. The narrower the wedge, the better the parallelization of the fibres. In addition to the diameter, the choice of doffer clothing, especially regarding the front rake relative to the main cylinder and the number of points and the space setting between the main cylinder and the doffer, is crucially important. The card output and the web quality are determined by the compression ratio $V_D$ as the quotient of the main cylinder speeds $v_T$ and $v_{Abn}$:

$$V_D = \frac{v_T}{v_{Abn}}$$

In mechanical web forming the ratio is not less than $5:1$ today. A greater than $5:1$ means that the web quality is improved. After the doffer come the take-up mechanisms which ensure transport from the doffer to the next unit.

The doffer comb is used mostly in those cases where flexible doffer clothing is still employed today. The maximum possible doffer speed is determined by the operation of the doffer comb blade, the number of strokes per minute and the requisite fibre-dependent overlap. Conventional doffer combs permit speeds of between 50 to 60 m/min. High production doffer combs allow speeds of up to 80 m/min, what is very important then is the way the doffer comb is fitted at the doffer. As far as possible, the assumed center line between the doffer comb and doffer must not be crossed. Higher doffer speeds can be attained below the assumed center.

**Take-up roller system**
These offer the advantage of high take-off speeds and low draft and compression. As a rule, the take-up roller systems consist of a take-off roller with card clothing and a further grooved, coated or smooth roller unit. Additionally a cleaning roller is located on the take-off roller to ensure constant long-term running. With these systems very high speeds of 300–400 m/min can be reached without any problems.

#### 4.1.2.2 **Plant technology**
The development of card technology is characterized by the demands of the nonwovens industry for higher carding rates, improved web uniformity and a low web areal density. This is associated with the processing of ever-finer fibres with a linear density equal to or greater than 1 dtex because a high web cover is achieved in this way. Care must also be taken to ensure that the fibres are distributed uniformly in the web and occupy the requisite positions. It has been shown that a high carding factor also makes for good web regularity. Years of practical experience show that the following empirical relationship between the fibre diameter to be processed and the minimum and maximum achievable web areal density may be specified as a limit value.

$$m_{Fmin} = 5 \cdot \sqrt{Tt_F}$$

$$m_{Fmax} = 3 \cdot m_{Fmin}$$

$m_F$ = areal density [g/m$^2$]
$t_{tF}$ = fibre fineness [dtex]

Currently, when processing polypropylene fibres, minimum web weights of 15 g/m$^2$ are obtained under practical conditions. The limit web forming rates are equal to and greater than 300 m/min. In addition working widths of up to 4 m are required.

High card outputs at wide working widths necessitate new, modern design developments and accurate card fabrication [17]. Table and overlap surfaces are ground to create spacings of a few tenths of millimetres between high-speed rollers and tables. The wide working widths result in a high sag value for the rollers, e.g. worker and clearer rollers at the same tube diameter, which gives rise to excessively large spaces between the working units. An increase in the diameter would lead to a reduction in the number of carding points and hence the carding factor. The sag can be considerably reduced by using carbon-fibre reinforced composite materials instead of steel for the same diameters. The heat produced at high carding intensity and high machine outputs, particularly when processing synthetic fibres, necessitates the use of direct conditioning equipment in order to maintain the fibre flow in the card at constant temperature and relative humidity.

Web doffing by the card deserves special attention. Low web weights and high speeds require independent solutions with regard to roller take-off systems. It is important that no excessively large drafts occur and that there is no web irregularity. One way to transport webs is on suction belts which fix the web on the conveyor belt.

Fig. 4-14 shows the principle underlying a universal high-production card from Spinnau GmbH for working widths of up to 3 m and webs speeds of up to 150 m/min. Automatically tilting compression rollers make it possible to obtain an MD:CD strength ratio of up to 4:1 in the case of heavy webs.

The high-production random web card HYPER-CARD HC 4-5 (Fig. 4-15) enables very light webs to be produced in maximum widths of up to 4 m at up to

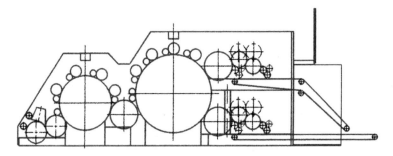

**Fig. 4-14** SUPER SERVO CARD SSC 4-5 universal high-production card

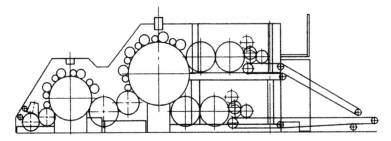

**Fig. 4-15**  HYPER-CARD HC 4-5 high-production random web card

300 m/min. The machine is fitted with take-off belts where the web is held and transported on the belt by means of a vacuum.

Today modern carding systems form an integral part of continuous web-forming and consolidation systems. Fibre preparation, web forming and nonwovens production form a single unit. The tendency is for opening of the fibre tufts and for uniformity of the webs goes more and more into the fibre preparation sector. Web forming and the arranging of the fibres in the web become the main task of the main roller and its working units. The INJECTION CARD from FOR, Biella/Italy [18] (Fig. 4-16) is a machine where the carding occurs at the main roller. With this system the clearers are replaced by a stationary unit which utilizes the air flow created by the rotation of the main cylinder. With this type of carding the workers remove scarcely any fibres from the main cylinder. The advantages over other forms of carding are the ability to process very fine fibres and the avoidance of neps in the web.

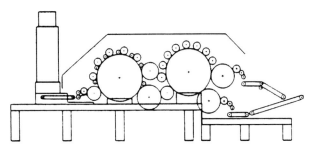

**Fig. 4-16**  INJECTION CARD MICROWEB from FOR [18]

### 4.1.2.3  Web forming
M. SCHÄFFLER

In web forming the card webs are stacked in several layers to form a web. The requirements relating to the webs are: mass retention, widthwise and lengthwise uniformity, fibre distribution and fibre positioning. Generally speaking, the fibres in the card web have a lengthwise orientation. In carding machines with randomizing mechanisms the fibres in the web have a non-emphasized position which, under ideal conditions, is random orientation.

The most important web-forming methods:
– parallel-laid webs and
– cross-laid webs

### 4.1.2.3.1  Parallel-laid webs

Continuous methods are preferably used here. This means that the card webs supplied by the sequentially-arranged or parallel cards are doubled on a common conveyor belt. In most cases the fibres in these webs have a lengthwise orientation. This means that the web strength is considerably greater in the lengthwise direction than the transverse direction. In the production of parallel-laid webs the web width is the same as the card web width. Widening is not possible. The speed at which the web is taken off the card determines the web-forming rate. The formation of parallel-laid webs permits the doubling of fibrous webs differing with respect to mass and fibre type, which makes a layer structure in the web possible. The number of layers in the web determines the number of cards needed.

### 4.1.2.3.2  Cross-laid webs

#### Principle of web laying

The web former (cross lapper) is located inside a web system beyond the card and takes up the card web at a specific rate. This web is laid in several layers on a take-off belt via a conveyor belt system with an oscillating carriage movement. This take-off belt moves at right-angles (90 degrees) to the carriage discretion. A multiplayer web can be produced using the speed ratio of the oscillating carriage movement and the delivery table (Fig. 4-17).

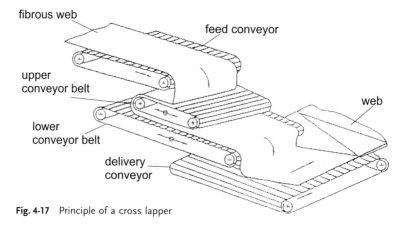

**Fig. 4-17**  Principle of a cross lapper

**Tasks of the web-laying machine**

- Increasing the web mass
  Using multilayer web laying, the web mass can be increased by a direct multiple of the card web mass to the required product weight. The smallest unit is a so-called double-layer, i.e. a web formed by the traversing of the plaiting-down carriage. For quality reasons, however, at least two double layers are used in practice.

- Increasing the web width
  Web laying machines can produce webs in widths (laying widths) of up to 7 m. Widths of up to 16 m are reached in special designs for papermakers' felts. Thus the web laying machine determines the maximum end-product width. However, the web laying machine can also be used to produce webs in infinitely variably smaller widths.

- Determining the web strength in the length and cross directions
  Depending on the design of the doffer system, the fibre orientation in the card web may vary. In conjunction with a web laying apparatus, attempts are made to produce a card web with largely parallel fibres, i.e. the majority of the fibres are in the lengthwise direction, thus having a longitudinal orientation. The orientation of the fibres on the take-off belt can be determined and the plaiting-down angle of the card via the laying width and the number of layers. The orientation of the fibre layer determines the lengthwise/transverse strength ratio in the end product.

- Improving the end product quality
  Distortions during web consolidation (mechanical, thermal, chemical) result in web width contractions which – viewed over the overall web width – are not equal, but are concentrated more strongly at the edges. The result is that the areal densities in the edge regions of the end product is higher than in the rest of it. This phenomenon of textile technology is also known in specialist technology as the 'bathtub' or 'smile' effect. Modern drive and control elements in the web laying machine permit irregular profiling to counteract this bathtub effect.

**Web-laying machine designs**

Web-laying machines can be divided into two categories: the camelback and the horizontal laying machine (plaiting-down mechanism) (Fig. 4-18).

While the camelback (Fig. 4-19) has a three-dimensional design and the variable laying width is included in the height which is dependent on it, the horizontal laying machine only operates two-dimensionally, seen from the abstract viewpoint. The height of the machine remains constant and is not dependent on the laying width.

Whilst the machine web throughput is always the same in the case of the camelback, this varies in the horizontal machine as a function of the direction of carriage movement.

Horizontal laying machines can in turn be subdivided into counter-current and synchronous categories (Fig. 4-18).

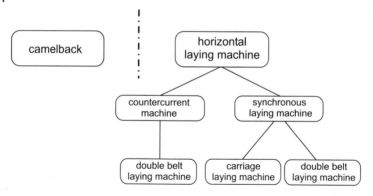

**Fig. 4-18** Types of laying

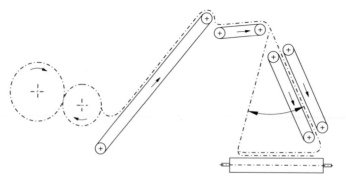

**Fig. 4-19** Camelback

Each horizontal laying machine has a so-called upper carriage and a laying carriage. These two carriages carry out an oscillating movement and thus transfer the web via the conveyor belts onto the take-off belt. If the upper carriage and laying carriage move in opposite directions, we speak about countercurrent types (Fig. 4-20). If the upper and laying carriages move in the same direction we refer to synchronous types (Fig. 4-21).

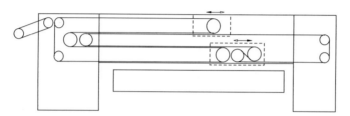

**Fig. 4-20** Countercurrent double belt laying machine

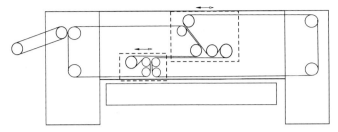

**Fig. 4-21** Synchronous double belt laying machine

With a countercurrent horizontal laying machine two web deflections of 180 de-grees each are needed for passing it through. With a synchronous machine has a single 180 degree deflection in the upper carriage.

Whilst the countercurrent machine is only known in the double belt laying ma-chine version, synchronous machines can be subdivided into carriage laying ma-chines and double belt laying machines.

The carriage laying machine (Fig. 4-22) has a stationary web feed system, an upper carriage and a laying carriage with enclosed conveyor belts. The web feed through the laying mechanism is open, likewise web deposition on the take-off table.

With the double belt laying machine (Figs. 4-20 and 4-21) the upper and laying carriage are connected with the two roller frames at the card and at the rear side via continuous conveyor belts. As a result the web is allowed to pass between the belts (hence also the designation 'sandwich layer') and plaited down on the take-off belt. In addition these belts act as a cover for the plaited-down web and thus give protection against air turbulence.

The synchronous double belt laying machine permits high laying rates and high web quality, and represents the current state of technology.

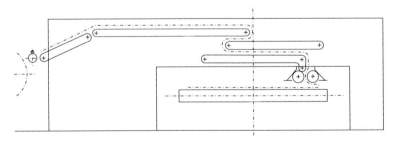

**Fig. 4-22** Carriage laying machine

**Machine and control technology**

Within a nonwovens system the web laying mechanism is the sole mechanism which has no uniform or synchronous throughput like the card, needling machine or dryer, for example, but the web doffing process occurs in two opposite directions. The result of this is that the moving masses have to be continuously accelerated and braked.

The machine design and drive and control technology also have to be looked at from this viewpoint.

If the cross lapper could be viewed as merely an adjunct to the card at the beginning of its development – also driven by the main card motor – over the course of the decades it has developed into an independent machine unit which controls productivity and quality within the nonwovens system.

The important components within a modern web laying system are:

- Laying belts
  In addition to transporting the card web, another job of the laying belts in a sandwich laying machine is to accelerate and decelerate all the moving masses within the laying mechanism. It must therefore have high longitudinal tenacity and transverse stiffness at low fabric weight. A smooth surface with as low as possible a surface friction coefficient is essential for web transport and web doffing. The surface must be chemically resistant to spin finishes and exhibit anti-static behaviour vis-à-vis man-made and natural fibres.

- Laying rollers
  Because of the translatory and rotary movements the mass inertia must be kept as low as possible for high machine speeds. This is still just possible in conventional steel technology. Alternative materials such as fibre-reinforced composites (carbon-fibre-reinforced composites) can be used here.

- Drives
  Motors which operate accurately are required for precision web doffing. In a synchronous double belt laying machine a separate AC servomotor is used for each moving component. This means that both each laying belt and each carriage (upper carriage and laying carriage), as well as the take-off belt, has its own drive shaft. The linear carriage movements are monitored via absolute path detection systems to ensure high switching accuracy.

- Carriage movement
  The drive reversing accuracy is transmitted to the two carriages via low-extension, low-noise toothed belts. These do not need lubricating or require maintenance.

- Take-off belt
  The original wooden lattice belts have been replaced by plastic lattice belts. These lattices make it possible to configure the surface in such a way that some fibres are held back during doffing. This is particularly important at high doffing speeds.

- Control
  The five drive shafts present in a synchronous double belt laying machine are controlled extremely accurately on a complex mathematical basis. These calculations are carried out using a personal computer or programmed control systems. The laying mechanism is displayed visually on a monitor. As a result, all the machine parameters can also be assigned locally and the setting of the laying mechanism is thus operator-controlled.

Technological aspects and features which determine the quality: the web which comes from the card is a sensitive material, particularly to air, which has to be laid down via the web laying mechanism at as high a speed and accuracy as possible. The speeds which can be attained depend on the mode, but are also considerably influenced by the following factors:

– web characteristic (fibre orientation)
– fibre type (synthetic fibres, viscose fibres, natural fibres)
– fibre dimensions (fineness and length)
– fibre elasticity (crimp)
– web mass
– spin finish and ambient conditions (temperature and relative humidity)

Basically speaking, the following relationship holds good: the greater the fibre/fibre friction and the greater the parallelization of the fibres in the web, the greater the stability of the web and thus advantageous for the laying speed which can be achieved. The web mass on the take-off belt of the laying mechanism is determined by the number of web layers, that is, via the ratio of the speeds of the incoming web and the outward bound take-off belt. Here the following basic calculation applies:

$$\text{Web mass} = \text{Number of single layers} \times \text{web mass (g/m}^2)$$

When determining the web mass at the output of the laying mechanism it should be noted that this can only be a whole multiple of the web mass. In all the mass and output calculations it should be borne in mind that the number of single layers is always used. The laying edge (loop) visible at the laying mechanism take-off is always a double layer, i.e. two single layers.

Numerical example for calculating the web mass:

| | |
|---|---|
| Web mass: | $25 \text{ g/m}^2$ |
| Number of layers (single): | 16 |
| Web mass= | $25 \times 16 = 400 \text{ g/m}^2$ |

The production rate of a web laying mechanism p is calculated in terms of the effective card web width, the web mass or weight and the web feed rate at the laying mechanism in accordance with equation (9):

$$p = \frac{AB_{eff} \cdot m_F \cdot v_F \cdot 60}{1000} \left[\frac{kg}{h}\right]$$

Example:                    effective web width $AB_{eff}$ 2.40 m at a nominal
                            working width of 2.5 m
Web mass $m_F$:              $40 \text{ g/m}^2$ (measured at the laying mechanism spot)
Card web fed rate $v_F$:     110 m/min
Laying mechanism output    $= \dfrac{2.40 \cdot 40 \cdot 110 \cdot 60}{1000} = 633.6 \text{ kg/h}$

To ensure optimum web quality it is essential to set the optimum so-called layer closure. By this we mean the point where the start of the topmost web layer is connected to end of the lowest web layer. This layer closure is defined in terms of the number of layers by the web feed rate, the laying width and the take-off speed.

$$\text{Number of layers} = \frac{\text{effective web width} \cdot \text{web feed rate}}{\text{effective laying width} \cdot \text{web takeoff rate}}$$

In practice the number of layers is determined on the basis of the desired end mass and the specified web card mass; the resultant take-off rate $v_{outlet}$ is also calculated on this basis.

$$v_{outlet} = \frac{AB_{eff} \cdot v_F}{LB_{eff} \cdot Z_{simple}} \left[\frac{m}{min}\right] \tag{10}$$

Numerical example:

| | |
|---|---|
| Effective card web width $AB_{eff}$: | 2.4 m |
| Web feed rate $v_F$: | 110 m/min |
| Effective laying width $LB_{eff}$: | 3.8 m |
| Number of single layers $Z_{simple}$: | 16 |

$$v_{outlet} = \frac{2.4 \cdot 110}{3.8 \cdot 16} = 4.34 \text{ m/min}$$

Depending on the user, the number of layers and the stretch during the subsequent consolidation, a certain amount of overlap is used in the cm range to maintain the maximum possible weight uniformity in the running direction in the end product. Web laying mechanisms with modern control systems have integrated this calculation process as a rule, i.e. the operative only needs to input the requisite number of layers; the corresponding take-off speed is automatically specified as a result. With conventional web laying machines there is a problem in that the web coming from the card at a constant rate is doffed at a constant rate even during the reversal of the laying carriage, i.e. more web is doffed in the laying carriages reversal sector than in the middle section. This results in the so-called 'bathtub effect', i.e. a web mass profile where the edges are clearly heavier. On the other hand, in modern web laying machines the web is stored during carriage reversal and then laid out again in a controlled manner. An example of this is the Topliner CL 4000 from AUTEFA Maschinenfabrik GmbH featured in Fig. 4-23.

As a result it is possible to doff the web at constant speed from the card and also to plait it down in synchronism with the variable-speed laying carriage movement during the switching process. This enables a considerably more uniform web mass distribution to be achieved over the laying width. Despite a uniform web feed from the laying mechanism, no optimum mass distribution is achieved in the end product. The main reasons for this are the distortions which take place

**Fig. 4-23** Topliner CL 4000 web-laying machine

during consolidation, which cause widthwise shrinkage in the product, leading to irregular mass distribution and a decrease in quality. Since a web laying machine determines the output profile for the later web consolidation process, the mass distribution can also be deliberately configured to be irregular to ensure as uniform a profile as possible after consolidation. This patented method was made possible by the introduction of preprogrammable control systems and servodrive technology. Using this control system – also known as 'profiling' – the web doffing can be monitored over the whole laying width [19], as shown in Fig. 4-24 in the examples for consistent fibre blends.

As ITMA'99 showed, the web laying machine with a profiled control system has made a breakthrough and is now being supplied by all the market leaders (Autefa Automation (Germany), Thibeau/Asselin (France), Octir/HDB, Thatham (UK)). The web feed rates have now already in excess of 150 m/min [20].

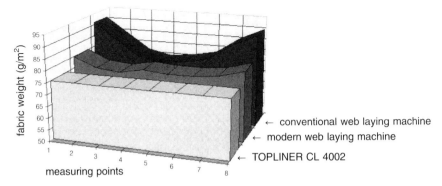

**Fig. 4-24** Ways of optimizing the quality using the profiled control system patented by AUTEFA ('profiling')

## 4.1.2.4 **Web drafting**

W. KITTELMANN, S. BERNHARDT

The purpose of web drafting is to reorient the fibres in the web in the direction of draft and to improve web quality. The crosslaid webs formed on the laying mechanisms usually have high transverse strength, but their longitudinal strength is relatively low. In many cases where nonwovens are used it is desirable that there is a balanced strength ratio in both directions. To achieve this, the fibres must be reoriented in the machine direction. This considerably improves the MD:CD strength ratio.

The production rate of a nonwovens machine can also be increased by forming heavier webs during laying which are made finer during drafting as a result of the higher speeds and thus lead to lower web weights. The higher delivery rate at the drawframe leads to a greater fabric production per unit time.

Web drafting is determined by the relationship between the length and strength of the drafted web and original dimensions. This is known as the draft. The following equation is valid for the draft.

$$V = \frac{v_o}{v_i} = \frac{m_{vi}}{m_{vo}}$$

$v_o$ = web speed at outlet
$v_i$ = web speed at inlet
$m_{vi}$ = web mass at inlet
$m_{vo}$ = web mass at outlet

The total draft in web drafting is equal to the product of the partial drafts. Care should be taken in web drafting to ensure that widthwise web shrinkage is

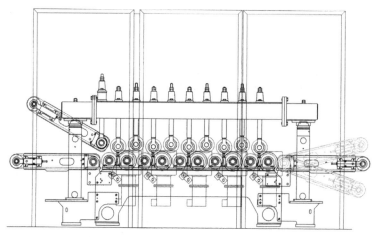

**Fig. 4-25** Section through the VST 4 web drawframe from Spinnbau GmbH

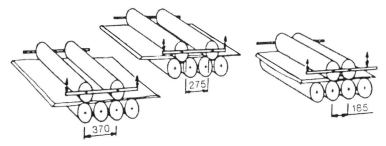

**Fig. 4-26** Possible roller arrangement for the drafting zones

avoided as far as possible during drafting. The web inlet/outlet weight ratio in drafting is inversely proportional to the relevant speeds. This also alters the number of fibres and their position in the web cross-section. It is important that the proportion of floating fibres, i.e. of those whose speeds cannot be controlled, is kept small during drafting to minimize irregularities. As a rule the web draw-frame has three to four drafting zones which are fitted with individual servo-drives, as a result of which the individual drafts have infinitely variable adjustment. The roller diameters are of the appropriate size, which means that the sag is low even with working widths of up to 7 m. The individual rollers have saw-tooth clothing to prevent web shrinkage. Their height adjustment is infinitely variable. Depending on the fibre length, the nip-point distances should be kept as short as possible and have a variable configuration.

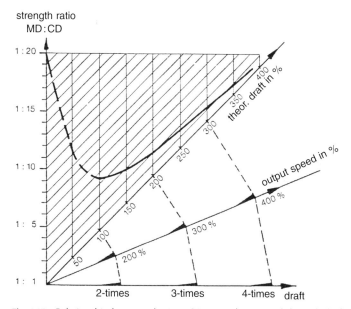

**Fig. 4-27** Relationship between the MD:CD strength ratio and the web draft

Fig. 4-25 shows a section through the VST 4 drawframe from Spinnbau GmbH.

Fig. 4-26 features the possible roller arrangements for drafting zones with nip-point distances.

Fig. 4-27 shows the relationship between the MD:CD strength ratio and the draft on the basis of fibre reorientation.

A favourable MD:CD ratio is particularly dependent on the external fibre properties, the web construction and the draft, which may be of the order of 1 to 4.

As regards web drawframes, we should like to refer to the VE drawframe developed by J.P. Dilo (see Section 6.1.6).

### 4.1.3
### Fibre webs following the aerodynamic procedure
A. Paschen, F. Leifeld, B. Wulfhorst

#### 4.1.3.1 Aim of the procedure
Aerodynamic web forming is a dry procedure to form a web out of fibres (short- or long-staple fibres). For this, the fibres are opened more or less intensively prior to the web forming process. The opening degree of the fibres is determined by the selection of the preceding openers or roller cards. These principles are described in Section 4.1.1 "Fibre preparation".

The desired characteristics of the web, as e.g. evenness and tuft mass of the web, are influenced to a high degree by the quality of the opening. Thus a thin, even web requires a higher opening degree of the fibres than really heavy webs.

Aerodynamic web forming makes it possible to produce a web possessing the same features regarding fibre orientation, elasticity, workability and strength both in longitudinal and cross direction. There are also advantageous features vertical to the web surface, as a favourable orientation of the fibres is reached even in this direction. This leads to voluminous webs which are flexible to pressure and have a relatively low volumetric weight. Due to their one-layer structure, there is no danger of web layer splitting, as is often the case of webs produced with web laying machines.

Apart from the technological aspects, the aerodynamic procedure offers economic advantages resulting from the investment volume and the operating cost for the installations. This allows a reasonable processing of different fibre materials at a high productivity.

#### 4.1.3.2 Description of the procedure
The idea to form webs by an aerodynamic procedure orientates itself by an ideal. According to this, the fibres in the plane or volumetric form produced are oriented in all directions. The distribution of directions and fibres is purely by chance, but is similar and regular in all zones. Density, thickness and structure are the same for small as well as for large checking areas. The more ideally this pattern is reached, the higher the technological and economic advantages of this procedure, as material input can be minimized if a favourable structure is obtained.

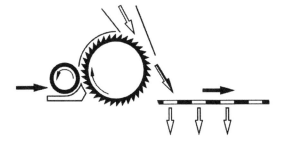

**Fig. 4-28** Basic principle of aerodynamic web forming

The procedures of aerodynamic web forming available on the market differ from each other and are more or less close to the ideal requirements. All procedures have in common that after the opening process the material is transferred into an air flow by the last opening roll. The air-fibre mixture comes to a continuously moving screen surface. This screen is under suction. The fibres deposit on the screen surface. When flowing through the deposited fibres, the air condenses the so-formed web. This basic principle is shown in Fig. 4-28 [21, 22].

The degree of condensing $V_d$ depends on the air speed and the air mass put through. According to the principle of linear momentum it is proportional to the product of air speed v and air mass m:

$$V_d = f(v \times m)$$

The procedures and installations used in practice mainly differ in obtaining different degrees of opening prior to web forming and in producing different air speeds and air masses flowing through the screen during fibre depositing. As a rule, the air speeds and the deposit surfaces sucked off are combined in such a way that with high air speeds the web forming zone consists of only narrow slots, in extreme cases sucked by vacuum pumps. With low air speeds the suction surfaces and thus the web forming zones are growing larger. The web forming zone that is not sucked can be considered as the borderline case. Here, the fibres are thrown out of the opening roll and finally arrive on the delivery belt mainly in free fall after having described a certain trajectory. Here, the flight and the formation of the fibres during their depositing are also influenced by the air stirred up in the deposit zone by the opening roll.

Screen drums and screen belts are used as deposition surfaces, in practice the screen belts predominate. There are several principles applicable to produce the suction at the screen surface. These are overpressure or subpressure systems or combinations of both, respectively. Fig. 4-29 shows the principle of such systems [21, 22].

The characteristics of the web produced can be described by geometrical, physical, technological and statistical values. If these values are scaled, quality criteria can then be determined and laid down. Thus it is possible to measure to what extent the web fulfils the requirements.

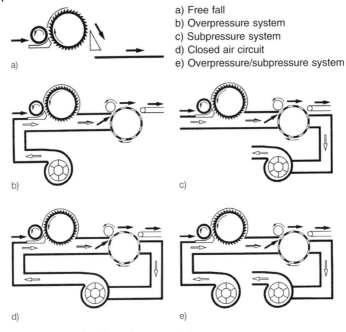

a) Free fall
b) Overpressure system
c) Subpressure system
d) Closed air circuit
e) Overpressure/subpressure system

**Fig. 4-29**  Principles of aerodynamic web forming

Due to the variety of values and to their various types the problem of quality description and quality assessment is very complex, expensive and difficult, especially if completeness is demanded. This should be borne in mind when reading the following descriptions. It is thus tried to list some important influencing factors for the quality of the web and to put them into a relationship. Therefore, this can only be done in a coarse manner without figures.

The characteristics of the web produced depend upon the fibre material used and upon the installation employed (Fig. 4-30). Here, first of all the fibre material itself with all its characteristics in the supply condition must be considered and assessed. After that, its condition, particularly the orientation of the fibres towards

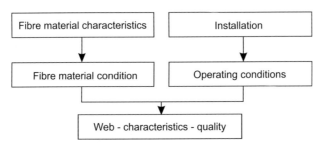

**Fig. 4-30**  Influencing factors for the web characteristics

each other, is changed. The result that can be obtained is first of all determined by the installation chosen with its special possibilities. Furthermore the operating conditions chosen at last determine the final result within these possibilities.

The most important influencing factors of the web quality are:

- **Characteristics of the fibre material**
  Fineness, cross-sectional form of the fibres, fibre length, fibre length distribution, curling, surface (e.g. roughness, friction factor fibre-fibre, friction factor fibre-metal, softness), electrostatic charging capability, strength, elasticity, flexural strength.

- **Condition of the fibre material**
  Opening degree, humidity, distribution in space, electrostatic charging condition, mixing condition, linking of fibres to each other.

- **Installation**
  Selection of individual units, arrangement of the units to each other, number of opening rolls, equipment of the units (e.g. opening unit: clothing, form and number of spikes), circumferential speeds of the rolls, deposit space, flight path, belt-cylinder, geometry of the screen surface, deposit zone and size, attainable air speeds, attainable air volumes, tendency towards turbulence formation.

- **Operating conditions**
  Throughput, circumferential speeds chosen (opening), screen surface speed chosen, air speeds chosen, air throughputs chosen, web mass chosen, adjustment of guiding elements.

For obtaining the desired web quality the installation and operating parameters must be optimally adapted to the fibre material. They influence the conditional characteristics of the fibre material in the transport air as well as the type and intensity of the air movement. What results from these influences is the landing condition of the fibre on the screen surface or on the web developing. This landing condition as well as the material condensing accompanying the air flow result in the interlocking of the fibres with each other and thus in their cohesion. This has a decisive influence on the mechanical characteristics of the fibre web regarding strength and elasticity.

In spite of the process parameters which are optimally adapted to each other, the aerodynamically formed fibre webs cannot reach the same evenness as mechanically folded webs. This difference becomes more and more recognizable with the lighter webs. In practice, the limit must be seen at approx. 100–200 g/m², this value is strongly correlating with fibre fineness. With the aerodynamic machines, the processing of fine fibres suitable for the lighter webs leads to problems caused by nep and lap formation as well as blockages of the screen surfaces. Here the practical lower limit of fibre fineness must be put at approx. 3 dtex [23].

Because of these reasons, the traditional application field of aerodynamic web forming lies in the area of rather heavy web masses made of coarse fibres up to 330 dtex that cannot be processed with mechanical procedures. The advantages here are the higher efficiency of this procedure and the possibility of processing larger material ranges, while the evenness problems take a back seat [23].

Regarding the processing of fibre blends of components with strongly differing physical features (fibre fineness, fibre density and fibre length), a separation of the blend components may occur in the air stream between fibre opening and screen surface. Different flight curves and corresponding landing points on the screen surface result from different aerodynamic characteristics, thus layers can develop. To avoid this effect so-called deflector shields can be used which redirect the heavier fibres and are thus supposed to compensate the separation. Such a shield is presented in the following descriptions of the machines [23].

### 4.1.3.3 Machines for aerodynamic web forming

The principles of aerodynamic web forming are realized in different manners by the machines currently available on the market. The corresponding manufacturers are listed in the manufacturers' index. The operating method of this procedure shall be explained by the examples of some established machines.

In Fig. 4-31 a typical installation configuration of Dr. E. Fehrer is represented, consisting of the pre-web forming machine V21/R and the random card K12. Via a vibration chute feed with filling trunk (1) and an opening device (2), the material is directly fed to a sucked screen belt (3) and is compressed by a pressure roll (4). This pre-web ls fed to the random card K12 (5), which is shown more in detail in Fig. 4-32.

The supplied material is fed via a feed roll (1) and a tray table (2) to a rotating card cylinder (3) and is thus pre-opened. Fibre opening is intensified and levelled by two pairs of working/clearer rollers (4) mounted at the cylinder. The fibres are detached from the cylinder by means of the centrifugal force. The fibres are taken by a laminar air flow and transported to a screen belt (5) under subpressure. It is here that the actual web forming takes place. The web is condensed by a web forming roll and delivered via the screen belt.

In the version presented in Fig. 4-32, this web forming roll is replaced by a high-loft roll (6). This is a screen drum with suction insert. By horizontal and vertical adjustments of the screen drum and the individual control of the air flows,

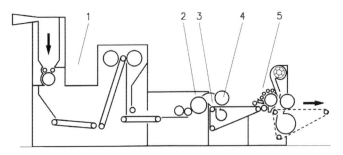

**Fig. 4-31**  V21/R – K12 of Dr. E. Fehrer

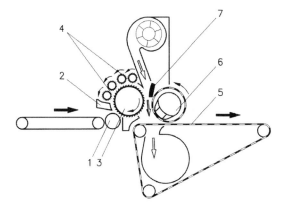

**Fig. 4-32** Random card K12 of Dr. E. Fehrer

web forming can be influenced with regard to the fibre used and the web mass or the web height desired. A web forming zone with suction from two sides develops, thus causing an increasingly vertical arrangement of the fibres. By this, web thickness can be increased by 80% (acc. to the manufacturer) [24, 25]. High loft webs are e.g. used in the clothing industry and in furniture manufacturing. In Fig. 4-32, the above-mentioned deflector shield (7) is shown, too.

The compact installation design as shown here is possible for a working width up to 4.8 m. Larger working widths can only be realized using a special width distribution device. The installation is suitable for processing all types of man-made fibres as well as cotton, reclaimed and recycled fibres of 1.7 to 200 dtex for producing interlinings, carrier material for coatings and artificial leather, geotextiles, filter material, web covers, wall and floor coverings as well as technical textiles in the finer web mass range (20 to 2000 $g/m^2$) [25, 26]. Current applications are in the field of the automotive industry. So, for example, formed parts made of flax or jute blends with polypropylen in the gsm range of up to 2000 $g/m^2$ or insulating mats of blends of recovered wool, cotton, polyester and foam rubber cubes are produced.

The pre-web former V21/R is a further development of the web forming machine V12/R, which is still in supply. It can be used for the heavier web mass range of 400 to 3000 $g/m^2$ and is suited for processing natural fibres and their reclaimed waste for producing insulating and mattress fabrics, filling material and mattress covers, under-carpets as well as wadding material for upholstery and automotive industry [25, 26]. It is also built in working widths up to 4.8 m.

Particularly for applications in the field of lighter webs (20 up to 130 $g/m^2$), Fehrer offers the high-production random card K21, which is represented in Fig. 4-33. The K21 can follow a roller card or a combination of roller card and horizontal web laying device. It is, however, also offered by the manufacturer as an installation together with the pre-web forming device V21/R.

The characteristic feature of the K21 is the use of four successive card cylinders which are each equipped with a pair of working/clearer rollers. A pre-web is fed via a tray feeding device (1) to the first card cylinder (2). Due to the centrifugal

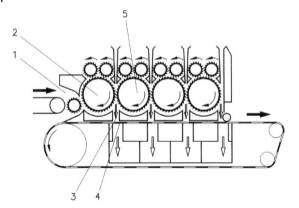

**Fig. 4-33** High-production random card K21 of Dr. E. Fehrer

force, this cylinder throws part of the fibre amount through a fibre guiding channel (3) onto the suction screen belt (4), the majority of the fibres is transferred to the second card cylinder (5). This is done in an analogous way when transferring the material to the following card cylinders. Thus a four-layer web is produced, the layers of which are inseparably linked to each other due to the three-dimensional depositing of the fibres.

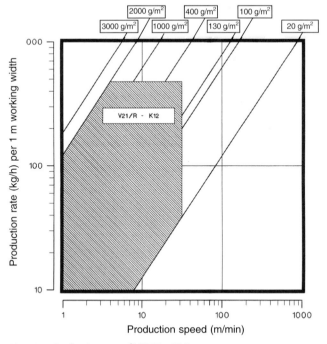

**Fig. 4-34** Production rate of V21/R – K12

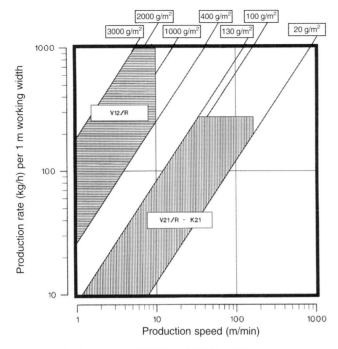

**Fig. 4-35** Production rate of V12/R and V21/R – K21

It is confirmed that this procedure results in a good random orientation of the fibres and an in some areas adjustable, really balanced strength ratio MC:CD [27–30].

Fibres from 1.7 to 3.3 dtex can be processed on the K21 to produce webs for hygienic and sanitary applications, for interlinings and basic material for cleaning cloth as well as coating carriers.

In Figs. 4-34 and 4-35, the production rates per 1 m working width of the presented Fehrer machines are shown depending on the possible production speeds and gsm values according to the manufacturer's statements. The working point that is to be chosen in reality will always depend upon the fibre material used and the web quality desired.

The company of Dr. O. Angleitner offers several web forming machines according to the aerodynamic web forming procedure. Their basic principle is shown in Fig. 4-36. A pre-web is transferred from the feed roll arrangement (1) to an opening roll (2). The fibres which are opened there are then guided via an air stream to a pair of screen drums (3) sucked from below where the fibre web develops.

Fig. 4-37 a shows the machine 1004 in its simplest execution. The fibre material, which has been specially prepared on continuously working fibre blending installations, is fed via a feed trunk (1) and forms the pre-web on a conveyor belt (2).

To improve fibre opening and web evenness, it is possible to use an execution with two cylinders (machine 1044) which is represented in Fig. 4-37b. Here an

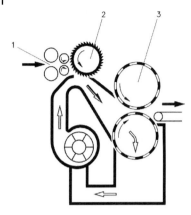

**Fig. 4-36** Principle of aerodynamic web forming of Dr. Otto Angleitner

aerodynamically formed web is fed to another opening roll, is separated again and is then fed to a second pair of screen drums with the air stream [31].

The machines can be equipped with screen drums of different sizes. Furthermore the filling trunk can be preceded by a condenser unit. It is also possible to add powder or foam tuft scatterers between filling trunk and web forming zone and thus to prepare the fibre material for an adhesive or cohesive bonding. Continuously working weighing units can serve to ensure an exact control of the web mass.

By positioning the upper screen drum as well as the fibre guiding area between opening roll and web forming zone, the collecting area of the fibres, and thus the web structure, can be selectively influenced. It is also possible to obtain "high loft" structures in this way [31].

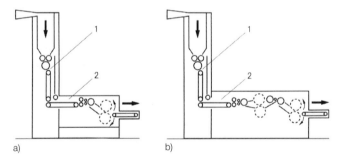

**Fig. 4-37** Web forming machines 1004 and 1044 of Dr. Otto Angleitner

The web forming machine 008-0445 (Fig. 4-38) of the French manufacturer Laroche S.A. can process natural fibres or blends with man-made fibres as well as fibreglass, recycled or waste fibres to produce webs in the heavier gsm range of 400 up to 1500 g/m². The prepared fibre material is fed via a rotating condenser (1) to a feed trunk (2). Via a spiked conveyor belt (3) it comes into a volumetrically working regulating trunk (4). From there the material is fed via two pairs of feed

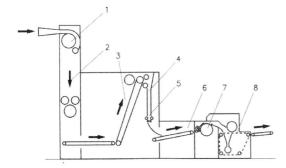

**Fig. 4-38** Web forming machine 008-0445 of Laroche S.A.

rolls (5) to a conveyor belt with continuous weighing device (6). It is then fed to an opening roll (7) which drops the fibres like snowflakes onto a screen belt (8), by this the web is formed.

The fibre webs can then be processed to become bed covers, mats, upholstery and insulation material, carrier material for carpets, industrial and geotextiles as well as furniture textiles – depending upon the fibre material used [32].

The Rando webber offered by the Rando Machine Corporation represents one of the oldest aerodynamic web forming procedures which is still in use today. The principle is shown in Fig. 4-39. A pre-web formed in different procedures is fed via a feed table (1) and a feed roll (2) to a fast rotating cylinder (3). Fibre take-off from the cylinder is supported by a blowing air stream; the air speed can be controlled by an excentric roll (4). In this way the fibre material is fed through a venture nozzle-type fibre channel (5) onto a sucked screen belt (6) where the web is formed [33].

Usually, the Rando webber is used in the gsm range up to $2500 \, \text{g/m}^2$ and processes blends of short- or long-staple primary or recycled fibre material to become webs for most different applications [33].

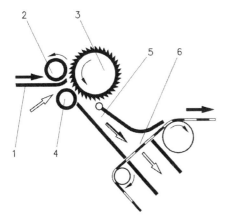

**Fig. 4-39** Principle of aerodynamic web forming of Rando Machine Corporation

The company of H. Schirp manufactures web forming machines for the gsm range between 100 and 3000 g/m². The principle is shown in Fig. 4-40. A pre-web is fed via a pair of feed rolls (1) to a cylinder (2), from where the individual fibres are separated by a blowing air stream. The fibre material is transported by the air stream to another pair of rolls (3), one of which is sucked from the inside and thus the web is formed [34].

This procedure is mainly suited for processing recycled fibre material to become webs for mattresses and blankets, for automotive form parts, needled webs and carpet underpaddings [34].

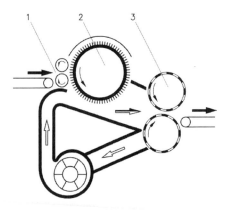

**Fig. 4-40**  Principle of aerodynamic web forming of H. Schirp

### 4.1.4
### Perpendicular laid fibrous highlofts STRUTO and ROTIS
R. Krčma †

The fibrous highlofts with fibres situated perpendicular to the fabric plain gained their market position mainly as filling and thermal insulating materials. The main reasons were as follows:

– increasing demand for filling and thermal insulating materials due to growing use and price of energies
– developing knowledge of the structure of nonwovens and of the relations of fabric properties and structural characteristics

In 1994, the production of filling thermal insulating materials reached two billion tons [35] and the demand is steadily increasing. These materials comprise high-loft fibre materials of textile, glass and ceramics fibres as well as polymeric, namely polyurethane foams. Latest developments lead to the use of textile high-loft materials as substitutes for non-recyclable, non-hygienic and environmentally non-friendly polyurethane foams, namely in automotive industry, in mattresses and upholstered furniture.

As it was shown in [36, 43], the compressional properties of fibrous highlofts depend mainly on

– properties of fibres and on their positions in the material
– properties of bonding agents, their content and distribution

The most important properties of fibres for production of thermal-insulating and filling materials are fineness of fibres (dtex) and fibre texturation. The influence of some fibre properties will be shown later. Most of highlofts are produced by cross-layering carded web or by air-layering. The fibres in these fabrics are situated mostly parallel towards fabric area. This gives the fabrics rather poor compressional properties, namely compressional resistance and elastic recovery after repeated or long-term loading.

Thickness of fabrics, the ability to keep thickness during loading as well as that after repeated and long-term loading are the most important properties of filling and thermal insulating fabrics. These properties are significantly dependent on the positions of fibres in the fabrics as will be shown later. In fact, this dependence was known and utilized many years ago in production of carpets Boucle and Neko [36].

Among the binder agents for highlofts, the bi-component bonding fibres are a top form of binder creating fabric structures with high number of point-like bonding sites and with rather long site-to-site free, movable parts of fibres. This structure gives the fabrics more elasticity and elastic recovery than other forms of binders, namely mono-component bonding fibres, powders and lattices applied by spraying.

### 4.1.4.1 Technologies of production perpendicular laid highlofts

The STRUTO technology [37] consists of

– production of carded web of the blend of base fibres and bonding mono-component or bi-component fibres
– forming a fibre layer using a special lapper producing upright situated folds of carded web on the conveyor belt of the through-air bonding chamber
– bonding the fibre layer by passing hot air inside bonding chamber, its solidifying by cooling

Any kind of fibres can be processed by the STRUTO technology provided a consistent carded web can be produced in the carding machine and that the fibres are suitable for thermo-bonding process. Carded web is transported from the card towards the vibrating or rotating perpendicular lapper situated over the conveyor belt of bonding chamber. A fibre layer of required thickness and density is produced on the conveyor belt and bonding process is then accomplished inside the chamber. After cooling the fabric and cutting the edges the fabric is wound up. A schematic diagram of a STRUTO production line is shown in Fig. 4-41.

The technology ROTIS [38] – see Fig. 4-42 – is the same as STRUTO as far as the carding and perpendicular layering procedures concerns. The perpendicular

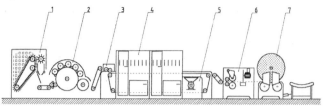

1 Fibre feeding
2 Carding machine
3 Perpendicular lapper STRUTO
4 Through air thermo-bonding oven
5 Cooling device
6 Cutting device
7 Take up mechanism

**Fig. 4-41**  Principle of the STRUTO production line

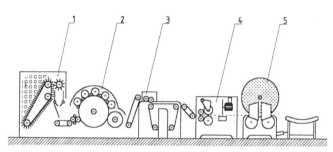

1 Fibre feeding
2 Carding machine
3 ROTIS unit
4 Cutting device
5 Take up mechanism

**Fig. 4-42**  Principle of the ROTIS production line

laid fibre layer is then lead together with one or two reinforcing nets into a solidi-fying mechanical device with a number of upright situated rotating elements. The elements create yarn-like bundles of the fibres at the surface of fibre layer The bundles act as bonding elements of the layer and simultaneously link the layer with reinforcing net or nets into a reinforced composite voluminous material.

Further, both the vibrating and rotating perpendicular lappers STRUTO as well as the ROTIS machine will be described in more detail.

A vibrating perpendicular lapper is shown in Fig. 4-43. This lapper produces a fibre layer (6) on a conveyor belt of through-air bonding chamber by folding carded web (2). Reciprocating comb (1) and reciprocating presser bar (4) are two main working elements of the lapper. The comb (1) pulls the carded web towards

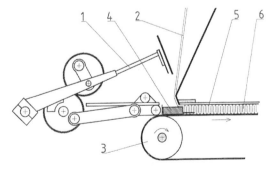

1 Forming comb
2 Carded web
3 Conveyor belt of thermobonding oven
4 Presser bar
5 Wire grid
6 Perpendicular laid fibre layer

**Fig. 4-43**  Reciprocating (vibrating) perpendicular lapper

the conveyor belt (3) in every working stroke. Then, the needles of the presser bar (4) keep the web close to conveyor and move it forward between the conveyor and the wires (5) of a grid. Thus, a perpendicular laid fibre layer is created between the conveyor and the grid. Then the layer is transported through the bonding chamber where it is bonded.

Thickness of the fabric is controllable by setting of the lapper, namely the distance between the grid and conveyor and the dimension of presser bar. Fabric density is easily controlled by the velocity of conveyor belt (3).

The base parameters of the device and produced fabrics are as follows:

Area weight of processed carded web:  7–250 g/m$^2$
Fabric thickness:                     18–35 mm
(when processing special fibres, the minimum thickness of ca. 10 mm and the maximum thickness of ca. 40 mm can be reached)

Fabric density:                       7–50 kg/m$^3$
(some fibres allow to produce densities as low as 4 kg/m$^3$, some fibres as high as 80 kg/m$^3$)

Area weight of fabric:                120–1700 g/m$^2$ (special fibres 60–2500 g/m$^2$)

Production rate of STRUTO machine (working width 2.5 m):
– input speed of carded web: up to 70 m/min
– production: up to 500 kg/h

A rotating perpendicular lapper is shown in Fig. 4-44. A carded web (1) is fed by a set of feeding disc (2) between the teeth (3) of working discs so that folds of carded web are created which then form a perpendicular laid fibre layer (6) between the conveyor belt (4) of bonding chamber and the wires of a grid (5).

Thickness of the produced fabric is controllable by setting the lapper, namely
– distance between the wire grid (5) and conveyor belt (4)
– ratio of the feeding discs (2) and working discs (3) velocities

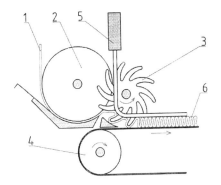

1 Carded web
2 Feeding discs
3 Forming discs with teeth
4 Conveyor belt of thermobonding oven
5 Wire grid
6 Perpendicular laid fibre layer

**Fig. 4-44**  Rotating perpendicular lapper

Density of the fabric is controlled in the same way as described at the vibrating lapper – by the velocity of conveyor belt.

The base parameters of the rotating lapper and fabric:
Area weight of processed carded web: 7–250 g/m$^2$
Fabric thickness:               8–50 mm
Fabric density:                 8–50 kg/m$^3$
Area weight of fabric:         60–1300 g/m$^2$

Production of machine (working width 2.5 m):
– input speed of carded web up to 120 m/min
– production: up to 1000 kg/h

Advantages of the rotating lapper when compared with the vibrating one:
– higher thickness of fabrics
– easier control of fabric thickness
– easier attendance and maintenance
– higher performance

Disadvantages:
– positions of fibres are not completely upright in some fabrics
– grooved surface of the fabrics due to action of teeth of working discs

To solidify the fibre layers, the through-air bonding chambers are used in the STRUTO technology. The performance of the chamber depends on required bonding temperature and on the working length of the chamber. Provided that the working width of the chamber is 2.5 m, following approximate performance can be expected:

| Length of heated zone (M) | Number of section | Ca. performance |
|---|---|---|
| 1.4 | 1 | 150 |
| 2.8 | 2 | 350 |
| 4.2 | 3 | 600 |

The ROTIS device is shown together with the rotating lapper in Fig. 4-45. Perpendicular laid fibre layer (3) is lead by an upper (5) and an lower (4) conveyor – both conveyors made of narrow longitudinal belts – to the rows of upper (9) and lower (8) rotating elements. The rotating elements (8) and (9) create yearn-like bundles of fibres located at the surface of fibre layer. The reinforcing nets (6), (7) can be brought to the lower and/or upper surface of the fibre layer before it reached rotating elements. In this case, the elements link the fibre layer with reinforcing nets during solidifying process.

Typical properties of ROTIS fabrics:
Fabric thickness:               10–50 mm
Fabric density:                10–35 kg/m$^3$
Area weight:                   150–1200 g/m$^2$
Input speed of carded web:  up to 120 m/min

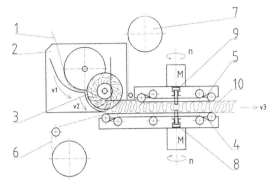

1 Carded web
2 Rotating perpendicular lapper
3 Perpendicular laid fibre layer
4 Lower belt conveyor
5 Upper belt conveyor
6 Lower reinforcing net
7 Upper reinforcing net
8 A row of lower rotating
   elements
9 A row of upper rotating
   elements
10 Both sides reinforced fabric

**Fig. 4-45** Principle of the ROTIS unit

Advantages and disadvantages of ROTIS technology when compared with STRU-TO: Mechanical bonding process does not require bonding agent, blending fibres with bicomponent or monocomponent bonding fibres, through-air bonding chamber as well as the energy to heat the chamber. The process is very simple and controllable. On the other hand, mechanically bonded bulky layers do not show excellent compressional resistance and elastic recovery which is the main advantage of STRUTO fabrics.

### 4.1.4.2 Properties of perpendicular laid highlofts

Thermal insulating and filling properties of highloft materials show a strong dependence on fabric thickness. Therefore, the properties of the fabrics which are submitted to compression during their end use depend on their compressional resistance.

The heat flux through an insulating material Q/t/A is equal to:

$$\frac{Q}{t \cdot A} = \lambda \cdot \frac{\Delta T}{d}$$

where   Q     is heat [J] passing through area A [m$^2$] in time t (s)
         $\lambda$     thermal conductivity [W $\cdot$ m$^{-1}$ $\cdot$ K$^{-1}$]
         $\Delta T$    the temperature difference between surfaces of insulating
               material [$^\circ$K]
         d     thickness of material [m]

Then, the total thermal resistance R [W$^{-1}$m$^2 \cdot$ K] is:

$$R = \frac{d}{\lambda}$$

Thermal conductivity $\lambda$ shows only a slight dependence on fabric density which is generally growing with compression. Thus, the insulating ability of material is an almost linear function of its thickness.

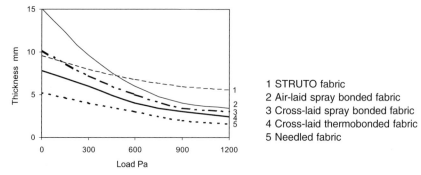

**Fig. 4-46** Dependence of the fabric thickness on compression. Area weight of fabrics $150 \, g/m^2$

Perpendicular laid highloft fabrics show higher compressional resistance than those made of fibres oriented parallel to the fabric area. Therefore, when compressed, they are able to sustain 30–60% bigger thickness and thermal resistance. Due to much better elastic recovery after repeated and long-term loading, perpendicular laid fabrics keep their functional properties during the use. In this respect is their behaviour close to that of polyurethane foams.

Change in thickness of various kinds of highloft fabrics during their compression is shown in Fig. 4-46. The STRUTO fabrics (1) show high compressional resistance when compared with air-laid (2), cross-laid spray bonded (3), cross-laid through-air bonded (4) and needled (5) fabrics.

Elastic recovery of various bulky materials is compared in Fig. 4-47. The materials were submitted to 25,000 loading cycles, whereas they were compressed to 50 per cent of their original thickness in every loading cycle. The curves in Fig. 4-47 show the time-dependent recovery of thickness expressed in per cent of original thickness.

A complex influence of fibre decitex, fibre position and fabric compression on total thermal resistance R of bulky materials is shown in Fig. 4-48.

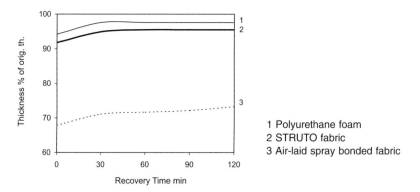

**Fig. 4-47** Process of elastic recovery after repeated loading (25,000 loading cycles, compression of 50 per cent of thickness in every loading cycle)

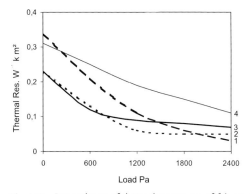

4 Perpendicular laid, fibres 17 dtex
2 Cross laid, fibres 1.7 dtex
3 Cross laid, fibres 17 dtex
1 Perpendicular laid, fibres 1.7 dtex

**Fig. 4-48** Dependence of thermal resistance of fabrics on compression

Fine fibres give the textile materials lower thermal conductivity $\lambda$. On the other hand, coarse fibres bring high compressional resistance which influences thermal resistance dramatically. Thus, coarse fibres show advantage in fabrics submitted to loading during end use. Beside this, better properties of perpendicular laid fabrics are demonstrated in Fig. 4-48.

Influence of fibre positions on the compressional resistance of fabrics is shown in Fig. 4-49 in more detail. Perpendicular laid fabrics containing fibres located in various positions against the fabric plane were produced. Their compressional resistance show strong dependence on angles $a$ between predominant fibre positions and the fabric plane.

The knowledge of conditions during the end-use and that of required fabric properties is very important for the designer of highlofts. Then the kind and positions of fibres can be chosen to optimize the fabric design and performance. The fabrics containing perpendicular positioned fibres bring considerable improvement to highloft properties.

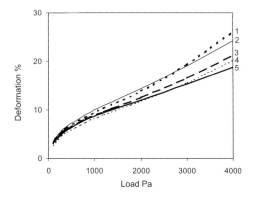

**Fig. 4-49** Dependence of fabric thickness on compression. Perpendicular laid fabrics with different positions of fibres. $a$ = angle between fibres and fabric plane. (1): $a = 69°$, (2): $a = 71°$, (3): $a = 76°$, (4): $a = 77°$, (5): $a = 80°$

**4.2**
**Extrusion nonwovens**
W. KITTELMANNN, D. BLECHSCHMIDT

For more than 50 years, high polymers have been melt-spun. Melt-spinning has led from grid spinning to extrusion spinning with its diversified production lines, which are distinguished by specific polymers, specific equipment or specific assortments of products. Using the extrusion spinning process, filament and fibre yarns are produced and then made into textile fabrics. The processes available to produce extrusion nonwovens (Fig. 4-1) allow the conversion of extruded polymer melt flows into textile fabrics in a continuous direct way. Basic developments in this field took place between 1950 and 1970 [44–48]. Extrusion nonwovens comprise filament and fibre spunbonded nonwovens as well as nonwovens made from extruded films of fibre-type structure. Within the last 10 years, the development of extrusion nonwovens has seen much progress worldwide. Novel processes and assortments of products have come up which are, in particular, based on polypropylene to achieve low mass per unit areas. They are used to produce hygienic and medical articles as well as agricultural and building textiles. Table 4-1 contains product developments within the last 10 years concerning nonwovens in general and spunlaid nonwovens in particular (according to EDANA).

Table 1 1   Production of nonwovens in Western Europe in 1,000 tonnes (according to EDANA)

|  | *1989* | *1991* | *1993* | *1995* | *1997* | *1998* | *2000* |
|---|---|---|---|---|---|---|---|
| Nonwovens in total | 414.0 | 480.6 | 554.5 | 646.4 | 759.5 | 836.0 | 1025.9 |
| Spunlaid nonwovens | 143.6 | 197.3 | 227.3 | 267.9 | 318.0 | 356.2 | 409.1 |

Within the period described above, production has doubled, the output of extrusion nonwovens growing by 250%.

In 2000, $409.1 \cdot 10^3$ t were produced, which can be distinguished by mass per unit areas as follows:

| | |
|---|---|
| $\leq 25$ g/m$^2$ | $159.3 \cdot 10^3$ t |
| $<25$–$70$ g/m$^2$ | $74.2 \cdot 10^3$ t |
| $<70$–$150$ g/m$^2$ | $87.5 \cdot 10^3$ t |
| $<150$ g/m$^2$ | $88.1 \cdot 10^3$ t |

Polypropylene is the most important polymer for extrusion nonwovens due to its properties and its low polymer density. In addition, it is relatively low cost. Frequently in German, English terminology is found with the different kinds of extrusion nonwovens. Table 4-2 gives an overview of both the German and English terms.

**Table 4-2** Nonwoven terminology

| German | English | Basic components |
|---|---|---|
| Extrusionsvliesstoff | Extrusion nonwoven | |
| Filamentspinnvliesstoff | Spunlaid nonwoven spunbond fabric | Mono- or bi-component filament |
| Faserspinnvliesstoff schmelzgesponnener Mikrofaservliesstoff | Meltblown nonwoven | Mono- or bi-component micro-fibres |
| Verdampfungsspinnvliesstoff, verdampfungsgesponnener Spinnvliesstoff | Flashspun nonwoven | Mono-components |
| Spinnvliesverbundstoff Spinnvlies (S) Mikrofaser (M) | Spunbond composite spunlaid (S) meltblown (M) | Filament fibre composite |
| Elektrostatikvliesstoff | Electrostatic nonwoven | Micro-fibres |
| Folienvliesstoff | Film nonwoven | Fibrillated film fibres |

## 4.2.1
## Use of polymers

With extrusion nonwovens, the choice of material is closely connected with the requirements to be met by the final product. In addition, the process parameters of the extrusion process also correlate with filament draw or film draw and the bonding process. With regard to extrusion nonwovens, Section 2.2 is supplemented in this place.

Spunbonded nonwovens are preferably made from the thermoplastic high polymers polypropylene (PP) and polyester (PET). To a small extent, other polyolefins are found such as polyethylene of high density (HDPE) and linear polyethylene of low density (LLDPE) as well as a variety of polyamides (PA), mainly PA 6 and PA 6.6.

High quantities of PP are used due to its low price and advantageous properties such as chemical resistance, hydrophobicity, sufficient or even better strength.

Raw material parameters:

- Melt index MFI [49]    20–40 g/10 min (with spunlaid nonwovens)
                          100–1,600 g/10 min (with meltblown nonwovens)

- Polydispersity $\dfrac{M_W}{M_N}$    3.5–7 unit to measure mole weight distribution

  $M_W$ mol weight    average weight
  $M_N$ mol weight    arithmetical average

- Atactic share    $\leq 2.5\%$

In recent times, novel PP raw materials have become known which are being developed or launched on the market. They are based on metallocene catalysts [50]. Contrary to common PP materials, metallocene PP show a tighter mole weight distribution, a lower atactic share and a 10–15 °C lower crystal melting point. Equal process parameters given at drawing, higher filament velocities and lower filament finenesses are achieved. Filament strength rises by 30–50% as compared to common PP, maximum elongation dropping by the same value. With spunbonded nonwovens, a similar improvement in strength could be given proof of with mass per unit areas ranging from 10–130 g/m$^2$ [51, 52].

Spunlaid nonwovens from HDPE and LLDPE show great softness, which is taken advantage of to make coverstock nonwovens for hygienic uses [53].

It is possible to use assorted, clean secondary granulates in spunbonded nonwovens, the degradation of the mole weight or, respectively, the increase of the melt index MFI and soiling being kept in limits. Equal processing parameters given, mechanical properties of spunbonded nonwovens can be achieved which compare well with those made from primary granulates [54].

After PP, PET is the second most important polymer to make extrusion nonwovens. As compared with PP, PET is particular suitable for a variety of uses that require high thermal resistance and low shrinking. Such uses are, for instance, bituminated roof covering, backing material for coating purposes and backing for carpets as well as tufting backing.

Raw material parameters:

– intrinsic viscosity IV ≤0.64
– low share in COOH-groups
– high crystallinity
– water content ≤ 0.004%

Spunlaid and meltblown nonwovens are exclusively made from crystalline PET. The same applies to film nonwovens, which require a higher intrinsic viscosity ranging from 0.7 to 1.0. Crystallinity influences pre-drying and extrudability as well as filament drawing orientation, which is basic to make products that meet the requirements and that are of proper strength. Pre-drying is inevitable as PET at thermal strain is subject to hydrolytic degradation when extruded. In addition, low water content avoids air pockets in the melt that might cause filament breakage.

Frequently, requirements can only be met by means of polymer modification. Except for the mechanical properties, UV-resistance and flame-retardancy are important with technical applications. UV-stabilization can be achieved by blackening or by means of substances that slow down warp degradation. Such nonwovens are used as agrotextiles, to give one example. By means of light-reflecting particles and pigments it is possible to achieve light-filtering effects for agricultural uses [55].

Nonwovens need to meet particular requirements concerning fire protection in the building and automotive industries. To reach flame-retardancy, PP materials are mostly equipped with substances based on halogen. With PET materials, they are found as well but, for environmental reasons, halogen-free additives are also used.

As compared with PP and PET, the polyamides PA 6 and PA 6.6 are of minor importance. PA used to be processed to a larger extent to make spunbonded non-wovens for application in the field of packing materials. According to [56], it is possible to make polyurethane (PUR) into spunbonded nonwovens. Insufficient mechanical properties and high prices are, above all, limits to a wider use of PUR as compared to PP and PET. Elastomer polyethylene (PE) is known as an alternative [57]. Products of this type are being developed.

A process different from extrusion is the processing of thermoplastic PE following the flashspinning process developed by DuPont [58]. In this process, a spun-bonded nonwoven known as Tyvek is made, which is water-proof and shows strong blocking behaviour. Under high pressure, PE dissolves in a solvent. The solvent evaporates after it has left the spinning solvent via the spinnerets.

Synthetic high polymers show, except for the advantages named above, a disad-vantage: They are not biologically degradable, which means they pollute the environment.

For this reason, several producers are trying to make degradable nonwovens or, respectively, films. The following raw materials are preferable for this purpose [59–61]:

– thermoplastic cellulose
– polylactide products
– polylactide/starch blends
– thermoplastic starch

Due to their processing properties, prospects are good for polylactide products (PLA). However, current prices are relatively high and need to be cut to PET level. At the time being, there is one spunbonded nonwoven based on PLA available on the market, which is used in agriculture [62].

Contrary to the manufacture of spunlaid nonwovens, with the meltblown process there is no or little orientation after extrusion. This is why a wide range of materials can be processed. Except PP and PET, the following polymers are used (to give examples):

– polyethylene of high density (HDPE)
– polyethylene of low density (LDPE, LLDPE)
– polyamides (PA 6, PA 6.6, PA 10)
– polystyrene (PS)
– polytrifluorochloroethene (PCTFE)
– polycarbonate (PC)
– polyurethane (PUR)

As compared with spunlaid nonwovens, meltblown nonwovens are characterized by very fine fibres of low strength.

Combining polymers of a variety of chemical bases and/or physical parameters, it is possible to achieve binder effects, crimp or split effects.

To achieve binder effects, copolymers that generally melt at lower temperatures, PUR or polyolefins are processed in connection with the correspondent homopolymers that melt at higher temperatures. The filaments can be extruded separately (bi-filaments) or as bi-component filaments (bi-co) [63, 64]. The single components of the bi-co filaments need to be sufficiently adhesive. Split effects want the opposite since disintegration into the single filaments is intended.

PP and HDPE are preferably used to make extruded film nonwovens. Depending on the process used, the melt index ranges from 2 to 10. As seen from the polymer, this allows higher substance strength than with spunbonded nonwovens.

Except for polyolefins, film nonwovens are made from PET and PA (to a smaller extent). Depending on the kind of material, these polymers do not split so well. To make them better to split, a mixture of incompatible components is added, which results in a matrix/island structure (e.g. polymer blends of PET and HDPE or LLDPE) [65].

## 4.2.2
### Generally, on the process technology and equipment to be used

Giving the example of spunlaid nonwovens, Fig. 4-50 shows the general process stages of the manufacture of extrusion nonwovens with the corresponding equipment. The main stages are spinning – drawing – web formation – web-bonding.

Spinning largely corresponds with the manufacture of synthetic fibre materials to the melt-spinning process. However, it is possible to spin a thread curtain across the full spinning width from several rows of nozzles. It took novel technological solutions to draw the filaments in a tensionally locked way and to make them into a nonwoven. Web-bonding is generally possible by means of the well-known physical and chemical processes described in Chapter 6. Contrary to the bonding of fibre webs, filaments are endlessly long, which needs to be taken into

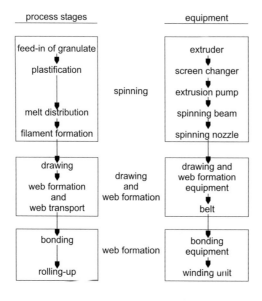

**Fig. 4-50** Schematic description of the process stages and the equipment required to make spunlaid nonwovens

consideration when determining the process variables of the bonding processes. With spunlaid nonwovens, calendering and thermal bonding are preferably used for lower mass per unit areas. For higher mass per unit areas needle-punching is recommendable.

Drawing and web formation equipment use the mechanical process of filament drawing by means of air, the filaments being laid down onto a perforated belt. Then, the loop-like filaments are fixed in their position by means of a suction current. Drawing the filaments by means of galettes (as is common in filament spinning) is difficult in practice since filament breakage may easily result in wind laps around the galettes.

Fig. 4-51 shows the basic process variants to make spunlaid nonwovens (according to Gerking [66] and Hartmann [67]). In variant a, air as hot as melting temperature emerges from closely to the nozzle holes, takes the filaments and draws them. The emerging air, at the same time, intermingles with the ambient air.

In all process variants, the web created on the perforated belt is held by means of a suction current and continues to travel. In variant b, the emerging air and the filaments are taken to a drawing channel. Blowing in additional pressed air, the drawing effect can be raised. In process c, cooling and drawing air are separated. The use of pressed air injectors means higher filament velocity and this way, a higher orientation of the filaments is caused with higher strengths and lower elongation values. Variant d is mainly the same as c. It is supplemented by a galette, which results in a higher drawing effect.

These basic variants to make spunlaid nonwovens serve, on the one hand, to develop marketable spunbonding machines. On the other hand, the manufacturers of spunbonded materials have created their own equipment so their know-how is used company-internally.

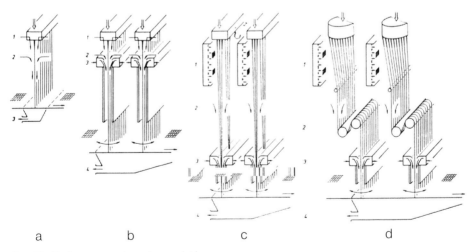

**Fig. 4-51** Basic processes to make spunlaid nonwovens

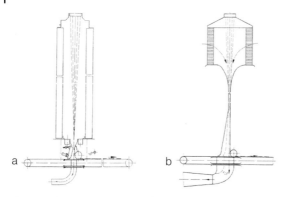

**Fig. 4-52** Basic processes to make spunlaid nonwovens to the vacuum process

Fig. 4-52 shows the variants available to make spunlaid nonwovens to the vacuum process.

The basic idea in variant a is that the force taking down the thread is punctually transferred to the filaments. Suction air is used to accomplish the transfer. This solution, however, is limited to a filament velocity of 800–1,000 m/min. Variant b means an advantage. Setting a lengthened drawing channel to its maximum, filament velocity can reach about 3,000 m/min [68]. This process uses the suction air to transfer the air friction forces to the filaments. Given an optimal transfer ratio of air velocity to filament velocity, filament velocity may reach up to 5,000 m/min.

The performance of the spunbonding machine per metre working width $P_{SP}$ in kg/h · m is determined by the throughput (depending on the raw material) per nozzle and the number of nozzles per metre of spinning width.

$$P_{SP} = \dot{m} \cdot n \cdot 0.06 \tag{1}$$

m=throughput per nozzle (g/min)
n =number of nozzles per metre of spinning width $(m^{-1})$

Throughput amount per nozzle and nozzle diameter determine the velocity at which the melt emerges from the nozzle.

$$\dot{m} = \frac{d_A^2 \cdot \pi}{4} \cdot v_A \cdot \rho_p \tag{2}$$

$d_A$ =nozzle diameter (mm)
$v_A$ =emerging velocity (m/min)
$\rho_p$ =polymer density $(g/cm^3)$
$\dot{m} = d_a^2 \cdot v_A \cdot \rho_p \cdot 0.785$
$\dot{m} = \dot{V} \cdot \rho_p \cdot 0.785$
$\dot{V}$ =volume flow $(cm^3/min)$

With solid polymers, the common polymer densities for polypropylene and poly-ester are applicable, a.o., for granulate. With the melt, the values are lower. Table 4-3 contains the polymer density for a variety of structures and their shares in the total structure.

The throughput amount kept constant, the nozzle diameter $d_A$ squares the velocity at which the melt emerges. At constant melt temperature, both values determine, as read at the nozzle mouth, the amount of heat transported by the filaments. Filament fineness is described as follows:

$$Tt_F = \frac{\dot{m}}{v_F} \cdot 10^4 \tag{3}$$

$Tt_F$ = filament fineness (dtex)
$V_F$ = filament velocity (m/min)

Filament fineness given – it can be determined from the filaments laid down – filament velocity can be calculated. This allows to measure at what velocity the filaments are laid down onto the perforated belt. The mass per unit area of the web created on the perforated belt is described by the equation (4).

$$m_V = \frac{P_{SP}}{v_T} \cdot 16.67 \tag{4}$$

$m_V$ = mass per unit area of the spunbonded nonwoven (g/m²)
$P_{SP}$ = kg/h · m
$v_T$ = speed of belt (m/min)

In the spunbonding process, the aerodynamic drawing of the filaments is responsible for filament attenuation between the point where it emerges from the nozzle and the point where it is laid down onto the perforated belt. At the same time, the forces acting on the filament cause structure formation and, thus, influence filament strength and elongation behaviour.

Creating a schematic model concerning the spunlaid process c in Fig. 4-51, Gerking calculated the aerodynamic drawing as well as the deformation strain acting on the filament [69].

Here is a summary of the results he came to: As Fig. 4-53 shows, the tensile force $F_Z$, due to the air flow in the injector channel, acts on the filament. As a

Table 4-3  Polymer densities for a variety of structures and their shares in the total structure

| Density (g/cm³) | Polyester | Polypropylene |
|---|---|---|
| Granulate | 1.38 | 0.91 |
| Crystalline share | 1.455 | 0.94 |
| Amorphous share | 1.336 | 0.85 |
| Melt | 1.18 | 0.75 |

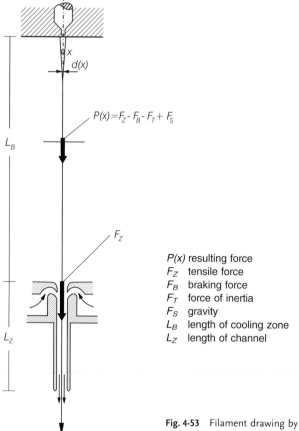

$P(x) = F_Z - F_B - F_T + F_S$

$P(x)$ resulting force
$F_Z$ tensile force
$F_B$ braking force
$F_T$ force of inertia
$F_S$ gravity
$L_B$ length of cooling zone
$L_Z$ length of channel

**Fig. 4-53** Filament drawing by means of injector: forces acting

consequence of the difference between air velocity $v_L$ and filament velocity $v_F$, this force is transferred onto the filament through air friction.

The filament undergoes considerable deformation after it emerges from the nozzle until it, having left the injector, reaches its final diameter $d_F$. Describing a variety of masses put through per nozzle, Gerking calculated those forces that act on the thread. The higher the deformation forces, the higher the molecular orientation of the filaments. Following Fig. 4-53, P(x) is the force resulting from the external forces at point x. These equations describe the external forces:

$$\text{Tensile force} \qquad F_Z = \pi d_F \int_{x=L_B}^{x=L_B+L_Z} \tau_{w2}(x)\,dx \qquad (5)$$

$\tau_{w2}$ = feeding tension in the channel

Braking force $\qquad F_B(x) = \pi \displaystyle\int\limits_{x=0}^{x=L_B} d(x)\tau_{w1}(x)dx$ $\qquad\qquad$ (6)

$\tau_{w1}$ = feeding tension in cooling zone

Force of inertia $\qquad F_T = \dot{m}[v_F(x_2) - v_F(x)]$ $\qquad\qquad\qquad$ (7)

Gravity $\qquad F_S = \dfrac{\pi}{4}g \displaystyle\int\limits_{x=L_B}^{x=L_B+L_Z} d^2(x)dx$ $\qquad\qquad\qquad$ (8)

For the case $v_L > v_F$, this connection can be taken for granted between the feeding tensions in the channel and in the cooling zone:

$$\tau_{w2} \approx \left[1 - \left(\frac{v_F}{v_L}\right)^2\right]\tau_{w1}$$

Equations (5) to (8) allow to determine the greatest force $P(x_2) = F_Z - F_B\,(x_2)$ acting on the filament. Force of inertia and gravity are not of much interest. Deformation tension $\sigma_v$ can be described as follows:

$$\sigma_v = \sigma_Z - \sigma_B = \frac{P(x_2)}{\dfrac{\pi d_F^2}{4}} = \frac{F_Z - F_B(x_2)}{\dfrac{\pi d_F^2}{4}}$$

Taking into consideration the resistance coefficients calculated, Gerking finds these results:

$$\sigma_Z(x) = 5.24 \cdot \rho_L \cdot v_L^{0.81}\left[1 - \left(\frac{v_F}{v_L}\right)^2\right](v_L - v_F)^{1.19}\frac{L_Z^{0.81}}{d^{1.62}} \qquad (9)$$

$$\sigma_B(x) = 5.24 \cdot \rho_L \cdot v_L^{0.81} \cdot v_F^{1.19}\frac{L_B^{0.81}}{d^{1.62}} \qquad\qquad\qquad (10)$$

Equation (9) takes $d = d_F$ and $v_F(x) = \dfrac{4 \cdot \dot{V}}{\pi \cdot d_F^2} \cong v_F$ for thread velocity.

With regard to the braking tension (10), a medium diameter $\overline{d} = 2d_F$ is assumed. The above given, the greatest deformation tension $\sigma_v$ follows this equation (11):

$$\sigma_v = 5.24 \cdot \rho_L \cdot v_L L^{0.81}\left\{(v_L - v_F)^{1.19}\left[1 - \left(\frac{v_F}{v_L}\right)^2\right] - \frac{1}{16}v_F^{1.19}\left(\frac{L_B}{L_Z}\right)^{0.81}\right\}\frac{L_Z^{0.81}}{d_F^{1.62}} \qquad (11)$$

$\rho_L$ = air density
$v_L$ = kinematic air tenacity

Fig. 4-54 shows, for a given proportion of $\dfrac{L_B}{L_Z}$, the deformation tension as depending on the filament diameter. Varying air velocity and volume throughput, a maximum is reached.

The maximum results in the smallest possible filament diameter. Gerking [69] calls the filament in question 'Grenzfaden' (minimum filament diameter). In this context, $v_L$ should be three times $v_F$. Falling below the minimum diameter means the spinning parameters are no more optimal.

Chen a.o. [70] have also looked into the dynamics of the spunbonding process. In laboratory tests focusing on pressed air drawing with injector, they determined the forces acting at a variety of pressures. The investigation carried out by them into the structures of PP filaments confirms that spinning tension influences double refraction and crystalline orientation. Here, parallels are seen to drawing as it is known from the classic spinning process.

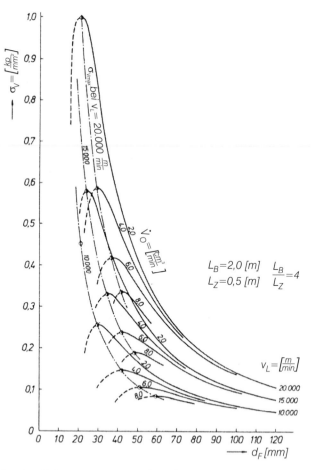

**Fig. 4-54** Deformation tension as depending on filament diameter, air velocity varied

Beyreuther [71] looked into the filament fineness achievable in melt-spinning polyamide and polyester. He found the wide variety of secondary parameters incurred in the form-fitting spinning process or in the tensionally locked spunbonding process need to be taken into consideration, too. This is also applicable with the full variety of technological solutions to make spunbonded nonwovens (filament drawing by means of pressed air or suction air). In the laboratory, Beyreuther and Brüning [72], following the suction air process, extended the thread formation model with regard to the manufacture of spunlaid nonwovens (see Fig. 4-52). They then used the model to optimize technology focusing on a variety of process parameters. The model developed by them is based on the proportion of melt flow $\dot{m}$ to filament fineness $Tt_F$ as described in equation (3), for filament force F as well as the change to filament temperature (which depends on the distance from the nozzle) as described in equation (12).

$$\frac{dT_F}{dx} = Nu \cdot (T_F - T_L) \frac{\pi \cdot \lambda_L}{\dot{m} \cdot cp} \tag{12}$$

$T_F$ = filament temperature
x = distance from nozzle
T = air temperature
Nu = Nusselt number
$\lambda_L$ = thermal conductivity of the air
$c_p$ = thermal capacity of the polymer

The Nusselt number is a dimensionless value that describes how the heat transfer coefficient $a$ between filament and air depends on the thermal conductivity $\lambda_L$ (which influences it) as well as the filament diameter $d_F$.

$$Nu = \frac{a \cdot d_F}{\lambda_L}$$

From the energy equation describing the heat exchange at the filament, Gerking [73], with reference to Schöne, comes to this equation (13):

$$\frac{T_F(x) - T_L}{T_S - T_L} = e^{-\frac{x}{x_o}} \tag{13}$$

$T_S$ = melting temperature at the nozzle exit

$$x_o = \frac{\overline{c}_p \cdot \dot{m}}{\pi \overline{\lambda}_L \cdot \overline{Nu}} \tag{14}$$

Following Schöne's equation, the average Nusselt number $\overline{Nu}$ is

$$\overline{Nu} = 0.91 \, \dot{m}^{0.21} \cdot d_F^{0.29} \cdot v_F^{0.05}$$

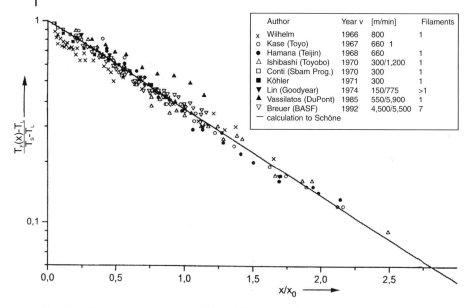

**Fig. 4-55**  Filament temperature profile below the nozzle (from [73])

Definitely, there is a correlation between the temperature profile and the length of the filament [73] (see Fig. 4-55).

The values below are approximately true for

|  | polypropylene | polyester |
|---|---|---|
| $\bar{c}_p$ in kJ/kg K | 2.7 | 1.7 |

Cooling parameters and forces acting on the filament given, it is possible to look into ways to optimize filament fineness as well as strength and elongation behaviour. Based on a mathematical model describing the Reicofil spunbonding process, Misra and Spruiell [74] investigated into the influence of material and process parameters on filament structure and the structure of the spunbonded nonwoven. Spinning parameters generally being constant, they found that rising extrusion temperature results in the filament diameter slightly decreasing. At the same time, final crystallinity (%) and double refraction decrease. Changing the temperature of the cooling air from 20 °C to 150 °C, the filament diameter is reduced. Crystallinity, at the same time, decreases (see Fig. 4-56 a and b). From the picture it becomes obvious that an increase in temperature delays diameter attenuation in the spinning direction. Crystallinity occurs after a larger distance below the spinneret.

Simulation based on the model describing the influence of the polymer and process variables on filament properties allow to evaluate the individual values. As to the material, these values are elongation viscosity and crystallization kinetics. As to process parameters, they are cooling air temperature as well as air flow quotas and the particular design of the pneumatic drawing equipment.

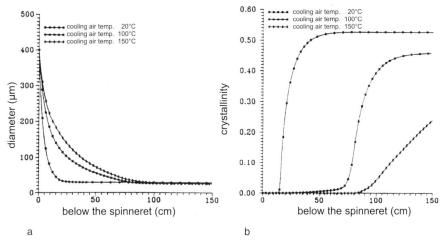

**Fig. 4-56** Influence of change to cooling air on (a) filament diameter profile and (b) crystallinity

In the zone where the web is created, the filaments crash onto the belt at high speed (1,000–5,000 m/min).

In most cases, filament air radiation causes a stagnation point flow which occurs after the filaments emerge from the drawing zone. Except for filament arrangement in the drawing zone, this air flow is essential for the filaments being laid down in a web. The belt velocity at which the filaments are laid down and transported is essential for the mass per unit area of web. Depending on the mass per unit area, the ratio of belt velocity to filament velocity may vary considerably. With mass per unit areas $\leq 10$ g/m$^2$, modern machines allow a belt velocity of up to 600 m/min. Generally, it is important the filaments are not laid down in bunches.

Theoretically, the filament is laid down continuously and spirally. On the conveyor belt, this results in a projected elliptic shape, which, via its diameters, defines the strength of the nonwoven in both MD and CD directions. Spunlaid nonwovens show higher strength preferably in MD direction. A transverse flow occurring when the filaments are laid down at the stagnation point, higher transverse strengths are achieved. Varying filament fineness may cause varying spinning, drawing and uncontrolled laying-down parameters, which results in fluctuation of the mass per unit area of the nonwoven and in nonwoven cloudiness. With low masses of nonwoven this fluctuation is visible. Generally, the web is created through the filaments being laid down in layers. The visible random-type arrangement of the laid down filaments can be described by an exponential function. Poisson distribution [66] applies to equally long distances of equally oriented filaments. Laying down can be influenced by means of swivelling – e.g. of the injectors towards the level of web formation. Three parameters are essential for the nonwoven to be created: the mass per unit area wanted, filament fineness, total length of filament per area unit.

This is described in the equations below:

$$d_F = 11.3 \cdot \sqrt{\frac{Tt_F}{\rho_P}} \qquad \begin{array}{l} d_F \text{ in } \mu m \\ Tt_F \text{ in dtex} \end{array}$$

The specific filament surface $S_F$ in cm²/cm is

$$S_F = \frac{40{,}000}{d_F}$$

Frequently, the specific filament surface $S_{FO}$ is given in cm²/g:

$$S_{FO} = \frac{40{,}000}{\rho_P \cdot d_F} \tag{15}$$

The length of filament per m² of nonwoven $L_{Ges}$ is described by

$$L_{Ges} = \frac{m_V}{Tt_F} \cdot 10^4 \qquad \text{in } m/m^2 \tag{16}$$

$m_V$ = mass of nonwoven in g/cm²

The specific nonwoven surface $S_{mV}$ is

$$S_{mV} = S_{FO} \cdot m_V \tag{17}$$

Filaments from polypropylene and polyester can be described this way:

Polypropylene:        Polyester:

$d_F = 11.85 \sqrt{Tt_F}$        $d_F = 9.62 \sqrt{Tt_F}$

$$S_{FO} = \frac{3709}{\sqrt{Tt_F}} \qquad\qquad S_{FO} = \frac{3013}{\sqrt{Tt_F}}$$

at $\rho_{PP} = 0.91 \ g/cm^3$        $\rho_{PES} = 1.38 \ g/cm^3$

Fig. 4-57 shows the specific fibre surface as depending on filament fineness. It is inversely proportional to filament fineness. The differences between polypropylene and polyester are caused by their densities. Spunlaid nonwovens of finenesses of $\leq 1.0$ dtex have been developed, the specific surface drastically rising with finenesses as low as that. With low mass per unit areas, this results in a more compact surface. In addition, finer filaments show lower bending strength, which means, equal bonding intensity given, softer nonwovens. Analogous to the specific surface, the light reflecting behaviour of the filaments depends on their fineness. This is why nonwovens made from fine filaments, generally, appear lighter than those of coarse filaments.

Fig. 4-58 shows the length of filament per square metre of nonwoven, filament fineness varying. From the graph it becomes obvious that at constant length of fil-

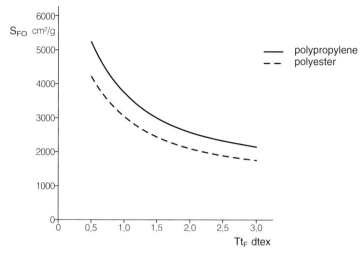

**Fig. 4-57** Specific filament surface $S_{FO}$ as depending on filament fineness $Tt_F$ for polypropylene and polyester

ament per area unit, filament attenuation may go together with a reduction in the mass per unit area of the nonwoven. This means less material is needed. At the time being, this tendency is predominant in the development of light-weight spunlaid nonwovens.

From Fig. 4-59 one can conclude how the specific surface of web rises with filaments becoming smaller. The current tendency towards the manufacture of very light-weight spunlaid nonwovens made from fine filaments ≤1 dtex entails high-

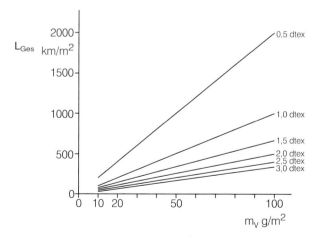

**Fig. 4-58** Length of filament $L_{Ges}$ per m² of web as depending on the mass of web, filament fineness varying

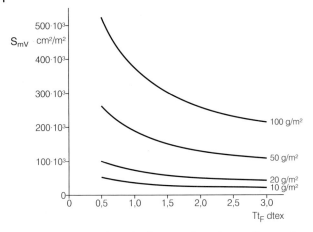

**Fig. 4-59**   Specific surface of web $S_{mv}$ as depending on filament fineness, mass of web varying

performance machines with working widths up to 5 m. Optimizing the process and making it computer-controlled, good-quality nonwovens can be produced.

### 4.2.3
**Processes to make spunlaid nonwovens and composites**

Focussing on the manufacture of extrusion nonwovens, this Chapter describes the variants of nonwovens known as spunlaid nonwovens, their properties and uses.

Process variants, on the one hand, show a number of common features. On the other hand, they are very diversified since they are influenced by a wide range of requirements to be met by the products. Requirements may even appear as pairs of opposites, such as

– long life/fast degradability
– mass per unit areas of 5–1,000 g/m$^2$
– softness/stiffness
– good permeability/blocking effect
– high resistance to deformation/easy deformability

The manufacture of spunlaid nonwovens is a one-stage process from the granulate to the final product. This entails economical production, which is high throughput or respectively, output of square metres of fabric. Since the beginnings in the 60s, parallel to the technical progress, the efficiency of the process has been raised to 5 to 10 times as much. This has become possible through

– a higher throughput performance of the extruders
– a wider working width (now up to 5 m)
– a higher working speed, which is belt velocity while the web is being created

**Table 4-4** Process stages of the manufacture of spunlaid nonwovens

| Stage | Characteristics | Notes |
|---|---|---|
| Granulate drying, melting the granulate/the blend of granulates on, adding additives/master batches, solving the polymers | mechanical, thermal | Heating, blending, homogenizing, injecting to dye, UV-stabilisation, flame-retardant equipment adding solvent |
| Filament spinning | mechanical, thermal, physico-chemical | heating, homogenizing, filtering, formation, evaporating the solvent |
| Cooling the filament, heating the filament | aerodynamic | cooling, exposition to hot air |
| Filament drawing | aerodynamic, mechanical | pressed air, suction air, galettes |
| Web formation | aerodynamic, electrostatic | laying down onto perforated belt or drum |
| Web-bonding | thermal mechanical hydrodynamic | calendering, thermofusion needle-punching meshing hydroentangling |
| After-treatment | thermal, mechanical aero-/hydrodynamic electrostatic | thermofusion, drawing the fabric spraying with lubrication, flame-retardancy, antistatic behaviour, hydrophilizing, water-repellent finish pretreatment for printing |

Working speed depends on the mass per unit area and may currently reach 600 m/min [75]. Higher strength or voluminosity (a.o.) may be achieved in a discontinuous additional process, however, this should be exceptional as it raises costs.

Table 4-4 shows, in a general way, the continuously connected stages of the process. As mentioned above, extrusion and filament spinning are largely based on filament yarn spinning and synthetic fibre spinning. This also applies to the bonding of spunlaid nonwovens, which uses the know-how available from the bonding of meltblown nonwovens. Filament drawing and web formation, however, are characterized by novel independent processes.

**Extrusion and filament formation**
The first stage of the process is analogous to the meltblown process and to the manufacture of film nonwovens: the polymer granulates or blends are melted on, are homogenized and then, the melt is transported to the extrusion tool. For this

purpose, it is common to use one-worm extruders with worm diameters D of 45–120 mm and worm lengths L of 28–35 D [76].

Strong spunbonding machines use extruders with a worm diameter of 160 mm. Dimensioning the extruder depends on the working width of the spunbonding machine and the mass per unit area of the nonwoven. For better machine flexibility in case of a change to the polymer used, e.g. from polypropylene to polyester, double-worm extruders with a variety of worm configurations are found [77]. Toothed wheel pumps are well-proven to achieve a constant dosing of the melt. Emerging from the extruder, the melt is filtered by means of a continuously working screen exchanger, the melt flow not being interrupted. Turning platforms KSF [78] and double-piston screen changers [79] are best known. Both of them allow a continuous melt flow with only a small drop in pressure.

Thoroughly filtering the melt is especially important when fine filaments are spun in order to avoid filament breakage, which would, due to faults in the nonwoven, spoil quality. Melt distribution in the spinning beam and the spinnerets depends on to what extent the filaments are spun in bunches, in partial curtains or as curtains all across the spinning width. Spinning bundles of filament or curtains requires circular or square nozzles, the melt being transported to them zone by zone via a spinning pump. It is important the melt reaches every point of spinning at constant temperature and pressure. This variant is used in case taking down and drawing the filaments is done by means of round injectors or slot injectors.

Spinning curtains of filament all across the spinning width requires the nozzles are arranged in lines. They are either one-piece or composed of several pieces. In this case, the filaments are taken down over the full width and drawn (see Fig. 4-51 b).

Cross-sections of the single nozzles are mostly circular, their diameters generally ranging from 0.3 mm to 0.8 mm. With fine filaments, nozzle diameters of 0.15 to 0.3 mm are found. Except for circular diameters, trilobal, dogbone or randomly designed nozzles are used.

Except for the nozzle diameter, the length of the borehole contributes to filament formation. In general, it is twice to fourfold the nozzle diameter. For careful filament spinning at high throughput a short borehole is recommendable. Currently, spunlaid nonwovens largely consist of mono-filaments. Multi-component filaments, in particular bi-component filaments, are becoming more and more important. These allow – depending on the requirements to be met by the nonwovens – binder effects, crimp as well as split effects. Spinning multi-component filaments requires two or more extruders, which, via separate melt pipes, transport one polymer component at a time to a multi-component spinning beam. The melt flows unite directly at the nozzle head. On their surfaces, both the components are adhesively connected with one another. The most important variants of filament cross-section are "side by side" and "core and sheat".

The manufacture of spunbonded nonwovens based on bi-component filaments is described in [80, 81]. One example is well-known as Colback nonwovens [82].

With machines used in the spunlaid process, one important criterion to measure cost-effectiveness is the throughput per hour and metre of spinning width

**Table 4-5** Polyester and polypropylene throughput

| Polymer | $\dot{m}$ [g/min, nozzle] | $\dot{m}_{preferable}$ [g/min, nozzle] | $P_{sp\ max}{}^{1)}$ [kg/h, m spinning width] |
|---------|---------|---------|---------|
| PP | 0.15–1.10 | 0.50–0.80 | 180 [83] |
| PET | 0.30–2.00 | 0.30–1.70 [84] | 300 [85] |

1) The maximum throughput $P_{sp\ max}$ is valid for one extrusion system.

(see equation 1). It is directly proportional to the throughput per nozzle and the number of nozzles per metre. Table 4-5 gives the throughput with reference to PP and PET, which currently are the most often used polymers. For more than 20 years, spunbonding machines have been known with double or multiple spinning beams. They can be fed by one or more extruders (see Fig. 4-60).

In the Lutravil process, spinning beams standing at an angle to the working directions and/or swivelling drawing units improve the evenness of filament lay-down [86]. Using several extruders and several spinning beams arranged one after another in working direction, a considerable increase in the throughput is possible.

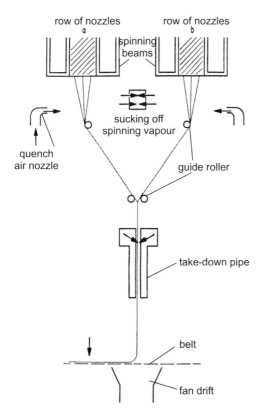

**Fig. 4-60** Equipment to manufacture spunbonded nonwovens with double spinning beam to DE 32 28002 A1

According to [83], it is 540 kg/h per metre of spinning width with e.g. the Rei-cofil 3 process (3×180 kg/h per metre of spinning width).

Using polymers that melt at different temperatures and with the help of thermal bonding, it is possible to achieve binder effects [87].

### Drawing and web formation

The process stages following filament extrusion are
– cooling the filament
– filament drawing
– web formation

They have great impact on the geometric and physical properties of the filaments and of the spunbonded nonwovens. These are

– filament fineness
– strength and elongation behaviour of the filaments
– mass per unit area
– thickness/density
– evenness

Filament fineness depends on the throughput per nozzle as well as on filament velocity. It largely influences such properties as softness/touch, pile structure, inner surface and permeability of the nonwovens. The finer the filaments, the higher are softness and inner surface of the nonwovens. Width of pore opening and permeability decrease with lower fineness. With fine filaments, strength as related to the mass per unit area increases [88]. The throughput per nozzle can be gravimetrically determined. In general, filament fineness is defined optically by means of oscillation (see Section 18.1.1). Filament velocity is received from the throughput per nozzle and filament fineness as described in equation (3). Filament velocity at constant throughput per nozzle is inversely proportional to filament fineness. Cooling the filament is accomplished by means of quench air, air suctioning or cooling air. It results in the melt solidifying at the point of bonding, which is at a certain distance below the nozzle. The distance much depends on the throughput per nozzle and the temperature of the cooling air.

Intensity and speed of cooling act on the formation of the crystalline structure and, in connection with the drawing process, on filament strength and elongation behaviour. To measure the height of drawing, the draw ratio is of interest:

$$V_R = \frac{Tt_D}{Tt_F}$$

$V_R$ = draw ratio
$Tt_D$ = filament fineness at the nozzle [dtex]
$Tt_F$ = filament fineness in the raw web [dtex]

Filament drawing is effected either in a tensionally locked way by means of air and/or in a form-closed way by means of drawing rollers or galettes. Currently, aerodynamic drawing is predominant as compared to galette drawing. The reason why is process security, as filament breakage can be minimized this way.

Theoretical calculations and models describing filament drawing have been dealt with in Section 4.2.2. With aerodynamic drawing, the effective air velocity (which depends on the drawing principle) acts on filament velocity. To measure the effectivity of the air acting on the filaments, it is important to look at transfer effectivity, which results from the ratio of filament to air velocity:

$$T = \frac{v_F}{v_L}$$

T = transfer effectivity
$v_F$ = filament velocity [m/min]
$v_L$ = air velocity [m/min]

Air velocity can be measured [89].

With galette drawing, the circumferential speed of the galettes determines the draw ratio and filament velocity [90]. However, with a combined air and galette drawing system, the total draw ratio results from multiplying the single draw ratios. Such systems are described in [91].

In recent years, filaments have been attenuated. This trend should be seen in the context of the requirements nonwovens need to meet, such as great softness, a mass per unit area as low as possible (e.g. between 5 and 7 g/m$^2$) and proper blocking behaviour to both gaseous and liquid media. To produce spunbonded nonwovens from fine fibres and fine filaments, there is a variety of ways available:

- Combination of meltblown nonwovens of finenesses >0.5 dtex and spunlaid nonwovens of finenesses >1 dtex (see Section 4.2.4).

- Manufacture of multi-component filaments of relatively coarse fineness and subsequent splitting into single filaments.

- Attenuation of the single filaments of finenesses ≤1 dtex reducing the through-put per nozzle and/or increasing filament velocity.

The great importance of filament attenuation can be seen by the fact that the amount of nonwovens from fine filaments and multilayer spunbonded composites (SM, SMS) has continuously grown so in 1997, it amounted to 88% of the total production [92].

Very light-weight spunbonded nonwovens were originally made to the melt-blown process. Later on, multi-component filaments, preferably bi-component ones, were processed. Meltblown nonwovens are characterized by an ultra-fine structure and a very soft touch, however, they show very low tensile strength [93]. Another disadvantage is the production process requires much more energy than the manufacture of spunlaid nonwovens. According to [94], the meltblown technology takes 7 kWh/kg polymer as compared with 2.5 kWh/kg polymer in the spunlaid process. In a novel suction-air process, the specific energy consumption is only 1.1–1.5 kWh/kg polymer [84].

Above this, fine-titer spunlaid nonwovens can be made on the base of multi-component filaments or to the common spunlaid process [95]. The multi-component process also allows to spin cross-sections known as *segmented pie* or *islands in the sea* [96]. In this case, element fineness is equal to filament fineness divided by the number of segments. This process requires sophisticated equipment. Splitting by means of chemically leaching out entails intensive after-treatment [97]. In the thermal process, one component is melted out, which causes a loss in material of 20–30%. Two mechanical splitting processes are known: hydroentangling and needle-punching.

Making fine filaments to the common spunlaid processes, spinning security decreases with finenesses <0.8 dtex, which means efficiency decreases [98]. In ref. [99], a fineness of 1.75 dtex is called the limit with regard to cost-effectiveness. This is why manufacturers of equipment and producers of spunbonded nonwovens are trying to enhance the spunlaid process. Both European and Japanese enterprises have come out with suitable solutions [75, 85]. At the time being, fine-titre spunlaid nonwovens are competing with SM composites. In [85], the high working speed is called the main advantage of the SM process while low flexibility in case a producer intends a change to titer is looked upon as the main disadvantage [100]. The mode of filament spinning also influences the process parameters with regard to filament drawing and web formation. Numerous solutions have been developed for these partial processes, both in design and technology.

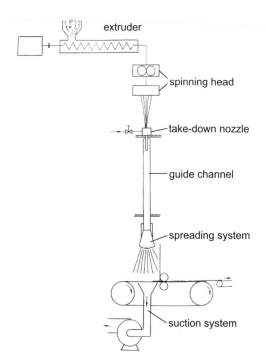

**Fig. 4-61** Process to make a variety of random-fibre nonwovens at Metallgesellschaft AG to DE 2014240

Currently, industrial enterprises cool the filaments, draw the filaments and create the web by means of aerodynamic processes. The individual flows of air are called primary, secondary, tertiary or process air [101].

Primary and secondary air are used to cool and draw the filaments while tertiary air serves to take down the filaments in bunches or as a curtain. Depending on the process chosen, the air is blown into the systems at excess pressure and/or sucked through at low pressure. The equipment is partially open or, with systems purely based on pressed air, hermetically closed. For better cost-effectiveness, producers often use air cycles.

One of the first industrially used spunlaid processes is the Lutravil process. In this process, the filaments are continuously cooled with conditioned air, which takes place below the nozzles. Then, using the adjoining quench ducts, the filaments are exposed to air at controlled room temperature.

The filaments then travel through a channel where they are drawn by means of tertiary air and at high pressure [101]. The Docan process is different in as much as the filaments are taken down via take-down nozzles (see Fig. 4-61) [101]. Analogous to the Lutravil process, the filaments are exposed to cooling air and at the same time, the cool air box serves to draw the filaments. Between the cool air box and the take-down nozzles the filaments travel through ambient air.

A variant to the Docan process developed by the same enterprise is known as New Spunbond Technology (NST) (see Fig. 4-62). Fig. 4-62 also shows filament and fibre filament technologies can be combined (SMS technology). Operating the

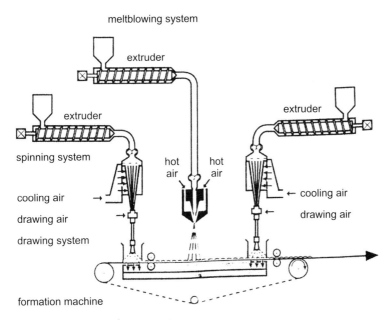

**Fig. 4-62** NST system of Zimmer AG

take-down nozzles or, respectively, injectors used in the pressed air process requires relatively much energy so both manufacturers of equipment and producers of spunbonded nonwovens focus on developments to reduce air pressure and energy consumption. An injector supposed to consume relatively little energy is described in [102].

To make no-shrink spunlaid nonwovens from polyester, filament velocity should reach 5,000 to 6,000 m/min. A drawing and web-formation system suitable for this purpose is shown in Fig. 4-63. This system uses a hermetically closed excess pressure box to draw the filaments and lay them down as a curtain. The channel is designed to set its limiting sheets zone by zone in order to fine-tune pressure and air flow.

Similar to the processes described above, the Reicofil process uses conditioned air below the nozzles, the temperature of which is adjusted to the polymer to be processed. In a second quench air zone, the filaments are exposed to ambient air. With the blended air, the filaments are sucked through the suction channel into a box where the web is laid down.

Drawing and laying down the filament takes place all across the width. Fig. 4-64 shows the latest variant of the REICOFIL 3 process. This process allows to make fine-titre spunlaid nonwovens (equipment: MF) and, in addition to polypropylene, to process polyester (equipment: PET).

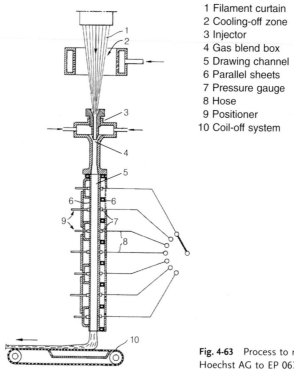

1 Filament curtain
2 Cooling-off zone
3 Injector
4 Gas blend box
5 Drawing channel
6 Parallel sheets
7 Pressure gauge
8 Hose
9 Positioner
10 Coil-off system

**Fig. 4-63** Process to make spunlaid nonwovens at Hoechst AG to EP 0674036A2

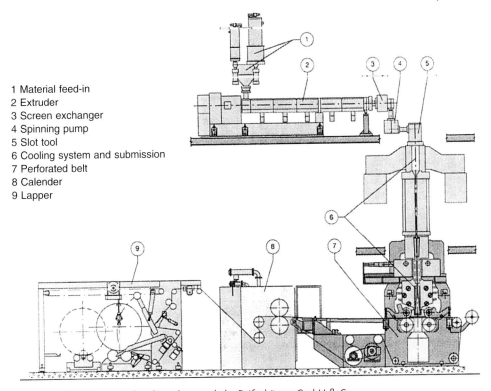

1 Material feed-in
2 Extruder
3 Screen exchanger
4 Spinning pump
5 Slot tool
6 Cooling system and submission
7 Perforated belt
8 Calender
9 Lapper

**Fig. 4-64**   REICOFIL 3 spunbonding plant made by Reifenhäuser GmbH & Co.

As early as the 1960s, a process was developed in which drawing and the web formation of a filament curtain are exclusively effected by means of suction air [103]. That was the technological base for the REICOFIL 1 process. The process described in [103] was industrially used until 1990 and then optimized with regard to both equipment and technology. The advanced version contains a change to the cooling-off zone. Suction channel and diffuser were lengthened. The basic process is used at the laboratory of Sächsisches Textilforschungsinstitut e.V. Chemnitz. By the help of this process, filament velocity could first be raised from about 1,000 m/min to 3,000 m/min and, then, to 5,000 m/min. This is how filament finenesses of <1 dtex can be achieved. To give a number of examples, research at this laboratory contains

– tests to novel raw materials [51]
– the processing of recycled materials [54]
– process optimization based on novel raw materials
– processes in combination with mechanical bonding technologies [98]
– the manufacture of novel fibre/spunbonded nonwoven composites [104]

To give two examples, Figs. 4-65 and 4-66 show results of investigation into the granulates Hostalen (common PP) and Hostacen (metallocene PP). Fig. 4-65 de-

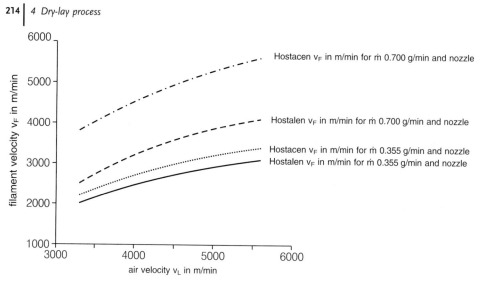

**Fig. 4-65** Filament velocity as depending on air velocity with regard to PP Hostalen and Hostacen

scribes the filament velocity reached as depending on air velocity. This also explains the filament attenuation found with Hostacen as compared to Hostalen. Fig. 4-66 shows the greater strength and better elongation behaviour of the spun-laid nonwovens based on Hostacen as compared to Hostalen, mass per unit areas being equal.

A novel process using suction air allows, due to a particular design of the take-down channel and the diffuser, high filament velocity with polyester, too [105].

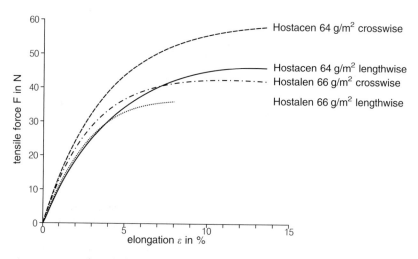

**Fig. 4-66** Ratio of tensile force to elongation with regard to PP Hostalen and Hostacen

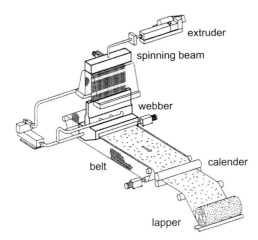

**Fig. 4-67** Spunbonding machine to the INVENTA Fischer process

The belt travels through the diffuser and is hermetically closed in this area, no loss in pressed air occurring. Fig. 4-67 shows the basic process.

With regard to two types of polyester, Fig. 4-68 describes filament strength as depending on filament velocity.

The change to the edge found when taking down filament curtains through channels does not occur. Making spunlaid nonwovens, one main question is how to achieve proper evenness. Answering the question which of the different spun-bonding technologies is most competitive, evenness is as essential as cost-effec-tiveness.

Insufficient evenness is seen, in particular, with low mass per unit areas, which show a "cloudy" nonwoven and high variation coefficients of mass per unit area, mechanical properties and permeability. Unevenness of spunlaid nonwovens is usually caused by aerodynamic filament drawing and web formation, mode of bonding playing a minor role. This means insufficient evenness of the raw web

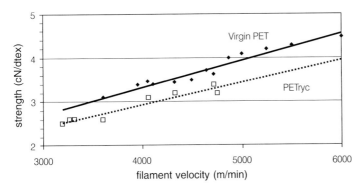

**Fig. 4-68** Filament strength as depending on filament velocity with regard to polyester (to the INVENTA Fischer process)

cannot properly be balanced out. Extrusion and spinning the filaments have little impact on evenness.

Process parameters being optimally set (e.g. temperature programme, worm revolutions per minute, melt pressure), the extruders available, the melt pumps, spinning beams and nozzles will transport the melt continuously. Consequently, the efforts of the manufacturers of equipment and the producers of spunlaid nonwovens mainly focus on better filament take-down and web-laying. One variant of solution is the spinning of yarn sheets, which are laid one above the other onto the belt [106].

Another variant of solution is known as suspended Coanda rollers. They move like a pendulum, thus improving the spiral arrangement of the filaments on the belt [107]. Taking the filaments down in bunches, it is important to spread the bunches after they have left the injector so as to avoid stripes appearing in the raw web. Fig. 4-69 shows the spreading occurs both in working direction and all across the width of the web. One technological variant of solution is electrostatic spreading as described in [108]. Spreading, however, can also be accomplished by means of bounce-blades, some of which may be perforated [109, 110].

Taking down filament curtains via channels all across the web width, the spaces of laying down were designed adjustably so as to enable flow-control. Basically, spinning and laying down filament curtains requires a steady thread flow (thread stand) with regard to trouble-free drawing so filament lay-down can go on as smoothly as possible to form a web.

With spunlaid nonwovens, one important quality parameter is the mass per unit area, which is largely determined by the later use. The mass per unit area de-

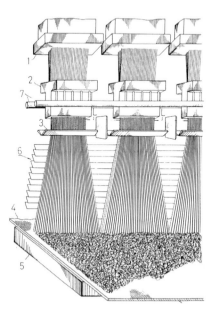

1 Nozzle
2 Air suction system
3 Filament spreading
4 Belt
5 Suction box
6 Flow-control sheets
7 Pressed air pipe

**Fig. 4-69** Process to make spunbonded nonwovens at Toyobo Boseki K.K. Osaka, Japan

pends on the throughput and the speed of the belt, which is many times smaller than filament velocity. The mass per unit area is calculated as described in equation (4). The number of openings and spinning width are constant, width of web is supposed to be analogous to the spinning width.

### Web-bonding

Except for the raw material parameters, the filament properties and the mass of web, the bonding process and its process parameters define the quality of the spunlaid nonwoven to be produced.

In the following, the bonding process is looked upon with regard to the peculiarities spunlaid nonwovens show. For more, see Sections 6.1 to 6.5. Several processes available, economic criteria such as throughput performance, working speed, working width are of interest. Giving a variety of bonding processes, Table 4-6 describes the most common mass per unit areas. With spunlaid nonwovens, the most frequently used bonding process is, due to the high share in products with low mass per unit areas, thermal bonding by means of calenders. This process is very well suitable to bond multilayer SM nonwovens.

Calendered with combinations of rollers smooth/engraved, spunlaid nonwovens are mostly punctually bonded, the surfaces to be bonded being between 5 and 25%. With lower mass per unit areas, such nonwovens are relatively soft, stiffness growing with growing mass per unit area. With mass per unit areas $>120 \text{ g/m}^2$, there is a risk layers may unbind in case calender engraving did not sufficiently penetrate them. Recent research results show such nonwovens can be calendered in a depth-engraving way when engraving speed is low and/or the material is preheated [111]. Spunlaid nonwovens bonded by means of calendering show the following properties:

– low thickness/voluminosity
– high density of the nonwoven
– medium strength
– low air permeability
– relatively poor suction behaviour
– medium water and water vapour permeability

**Table 4-6** Mass per unit areas with regard to a variety of bonding processes

| Mode of bonding | | Mass per unit area [g/m²] | Preferable mass per unit area [g/m²] |
|---|---|---|---|
| Thermal | calendering | 5–120 | 5–80 |
| | thermofusion | 60–500 | 80–400 |
| Mechanical | needle-punching | 60–1,000 | 100–400 |
| | stitch-bonding/meshing | 80–1,000 | 100–400 |
| Hydrodynamic | | 20–400 | 30–200 |

Owing to their properties, such nonwovens, which are preferably made from poly-propylene, are used for hygienic and medico-textile purposes.

With higher mass per unit areas, thermofusion is frequently used in connection with needle-punching [112]. Higher mass per unit areas require needling rather than calendering. As compared to products bonded by means of calendering or thermofusion, needle-punched spunlaid nonwovens are softer, more voluminous and more absorbent. Except for the choice of stitch-bonding process and the kind of needles, stitch density and stitch depth largely influence strength and elonga-tion behaviour.

Investigation into the needle-punching of spunlaid nonwovens shows that with mass per unit areas ranging from 150–300 g/m$^2$, the optimal stitch density is be-tween 200 and 300 stitches/cm$^2$. Higher stitch density entails the risk the fila-ment is done damage to [88]. Spunlaid nonwovens are continuously needle-punched at speeds ranging from 10–40 m/min. For this purpose, several serially connected stitch-bonding machines are used in order to reach the stitch densities necessary.

Additionally bonded by means of thermal treatment, needle-punched spunlaid nonwovens show high punching strength. In particular, they are applicable in the building industry and as geotextiles [112]. For examples of products see [113]. Both stitch density and stitch depth influence the width of pore opening as well as horizontal and vertical water permeability or, more generally said, the filter properties [88]. So far, meshing spunlaid nonwovens is of minor importance since, due to the low working speed of the stitch-bonding machine, the process has to be discontinuous.

An advanced variant of hydroentangling webs by means of high-pressure water beams allows an adaptation to the spunlaid process. Up to recently, hydroentan-gling has exclusively been used with fibrous nonwovens. Hydroentangling means the filaments are entangled horizontally and partially, vertically in the nonwoven cross-section.

The hydrodynamic bonding of spunlaid nonwovens and multilayer composites is described in a number of patents [114, 115]. Well-known manufacturers of hy-droentangling machines offer the necessary equipment [116]. In particular, enter-prises that are active on the market as both manufacturers of equipment and of nonwovens and, consequently, have the complex know-how give hope a continu-ous spunbond/spunlace system will be launched on the market soon [117].

To this end, an increase in the working speed of hydroentangling machines up to 600 m/min and a rise in tool pressure up to 600 bars are essential [116].

Recent research results have shown that hydroentangled spunlaid nonwovens from polypropylene with mass per unit areas between 35 and 120 g/m$^2$ are char-acterized by higher strength and better elongation behaviour and, consequently, better shaping behaviour as compared with thermally bonded and needle-punched spunlaid nonwovens [98]. With mass per unit areas ranging from 90 to 200 g/m$^2$, punching strength has been improved as compared with needle-punched spunlaid nonwovens. The same applies to softness. With higher mass per unit areas, re-cent results show the specific energy input is not yet sufficient. With low mass per unit areas, prospects are good to hydroentangle spunlaid nonwovens soon. Here, the qualities wanted are easier to achieve.

Speaking about spunlaid nonwovens, chemical bonding with baths of binding agents is of minor interest. It is still found where spunlaid nonwovens are used to make garments. Mechanical, thermal and chemical bonding can also be used in combination, in particular, in order to meet certain requirements with regard to strength and natural stability. To give two examples: roofing sheets and insulation material made this way are known to the building industry (see Section 15.3.2) [118, 119].

Strength and elongation behaviour in MD/CD direction can be changed so as to reach isotropic strength [120]. For this purpose, the spunlaid nonwovens can be drawn biaxially, after-treatment following. This entails a reduction in the mass per unit area, which means less material is used. For ways into finishing and equipping the nonwovens see Part III.

### Nonwoven composites

The manufacture of extrusion nonwovens also allows to make nonwoven composites to continuous processes. Within the recent 15 years, this has been of considerable importance with regard to novel uses of spunbonded nonwovens. Combining processes, nonwoven composites can be made whose single layers all contribute with their individual properties to meet the requirements.

Except for hygienic and medical uses, such composites have found a market as geotextiles, filtration textiles and protective textiles. In particular, this requires proper strength and stability as well as the porosity, permeability and blocking performance wanted.

As shown above (describing the processes of the manufacture of spunlaid nonwovens), processes are combined if that is helpful to spin the filaments, to draw them or to make them into a web.

Taking spunlaid nonwoven material for a basic component, the following combinations with other materials are known:

- one or multi-layer spunlaid nonwovens (SS, SSS)
- one or multi-layer meltblown nonwovens (SM, SMS, SMMS) [121]
- films (SF) [85]
- meltblown nonwovens and films (SMF) [85]
- short fibre pulp with or without meltblown nonwovens
- fibrous nonwovens

Spunbonded nonwovens are also made into composites using claw mats, plastic grids, thread composite, cloth or warp-knitted fabrics. To give an example, Fig. 4-70 shows a plant with two spunbonding units laying down the web onto one belt (REICOFIL process).

Fig. 4-62 shows an example of the combination SMS. Such lines have also become known as Multidenier process [122]. Spinning filaments or fibres of a variety of finenesses, it is possible to reach a gradient of porosity or a graded barrier which allow to use these nonwoven composites as depth filters [123]. The combination of spunlaid

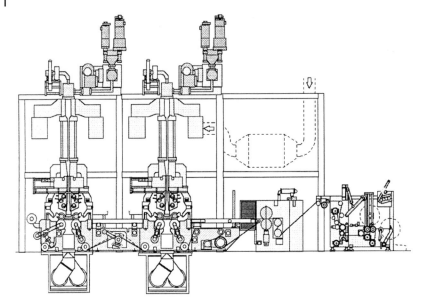

**Fig. 4-70** Double REICOFIL spunbonding machine
at Reifenhäuser Maschinenfabrik GmbH & Co.

nonwoven material, coating material and composite drawing is a new technological
solution. Fig. 4-71 shows the process elements with regard to the VAPORWEB composite. Directly after the thermal bonding of the spunbonded nonwoven, the composite is coated and then biaxially drawn [124]. Inserting calcium carbonate into the
polymer melt used to coat the composite (which takes place after drawing), it receives
a micro-porous water vapour blocking structure. The composites are well suitable for

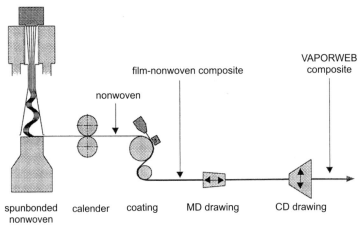

**Fig. 4-71** Process elements of the VAPORWEB technology

use as coverstock materials in hygienic articles (see Chapter 10). Combining spunbonded nonwovens with short fibre pulp, absorbency can be influenced [125].

Combinations of spunbonded nonwovens with claw mats, plastic grids or other structures of thread are widely used in the building industry and as geotextiles. The same as they may reinforce spunbonded nonwovens, fabrics of threads are also suitable to reinforce composites of spunlaid nonwovens. They are, for instance, used as bituminated roofing sheets or in the automotive industry. Lightweight spunlaid nonwovens are also found as nonwovens covering carbon-fibre reinforced composites.

Section 4.2.1 on raw materials already referred to the manufacture of spunlaid nonwovens from biologically degradable polymers. The production of degradable spunbonded nonwovens from these polymers will certainly lead to novel nonwoven composites whose purpose is to serve the user for no more than a limited period of time. It is here that certain filter materials [126] as well as agrotextiles and geotextiles [62] belong.

## 4.2.4
### Processes to make fine-fibre spunbonded nonwovens

The extrusion of thermoplastic polymers also allows to make spunbonded nonwovens with a fibrous structure. There is no strict differentiation between the terms fine fibres and finest fibres. With regard to polypropylene, the example following below shows what filament diameters should be assigned to the individual finenesses.

| | |
|---|---|
| 1.0 dtex | $\cong$ 11.9 µm |
| 0.1 dtex | $\cong$ 3.8 µm |
| 0.01 dtex | $\cong$ 1.3 µm |
| 0.001 dtex | $\cong$ 0.4 µm |

With spunbonded nonwovens of low mass per unit areas, it is advantageous to make very fine filaments of finenesses ≤1 dtex, e.g. for hygienic articles, agrotextiles and filter materials. To spin filament finenesses of 0.5 dtex means a great challenge to both process equipment and technology (speaking of polypropylene, this corresponds with a filament diameter of 8 µm).

Consequently, the common process equipment used in mono-component spinning from nozzles reaches its limits. This is certainly the reason why, together with the development of spunlaid nonwovens, research was carried out to find novel processes that allow to produce fibres of very great fineness and to process them into spunbonded nonwovens. They comprise

– the meltblowing process
– the flashspinning process
– electrostatic spunbonding

The first two of these processes are now industrially used.

### 4.2.4.1 **Meltblowing process**

McCulloch describes how the production of nonwovens from thermoplastic finest fibres to the meltblowing process developed [127]. The beginnings date back to Exxon at the end of the 60s, a one-stage process being the first step. Licences were granted to Kimberley-Clark, Johnson & Johnson, James River, Web Dynamics, Ergon Nonwovens a.o. Reifenhäuser GmbH, too, took a licence from Exxon to develop and build spunbonding machines.

In the meltblowing process, finest fibres of diameters of 1–5 μm are made from a polymer melt [49], the melt being exposed to hot air directly at the nozzle mouth. This causes a fibre-air mixture. Fibre drawing is accomplished at high air velocity of 6,000 to 30,000 m/min, which depends on melt condition and temperature as well as the fibre shape wanted. Air temperature is adapted to melt temperature in order to draw the polymer at high temperature. Below the mouth of the nozzle, air turbulences are created leading to fibre break-down and, thus, to an air-fibre mixture. The air intermingles with the ambient air so it cools the fibres. The fibres are now very fine and are laid down in random arrangement onto a perforated belt. The air is sucked off. Making very fine fibres requires the melt at the nozzle exit is of very low viscosity. The higher the melt temperature, the lower its viscosity. At the same time, air velocity can be reduced. Air velocity and air temperature largely define fibre properties and web quality. Higher velocity causes a greater drawing effect at the fibre and raises the fibre-air turbulences at lay-down. Not far from below the nozzle, fibre formation is complete. This is why, contrary to the spunlaid process, the distance between nozzle and belt can be kept small.

In the meltblowing process, the design of the melt blow head is of particular importance. Fig. 4-72 describes the basics [128].

Most important is the design of the pipes that guide the melt and hot air directly at the nozzle mouth of the blow head. Generally, the nozzles are arranged in a row, the hot air emerging from air slots at high velocity. With regard to the manufacture of meltblown nonwovens, the European patent 067 4035 A2 of Kimberley Clark Corp. describes both the process and the nozzle head with the pipes that guide the melt and hot air [129]. Fig. 4-73 shows the basic equipment and Fig. 4-74 a cross-section of the blow head.

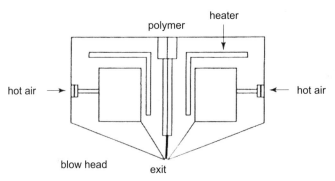

**Fig. 4-72**  Basic meltblowing nozzle [128]

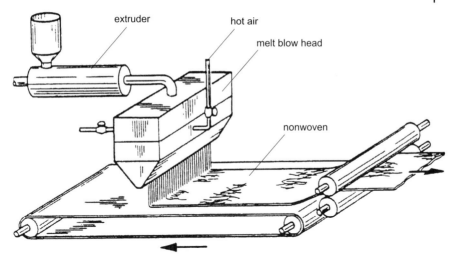

**Fig. 4-73** Meltblowing process [129]

With a nozzle head to Kaun [130], the slots of the air nozzle are found on both sides of the polymer nozzle. The nozzle slots are made up of an air lip on each side. Together with the lip edges opposite the wedge-shaped polymer nozzle, they form the slots of the air nozzles. Gerking [131] suggests a variant of design which embeds each melt-supplying borehole of the spinning nozzle in a ring-shaped blow-nozzle. The incurred concentric flows of gas support drawing. Frey [132] uses the nozzle plate temperature as a process variable as well as slot width at the point where the air-fibre blend escapes and the air-flow resistance at the lay-down system to control both process and/or the mass per unit area of the web. Follow-

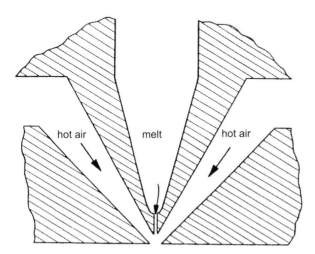

**Fig. 4-74** Cross-section of a meltblowing nozzle [129]

ing Balk [133], the meltblowing process is also well suitable to make finest fibres from two or more polymer components and to make them into webs. Adhesion between the single components being small, the meltblown fibres can be split into their single components and laid down as a web. This allows to vary the design of the nonwoven of finest fibres and to adapt it to the properties wanted. Wadsworth and others [134] speak about test results with regard to nonwovens from bi-component fibres made to the meltblowing process on a novel REICOFIL bi-component plant. Reifenhäuser Maschinenfabrik GmbH & Co. [135] also use the melt-bowing process to make spunlaid nonwovens. Process parameters are set so as to create filaments of a diameter of below 100 µm. Afterwards, the thermo-bonded web can be heated up to drawing temperature and biaxially drawn (100 to 400%). Then, the spunlaid web is stabilized by means of thermosetting.

### 4.2.4.2 Flashspinning process

This process was developed by DuPont. The Tyvek spunbonded nonwovens made to the process are produced exclusively by DuPont at two locations. The annual output in 1997 was 55,000 t. The process allows to make very fine fibres ranging from 0.5 µm to 10.0 µm [136]. Following this process, polyethylene of high density is heated in the autoclave to make a boiling solution of more than 200 °C hot. Tri-chlorofluoromethane or freon are suitable solvents. Dissolving is accomplished at high pressure (4,000 kPa to 7,000 kPa) [128]. Fig. 4-75 shows the basic process. At pressure as high as this, the solvent, in a controlled way, is set free. The solvent

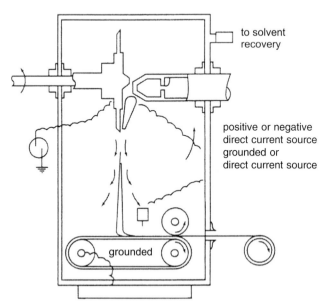

**Fig. 4-75**   Basic flashspinning process [128]

evaporating, a network of very fine fibres is created of diameters as described above. Following this, the fine-fibre web created undergoes after-treatment.

Depending on what equipment is intended, very dense and solid nonwovens or, if preferred, soft ones can be made. The nonwovens developed by DuPont for particular uses are well-proven e.g. in the filter-producing industry and in the manufacture of protective clothing (see Chapter 14.2).

### 4.2.4.3  Electrostatic spunbonding

Electrostatic spunbonding means a polymer solution or polymer melt is, at very high voltage, exposed to an electric field, broken down into very fine fibres and then made into a web. Fig. 4-76 shows the basic process.

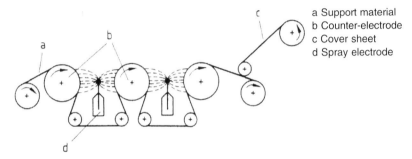

a Support material
b Counter-electrode
c Cover sheet
d Spray electrode

**Fig. 4-76**  Electrostatic spunbonding process [137]

At high voltage and in the electric field between counter-electrode and spray electrode, the melt or solution is sprayed onto a support material to create a fibre. The web created is covered by means of a cover sheet [137, 138]. The diameters of the fibres made at high voltage of between 5 kV and 20 kV range from 500 nm and 40 nm, which means they show a great surface-volume ratio. Thus far, this interesting process is not used in industry yet.

However, prospects are good for its worldwide use in the manufacture of protective clothing since web structure is well suitable for this application and the materials made are easy to equip. This technology is also suitable to spray the nonwoven onto three-dimensional shapes so it may be the base to make moulded articles from spunbonded nonwovens in a direct process.

### 4.2.5
### Processes to make film nonwovens

Film nonwovens are nonwovens based on the extrusion of films in one-stage or multi-stage continuous processes. The films are converted into fibre net-like fabrics, then made into a web and mechanically bonded to create film nonwovens.

Following the patent and specialist literature, there are a number of technological solutions. However, as compared to the production of spunlaid and spunbonded nonwovens, they are not widely used in industry yet. The following focuses on technologies which are either being developed or already used in practice.

Making film fibre materials, the starting point is the extrusion of flat or blown tubular films. The polymer granulates are melted on in a way similar to spunlaid nonwovens, using one or several extruders with worm diameters D ranging from 45 to 160 mm. Then, the polymer melt or, respectively, melt blend is, mostly by means of spinning pumps, transported to the flat or blow film nozzles. The films are extruded using flat or round slots, which are preferably 0.4 to 0.8 mm wide. These nozzle slots may be smooth or show a profile [139]. Except for the manufacture of one-layer films, multi-layer films have been made for a time. With them, analogous to filament spinning, several flows of melt supplied by several extruders are united in the nozzle tools [140]. Film design corresponds with the *side by side* design known from filaments. The polymers used to extrude the films are characterized by higher viscosity or, respectively, a lower melting index MFI than the types of raw material known from filament spinning (see Section 4.2.1). The manufacture of film nonwovens started as early as the end of the 60s. At that time, the raw materials used were preferably polypropylene and polyethylene. Polyamide and polyester were also processed [139]. Later on, these polymers were supplemented by polymer blends, which allowed to directly influence the processing properties of the films. Polymer blends may be distinguished by chemically compatible and incompatible systems. Chemically compatible blends, e.g. of different polyolefins, serve to improve drawing behaviour and to raise strength. Chemically incompatible polymers cause a matrix-island structure, which is helpful in the splitting process known in the manufacture of film nonwovens. They cause a material-specific uncontrolled inclination to split. Examples can be found in [141] and [142]. The main components used were polypropylene, polyamide 6, polyamide 6.6 and polyester. These polymers were blended as well as the main components, such as polystyrene with other polymers [141]. It was found strength decreases at a share in secondary components of more than 10%. Inclination to split, however, is best at a share in secondary components ranging from 20% to 80%, 50/50 blends being the optimum. Later on, polyester/polyethylene blends were industrially used in the connection of splitting by means of needle rollers [143], which the following will say more about. For polyester components, recycled materials were used.

Making film nonwovens from multi-component (preferably bi-component) films, the same polymers or combinations of polymers are found as with one-layer films. One well-proven process is known as the Baroflex process [141]. In ref. [144] a corresponding production plant is mentioned, which will be described further below. The bi-component process – similar to bi-component filaments – focuses on the attenuation of the individual fibres and/or, with thermal bonding, on a binder fibre effect or a crimp effect.

With film nonwovens, for instance, a crimp effect is most wanted, which is caused after the films have been split [141, 145]. It is important the components can easily be split. At the same time, they should stick to each other so they can crimp due to their different shrinking behaviour. This can be influenced by the choice of polymers used [141]. The films emerging from the nozzles, they are cooled and drawn. Film thickness is usually between 0.05 mm and 0.2 mm while the drawing ratio is about 1:4 to 1:12. Similar to filament spinning, crystallization and orientation takes place. Strength grows several times as compared with extruded films not drawn.

Contrary to the spunlaid process, drawing is accomplished as much as exclusively by means of galettes or roller drawing systems. Given a wide drawing slot, a contraction in width is found and a decrease in thickness, which can be described like this:

$$b_V = \frac{b_U}{\sqrt{V_V}}$$

$b_V$ = width of drawn film [mm]
$b_U$ = width of extruded film [mm]
$V_V$ = drawing ratio

$$d_V = \frac{d_U}{\sqrt{V_V}}$$

$d_V$ = thickness of drawn film [mm]
$d_U$ = thickness of film not drawn [mm]

Drawing the film in a tight drawing slot, the width of the film is maintained and the drawing ratio is fully effective with regard to the reduction in thickness. Extensively, [139] and [146] talk about this matter, e.g. in the context of the production of film ribbons. In this place, no more shall be said.

Presently, a number of processes is preferably used to make film nonwovens:

– splitting the films by means of needle rollers [143, 144, 147]
– embossing the films and then, splitting them by means of monoaxial or biaxial drawing [148, 149]
– loaming the films and splitting them after extrusion (caused by additives inserted prior to splitting)

The process most often used is splitting, which is also known as fibrillating. It uses needle rollers whose circumferential speed is higher than the speed of passage of the film. The proportion between the two speeds is the fibrillation ratio $V_F$ as described in the equation

$$V_F = \frac{V_{NW}}{v_F}$$

$V_F$ = fibrillation ratio
$V_{NW}$ = circumferential speed of the needle roller [m/min]
$v_F$ = speed of passage of the film [m/min]

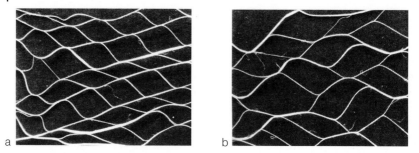

**Fig. 4-77** Net-like structure of films. Fibrillation ratio: a) 1.4:1; b) 1.6:1

The fibrillation ratio is usually 1.3:1 to 8:1. [144] gives a fibrillation ratio of 5:1 to 8:1.

Except for the fibrillation ratio, the following parameters influence the splitting effect:

– diameter of the needle roller
– needle density per cm of width
– angle of attack of the needles
– way in which the film is exposed to needle roller operation
– number of needle bars all over the circumference

The splitting process causes net-like structures of varying finenesses of links and lengths of slots (Fig. 4-77). Frequently, link finenesses are called single fibre finenesses even if the networks are endless. The cross-section of the single fibres is square and wider than with aerodynamic drawing or with the galette drawing of filaments [145, 146]. The average of the single fibre unit ranges, generally, from 15 to 70 dtex, which is higher than spunlaid nonwovens [151]. Using bi-component films the components of which crimp differently, a crimp effect can be achieved after splitting and exposing the films to heat (Fig. 4-78).

With direct processing, the split web of film is the same as a raw web in the spunlaid process, laid down onto a belt and transported to a rigger, which cross-laps the

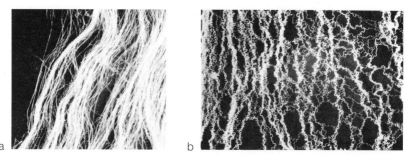

**Fig. 4-78** Fibre net of films, a) smooth, b) crimped

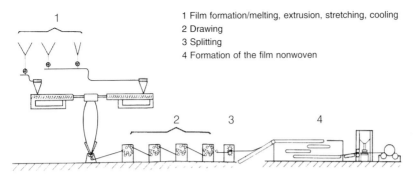

1 Film formation/melting, extrusion, stretching, cooling
2 Drawing
3 Splitting
4 Formation of the film nonwoven

**Fig. 4-79**  Process to make a film nonwoven (schematic)

raw web like a carded web and takes it, for instance, to a needle-punching machine so it can be mechanically bonded. Fig. 4-79 shows the basic process (schematically).

In another novel process, the film is drawn and split as is common. However, contrary to cross-lapping as described above, the web of film is directly taken down via an air channel and laid down onto a conveyor belt. In this process, the raw web is entangled in a diffuser (Fig. 4-80). In the process shown in Fig. 4-80, the net-like arrangement of fibres is laid down in loops onto the belt, the crosswise links created by splitting the film being maintained. This results in an advantageous MD:CD ratio of strength. The inventors of the process described are applying for patent [152].

In another process, splitting the film is accomplished by means of a stitch-bonding machine and an integrated needle roller [143, 147]. Subsequently, this machine bonds the film nonwoven inline, which means cover-seaming. This is why these nonwovens usually show greater longitudinal strength. A number of enterprises in Great Britain and the USA have been making film nonwovens based on embossed films [148, 149]. The films from preferably polyethylene of high density are embossed after extrusion and prior to drawing. The films, alternatingly, show thicker and thinner areas. The latter, as predetermined breaking

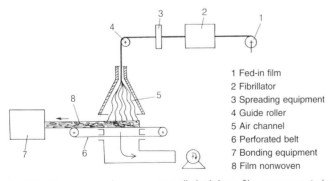

1 Fed-in film
2 Fibrillator
3 Spreading equipment
4 Guide roller
5 Air channel
6 Perforated belt
7 Bonding equipment
8 Film nonwoven

**Fig. 4-80**  Process to make a pneumatically laid down film nonwoven (schematic)

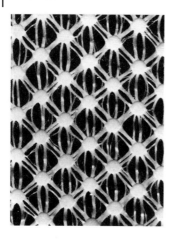

**Fig. 4-81**  Embossed film net structure

points, cause splitting or defibring. Fig. 4-81 shows such a grid-like film nonwo-ven. Further processes to make nonwovens of this kind are found in [153]. They are based on foliated extrusion or grid extrusion.

Film nonwovens made by means of embossing are mostly used as bonding layers between other fabrics. Their low melting point, under the impact of heat and pres-sure, makes the thinner areas melt, the thicker ones act as points of bonding. Film nonwovens can also be made to what is known as the burst process [150]. A foaming reagent is added to the polymer melt which sets free a gas when being extruded. Emerging from the radial nozzle, the film is air-cooled and a film of foam is cre-ated. The biaxial drawing process that follows results in a tubular web.

A similar material is created extruding a film of 70% polyethylene and 30% polyester with a matrix-fibril structure. The crimped film is defibred by means of an aerodynamic webber [154].

A two-stage extraction with xylene leaches the polypropylene as a matrix compo-nent out of the web. Depositing a cover layer, it is possible to make the web into synthetic leather.

Following [142, 144], fields of application are, for instance:
- oil separators
- geotextiles
- permeable to water membranes
- textile tapestry

These nonwovens are also well-known to concrete reinforcement. The net-like materials made from embossed films are mostly used as coverstock nonwovens or bonding components to make multi-layer composites. With nonwovens made of polymer blends that mainly consist of polyester, a higher resistance to heat as compared to polyolefins is specifically used. So far, they are frequently found on the market as roofing sheets, carpet backing and as components contributing to flocked floor covering [143].

# References to Chapter 4

[1] Leifeld F (1982) Fortschritte beim Öffnen und Kardieren, Vortr 20. Internat Chemiefaser-tag 25.09.81 Dornbirn, Melliand Textilber 63

[2] Leifeld F (1983) Neue Technologie für die Vorbereitung von Krempelvliesen, Vortr 22. Internat Chemiefasertag 08.06.83 Dornbirn, Melliand Textilber 64

[3] Leifeld F (1986) Meßmethoden zur Ermittlung der Gleichmäßigkeit von Karden- und Krempelspeisevliesen, Internat Textil-Bull, Weltausgabe Garnherstellung 1

[4] Schlichter S (1986) Maßnahmen zur Verbesserung der Gleichmäßigkeit des Vorlagevlieses an Baumwollkarden, Internat Textil-Bull, Weltausgabe Garnherstellung 1

[5] Schlichter S (1989) Aktuelle Entwicklungen in der Vorbereitung von Fasermaterial für die Vliesstoffherstellung, Vortr 3. Hofer Vliesstoffseminar

[6] Leifeld F (1993) Perfektes Mischen vom Ballen bis zum Band, Vortr Beltwide Cotton Con-ferences 10.–14.01.93 New Orleans

[7] Schlichter S (1993) Moderne Faservorbereitung, Vortr Index 93 Kongreß Genf

[8] Leifeld F (1995) Abfall-Recycling in der Spinnereivorbereitung, Vortr Ecotextile Conference 11.–14. April 95, Internat Textil-Bull 3/96

[9] Leifeld F (1997) Qualität beginnt mit der Vorbereitung – Neueste Entwicklung für die Herstellung von Krempelvliesen, Vortr XXV Internat Vlies-Kolloquium 22.–23. Oktober 97 Brno/CS

[10] Leifeld F (1998) Vliesprofilregelung – Steigerung der Vliesgleichmäßigkeit von aerodyna-misch gebildeten Vliesen und Krempelvliesen, Vortr Nonwovens Symp 4.–5. Juni 98 Paris

[11] Prospekt der Spinnbau GmbH Bremen (1999)

[12] Damgaard T (1996) Garniturkonzepte für Krempel und Öffner im Vliesstoffbereich, 11. Hofer Vliesstoffseminar Nov 96

[13] Grothe H (1876) Technologie der Gespinnstfasern, Vol 1: Die Streichgarn-Spinnerei und Kunstwoll-Industrie, Springer-Verlag Berlin

[14] Bernhardt S (1986) Grundsätzliche Funktionen zum Krempelprozeß, Hofer Vliesstoffse-minar Nov 86

[15] Bernhardt S (1990) Optimierung der Faserlage MD:CD, INDEX Kongress 1990

[16] Krusche P (1999) Grundlage des Krempelns, Internes Material der Spinnbau GmbH

[17] Bernhardt S (1999) Qualitätsbeeinflussende Merkmale an Hochleistungs-Vlieskrempeln, Vortr Vliesstoffmaschinen Symp des VDMA, April 99

[18] Firmenprospekt INJECTION CARD der FOR Biella/Italien (1999)

[19] Schäffler M (1999) Moderne Vlieslegertechnologie, ITB Vliesstoffe-Techn Textilien 2: 56–58

[20] Fuchs H (1999) Anlagen zur Bildung von Faservliesen, ITB Vliesstoffe-Techn Textilien 3: 49–50

[21] Knaus R (1974) Die aerodynamische Vliesbildung, Chemiefasern Text Ind 24/96: 209–212, 215–218

[22] Knaus R (1976) Die aerodynamische Vliesbildung, Textilbetrieb 94, 3: 36–42

[23] Lünenschloß J, Albrecht W (1982) Vliesstoffe, Georg Thieme Verlag Stuttgart/New York

[24] Jakob H (1993) Neue Entwicklungen in der aerodynamischen Vliesbildung und der Vernadelungstechnologie, Melliand Textilber 74: 291–293

[25] Vliesmaschinen und Wirrvlieskarden, Firmenschrift Dr E Fehrer AG Linz/A (1995)

[26] Web forming and random carding, Firmenschrift Dr E Fehrer AG Linz/A (1997)

[27] Jakob H (1990) Aerodynamic web-forming of lightweight nonwovens, Text Technology Internat 7: 138–141

[28] Jakob H (1989) Random web cards – New techniques and opportunities in the aerody-namic web-forming sector, Vortr XXI Internat Vliess-Kolloquium Brno/CS

[29] Jakob H (1989) Erfahrungen mit der Wirrvliestechnologie, Melliand Textilber 70: 185–189

[30] Pühringer S (1992) Wirrvlieskarden und Nadelfilzmaschinen, in Loy W (1992) Taschen-buch für die Textilind, Fachverlag Schiele & Schön Berlin: 367–381

[31]  DOA, Firmenschrift Dr Otto Angleitner GmbH & Co. KG

[32]  Laroche, Firmenschrift Laroche S. A.

[33]  The Front Runners, Firmenschrift Rando Machine Corporation

[34]  Innovative textile Technologien für Recycling & Nonwoven, Firmenschrift H. Schirp GmbH & Co. KG

[35]  Holliday D (1995) Heighloft Nonwovens Update, Heighloft's 95, Charlotte NC (1995)

[36]  Krčma R, Jirsák O (1990) Heißluftgebundene Textilien mit neuer Struktur, Chemiefasern Text Ind 40/92

[37]  Krčma R (1976) Neue Erkenntnisse über die Struktur, Parameter und Eigenschaften von Vliesstoffen, Textiltechnik 26, 6: 347–351

[38]  Krčma R (1963) Textilverbundstoffe, Fachbuchverlag Leipzig: 241

[39]  Krčma R, Jirsák O, Hanus J, Saunders T (1997) Nonwovens Industrie 28, 10: 74–78

[40]  Krčma R., Jirsák O, Hanus J (1997) Fortschritte bei der Entwicklung hochvoluminöser senkrecht verlegter Textilien, Techn Textilien 40, 1: 32–34

[41]  Jirsák O, Hanus J, Plocarová M (1992) Dresdner Textiltagung 92

[42]  Krčma R, Jirsák O, Hanus J, Macková J, Plocarová M (1994) 75th World Conference of the Textile Institute, Atlanta

[43]  Jirsák O, Hanus J, Lukás D (1996) INDATEC'96, Washington DC–Crystal City

[44]  Guandique E, Katz M (1964) Non-Woven-Fabric, US-Patent 3117055

[45]  Kinney GA (1967) Process for Non-Woven Filamentary Structures from Fiber Forming Synthetic Polymers, US-Patent 3338992

[46]  Hartmann L (1970) Process for Producing Non-Woven Fabric Fleeces, US-Patent 3502763

[47]  Dobo EI (1971) Lightweight Nonwoven Web of Increased Opacity, US-Patent 3630818

[48]  Stevensen PI (1971) Process for Forming Lightweight Non-Woven Web, US-Patent 3607543

[49]  Cheng CY, Permentier D (1999) Effects of resin properties on processing and properties of PP nonwovens, Chem Fibers Intern 10: 384–387

[50]  Scott ND (1996) Metallocenes and Polypropylene – New Solutions of old Problems? Paper 3 World Textile Congress on Polypropylene in Textiles Huddersfield/UK

[51]  Blechschmidt D, Fuchs H, Vollmar A, Siemon M (1996) Metallocen-catalysed Polypropylene for Spunbond Applications, Chem Fibers Intern 10: 332–336

[52]  Techn Datenblatt Pegatex-MICRO S, Pegas a. s. Znojmo/CS (1999)

[53]  Techn Datenblatt ASPUN Fibre Grade Resins Dow Europe SA/CH (1999)

[54]  Ermittlung der Herstellungsbedingungen und Eigenschaften von PP-Fasern und PP-Spinnvliesstoffen für Geotextilien, die aus Mischungen von Originalrohstoffen und Recyclaten bestehen, Schlußber AiF 10117B Sächs Textilforschungsinst eV Chemnitz (1996)

[55]  Kauschke M (1992) Verwendung eines photoselectiven Spinnvliesstoffs zum Bleichen lebender Pflanzen, Vortr 4. Internat Techtextil-Symposium Frankfurt am Main

[56]  Hartmann L (1973) Spinnvliestechnik und industrielle Anwendungen von Spinnvliesen, Melliand Textilber 6: 605–608

[57]  Hoffmann M, Beyreuther R, Vogel P, Tändler B (1999) Thermisch-mechanisches Verhalten schmelzgesponnener Ethylen-Octen-Copolymerfäden, Vortr 16. Stuttgarter Kunststoff-Kolloquium

[58]  Nonwovens Markets, Internat Company Profiles Miller Freeman Inc/USA (1997)

[59]  Bogaert JP, Coszach P (1999) Polylactid Acids: New Polymers For Novel Applications, Paper Index 99 Congress Geneva/CH

[60]  Fritz H-G (1999) Die Rolle der Stärke bei der Generierung innovativer Polymerwerkstoffe, Vortr 2. Internat Symposium Werkstoffe aus nachwachsenden Rohstoffen Erfurt

[61]  Yamanaka K (1999) "LACTRON", a biodegradable fiber, its development and applications, Paper 38th Internat Man-Made Fibres Congr Dornbirn

[62]  Ehret P (1996) Biologisch abbaubare Vliesstoffe, ITB Vliesstoffe-Techn Textilien 03: 29–30

[63]  Spunbond-Anlagen (1999) ITB Vliesstoffe-Techn Textilien 03: 61–62

[64] Ward DT (1998) Immer breiteres Angebot an Spunbond-Anlagen, ITB Vliesstoffe-Techn Textilien 02: 27–33

[65] Jakob W, Michels C, Franz H, Berger W (1974) Die Spaltung von Polymermischungs-folien – ein neuer Weg zur Herstellung von Foliefaserstoffen, Faserforsch u Textiltechnik 25, 6: 229–234

[66] Gerking L (1974) Spinnvlies-Herstellungsverfahren und Besonderheiten, Melliand Textil-ber 4: 326–330

[67] Hartmann L (1974) Widening applications of Spunbonds, Textile Manufacturer 9

[68] Kittelmann W, Jossa W, Lindner R (1994) Möglichkeiten und Grenzen zur Erhöhung der Filament-Geschwindigkeit bei der Herstellung von Spinnvliesstoffen nach dem Unter-druckverfahren, text praxis internat 3: 149–155

[69] Gerking L (1976) Berechnung der aerodynamischen Verstreckung beim Spinnvliesprozeß, Verfahrenstechnik 12

[70] Chen CHH, White JL, Spruiell JE (1983) Dynamics, Air Drag and Orientation Develop-ment in the Spunbonding Process for Nonwoven Fabrics, Text Res Journ 1: 44–51

[71] Beyreuther R (1991) Grenzen der Elementarfadenfeinheit beim Schmelzspinnen, Mell Textilber: 795–799

[72] Beyreuther R, Brünig H (1994) Abschlußbericht des Inst für Polymerforsch Dresden eV für STFI Chemnitz 30.01.94

[73] Gerking L (1993) Änderung der Fadeneigenschaften vom Polymer her und in der Spinn-linie, 32. Internat Chemiefasertagung Dornbirn

[74] Misra S, Spruiell JE (1992) Investigation of the Spunbonding Process Via Mathematical Modelling, INDA Journ of Nonwovens Research 5: 13–19

[75] ITS-Charts: Spunbond-Anlagen, ITB Vliesstoffe-Techn Textilien 02: 30–31 (1998)

[76] Fourné F (1993/1995) Polypropylen-Fasern: Herstellung, Eigenschaften, Einsatzgebiete, Chemiefasern Text Ind 43/95: 811–820

[77] Tokunaga A (1998) Stand der Technik bei der Spinnvliestechnologie, ITB Vliesstoffe-Techn Textilien 04: 21–24

[78] Gneuß D, Filtrieren von Kunststoffschmelzen, 2nd edition, Feist & Lehbrink Hiddenhau-sen

[79] Siebwechseleinrichtungen zur Verarbeitung von Thermoplasten, Prospekt Maschinenfabr J Kreyenborg GmbH & Co Münster-Kinderhaus

[80] Patterned Embossed Nonwoven Fabric, Cloth Like Liquid Barrier Material and Method for Making Same, US Patent 55 99 420 (1997)

[81] Navone M (1999) Faré: new machines for the production of staple fibers and bicomponent fibers, Chem Fibers Intern 49: 238

[82] Brown DL, Wagner CS (1992) Effect of bicomponent filament bonding of colback nonwo-vens in industrial application, Konf Einzelber: INDA-TEC 92, The Internat Nonwovens Technical Conf, Ft. Landerdale USA: 233–248

[83] Balk H (1999) 100 Spinnvliesanlagen – ein besonderes Jubiläum bei Reifenhäuser, Reifen-häuser-Nachrichten, Ausgabe 27, 11

[84] Weger F (1999) New spunbond line for virgin and recycled PET, Chem Fibers Intern 49, 5: 442–443

[85] Kunze B (1998) Vliesproduktion mit Spunlaid-Technologien, ITB Vliesstoffe-Techn Texti-lien 03/98: 41–45

[86] Vorrichtung zur Herstellung von Spinnvliesen mit Hilfe von Längsdüsen, DE AS 15 60 790 (1971)

[87] Tuftingträger aus Spinnvliesstoff, DE 36 42 089 A1 (1988)

[88] Mechanisch verfestigte Verbundstrukturen für das textile Bauwesen, vorzugsweise unter Verwendung von Filamentvliesstoffen, Schlußbericht AiF-Forschungsvorh 10544B Sächs Textilforschungsinst eV Chemnitz (1998)

[89]  Untersuchungen über den Einfluß der Filamentgeschwindigkeit auf die Vliesbildung und die Isotropie der Festigkeitseigenschaften nach dem Unterdruckverfahren, Schluß-bericht BMWi-Forschungsvorh 181/93 (1994)

[90]  Fourné F (1988) Spinnvlies: Herstellungsverfahren und Anlagen, Chemiefasern Text Ind 38/90, 8: 691–696

[91]  Verfahren und Vorrichtung zur Förderung und Ablage von Scharen endloser Fäden mittels Luftkräften, EP 05 98 463, Karl Fischer Industrieanlagen GmbH Berlin (1994)

[92]  Butler J (1999) Worldwide Prospects for Spunbonds, Nonwovens World 8/9: 59–63

[93]  Wehmann M (1990) Neue Produkte durch Spinnvlies- und Melt-blown-Systeme, Chemiefasern Text Ind 40/92, 7/8: 756–757

[94]  Malkan SR (1994) An overview of spunbonding and meltblowing technologies, Konf Einzelber Nonwovens Conf Proc, Orlando/USA: 31–37

[95]  Harris WS, Moller A (1998) The possibilities for really Soft nonwovens using 0,5 denier Bicomponent spunbonds, Konf Einzelber Insight 98, Fiber and Fabric Conf Orlando/USA: 5.1–5.9

[96]  Watzl A (1999) Mikrofasern, Allgem Vliesstoffrep 4: 37–40

[97]  Matsui M (1998) Bikomponenten- und Mikrofasern, Kettenwirkpraxis 02/98: 13–18 and 03/98: 23–28

[98]  Blechschmidt D, Brodtka M, Lindner R (1999) Die Wasserstrahlverfestigung von Spinn-vliesen – eine innovative Verfahrenskombination, Vortr 14. Hofer Vliesstofftage

[99]  Mothes A (1995) Spunlaid technology combining fine count fibers. New web laying systems and their competitiveness, Konf Einzelber Internat Nonwovens Symp Bologna/It: 1–8

[100]  Tokunaga A (1999) Spunbond Nonwovens Plant: Risk Management and Analysis, Nonwoven World 08/09: 64–69

[101]  Malkan SR, Wadsworth L (1992) Ein Rückblick auf die Spinnvliestechnologie, INB Nonwovens 03/92: 4–12 and 04/92: 24–34

[102]  Verfahren zur Herstellung eines Vlieses aus Endlosfäden und Vorrichtung zur Durch-führung des Verfahrens, DE 35 41 128 A1 (1987)

[103]  Reif KA, Lasch G, Heinze A (1973) Entwicklung eines Elementarfadenvliesstoffs, Textiltechnik 23, 2: 82–87

[104]  Blechschmidt D, Lindner R, Lieberenz K (1998) Verbundstrukturen aus Filament- und Faservliesstoffen für technische Einsatzgebiete, Vortr 13., Hofer Vliesstoffseminar

[105]  Weger F (1995) Neues Saugluftverfahren zur Herstellung von Spinnvliesstoffen, ITB Vliesstoffe-Techn Textilien 03/95: 52–55

[106]  Kunze B (1997) Das Programm der „Reicofil"-Spinnvliesanlagen, Reifenhäuser-Nachrichten, 25, 09

[107]  Verfahren zur Herstellung von Spinnvliesen mit erhöhter Gleichmäßigkeit, DE 35 42 660 A1 (1987)

[108]  Verfahren zur Herstellung von gleichmäßig verteilten Filamenten aus einem Spinnfilament und Herstellung des Spinnvlieses, EP 04 53 564 B1 (1995)

[109]  Verfahren zur Steuerung der Anisotropie von Spinnvliesen, DE 44 28 607 A1 (1995)

[110]  Vorrichtung zum gleichmäßigen Ablegen der Elementarfäden beim Herstellen eines Vliesstoffs im Direktverfahren, DD-PS 261 179 B5 (1988)

[111]  Untersuchungen über den Einfluß der Kalandrierungsbedingungen einschließlich der Bindepunktgestaltung und des Flächengewichts auf die Eigenschaften von Bautextilien aus Filamentvliesstoffen, Schlußbericht BMWi-Forschungsvorh 547/98 Sächs Textilforschungsinst eV Chemnitz 12/99 (1999)

[112]  Watzl A (1997) Verbesserung der Eigenschaften von Geotextilien durch Thermofixierung, Vliesstoff Nonwoven Internat 127: 230–232, 251–253

[113]  Geotextilien Polyfelt TS, Prospekt Polyfelt GmbH Linz/A

[114]  Verbund Spinnvlies mit Wasserstrahlen verfestigt und das Herstellungsverfahren, EP 05 57 678 B1 (1996)

[115]  Composite nonwovens web material and method of formation thereof, EP 05 77 156 A2 (1994)

[116]  Fechter TA (1999) Die Wasserstrahlverfestigung von Spinnvliesen – Anlagentechnik und Umsetzung, Vortr 14, Hofer Vliesstofftage

[117]  Völker K (1999) SPUNjet – Ein neues Spinnvlieskonzept von IBCT Perfojet, Allgem Vliesstoffrep 5/6: 13–14

[118]  Fischer K (1996) Polyester-Spunbond mit Polymerbinder für Dachbahnen, Techn Textilien 39: 60–65

[119]  Trägerbahn für Dachunterspannbahnen, DE 38 31 271 A1 (1990)

[120]  Verfahren zur Herstellung von Spinnvliesen, EP 00 13 355 (1980)

[121]  Madsen JB, Axelsen MS (1998) Improving nonwovens performance by using multiple spunmelt layers in advanced composite structures, Konf Einzelber EDANA 1998 Internat Nonwovens Symp, Technol, Fibers & Webs, Hygiene, Treatment & Bonding Paris/F: 12.1–12.14

[122]  Fahmy T (1995) Multidenier – Beispiel eines fortschrittlichen Spinnvlies-Verbundstoffes, Vliesst Nonwovens Internat Vol 10, 10: 288, 291

[123]  Neue Filtermedien von BBA Nonwovens, Vliesst Nonwovens Internat 131: 122 (1997)

[124]  Kunze B (1999) A new approach to breathable structures, VAPORWEB – leightweight breathable nonwoven film product in line, Vortr INDEX'99 Congr Geneva/CH

[125]  Composite nonwoven web material and method of formation thereof, EP 05 77 156 A2 (1994)

[126]  Flächiges, biologisch abbaubares Trägermaterial für Denitripikanten in biologisch betriebenen Klärstufen, DE 42 20 795 A1 (1994)

[127]  McCulloch W, John G (1999) Nuclear origins of sub-micron extrusion, Nonwoven report intern 11: 29

[128]  Nonwovens Markets 1996, International Factbook and Directory, Miller Freeman, Inc United News & Media San Francisco/USA (1996)

[129]  Mc Dowall DJ (1995) Polyethylen meltblown fabric with barrier properties, EP 0674035 A2

[130]  Kaun A (1994) Düsenkopf für eine Anlage zur Spinnvliesherstellung nach dem Meltblown-Verfahren, DE 4238347 A1

[131]  Gerking L (1991) Verfahren und Vorrichtung zur Herstellung von Feinstfäden, EP 0455897 A1

[132]  Frey D (1994) Verfahren und Vorrichtungen zur Herstellung eines Spinnvlies-Flächenproduktes DE 4312309 A1

[133]  Balk H (1991) Verfahren zur Herstellung von einem Vlies aus Spinnfasern aus thermoplastischem Kunststoff, DE 3927255 A1

[134]  Wadsworth LC, Sun Q, Zhang D, Zhao R, Kunze B (1999) Enhanced barrier performance of bicomponent Fiber meltblown nonwovens, Nonwovens World 8/9:40-46

[135]  Joest RH, Geus HG, Balk H, Kunze B, Schulze H (1996) Verfahren zur Herstellung einer Vliesbahn aus thermoplastischen Polymerfilamenten DE 19501123 A1

[136]  Nonwoven Markets (1997) International Company profiles, Miller Freeman Inc. United News & Media, San Francisco/USA

[137]  Weghmann A (1982) Production of Electrostatic Spun Synthetic micro fiber nonwovens and applications in Filtration, Nonwovens Ind 11: 24–32

[138]  Gibson P, Schreuder-Gibson H (1998) Electrospinlacing of fibers, Industrial Textiles, Univ Clemson USA, Jan 28–29

[139]  Peuker H (1969) Fäden aus Folie, Textilindustrie 71,) 2: 82–88

[140]  Krejci M (1986) Technologie und Anlage zur Herstellung von Foliefasern und deren Einsatzgebiete, Textiltechnik 36, 11: 601 ff

[141]  Gayler J (1974) Einige neue Aspekte der Splitfasertechnologie, Vortr 13, Internat Chemiefasertag Dornbirn/A

[142] Berger W, Kammer HW (1977) Physikalisch-chemische Grundlagen der Fibrillierung, Vortr 16, Internat Chemiefasertag Dornbirn/A

[143] Merkel K, Schaller G (1978) Das Florofol-Verfahren – ein effektiver Weg zur Herstellung von Foliefaserstoffen, Textiltechnik 28, 6: 359–361

[144] Jezek H, Swoboda L (1991) Polypropylen-Bikomponentenfasern für textile Wandbespannungen und andere Heimtextilien, Chemiefasern Text Ind 41/93: 1191–1194

[145] Badrian W (1971) Polypropylene Film Fibre Applications and Developments, Vortr Konferenz Fasern aus Folie, Trencin/CS

[146] Grundlegende Untersuchungen des Zusammenhanges zwischen der Gleichmäßigkeit fibrillierter Folienfäden und der Fadenbruchhäufigkeit bei der Verarbeitung zu technischen Geweben, Schlußbericht zum AiF-Forschungsvorh 247D, Sächs Textilforschungsinst eV Chemnitz (1994)

[147] Frenzel B, Schaller G (1988) Folieverarbeitung auf Nähwirkmaschinen zur Herstellung textiler Bodenbeläge, Melliand Textilber 5: 338–342

[148] Lennox-Kerr P (1968) Textilien aus Folien, Textilindustrie 70, 5: 331–334

[149] Roussin-Moynier Y (1987) Manufacturing Process for Nonwoven Webs and Associated Composites, Konf Einzelber Index'87 Kongress: p 1 ff

[150] Malkan SR, Wadsworth LC (1992) Ein Rückblick auf die Spinnvliestechnologie Teil III, INB Nonwovens 04/92: 24–34

[151] Lünenschloß J, Albrecht W (1982) Vliesstoffe, Georg Thieme Verlag Stuttgart/New York, p 37

[152] Verfahren und Vorrichtung zur Herstellung von Vliesstoffen aus spleißfähiger Folie, PA-Anmeldung 199 39 084.B (1999)

[153] Rasmussen OB (1989) Fibriller network coverstock material, Konf Einzelber 2nd Internat PIRA conference, Nonwoven Absorbency 89, Aarhus/DK, Vol 2: 1–13

[154] Berger W, Gräfe F (1992) Herstellung ultrafeiner Fasern mittels Folientechnologie: Neue Wirkungsprinzipien – neue Produkteigenschaften, Vortr 4, Internat Techtextil-Symp, Neue Perspektiven für technische Vliesstoffe – Aktuelle Entwicklungen im Materialbereich Frankfurt/Main

# 5
# Wet lay method
H. PILL, K. AFFLERBACH

The wet lay method can be used to form a web from all fibres which can be dispersed in fluids. The following are characteristics of the wet lay method [1]:

– very good product homogeneity
– versatility in the product scale
– high production

The most important products which are produced by the wet lay method are listed in Table 5-1.

The use of machine groups which resemble or are similar to papermaking machines in the sectors ranging from material preparation to batching shows that, because of the close relationship in their manufacture and the end product, their external appearance and composition, these wet-laid web products can be designated as much as webs as 'long-fibre paper'. The concept 'filter paper' is frequently used in practice.

**Table 5-1** Wet-laid webs [2]

| Special papers | Industrial nonwovens | Nonwovens resembling textiles |
|---|---|---|
| Synthetic fibre paper | Nonwovens for roofing | Medical underwear |
| Air filter paper | Nonwovens for carpets | Bedlinen |
| Liquid filter paper | Nonwovens for filters | Serviettes |
| Cigarette paper | Nonwovens for separators | Hand towels |
| Overlay paper | Cleanroom filters | Hygiene products |
| Sausage wrapping paper | Webs for PCBs | Textile interlinings |
| Screen (stencil) paper | Nonwovens for coating | Nonwoven furnishing fabrics |
| Vacuum cleaner bag paper | | |
| Teabag paper | | |

## 5.1
## Principle of the method

The manufacturing of nonwovens by the wet lay method is derived from paper-making. The following stages in the process are typical:

– dispersion of fibres in water
– continuous web forming on a wire cloth through filtration
– consolidation, drying and batching up the web

Fig. 5-1 shows the individual production stages.

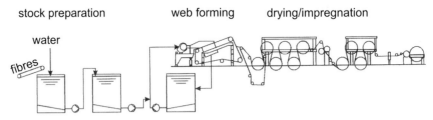

Fig. 5-1   Production stages in the wet lay process

## 5.2
## Development

The first ideas of producing a web on obliquely running belt are set down in US Patent No. BP 11394 dating from 1848 [3]. In the thirties the American company Dexter attempted to produce Japanese hand-made paper by mechanical means. After the long fibres required for the product had been found, laboratory rests found that a great deal of dilution was required for these fibres to be turned into uniform sheets. An attempt to use such a dilute suspension to produce paper on a conventional endless wire papermaking machine failed, however, there was success when the wire was set obliquely. In the course of the next few years several papermaking machines at Dexter were converted to inclined-belt machines, as were some under licence at Crompton in England. Many long-fibre papers were developed before the Second World War [4]. The developments at Dexter were their own, and were available to only a few companies under licence.

In the sixties of the 20th century various manufacturers of papermaking machines started to adapt the method for the mechanical production of paper and webs largely to the technical requirements of textile fibres. The developments went back to wish of the papermakers to produce also long fibres (abaca, sisal) and synthetic fibres (polyester, nylon) in addition to the fibres used at that time.

The use of long fibres requires special measures at all stages of the process which differ from conventional methods. Crimp-free or slightly crimped fibres with lengths of 5 to 30 mm are preferred when man-made fibres are employed.

Fibres of this type, and also natural raw materials such as unground cellulose pulp, cotton linters or hemp-based pulp, require much greater dilution than in the production of conventional paper for dispersal in an aqueous medium – both at the preparatory stage and during sheet production. Densities of 0.2 to 0.8 g/l are needed in order to produce homogeneous webs from long fibres. The relatively high dilution is the reason why inclined-belt machines are used for the processing of long fibres, and not conventional fourdrinier machines. If papermaking machines were to be used, forward ends with lip openings of 100–200 mm would have to be used. As controlled sheet formation is no longer possible with such large openings, conventional papermaking machines are not suitable for the processing of long fibres.

The road from the first open inclined-belt machines to the hydraulically enclosed forward ends led through various stages of development. The sheet formation pressure had to increase with an increase in the production rate, which caused the forward end to build up sheet formation pressure appropriate to the production.

The development stages were as follows:
– the suspension flows horizontally at the inclined belt
– a baffle is set up at the end of the sheet formation zone
– the sheet formation zone is screened off by a front wall and an air scoop
  is created to build up the pressure
– the inclined belt is sealed, and the air pressure increased, by an air cushion
– hydraulic sealing of the inclined belt

Until now the majority of inclined-belt machines were built by J. M. Voith, Heidenheim, and Voith Sulzer Papiermaschinen, Dueren, formerly Doerries. Their hydroformers can come in single or two-layer designs. Valmet Sandy Hill with the Deltaformer and Bunderhaus with the No-Wo-Former were further well-known suppliers of inclined-belt machines until recently..

## 5.3
## Raw materials and fibre preparation

On the basis of many years of experience and very many experiments using an inclined-belt laboratory machine from PILL WET LAY WEB TECHNOLOGY, using a very wide range of materials, it is possible to state that almost all fibre types can be laid to form a web. The dispersing behaviour of the fibres and the fibre length are important for the product. Up to now the following fibre types have been processed on this laboratory machine in different lengths, thicknesses, states of preparation and blends, in single or two layers:

- cellulose pulps          a wide range of materials
- man-made fibres        viscose rayon, polyester, nylon, polypropylene, polyacrylate
- binder fibres            PVA, copolyesters, bicomponent fibres

- high-tech fibres          ceramic, aramids, polyuron, PTFE, carbon
- mineral fibres             C+E glass, microglass, rock wool
- metal fibres               titanium, special steel, steel
- natural fibres            flax, jute, hemp, kenaf, bast, peat, cotton, coir, ramie
- waste and recycled fibres   textiles, leather, shearing dust, carpets

**Special aspects of fibrous raw materials**

The dispersibility of fibres in a fluid or liquid is very important for the wet lay method. The raw materials (fibre bundles, fibre staple, cellulose pulp plates) should be capable of being split up homogeneously into single fibres in the preparation mechanism before dispersion to form a suspension and remain uniformly distributed while being transported to the web forming stage.

Many fibres do not satisfy this requirement. They do not allow themselves to be separated or they become knotted together in the suspension to produce clumps. The result is an irregular distribution in the web with visible knots and a non-uniform sheet structure.

Fibre dispersing behaviour is essentially influenced by the following factors:

- the fineness ratio, calculated on the basis of the fibre length
  and the fibre fineness
- the fibre stiffness in a liquid medium (wet modulus)
- the type of crimp
- wettability
- fibre staple quality

The dispersing properties deteriorate with an increase in the fibre fineness ratio and a decrease in fibre stiffness. Furthermore, the fibre type and fibre dimensions are also criteria which determine fibre processing. Normally fibre groups containing non-crimped fibres are easier to turn into a homogeneous suspension than long, fine, fibrillated or crimped fibres. These are more inclined to form clumps. The points enumerated here mean that, with a few exceptions, only fibres with a length of less than 30 mm can be processed by the wet lay method. In the textile industry staple fibres of this length relate to the short-staple sector. The fibres used in similar papermaking methods have a maximum length of 5 mm.

Experience during recent years has shown that, as a rule, fibres with good wettability also have good dispersing properties and other fibres types tend to flocculate. With the large number of new fibres and various modified properties now available, these statements can no longer be looked at generally. The fibre surface, fibrillation, the viscosity of the liquid and the types of mechanisms used to produce the suspension are also crucial. The result is that synthetic fibres for the wet lay method are provided with spin finishes by the manufacturers which improve their wettability. For this reason wetting agents or other auxiliaries are added to the fibre suspension. In the case of many of the 'heavy' fibres such as a glass, metal and mineral fibres – and other, similar fibres – it is therefore essential to thicken the liquid so that the fibres remain constantly floating in the suspension until the web is laid.

One typical fibre fault is badly cut fibres. They include fibres, the length of which may exceed the actual staple length several times. Because the cutting blades have become blunt the fibre ends end up being hook-shaped and the free fibres on both sides form a 'double-knot'. In tow cutting many fibres in the tow may become welded together through squeezing at one end and then resemble a brush.

The demand for perfect staple comes about because of the realization that only a small number of these excessively long fibres can be the cause of lumps and knots in the suspension. These faults can be detected by using optical detectors on a laboratory inclined-belt machine. Thus fault-free staple quality can be ensured in the majority of cases. Filaments which have become stuck together can be largely separated using chemicals, but this solution cannot be applied to fused filaments.

## 5.3.1
### Fibre types

More and more high-tech fibres have been developed in the last few years as a result of new technologies. In addition the advantages of natural fibres have been recognized again, and they have become topical in that they are renewable resources.

The properties of fibre types available are described in Part I, only brief details will therefore be given of the types used in the wet lay method.

### Vegetable fibres from annual plants

In recent years the range of fibres which can be used in the wet lay method e.g. cotton, cotton linters, flax, hemp, and abaca, has been extended to the special types such as mulberry bast, sisal, ramie, jute, coir, kenaf and pulps derived from straw and esparto.

Of these raw materials Manila hemp and cotton linters are important for wet laying. Both types of fibres, finished by kier-boiling and a subsequent bleaching process, are used. Their dispersing properties are good.

Cotton plays only a subordinate role, as it cannot be easily dispersed because of its length and, in some cases, pronounced crimp. It is occasionally found in much shortened form in wet-laid nonwoven fabrics. In this case these fibres lend bulk to the fabric and impart elasticity and a soft surface, high density, high absorbency, good opacity, good dimensional stability and resistance to ageing in the nonwoven. They are well suited for filter papers and blotting paper.

Flax and hemp papers are used both in the production of paper and wet-laid nonwoven fabrics. They are easy to fibrillate and are suitable for the production of tough and crease-resistant nonwoven fabrics or composite materials for door linings, spoilers and dashboards in vehicles.

Abaca fibres or Manila hemp are light yellow to white in colour, with a silky lustre, and can be dyed, they are very strong, are light and highly resistant to moisture. Strong, tough nonwovens are produced by using these fibres. They are used for fine, thin nonwovens, e.g. for teabags, wrapping paper and fuse cord paper.

Mulberry bast fibres (*Broussonetia papyrifera* – paper mulberry tree) fibrillate very easily. The fibres occur in two forms – thick fibres with pointed cell ends and ribbon-like, frequently twisted flat types. These fibres produce strong and tough nonwoven fabrics with a textile-like handle, e.g. for decorative Japanese paper, gift wrapping paper, paper used by florists, and nonwovens for teabags.

Ramie fibres consist of almost pure cellulose (approximately 72%) and fibrillate very easily. These fibres produce strong, tough nonwovens with low static charge levels.

Jute fibres can be fibrillated, but are not suitable for high-grade paper webs. They have a high density, are not easy to bleach, but otherwise are very similar to flax fibres.

Kenaf fibres resemble jute, but are stronger and have a higher lustre. They are used relatively rarely in Europe.

Straw pulp is mostly only used for paper because of the compact paper surface, good writing qualities and resistance to erasing. Straw fibres are added to produce decorative effects.

Esparto pulp (alfa grass) is mostly used for paper on southern Europe only to give high bulk, softness and absorbency.

Coir fibres have a smooth surface and are very resistant to mechanical wear, moisture and rotting. They float in water due to the high level of air inclusions. They are used for open-pore nonwovens which are subject to high levels of stress.

**Animal fibres**

Wool is resistant to acids, but not to alkaline solutions. Dyed fibres are used as melange fibres. They are further used for blotting paper, roofing felts, calendar bowl paper or wool felt board.

*Leather fibres*

Dry leather waste is cut up into $1 \text{ cm}^2$ pieces and then caused to disintegrate in wet condition at approximately 5% stock density using disc refiners (Asplund refiner RPO). This raw material is frequently blended with 10% other fibres. Blending with a very wide range of fibres is possible. Long dwell times are necessary for sheet formation and drying, because it is very difficult to extract water from the fibres. When drying it should be noted that the fibres cannot be heated to above 70°C, as the leather starts to decompose then and the material becomes brittle. The products are used in the production of shoes and pursemaking fabrics.

**Vegetable fibres based on wood**

Like all fibres of vegetable origin, cellulose pulps are a mixture of fibres having an irregular length distribution, depending on the end-use, different types of pulp are used in the wet lay method.

The basic substance in the fibres is cellulose. Depending on plant type, paper pulps further contain different fractions of lignin, hemicellulose and resin. Alpha fibres are almost free of these impurities.

The cellulose pulps available on the market are mostly conifer- or deciduous-based celluloses pulps. These two types differ with respect to fibre shape and dimensions. Conifer-based fibres are approximately 2–4 mm long and 0.02–0.07 mm thick. Fibres from deciduous trees are approximately 1 mm long and 0.03 mm thick. Nether type gives any dispersing problems and is easy to transport in suspension.

As a rule, cellulose pulps are processed together with other fibrous materials. In this case the cellulose pulp performs very different tasks:

– it is cheap material
– it acts as a carrier for binder particles which are bonded to the fibre surface
– as binder fibres (formation of hydrogen bonds) to achieve preliminary bonding which improves the wet strength and/or bonds the dry web together so well that it can be fully consolidated outside the machine in some cases
– to improve the water absorbency and opacity of nonwovens

Cellulose pulps can also be produced from the following and are used in accordance with the geographical location of the paper web manufacturer:

– long-fibre conifer cellulose pulp from pine, fir, spruce, mulberry bast
– short-fibre deciduous cellulose pulp from birch, beech, poplar, eucalyptus, abaca

**Cellulosic staple fibres**

Short-staple viscose rayon fibres are one of the fibre types which are most frequently used in the wet laying method. Use is mainly made of the non-crimped fibre type which has good dispersing properties. Because of their good properties they are also blended with other fibre types.

Cuprammonium rayon fibres resemble viscose rayon fibres. They are very fine, have a smooth surface and a circular cross-section, good dyeing properties and high absorbency.

**Staple fibres from synthetic polymers**

As far as the fibres belonging to the group are concerned, polyester and nylon fibres are primarily used to form webs, whereas polypropylene and various mixed polymer fibres are used to achieve heat-sealing properties, and polyvinyl alcohol, polyvinyl alcohol copolymer and bicomponent fibres are used for web consolidation.

Synthetic fibres have a low water retention capacity and require a special finish to improve their dispersing properties. If a good finish is applied, these are just as good as – or – because the fibre stiffness is somewhat greater – better than those of viscose rayon fibres.

Polyester and nylon are also made more available in crimped form for wet laying. Fibres with two-dimensional crimp have proved themselves useful. They are particularly suitable for the production of bulky nonwoven fabrics.

### Synthetic cellulose pulp

This raw material comprises fibres from low-pressure polyethylene. It is produced as early as the polymerization stage, resembles cellulosic fibres in shape, but is thermoplastic (melting point 130–135°C) and has worse wetting properties. However, its processing properties during wet laying are good when blended with other materials. It has, on one hand, the function of a binder fibre and, on the other, it is used in the production of heat-sealable products.

### Inorganic fibres

Glass fibres are at the leading end of this product range. They are used in, amongst other things, bitumen-coated roofing fabrics, weatherboards and separators are used to carry the coating in carpets: all products which have to be rot-resistant and dimensionally stable. In the main, E- and C-type glass are processed.

C-glass is an alkaline glass (alkali/calcium/borosilicate) with an increased boron trioxide content and a high chemical resistance.

E-glass is an alumino-borosilicate glass (calcium/aluminium/borosilicate) with 1% alkali. It is used in glass-fibre-reinforced composites and electrical technology.

The poor wettability necessitates special measures for fibre finishing. A fibre cannot be used unless it has been finished. As a rule the fibres which are available today have been finished in such a way that the fibre packets open up during dispersing. In many cases, however, the finish is insufficient to maintain this state up to sheet formation. The fibres accumulate again and form clumps. This can be prevented by adding wetting agents and dispersing auxiliaries to the suspension. The task of the latter is to form a protective colloid on the fibres and these from lying on top of one another again during the preparatory phase. Nonionic surfactants and carboxymethylcellulose (CMC), guar gum and hydroxyethylcellulose, amongst others, have proved their values as auxiliaries.

Microglass fibres are characterized by relatively high stiffness on one hand and poor wettability on the other. The former is crucial because the fibre lengths and fineness ratios which are permissible for the wet lay method are greater here than for man-made fibres. The filter webs produced are used for heat resistant (up to 700°C) and chemically-resistant fine filters, for liquid media (hydraulic oils, fuel), HEPA paper for sterile clean-room filters, separator paper, graph paper, gasket elements, thermal insulation, and capacitor paper. Dispersing is usually carried out in acidified water (up to pH 3). Microglass fibres can also be used in combination with other fibres (cellulose pulp, ceramic) to improve separation and the rate. When dry material is processed it should be noted that skin irritation may occur and that fine fibres may penetrate into the lungs.

Rock wool fibres are produced from liquid rock by the centrifuge method. A mixture of fibres having different lengths and thicknesses is produced. Generally speaking, however, the fibre dimensions are in a range which permits processing by the wet lay method without any special fibre preparation. They are used in insulating materials.

*Carbon fibres*

The nonwoven fabrics produced are used for heat resistant (up to 1500 °C) and chemically-resistant products, e.g. air and liquid filters, protective clothing, and insulation for electrical products. Blends with other fibres are possible.

Ceramic fibres are used for very highly heat-resistant nonwoven fabrics (up to about 1500 °C) thermal electrical insulation (very high electrical resistance) and in aerospace technology. Organic binders burn when first used. In dry processing it should be noted that skin irritation may occur.

Metal fibres are used for the screening of electromagnetic waves and radar, particularly in aircraft and aerospace technology [6]. Examples here are stainless steel and titanium fibres with lengths of 3–50 mm.

### 5.3.2
### Production of short-staple fibres

Special cutting machines are available for producing short-staple fibres. A distinction is drawn between two methods: the chopper and the circular cutter method.

With the chopper method a very precise cut is achieved via oblique means. Depending on the size of the machine, tows of flat, continuous material in the range of approximately 2–20 million dtex are cut to any specified length. The cutting rate is variable between 50 and 400 cuts/min, because of the nature of the system it is not possible to cut directly off the spinning machine.

On the other hand circular cutting machines, as the name already states, cut at high speeds directly off a spinning machine using a circular cutter. However, only specific lengths can be cut. These are determined by the pitch of the rim and, in some cases, also by the construction itself. Lengths below 6 mm are uncommon; 6 mm, 12 mm etc. are the lengths cut as a rule.

The advantage of these machines is that very high production levels can be achieved and that crimped materials can be processed, which is not the case with the chopper method.

### 5.3.3
### Binders

Binder fibres and synthetic dispersions can be used for bonding wet-laid webs. These are used either in preliminary bonding or at the full consolidation stage.

Preliminary bonding requires that the binder is used during the suspension preparation phase. It is a partial consolidation stage phase which imparts so much strength to the web that it can be subjected to further treatment stages when dry.

Full bonding is effected when the web requires its full strength for further processing. In this case the binder can be added to the stock and/or applied to the preconsolidated web.

## Binders for preliminary bonding

*Cellulose pulp*
To be effective as a preliminary binder, the pulp content of the fibres used must be 20–50%. Its use is therefore confined to products where a high pulp content is required or can even be tolerated. The low wet strength of the pulp-bonded non-woven fabric is a disadvantage.

*Cellulosic binder fibres*
Fibrillated-film and hollow fibres, as well as 'dispersion fibres', give an acceptable strength if they are processed without any further non-bonding components.

*Polyvinyl alcohol fibres*
Used in conjunction with cellulosic and synthetic staple fibres, it is enough to add 1–5%: with glass-fibre webs fractions of approximately 10% are necessary. The fibres can be prepared without any problems. Convective dryers are ruled out for drying purposes. Contact dryers, where light pressure is applied to the web during drying, have proved to be useful. The tensile properties and drapeability of the nonwoven are determined, on one hand, by the binder fibre fraction, and by the water content of the web during drying. If the web is too wet on reaching the fibre dissolution temperature (dry contents – below 25% – according to DIN 6730, ratio of the mass of a sample dried under specified test conditions to the mass before the start of drying), the fibres dissolve completely and a film is formed. If the dry contents are the higher the fibres are only slightly dissolved and retain shape. Punctiform bonding is established. Under these conditions the nonwoven is less strong, but softer. The wet strength of the consolidated nonwovens is thus low because the binder fibres are soluble in water. If *synthetic pulp* is used, satisfactory preliminary bonding can be achieved with a binder fraction of approximately 10%. It should be noted that, depending on the type, the softening point is relatively low (60–135 °C).

*Binder dispersions*
The synthetic dispersions added to the fibre suspensions are mostly acrylates or binders based on butadiene/styrene/acrylonitrile. These are anionic substances. Ethylene/vinyl acetate dispersions are also used. The dispersions can be added to fibres as follows:

- In combination with precipitants: with this method a cationic precipitant is initially added to the fibre dispersion. This is absorbed by the fibres, with the result that coagulation occurs at the fibre surface when the suspension is subsequently added. Waterproofing agents which are known in paper manufacture are suitable as precipitants. As these are preferentially absorbed by cellulosic fibres, coagulation mostly occurs on these fibres.

- As a precoagulated dispersion: in this case coagulation is carried out separately from the fibres in a special machine. The precipitants are alum or aluminium sulphate. The coagulate particles thus produced have a fibrous shape.

- As a perisuspension: these are produced by peripolymerization. The binder accumulates in the form of globular droplets and is prepared in this form by the manufacturer in an aqueous medium.

Adding the synthetic dispersion to the stock preparation unit of a wet web-laying machine ensures that the binder is distributed uniformly in the web. This is retained in the finished nonwoven, as coagulated binders do not migrate during drying. The level of nonwoven fabric strength which can be attained with the same binder content is much lower than when the binder is applied to the web. The surface strength is better if the binder is added to the fabric surface. The binder content is between 10 and 25%.

### Binders for full bonding
Melt-bonding fibres include nylon copolymer, polyester copolymer, bicomponent, heteropolymer and polyvinyl alcohol fibres.

With the exception of polyvinyl alcohol fibres, the fibres are used exclusively for the production of special types of nonwovens. They act as binders and also impart special properties to the webs. Because of their shrinkage properties, bicomponent and nylon copolymer fibres produce bulky products having adequate strength.

Synthetic dispersions and solutions can be applied to the finished nonwoven in the same way as in the production of nonwovens by the dry method.

Synthetic resin solutions based on urea formaldehyde, polyvinyl alcohol and polyvinyl acetate are preferably used for consolidating wet-laid glass-fibre webs. It is important that these fibres should guarantee high strength, resistance to high temperatures and also, in part, low solubility in water. To ensure the requisite end-use properties for the product, these binders are often used in combination. The binder add-on is of the order of 20% relative to the dry fibre content.

### 5.3.4
### Fibre preparation

As already mentioned in the case of the raw materials, many types of these fibres have to undergo mechanical or chemical pretreatment.

### Mechanical preparation methods
These are mostly the same methods as used in paper manufacture. The fibres are supplied in partially opened bundles: pulps come in pressed plates or in bales as one loose stock.

The task of the stock preparation mechanisms is to 'break down' these fibrous materials into single fibres in an aqueous medium, to mix together the individual raw material components, and to transport the resulting suspension to the forward end of the wet-laying inclined-belt machine while keeping it free of clumps and flakes. Excessive turbulence in the fibre suspension or the deposition of fibres on rough container walls and sharp edges in the machines may lead to lumps. Specific criteria are used for selecting the machine components. In the case of machines which process both pulp and long fibres, the stock preparation stage is divided up into a pulp short-fibre line, a synthetics long-fibre line and possibly also into a waste line, e.g. in the production of nonwovens for filters.

Pulpers are used for breaking down solid pulp sheets and bales with consistencies of 3–8% and for fibres which require strong turbulence. The fibre bundles are converted into a pumpable suspension in the pulper while water is added. Pulpers work continuously or discontinuously. They are large mixing vessels from acid-resistant steel. A sharp, frequently helical, conical rotor is used as the beater. The pulp fibres are fed into the next unit through a perforated plate in the floor of the container. Depending on the type of material, the consistency and the container size, the dissolution time is 10–30 min. Pulped fibres are never free from clusters.

Conical centrifuges or centrifugal strainers are used to remove solid material. They are used to deaerate and clean the pulp or to remove impurities from glass fibres. The equipment for removing solid material operates best at stock densities of 3–6%, whereas equipment for removing finer material operates best when this density is below 1%.

Cluster removal equipment is used to break any clusters present in the pulped slurry down into individual fibres. The construction of this equipment resembles that of a rotary pump. Instead of a runner/impeller, use is made of perforated discs or toothed rims with set clearances of 0.5–2.0 mm, cluster removal is carried out using very high turbulence levels without damaging the fibres and at stock densities of 3–6%.

*Grinding machines*

In many cases cellulose pulps must be ground before use. This grinding is a form of finishing the fibre and its purpose is to make the fibre suitable for turning into a nonwoven. To obtain the appropriate fibre lengths for nonwovens (long fibre paper) the pulp is only lightly ground:

| | | |
|---|---|---|
| up to 30°SR | coarse material | for coarse, open-pore webs |
| 31 to 50°SR | moderately ground material | for fine-pore webs |
| over 51°SR | wet-ground material | unsuitable for webs |

The degree of grinding (SR) is, according to Schopper-Riegler, a factor which characterizes the hydroextraction behaviour of a fibre suspension [7].

Refiners come in the following designs: flat and steep cone, one- and two-disc refiners.

Conical refiners have a cast-iron conical housing (stator) with a knife box. The bladed rotor which rotates in this housing is mounted on an axially-displaceable shaft. The gap is adjusted by displacing the cone. In contrast to beaters, the whole of the grinding surface is fully utilized in the case of refiners. Thus a considerably higher level of efficiency is reached for the same amount of manpower.

Disc refiners come in two versions: single- and double-disc. In the single-disc variety the stator disc is fixed and the rotor disc is axially displaceable. In the double-disc refiner one stator disc is firmly fixed to the housing and the second disc is axially displaceable. The rotor disc is not fastened to the shaft and is moved into position by the displaceable stator disc. The flow is parallel. For grinding lines several refiners are frequently connected in series or parallel.

Beaters, which were formerly common in the paper industry, are now being superseded by refiners. As before, beaters are ideal for natural fibres, as they fibrillate the fibres more than shorten them. A beater consists of an oval concrete or steel trough with a blade and a basalt or lava baseplate. A rotating height-adjustable bladed roller, similar to a water wheel, is mounted over this. The blades are made of bronze or acid-resistant steel.

Containers for dissolution, intermediate storage and mixing, so-called vat-type containers, are produced from special welded steel or concrete in different sizes and shapes. In a propeller built horizontally into the side wall sucks the fibre suspension up to the vat wall and then circulates back again. The propeller blades can be welded in or they may be adjustable. The shape of the vat, the propeller speed and its shape are crucial for the mixing or dispersing. Depending on the type of fibre, the stock density is between 0.5 and 4.0%.

Special vats are provided for the dissolution of the still-wet web in the case of a web break and for sprayed edges. These containers are usually located below the end of the inclined-belt end. Two or more propellers mounted on a shaft with bearings at both ends is used as the circulating unit.

With open systems screen water containers are located below the inclined-belt machine. The circulating water from them is taken into intermediate storage. The mixing pump pumps diluting water from them into the forward end. The water returning from the hydroextracting elements is recycled back into them.

## Pumps

*Eccentric worm pumps*
Because of the risk of lumps, eccentric worm pumps are frequently used as circulating pumps for transporting long-fibre suspensions. The pumps, which operate in accordance with the rotary displacement principle, have a regular, low-pulsation flow with a minor shear effect. These pumps are therefore also preferably used for the forwarding and feeding of chemicals and binders.

The pumps are self-aspirating up to a pressure of 0.9 bar and can be supplied for pressures of up to approximately 48 bar at delivery rates of up to approximately 6 m$^3$/min.

Mode of operation and feed principle: the single-gear eccentric worm rotates and oscillates within a fixed stator. The hollow spaces which alternately open and close in continuous sequence as the rotor rotates, continuously transport the goods from the suction to the pressure side. As a result of the geometry and the constant contact between the two feed elements, sealing bands are produced which provide a seal between the suction and pressure sides. This gives the pump its high suction capacity and facilitates a high pressure build-up, irrespective of the speed.

Circulatory pumps are the best known pumps in the pulp and paper industry. They are used for the transport of water-like media or cellulose pulp suspensions. Nowadays new type of impeller also enables long-fibre suspensions to be used. Their use in the pumping of glass-fibre suspensions started. Today this pump has been used in connection with more and more fibre types.

The main elements are the spiral housing and the open or enclosed impeller and the drive. The delivery head of the pump depends on the pump speed, the diameter and type of impeller.

Diaphragm or tube pumps are used for the gentle and non-destructive delivery of glue dispersions, acids, alkalis, and other sensitive substances.

The risk of clumps is the reason why scarcely any throttling or blocking elements are used in the main line.

Fibres which are difficult to disperse can/must spend adequate time in a swelling bath which is suitable for the fibre type before dissolution. A size can be applied to the fibres, a chemically modified surface can be produced, or another load is applied. Sometimes the swelling time is sufficient for deaerating the fibres, to ensure improved dispersing behaviour.

## 5.4
## Construction of wet lay systems

A large number of raw materials and combinations of raw materials can be processed on one and the same inclined belt. Here it must be borne in mind, however, that, although the inclined belt is the versatile core of a wet lay or special paper machine, it is, on the other hand, the design of the stock preparation unit which determines which fibres can be used and the layout of the drying unit which product properties can be achieved. Systems for various end-uses can also have very different designs. The production machines for the most important end-use factors are described below.

In addition to the many other applications, the following are currently the most important:

– glass-fibre machines (see Fig. 5-3)
– machines for teabag paper (see Fig. 5-7)
– filter paper machines (see Fig. 5-8)

### Machine for producing glass-fibre nonwoven fabrics
At the present time machines for producing glass-fibre webs form the majority of wet lay machines.

*Stock preparation: glass fibres*
The design of the stock preparation unit for such a glass-fibre web production machine is comparatively simple (Fig. 5-2). As a rule it consists of an opener, where the raw material is opened as gently as possible so that the fibres are not short-

continuous system

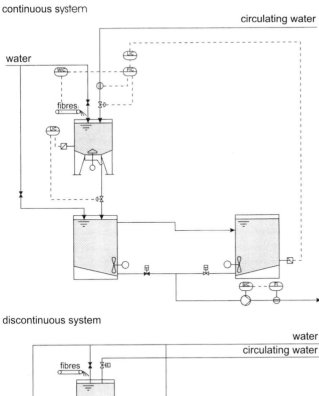

discontinuous system

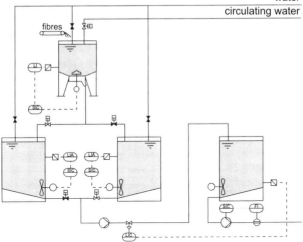

**Fig. 5-2** Continuous and discontinuous stock preparation from glass fibres

ened, and a series of vats for mixing and further dispersing the fibres. Units such as cleaners and separators are used on their own.

A distinction is drawn between continuous systems, where the raw material is continuously fed to the opener, and discontinuous systems, where batches of raw material are opened up in the opening unit.

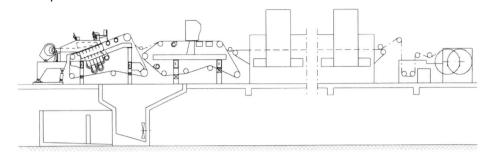

**Fig. 5-3**   Glass-fibre machine (Voith Sulzer Papiermaschinen GmbH)

The glass-fibre web machine (Fig. 5-3) consists of

– an inclined belt
– a mechanism for impregnating the web with a binder
– a tunnel dryer for drying the web and curing the binder and
– a batcher for batching up the finished nonwoven fabric

The impregnating unit can be set up directly behind the inclined belt because the web can be hydroextracted on the hydroformer to have a dry content of approximately 50% otro.

Glass-fibre web systems with a belt width of 2 to 4.4 m operate with a production rate of from 100 to 400 m/min. The capacity of such plants ranges between 10 and 230 t/24 h.

### Machines for producing teabag paper
Nowadays teabag paper comes in single- or double-layer versions. Double-layer teabag paper has a heat-sealable second layer which is activated to glue the teabag when the paper is processed further.

### Stock preparation for single-layer products
As a rule the stock preparation unit for such a system consists of a cellulose pulp line and a synthetic fibre line (Fig. 5-4). The fibres are fed in batches.

The pulp line comprises the machines and units which are used in the paper industry: fibre openers, cluster remover, refiner, sorters and cleaners.

Most of these machines are not present in the synthetic fibre line. Here the primary task is to disperse the raw material, which is ready in staple-fibre form, uniformly in the process water and to feed it to the machine in this condition. To avoid lumps and knots when long and fine fibres are used, the fibres are not opened in a conventional opener with high turbulence, but in an opening vat. The stirring mechanism in such a vat has a frequency-controlled drive, so that the fibre processing intensity can be adjusted in accordance with requirements. For the same reasons volumetric pumps are used for feeding the suspension through instead of centrifugal pumps.

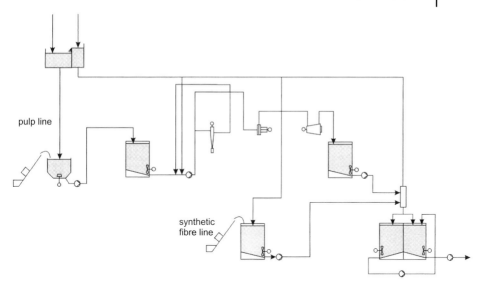

**Fig. 5-4** Stock preparation with pulp and synthetic fibre line
(Voith Sulzer Papiermaschinen GmbH)

Fig. 5-5 shows the stock preparation unit for conventional pulp and so-called long fibres. The latter require special measures during grinding, as they are strongly inclined to form knots.

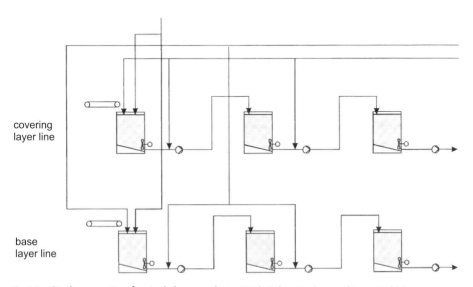

**Fig. 5-5** Stock preparation for single-layer products (Voith Sulzer Papiermaschinen GmbH)

*Stock preparation for multi-layer products*

Fig. 5-6 features the stock preparation unit for a machine for producing two-layer nonwovens.

It consists of a preparation line for the perforated layer and the covering layer in each case. The synthetic fibre fraction is fed directly into the mixing vat. The pulp is processed separately in a third line and added to the top layer or the perforated layer material as required.

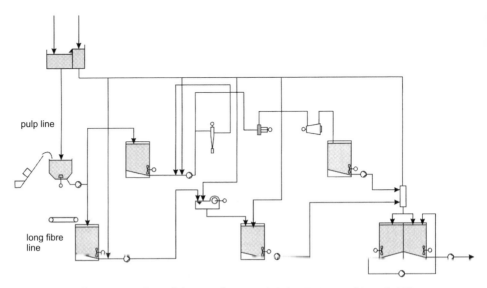

**Fig. 5-6** Stock preparation for multilayer products (Voith Sulzer Papiermaschinen GmbH)

*High-production machine with two circulating-air dryers*

The machine shown in Fig. 5-7 consists of

– an inclined belt for the production of two-layer products with an open or enclosed forward end
– a pick-up mechanism for transferring lightweight webs or a feed screen for transferring heavyweight products
– circulating-air dryer for pre-drying
– a glue press or impregnating mechanism
– an after-dryer for finish-drying the web and curing the binder
– end-group with moistening device, calender, measuring frame and batcher

Machines having this construction cover most of the nonwovens and special paper production range. They are used for the production of single- and multi-layer filter papers (for a wide variety of end-uses) and of nonwovens with fabric weight between 10 and 150 g/m$^2$ and operated at speeds of up to 500 m/min. Examples include sausage wraps, overlay papers and screen papers.

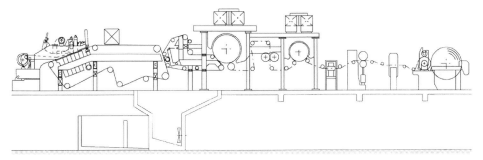

**Fig. 5-7** Two-layer teabag paper machine with two circulating-air dryers
(Voith Sulzer Papiermaschinen GmbH)

### Machines or the production of filter paper

The machine feature in Fig. 5-8 is a machine for special paper for producing low-
and high-porosity filter paper. Its basic components are:

– an inclined belt
– a press unit
– a pre-dryer with contact drying
– a glue press
– an after-dryer with contact drying and
– an end-group

In many cases filter paper machines are provided with contact dryers to produce
the specific requisite surface qualities. However, dryer combinations can also be
used, e.g. circulating-air and cylinder dryers, as well as circulating-air dryer systems on their own.

The porosity of the products is influenced by the selection and preparations of
the raw materials and the compression of the paper web. Low-porosity paper is
compressed, but not high-porosity paper.

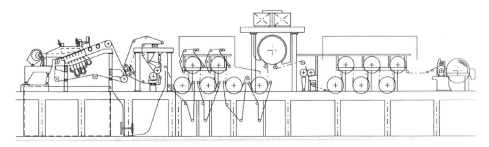

**Fig. 5-8** Machine for filter papers (Voith Sulzer Papiermaschinen GmbH)

5.4.1
**Web forming**

**Initial development stages on a laboratory machine for wet-laid webs**
These days it is no longer necessary to carry out expensive development trials on production machines. New products can be developed or optimized at considerably less expense on a laboratory machine (Fig. 5-9). Such machines need very little space or energy (25 m², 14 kW). The advantages of inclined-belt laboratory machines are as follows:

– new products can be developed on these at low cost
– the results are clear and visible
– the development results can often be transferred to production machines
– little raw material is required, as trials can be carried out with 200 g of fibres
– they are simple to operate and manpower and space requirements are low
– only 1.5 m³ water and little energy are required because the system is enclosed

**Fig. 5-9** NVLA 22 wet-lay laboratory system (Pill Nassvliestechnik GmbH)

**Further steps using a wet-lay pilot plant**
Pilot plant, which is operated at speeds similar to those of production machines, can be used to optimize important developments and to produce small batches.

**The inclined belt/screen** can be configured as a cylindrical sieve having an infinite radius. The sheet formation principle is the same for both machine types. The fibre suspension in a forward end is held back by a screen which is located above a hydroextraction unit. The fibres are laid on the screen via filtration to form a web, while the water from the suspension is drawn away by the hydroextraction units.

---

With a cylindrical sieve the sheet formation length is more restricted than when an inclined belt is employed. Because of the higher production rates and wide machine widths, inclined belts are used nowadays for the production of wet-laid webs in the overwhelming majority of cases.

The main features of an inclined-belt former are:

- the forward end
- the hydroextraction unit
- the screen unit and
- the screens/sieves

It is the job of the machine manufacturer to point the way to the optimum production of a product. For this reason the combination of pre-developed machine parts from different manufacturers is very important.

Using a filter web machine as an example, where three fibre lines are present, a machine design is discussed on the basis of the selection criteria mentioned in the previous chapter.

*Pulp line*
Pulp bales or sheets of the long-fibre cellulose pulp are fed into the pulper, broken down to approximately 5% and pumped into a vat; from here they are forwarded to the next vat via a machine which removes dense material and a machine which takes out clusters using a rotary pump. This suspension is now ground to the necessary state using refiners or beaters, pumped into a supply vat using a centrifugal pump, diluted and mixed with the requisite auxiliaries. From here the fibre suspension is pumped via cleaners and sorters in the suction line in front of the mixing line located in front of the front end. This leads to additional mixing.

*Synthetics line*
Since it may take longer than foreseen for the suspension of opened fibres to be ready and the machine vat is relatively small, two dissolving vats are required. The fibres and the requisite auxiliaries are introduced into the water in the dissolving vats. Propellers are used to open fibre bundles in the aqueous suspension and then pumped into the supply or storage vat in the long-fibre line. There they are circulated constantly to prevent unmixing. This long-fibre suspension is then pumped into the forward end feed line beyond the mixing pump using and eccentric worm pump.

*Waste line*
The sections which are not used when the web tears and the sprayed edge strips which are not used are taken out of the waste bin, thinned down and broken up with propellers. The long fibre/pulp mixture is forwarded to an intermediate vat using eccentric worm pumps and diluted to a specific concentration before it can be fed into the process again.

With many other systems this line can be constructed as a pulp or a synthetics line and supplied with other secondary fibres.

The job of the *forward end* or *breast box* is to feed the fibrous suspension to the sheet-forming zone. To obtain a homogeneous web the fibres have to be uniformly distributed over the width of the machine. Distribution systems take care of this.

- The flow at the pipe network and transverse distribution units is sideways on, they therefore need little space. These mechanisms can be tilted for cleaning. Part of the suspension (5 to 10%) is recirculated via a pressure-controlled pipe.

- With a pipe manifold system the suspension is distributed over the machine width via one or several rows of tubes.

- Transverse flow distributors have a conical access and a rectangular or circular cross-section. To even matters out it is connected to a turbulence unit with stepped bores.

- A circular distributor is a central tube where the flow approaches from below and which terminates in an extended mushroom-shaped chamber. Many tube outlets are radially arranged at this circumference, from which hoses of equal length lead to the centrifugal chamber of the forward end.

- The diffuser is a continuously-extending shaft which stretches with a maximum angle of 15° from the pipe cross-section to the machine width where the fibre suspension flows from the bottom to the top. These diffusers are suitable for processing any type of fibre. However, they need a relatively large amount of space and are used for low speeds and small machines.

- Nowadays the front ends are mostly fitted with transverse-flow or circular distributors.

Fig. 5-10 features the design of a modern, hydraulically compact hydroformer.

The hydro former consists of the following units:

- a conically parabolic transverse distributor
- a turbulence unit
- a forward end table
- side walls
- front wall

The suspension is distributed uniformly over the machine width in the transverse distributor. Pressure differences in the transverse distributor which hinder uniform distribution can be evened out by using a recirculation line. To produce a microturbulence a short distance in front of the web-forming zone, a turbulence pack is located behind the transverse distributor. Enclosed by the back wall, the forward end table and the side walls, the suspension flows in on to the belt as if in a tunnel. The web-forming zone is located on the ascending part of the belt. It extends over the whole length of the hydroextraction unit. The suspension is covered by a flat front wall in the web-forming zone. The adjustable angle between the front wall and the belt is set in such a way that the fibre suspension in the whole of the sheet-form-

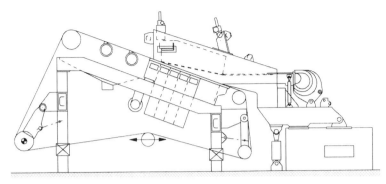

**Fig. 5-10** Hydraulically compact hydroformer (Voith Sulzer Papiermaschinen GmbH)

ing zone flows in the machine direction at approximately with the belt speed. The speed of the suspension can be controlled by adjusting the front wall angle, which has a direct effect on the strength ratio of the nonwovens.

The sheet-forming pressure in the forward end is created with the forward-end pump in the case of a hydraulically compact inclined belt or screen. The influence of the angle of inclination of the belt is evaluated differently by different manufacturers. The hydroformer was built for many years with an angle of $<10°$. As a result of ever-higher belt speeds, and hence higher sheet-forming pressures, the angle was increased to $<20°$ for design reasons. The angle of inclination of the deltaformer was adjustable between $<10°$ and $35°$. The No-Wo Former was designed with fixed angles between $<8°$ and $35°$.

The *hydroextraction unit* is located below the forward end. It consists of different hydroextraction compartments. Water can be drawn out of the compartments via control valves. Wide machines have connections to hydroextraction lines in the bottom part, on both sides or in the centre. Smaller machines have these connections on the drive side. Uniform water removal over the machine width ensures uniform fibre deposition over the machine width. At the tops of the units there are hydroextraction ridges which form suction slots and which are arranged transversely to the machine direction. The belt or screen passes over these ridges. A fixing suction apparatus is located behind the last hydroextraction unit. The web is hydroextracted further using several subsequent vacuum-assisted tubular suction sources. A web dry content of 15–50% can be attained, depending on the water retention capacity, the vacuum and the machine speed.

The job of the *screen unit* is to move the endless screens over rollers and to clean them.

Control, tensioning, and guide rollers are fastened to the supports. So that the endless belts can be changed easily in the case of wear, cantilever supports are used. These are so pretensioned on the drive side that intermediate sections can be taken out of the brackets on the guide side to change the screen. With the forward end raised, the endless belt can easily be drawn into the machine via all the rollers.

Spray tubes for cleaning the belt and the roller scraper are essential components of the screen/belt unit. The hydroextraction unit is fastened to the mounting of the latter.

*Two-layer forward end*
With a two-layer forward end to produce a two-layer web in one operation. The
two layers may consist of different fibres or fibre blends. Fig. 5-11 illustrates such
an inclined belt.

With these machines the two sheet-forming zones are located one behind the
other. They are separated by the front wall of the base layer which simultaneously
serves as the table for the covering layer. The stock for the base layer is fed in as
in the case of a single-layer forward end. The suspension for the covering layer ar-
rives in the second sheet-forming area via a separate distribution system.

The two nonwoven layers are firmly joined together, as some of the fibres in
the covering layer have formed a mixed layer with the fibres in the base layer.

Two-layer inclined-belt machines have proved themselves mainly in the production of fine webs, e.g. webs for teabags, from natural and synthetic fibres.

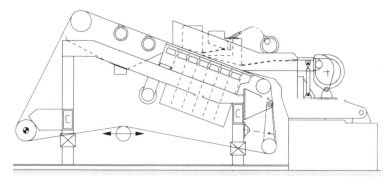

**Fig. 5-11** Two-layer forward end (Voith Sulzer Papiermaschinen GmbH)

**Water circulation systems**
The water circulation for a wet lay machine consists of two closed cycles.

The inner cycle (screen water cycle) contains the water supply which is directly
linked with web forming. Water is pumped from the container into the forward-
end. The water passes through the belt into the hydroextracting units positioned
below, and from there back into the belt water container via hydroextraction lines.

The fibre suspension coming from the stock preparation unit (thick material)
reaches the belt water cycle before or after the forward end pump and is thinned
down to the requisite sheet-forming stock density.

In the outer cycle the excess water in the belt water container is largely pumped
back directly into the stock preparation unit. The rest passes through cleaning
units and is used in spray tubes and as blocking water. Adequate water economies
are achieved by recycling process water.

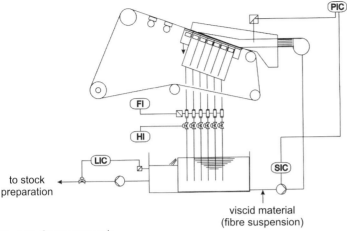

**Fig. 5-12** Open water cycle

In theory the water consumption of the plant only corresponds to the mass of water which evaporates when the web is dried. In practice however, the water consumption is higher, because the process water has to be changed when fabric types are changed or when there is strong contamination by auxiliaries.

Wet lay machines can be fitted with different belt water cycles:

– open cycle
– closed cycle

With an open cycle (Fig. 5-12) the vacuum needed for hydroextraction is built up via fall tubes which lead to an open water tank located below the inclined belt. The hydroextraction capacity depends on the geotedical height of the fal tubes.

With a closed cycle (Fig. 5-13) the hydroextraction lines lead to a closed water container. The forward end/breast box pump has two functions with this system.

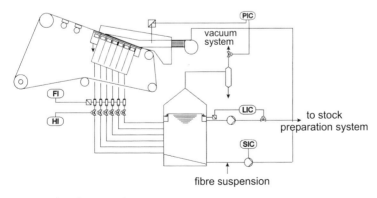

**Fig. 5-13** Closed water cycle

On one hand it supplies water to the inclined belt and, on the other, it acts as a suction pump for the hydroextraction system. The enclosed water tank is under a vacuum. Excess water flows into an adjacent extraction compartment with level control.

The enclosed water tank has the following advantages:

- the suction is not dependent on the difference in height between the level of the water tank and the hydroextraction zone
- better deaeration of the water in the tank is achieved as a result of the greater vacuum in the tank
- in addition denser nonwovens can be produced because of the higher vacuum levels
- the water tank does not have to stand under the inclined belt
- a rapid start is possible, as the pipes remain full when the machine is switched off

With an enclosed system the energy consumption is higher than with an open system, as an additional vacuum pump is needed for the water tank.

Two-layer inclined-belt machines need separate water cycles for the base and covering layers.

## 5.4.2
## Bonding

### 5.4.2.1  Adding the binder fibres
These fibres are fed to the mixing vat in the same way as the other materials. Special mechanisms are therefore not needed. Short fibres with a low melting point are opened in the opening unit before being added to the mixing vat. This is essential because many of these fibres are fused together at the ends and cannot be opened by the turbulence created by the vat propeller/impeller.

### 5.4.2.2  Adding binder dispersions to the pulp
Coagulation in the vat: this method also does not require any additional mechanisms in the stock preparation unit. The binder is added either to the mixing vat or the pulp vat. The prerequisite for optimum binder utilization is that the dispersion coagulates out to form particles a specific size and that these adhere firmly enough to the fibre surface. Since the particles adhere best to cellulosic fibres, the method is preferably used when such fibres are present in the breast box. If the cellulosic fraction is sufficiently high, the coagulation should occur in the pulp vat.

The particle size plays a part because, in practice, not all the coagulate particles are bonded to the fibres. In addition some of the bonded coagulate is shed by the fibres on the way from the mixing vat to the web former. If the particles are too small they pass through the belt with the water during web forming and into the circulation system. The result of this is a web with a reduced binder content on the belt. In addition there is a risk of the particles coalescing together in the cycle and contaminating this or passing into the drying unit together with the web.

The adhesion and particle size are influenced by the type of binder, the type of precipitant, and the concentration of the two components in the mixing vat. The size of the particles should not fall below 0.2–0.3 mm. The special formulations of binder manufacturers for different end-uses should be taken into consideration.

Precoagulation of the dispersion: this is carried out by the Ciago method in a special machine, the so-called coagulator. The dispersion is mixed in this with the precipitant and the coagulated binder is subsequently exposed to strong shear forces. The concentration of the two raw materials and the speed of the machine rotor which creates the turbulence determine the particle size and their shape. The aim is to produce a fibre shape and lengths between 2 and 4 mm.

The precoagulated binder is either fed in batches to the mixing vat or continuously to the stock flowing to the web former. The binder particles form a separate raw material component in the suspension. They can be processed with every type of fibre. High retention values can be expected if the above-mentioned fibre lengths are used.

### 5.4.2.3 Adding the binder to nonwoven web

Binder applicators for the wet lay method must ensure a uniform add-on at high machine speeds without the wet web which is not very strong being too heavily stressed. From the range of conventional methods, spray and pouring systems, doctors for the addition of foamed dispersion, printing, and impregnation in an immersion bath between two screens satisfy these requirements.

In many cases the binders are added to the pre-dried web (approximately 60% dry content). The web should be finish-dried while using partial binder application pressure.

Binder application to the still wet web before the dryer is the exception when the webs are from pulp, viscose rayon and/or synthetic fibres. In the production of glass-fibre nonwovens it is, however, the rule, as the web is hydroextracted in the suction zone to dry contents of 40–60%.

### Spraying method

During spraying it is advantageous if the binder is applied to the web without part of the machine coming into contact with it. Any excess is not squeezed off, but sucked through the web by subsequent suction mechanisms.

Sprayers are particularly useful for the full impregnation of lightweight webs with fabric weights in the range 10–60 g/m². Spraying is from one or both sides, depending on the product, using spray pistols located one behind the other. These move across the fabric and either operate with compressed air or in accordance with the airless system. The number, arrangement and traversing rate of the pistols can be adapted to the circumstances. In electrostatic spraying the binder droplets are charged in an electrical field and are attracted by the wet, earthed web. The use of electrostatics primarily helps to reduce spray losses.

- **Pouring on the binder dispersion:** With this method the binder dispersion is applied to the web from the breast box via an overflow edge. To avoid damage of the web surface, the web and flow speeds must be synchronized. Excess binder is applied. The excess amount goes back to the forward end via a suction element in a closed cycle.

- **Foam impregnation:** This method can be used in the same way as the spray method. The foam binder imparts a soft handle to the web. As the foam contains less water than the sprayed-on dispersion, there is less drying to do later on.

- **Printing on the binder:** Rotary printing has proved to be useful for printing on the binder. In this case the web runs over a perforated roller-form screen. The thickened dispersion is printed onto the web as printing paste via the sheath. The result is partial bonding or consolidation, particularly the web surface. Through printing of the web is only possible with very thin materials. The perforation pattern is crucial for the softness, flexibility and air permeability of the product.

- **Impregnation in the immersion bath:** This method of impregnation in an immersion bath between two screens can also be used to fully impregnate heavy-weight products. It is on slow-running machines.

- **Glue press:** In the case of filter papers a glue press is often inserted between two drying units. Here the pre-dried web is passed through a roller nip. The rollers have a horizontal or an inclined arrangement. The liquid binder (e.g. latex dispersion or foam) is fed into the nip in both sides of the running web. Pressure is applied to press the binder into the web. The rollers are sometimes also wetted using metering rollers.

### 5.4.2.4 Pressing

**Use of pressing devices**

Devices of this type are used to continue the hydroextraction of the web which started in the vacuum section of the inclined-belt machine. At the same time the nonwoven web is compressed during this process and thus also bonded. Mechanical hydroextraction is cheaper than the thermal version.

Whether pressing is used or not depends on whatever requirements relate to a particular product and also on whatever facilities exist for the subsequent finishing of the nonwoven. For example, a thin web which is produced in a given machine is pressed to give it adequate strength for passing through the pre-drying and impregnating units. The loss in bulk or softness can be evened out by subsequently compressing the nonwoven.

The presses used are double-felt presses where the web is fed through the nip between the take-off felt and the press felt.

**5.4.3**
## Web drying

This stage of the method is not only responsible for removing water from the nonwoven, but is also needed for bonding the web. Contact, circulating-air and radiation drying systems are used. The stock preparation and web-forming sections of a wet lay unit can be used for a large number of applications, depending on the design. On the other hand, the configuration of the binder applicator and the dryer unit is more product-specific.

### 5.4.3.1  Contact drying

Drying is effected through the transmission of heat from the surface of a cylinder heated by steam, electricity, oil to the web surface. This is either fed through freely or pressed to the cylinder surface by a felt or screen running along with the web. The thermal energy transmitted evaporates the water in the web. The water vapour is absorbed by the ambient air. To prevent the air becoming saturated with water vapour, numerous hot-air ventilation systems are installed for removing the water vapour. The evaporation efficiency decreases with web thickness and a decrease in porosity because this hinders the release of water vapour.

As the evaporation preferentially occurs at the boundary layer of the web facing the cylinder, this favours binder migration in this direction. The side of the fabric facing the cylinder is smoothed out as a result of the close contact between the web and the smooth cylinder surface. The method is used in the wet lay process for pre-drying the still binder-free web; additionally it is used for the finish-drying of webs with binder fibres which have to be pressed for triggering off the bonding during drying. Because of the risk of the web sticking to the cylinder surface, the latter is often coated with a 'Teflon layer'. For the same reason care is taken to see that the web is dried at a single contact surface. Because of this, use is frequently made of so-called Yankee cylinders with a diameter of 3–6 m.

### 5.4.3.2  Circulating-air dryers

In the wet lay process use is most frequently made of sieve-drum dryers which operate on the flow-through principle. Tensionless dryers are used in special cases. The web is fed over a perforated drum. The hot air flows from the outside through the fabric into the interior of the drum. The drums have either a perforated-roller or so-called honeycomb construction and nowadays have an up to 96% open surface. Several drums can be fitted in sequence.

Circulating-air drying is particularly thorough, as the surfaces of practically all the fibres come into contact with the air flow when the hot air passes through the web. The prerequisite for this is that sufficient air can pass through the web in the shortest time. The fabric air permeability thus plays a crucial role as regards the suitability of this drying method. The greater the air permeability, the more effective the drying.

From the point of view of the wet lay method it follows that this method of drying is particularly suitable for the production of lightweight nonwoven fabrics and thin paper, i.e. for production at high machine speeds. Since, in the case of the through-flow principle, the goods are heated uniformly over their whole cross-section, the risk of binder migration is relatively low here. To achieve a high specific evaporation rate, air temperatures of between 150 and 400 °C are used. This can be done without damaging the web, as the web, as long as it is still wet, cannot be heated to above 100 °C under the pressure conditions prevailing in the dryer.

### 5.4.3.3 Tunnel Dryers

Tunnel or belt dryers have proved themselves useful in the production of nonwoven fabrics from glass, mineral and leather fibres. Because of the brittle nature of the fibres, these materials require a path which is as deflection-free as possible. High vacuums must be avoided, especially during the initial phase of drying, to prevent the binder being sucked out of the web. The units used are therefore single belt dryers with a combined baffle and circulating-air drying system.

### 5.4.3.4 Radiation drying

This is a contactless drying method, where thermal radiation is created by electrically- or gas-heated radiant bodies and directed at the goods to be dried.

Infrared radiators are used in most cases. The wavelengths here are over 0.7 μm. The radiation temperatures are in the range 500–1800 °C.

Modern dryers are constructed on the module principle from standard individual units. In the wet lay process this method of drying is confined to individual cases of preliminary drying. We are aware of cases where radiation dryers are used to raise the temperature of a wet web containing a nylon copolymer component to approximately 80 °C for prebonding and to release the shrinkage. The web is passed freely and in as tensionless state as possible between two banks of radiated heaters. A further example of their use is for the coagulation of heat-sensitive binders in the impregnated web before the actual drying process. In this case the objective is to achieve prebonding through coagulation of the binder at 40–80 °C, especially at the surface, which should prevent the web sticking to the subsequent drying elements.

### 5.4.4
### Batching

A distinction is drawn between central and supporting drum rollers.

In the case of central batchers a batching roller with a cardboard tube is directly driven. The batching speed of the paper roller must be controlled accurately to ensure a constant peripheral speed and tension for the nonwoven web. Modern central batchers operate with an electrically controlled drive. They enable the batching tension to be set accurately as a function of the goods being batched. Their design makes them suitable for smaller machines and lower machine speed ranges.

With the supporting drum system the web is wound onto a steel drum. This runs a supporting drum and is driven by the latter through friction. Here the batching speed does not need to be controlled. The supporting drum roller design makes it suitable for high machine speeds and wide machine widths. The rolls it supplies are hard. However, because of its operating principle, its use is restricted to high-strength nonwoven fabrics.

To circumvent this disadvantage supporting drum rollers with electrical drives are available today. These are additionally fitted with an axial drive for the batching drum so that, when delicate nonwovens are being batched up, the roller can also be operated separately from the supporting drum.

## References to Chapter 5

[1] Albrecht W (1982) Vliesstoffe, Georg Thieme Verlag Stuttgart
[2] Scholz B (1989) Naßvliesherstellung auf dem Hydroformer®, Spezialpapiersymposium München
[3] USA Patent No 11394: 1848
[4] Fay H, Osborne, The History of Dexter's long fiber paper development
[5] Pill H (1998) Herstellung von Vliesstoffen nach dem Naßverfahren, Taschenbuch f d Textilind Schiele & Schön Berlin
[6] Lennox-Kerr P (1997) Technical fibre products – the fibre manipulators, Techn Text Internat 6, 9: 13–16
[7] Bestimmung des Entwässerungsverhaltens einer Faserstoffsuspension mit dem Schopper-Riegler-Mahlgradprüfer, Papiermacherschule Steyrermühl
[8] Hutten JM (1995) The concerns of forming wet lay nonwovens from long fibers, TAPPI Press Atlanta: 161–172

**Advertisement section**

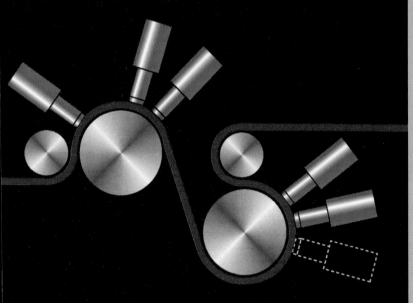

# *Nonwovens are* all the same -
## but *so changeable*

Nonwovens have not yet reached their true *innovative potential*. These new generations of textiles are the result of 50 years of *experience* and intensive *research*. Although they look the same, they're all different. *New production technologies* enable their characteristics to be exactly matched to end user needs. Today's nonwovens offer *customized solutions* for a broad range of applications. This improves *profitability during processing*, saving both time and money. The diversity of nonwovens is almost unlimited. *Working together with customers*, we develop nonwovens that are as individual as the product in which they are utilized. This opens up new markets - *worldwide*.

The Freudenberg
Nonwovens Group

 **Freudenberg**

Freudenberg Vliesstoffe KG · 69465 Weinheim · www.nonwovens-group.com

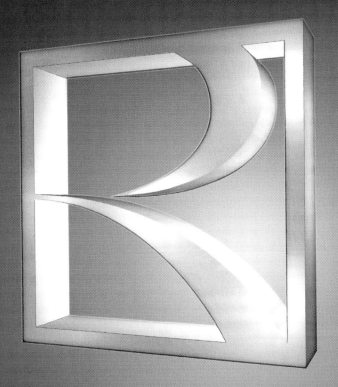

# edana

# Worldwide Acceptance

Since 1992 the internationally renowned Oeko-Tex Institutes have checked textiles for harmful substances. Safe textiles for the consumer. Worldwide.

**Oeko-Tex Standard 100 – a quality achievement.**

CONFIDENCE IN TEXTILES
Tested for harmful substances
according to Oeko-Tex Standard 100
Test-No. 00-0000 Institute

**The International Association for Research and Testing in the Field of Textile Ecology (Oeko-Tex)**

Institutes and Agencies in:
Austria, Belgium, China, Denmark, France, Germany, Hungary, Italy, Japan, Poland, Portugal, Slovakia, South Africa, South Korea, Spain, Sweden, Switzerland, Turkey, UK, USA.

**www.oeko-tex.com**

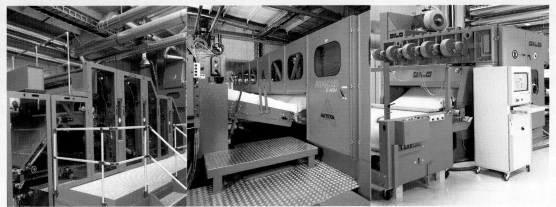

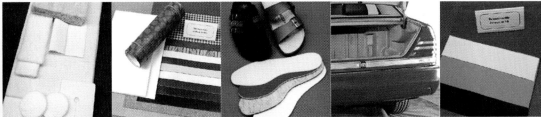

# BGB STOCKHAUSEN

**Textile Auxiliaries**

**Man Made Fibre Auxiliaries**

**Economical**

# AROUND
# TEXTILES

**Ecological**

BGB Stockhausen GmbH
Bäkerpfad 25
D-47805 Krefeld/Germany
Tel.: + 49 - 21 51 / 38 18 66
Tel.: + 49 - 21 51 / 38 18 22
Fax: + 49 - 21 51 / 38 10 05
www.bgbstockhausen.com
e-mail: info@bgbstockhausen.com

# 6
# Web bonding

During the web bonding process, the fibres are consolidated to ensure that the produced felts meet the necessary expected performance parameters for the intended end use. The degree of web bonding depends on the specific fibre characteristics, such as fibre geometry, tenacity and shape, as well as fibre location within the felt, felt mass and the specifics of the bonding process changes.

The transition of a fibre batt to a fabric is accomplished by either a physical, or chemical or thermal process. The bonding takes place over the entire surface or portions thereof. The entanglement of fibres or continuous filament is strengthened by the consolidation of the batt during the bonding process. Physical entanglement between fibres is based on friction, cohesion and adhesive surface forces [1]. The bonding processes between fibres are complex, especially when synthetic dispersions – also called binding agents – are used. Chemical bonds between fibres occur only when dissolved fibre surfaces meet or binding agents of the same polymer are applied. These fibre surface bonds can be obtained as a result of the chemical composition of the polymer's monomer unit.

In contrast to the physical/chemical fibre to fibre bonding, the technologies applied depend on the nature of the bonding (Fig. 6-1). Mechanical web bonding processes result in different structures unique to the type of process technology selected (Table 6-1).

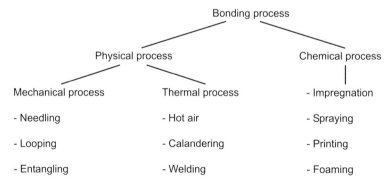

**Fig. 6.1** Description of the bonding processes according to ISO/DIS 11224 [2]

**Table 6-1** Structure elements of the fibres' orientation within the felt

| Mechanical process | Structure element | Fibre orientation within the felt |
|---|---|---|
| Felt needling | Plugs | Partly fibre bundles with reorientation to the vertical felt plane |
| Felt looping | Loops | Fibre bundling and horizontal transforming to loops |
| Felt entanglement | Balls | Fibre displacement and entanglement |

Thermal bonding requires fibres or powders with thermoplastic characteristics. In the chemical process polymer dispersions are used as binders so that the consolidated fibre batt can be considered a bi-component composite. The binding agent's ratio in the fibre batt and the type of application influence the kind of bonding process and binder distribution in the substrate. Through combination of different bonding processes, a variety of special bonding effects can be obtained, which are necessary for the respective product application.

## 6.1
**Needling process**
W. KITTELMANN, J. P. DILO, V. P. GUPTA, P. KUNATH

During the mechanical bonding of fibre webs via the needling process, the fibres or filaments are reoriented into the vertical plane in a way that tufts or stitching channels are formed. The applied compression between fibre sections results in a friction lock, which is essential to the degree of bonding in the felt.

The needling process was invented more than 100 years ago and later used industrially through the developments by the companies Bywater/GB and Hunter/USA in the processing of fibres unsuitable for natural felting. In the beginning only natural and regenerated fibres were used. With the advent of man-made fibres in the 40s and their use in nonwovens, there followed a focused development effort in the areas of equipment and process technology for the needling of fibre webs. This is characterized by improved needle design, increased needle density per working width, increased stroke frequencies, delivery speeds, working widths, and the optimization of known basic needling principles, as well as the development of new needling process technologies.

### 6.1.1
**The principle of needling**

The principle of needle felting is based on the reorientation of a portion of horizontally located fibres or filaments into the vertical plane in form of fibre tufts through the use of barbed needles. The tuft ends remain outside the fibre batt during return of the needles, are flattened by the horizontal movement, and are simultaneously

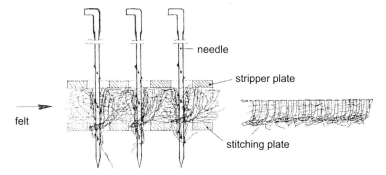

**Fig. 6-2**    Principle of needling a fibre web

interlocked through repeated needle penetrations. This results in a condensation of the reoriented fibres into bundles, which increases the friction between fibres so that the fibre consolidation is achieved through shape and friction as shown in Fig. 6-2. During fibre reorientation to the vertical plane, a fibre movement takes place within the felt as the fibres caught by the needles slide by each other and change their position. The majority of fibres or fibre sections which are not consolidated into fibre tufts remain in their horizontal position. The fibre movement within the fibre batt, caused by the needling action, results in length and width changes of the area mass influenced by the process conditions. While needling filaments, it is assumed that they are positioned randomly, elliptically and overlapping into a substrate, which change position during the needle penetrations, which can lead to fibre breakage. The forces generated through needling of the substrate are the result of the column stiffness of the fibres, the number of fibres within each tuft and depth of needle penetration, but also on the coefficients of friction between fibres and steel. By application of suitable fibre finish agents (avivage) it is possible to minimize friction between fibre and steel and to change intrafibre friction. This greatly assists in reducing needling forces and fibre damage.

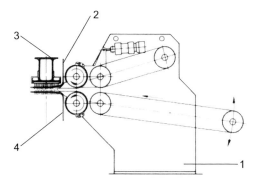

**Fig. 6-3**    Scheme needling technology
1 feeding system, 2 stripper plate, 3 needle beam, 4 stitching plate

The principle of needling is shown in Fig. 6-3, which shows the presentation of the fibre batt to the needle zone via a Compressive Batt Feed system. The felting needles in the needle board attached to the needle beam penetrate the fibre batt in the needling zone forming fibre tufts. As a result of the eccentric drive of the needle beam, the needles prescribe a vertical up and down motion. Stripper plate and bed plate are perforated in the identical pattern of the needles in the needle board. With the help of a delivery roll pair, the substrate is being transported through the needling zone and wound up, as well as edge trimmed, or crosscut if applicable.

The stitch density, i.e. the number of penetrations per area square ($cm^2$ – $inch^2$) of the felt is calculated according to the customary general equation (1).

$$Ed = \frac{n_h \cdot N_D}{v_v \cdot 10^4} \tag{1}$$

Ed = stitches per area $[cm^{-2}]$
$n_h$ = number of lifts $[min^{-1}]$
$N_D$ = number of needles by m working width $[m^{-1}]$
$v_v$ = web outlet speed $[m \cdot min^{-1}]$

The ratio $v_v$ to $n_h$ is equivalent to the material advance $L_v$ per stroke. Formerly one operated with intermittent advances, which meant a felt transport, was possible only while the felting needles were not in engagement with the material. Modern needle looms operate with continuing felt transport. For optimum needling efficiency, the following process solutions are of importance:

– batt feeding
– needling zone
– felt delivery

### 6.1.1.1  Batt feeding

The task of the batt feed is to lead the unbonded voluminous batt evenly to just in front of the first row of needles without compromising the following passage through the needle zone. This way uncontrolled drafts during batt transport are minimized. Uncontrolled draft during the feed into the needle loom is causing intra fibre migration, which leads to length dimensional changes during needling, resulting in uneven surface mass and/or felt thickness. A good batt feed is especially important during the first needling pass, also referred to as "pre-needling". Generally feed systems comprise batt transport apron, pre-compression apron and a pair of feeding rolls. The roll gap should correspond to the pre-compressed batt thickness.

Fig. 6-4 shows the feed system "CBF-Transfer" of Oskar Dilo KG. In this system, grooved rolls and plastic fingers ensure the batt transport to the first row of needles. Additionally, an intermittent transfer roller bridges the gap between aprons and pair of "finger roll." The FFS-Fehrer is similar and minimizes the draft at transport of voluminous batts to the needling zone; however, without the

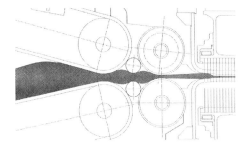

**Fig. 6-4** Batt feeding system "CBF-Transfer" of Oskar Dilo Maschinenfabrik KG

transfer rollers mentioned above. All manufacturers offer systems with lattice or smooth conveyor aprons. Systems without guide fingers are unable to deliver the batt directly to the needle zone. Another technical solution is the rotary pre-needling approach of Asselin (Fig. 6-5). The batt is fed between two driven perforated cylinders, which are each equipped inside with a needle beam, thus creating a needle zone. Pre-needling minimizes the potential fabric with shrinkage and draft. Only shallow penetration depth, as well as low stitch densities, are possible and, because of the particular needle patterns, needle markings are a real danger.

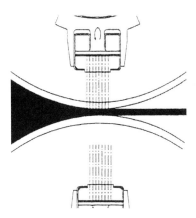

**Fig. 6-5** Cylindric pre-needle loom of Asselin

#### 6.1.1.2 Needling zone

The needling zone consists of a needle beam with the needle board equipped with the felting needles, the stripper plate and the stitching or bed plate. The needle beam is vertically driven by the main shaft via an eccentric shaft. While needling a batt with felting needles penetrating from top to bottom, the batt is pressed against the bed plate showing holes or slots. When the needles raise again, the felt is held back in its position by the stripper plate. The penetration depth of the needles in the felt is being changed by raising or lowering the bed plate. The distance between stripper and bed plates depends on the thickness of the felt. Appearance and degree of compression of a needle felt are mainly influenced by:

- needle arrangement in the needle board
- direction of needling (from top, from bottom or from both sides)
- needle parameters (gauge, form of barb, number of barbs)
- needling parameters (penetration depth and density, draft)

Formerly the needles were fixed in boards made of wood but nowadays primarily sandwich construction needle boards of e.g. aluminium and polyamide (Oskar Dilo KG) are being used. The horizontal and vertical arrangement of the needles within the board (needle arrangement) is important. It should be chosen to minimize markings or fibre accumulations in the felt due to advance per stroke and due to the possible draft in machine direction or the shrinkage in cross direction. The percentage of the mass changes in machine and cross directions ($\Delta L$ and $\Delta Q$) are basis for the calculation of the changes in the plane $\Delta A$ according to the following equation (2):

$$\Delta A = \Delta L + \Delta Q + \frac{\Delta L \cdot \Delta Q}{100} \ [\%] \tag{2}$$

This results also in a change of the average area of the needle felt according to equation (3):

$$m_{NV} = m_v \left( \frac{1 + \Delta A}{100} \right) \tag{3}$$

$m_{NV}$ = area weight of the needle felt g/m$^2$ after needling
$m_v$  = area weight of the needle felt g/m$^2$ before needling

The actual needle board surface $A_v$ is calculated using equation (4):

$$A_v = L_N \cdot L_B \tag{4}$$

$L_N$ = length of needle board equipped with needles [cm]
$L_B$ = width of needle board equipped with needles [cm]

The number of needles in the board $N_z$ is according to equation (5)

$$N_z = N_A \cdot N_R \tag{5}$$

$N_A$ = number of needles per row
$N_R$ = number of rows equipped with needles

When the needles penetrate the first time, penetration density E can be calculated according to equation (6)

$$E = \frac{N_z}{A_V} \tag{6}$$

The overall penetration density after a needling passage can be seen in equation (1).

In order to simplify comparisons, it is common practice to indicate the number of needles in the board $N_z$ per 1 m of working width. The needle boards are for example 200, 300 or 350 mm wide.

For many years, the "herring-bone" pattern has been known for the arrangement of the needles in the needle board. The arrangement of the needles in this pattern shows in the projected way that the needles deviate in the same amount in horizontal and vertical directions. Nowadays computer programs help develop needle patterns which take into account the processing parameters as well as the expected appearance of the needle felt [3, 4]. The number of needles in the needle board vary from 1500 to 5000 needles per meter and per board of 200 mm width.

Dr. Fehrer AG, Linz offers the needle pattern F9 [5]. In contrast to other known patterns this one is less dependent on the felt's advance per stroke. This has been achieved by assigning several needles to one slot in the base and stripper plates.

To prevent the needles from twisting, clamping grooves are milled in the softer polyamide layer of the needle board. Rapid clamp systems facilitate the quick changing of the needle boards at the needle beam.

Flat needling of felts is the most common kind of compression. During this process, the felting needles penetrate the felt vertically. Fig. 6-6 shows the possibilities. One has to distinguish between one-board, two-board or four-board needle looms (Fig. 6-6 a, b, d). Additional possibilities are the penetration direction either from top to bottom and vice versa (Fig. 6-6 a, b) or a combination of both (Fig. 6-6 c, d). With double-sided needling, the needle boards can be arranged opposite each other or offset.

These possibilities result in different fibre transport and dwell time of the needles in the felt and, therefore, in different compression conditions within the felt. When the needle boards are arranged "opposite to each other", two needling modes "simultaneous" and "alternating" are possible. The needling "simultaneous" mode means that both boards penetrate the felt at the same time. Therefore only half the needles can be inserted in alternating rows in each board to prevent confrontation of the needles. To reach a certain and higher compression, the necessary total penetration density is shared between several needle looms. The needle looms are arranged one behind the other in so-called needling lines.

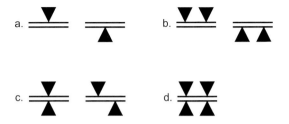

▼ direction of needle penetration through the felt

**Fig. 6-6**  Flat needling variants of felts

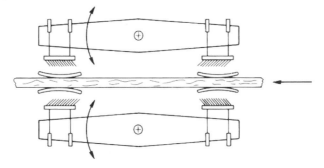

**Fig. 6-7**  Chatham procedure (Fibrewoven), according [8]

Process modifications for the needling technique [6–8] are intended to bind the fibres more intensely into the felt and, therefore, at better compression and solidification. In particular in the United States, the "Fibrewoven" process developed by Chatham has been known [9]. This process uses double-sided alternating needling with diagonally penetrating needles (Fig. 6-7). The possibilities of diagonal needling which have been realized formerly by DILO in the machine types OR, RON-TEX, DI-LOFT and SKR have been taken up by Dr. Ernst Fehrer AG, Linz/Austria in 1996 for their one-board and two-board needle looms and put into practice. In this process, the felt is fed during needling through a curved bed plate thus obtaining the diagonal penetration. The stitching plate's curvature determines the angular change of penetration which decreases from infeed to the curve's apex and increases toward the delivery end. Fig. 6-8 shows the different base principles of needling according to the conventional technology and the H1-technology.

The following differences between the H1-process of Fehrer AG and the conventional process are summarized:

- The diagonal penetration realizes a deeper penetration into the felt and thus diagonally arranged tufts with bigger compression effect (Fig. 6-9).
- Different penetration angles of the needles guarantee always different positions of the fibre tufts in the felt's cross section contributing to high felt compression.
- The increasing penetration angle at the delivery side results inevitably in a low-draft fibre reorientation from the cross direction to the machine direction.
- Steady felt transport between the stitching plates during needling.

This needling principle has been realized successfully by Fehrer AG (Fig. 6-10) for needling from top or bottom in pre- and finish-needle loom applications.

In contrast to the present single-axle movement of the needles, OSKAR DILO Maschinenfabrik KG, Eberbach, have developed an elliptical needle movement while still maintaining optimal needling patterns.

Originating from the known technical solutions of the proven DI-LOOM series, DILO was already able in 1995 to present a machine which had been developed primarily for needling spun-bondeds with high through-put speed. The needle

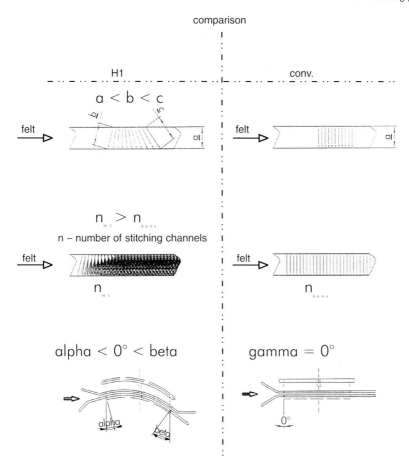

comparison

Fig. 6-8    H1-technology of Ernst Fehrer, Linz/Austria

beam of this machine type HSC Hyperpunch operates with a simultaneous vertical and horizontal stroke movement through an additional eccentric (Fig. 6-11).
Thus at high stroke frequencies the product's typical advance per stroke can be increased. This was achieved by the needles following the material flow at the point
of penetration until their exit from the felt as a result of their elliptical movement.

The technical solution considers primarily particularly high needle densities,
high through-put speeds at the highest machine speeds and the horizontal fixed
follow through of the needles to the adapted process.

Meanwhile, the newly developed and already available machine type HV is especially designed for needling staple fibre with as low a draft as possible. In a newer
design solution it is possible to adjust the horizontal stroke infinitely. With the
help of computer-developed proven random needle patterns, the movement of

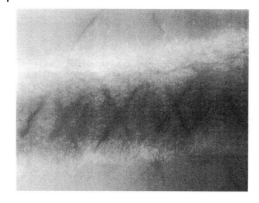

**Fig. 6-9**  Felt structure of a felt consolidated with H1-technology (photograph by courtesy of Ernst Fehrer, Linz/Austria)

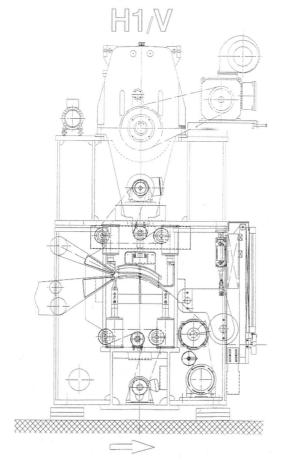

**Fig. 6-10**  Pre-needle loom for H1-technology of Ernst Fehrer (photograph by courtesy of Ernst Fehrer, Linz/Austria)

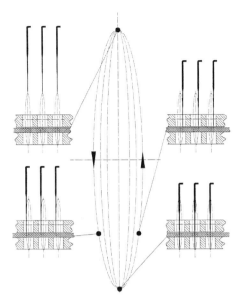

**Fig. 6-11** Elliptical needle movement of O. Dilo Maschinenfabrik KG (photograph by courtesy of O. Dilo Maschinenfabrik KG)

each needle takes place within the respective hole of the bed plate. Thus the needles do not only compress the material but provide in addition a transport infinitely variable to the material's delivery speed during their dwell time, resulting in a drastic reduction of previous prevalent drafts. During pre-needling, the elliptical needle beam kinematics provide an active felt transport through the needling zone. With this, the detrimental drafts are nearly completely avoided. Furthermore, shrinkage which has been caused by draft is minimized. The homogeneity of the needle pattern is enhanced as the homogeneity of the penetrations' distribution always gets worse with increased draft. In particular the felt's evenness regarding the area weight can be increased considerably by the horizontal advance as the draft's impact has been almost totally eliminated.

During finish-needling the main advantage is that the needles follow the material flow within the needle zone and do not hinder it, or stop it completely during penetration. In conventional needle looms the already very sensitive felt is being used as "crumple zone" during deceleration. This results in shifts within the felt layers in the base material causing a deteriorated surface and short term variations in thickness or area weight. The intermittent advance in conventional needle looms causes detrimental vibrations of all parts of the needle looms, in particular the material transport rolls and tension rolls. The damage to the base material which is to be expected is much higher in conventional needle looms as the needle rubs or saws with high friction at the fibres. No elongated needling holes are created, in fact are totally avoided by use of the Hyperpunch needle loom [8].

Apart from flat-needling, OSKAR DILO Maschinenfabrik KG, Eberbach, has also realized the production of circular needle felts [10, 11]. The principle for

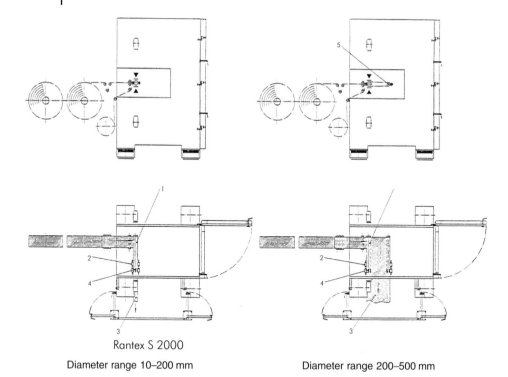

Rantex S 2000

Diameter range 10–200 mm          Diameter range 200–500 mm

**Fig. 6-12**  Circular needling principle RONTEX of O. Dilo Maschinenfabrik KG
(photograph by courtesy of O. Dilo Maschinenfabrik KG)
1 stitching tube, 2 pressure cylinders, 3 tube, 4 delivery, 5 tension roller

needling circular tubes developed by DILO is based on winding up lightly compressed fibre or filament needle felts around one or two adjustable rollers and
then needling them from top and bottom with two needling units. During this
process, the high compression is achieved by changing penetration angles and
cross stitching channels. The tube is continuously drawn away by the rollers
(Fig. 6-12). With the latest development of the "RONTEX S 2000" into a highly
productive process by DILO, the prerequisites have been created to open up new
technical applications for needled tubes [11].

This process allows the production of tubes with
– inside diameter of 10–500 mm
– wall thicknesses of 2–20 mm
– a tube weight range of 400–4800 g/m length
– density variations
– density gradient across the tube wall
– reinforcements
– a multi-layer wall construction by using fibre felts of different fibre types
  and fineness

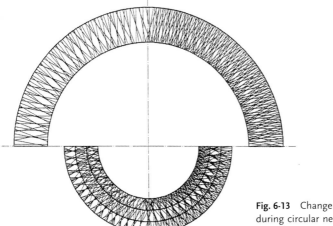

**Fig. 6-13** Change of stitching angle during circular needling, system RONTEX

The multitude of product variations allow not only the use of needled felted tubes as roll covers or filter cartridges but offer also the possibility to use them in the area of cable wraps, pipe renovation, drainage pipe and other technical felt products. Fig. 6-13 shows how the penetration angle changes during circular needling if the presented needle felt is guided on a fixed stitching tube. The possibilities for the construction of a tube cross-section in layers and thus of a thickness gradient are shown in Fig. 6-14.

The needle design type RONTEX S 2000 allows continuous tube production
– at stroke frequencies of max. 2500 $min^{-1}$
– with two needle boards and
– with up to 3000 needles/1 m of working width

The delivery speed of the tube depends on the required lead of the felt spiral which is being built up on the fixed mandrel as well as on felt mass and felt width and the desired tube thickness at a given infeed speed. An automatic length cutting device guarantees the production of reproducible constant felt lengths.

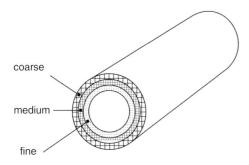

coarse

medium

fine

**Fig. 6-14** Construction of a tube diameter in layers

### 6.1.1.3 **Felt removal**

At the end of the needle zone the needle felt is guided between a pair of clamping rolls. Previously, at low stroke frequencies, the delivery of the needle felt occurred intermittently. The needle penetrated in the felt, which was not moved. As soon as the needles were disengaged, the needle felt was being advanced. Modern needling installations function with continuous advance. The contact pressure of the delivery rolls and the contact area can be adjusted to the product. In general, the needle felt is being guided positively and without deformations. The drive of the delivery rolls can either be synchronous to the stitching frequency of the needle beam or individually controlled.

If longitudinal cutting or individual cross cuts for tiles are desired, an accumulator, longitudinal and cross cutting device with separate winder and/or plaiter can be installed after felt delivery.

### 6.1.2
**High-performance needling technique**

During the past 20 years, the needling of felts has developed into a high-performance technique in the systems engineering field. Further developments of this process as e.g. the one- and two-sided needling of felts have increased the machines' efficiency. This efficiency is characterized by stroke frequencies of the needle beam of more than 3000 $min^{-1}$ and high needle densities per meter of working width. The high speed needling results at the same time in high dynamical needling forces per cycle. These forces are a function of the felt's resistance at the needles' penetration, the needle's gauge and its design.

An analysis of the needling forces has shown that the maximum penetration and pull-out forces as well as the recurring energies depend practically on the resistance of the fibre felt [12]. It is important to minimize existing friction or forces between fibres and between the fibres and the steel by using suitable fibre finishes. Optimum fibre lubrications ease the fibre movement at the tuft formation and have a positive effect on the material's appearance [13]. The high-performance needling technique of the leading installation manufacturers, e.g. DILO (D), FEHRER (A) and ASSELIN (F) are – depending on the construction – characterized by

– the use of standardized modules and elements
– modular main shaft system with rotating and oscillating masses, connecting rod and rocker arm guiding system (DILO), or alternatively gear box with piston guiding (FEHRER, ASSELIN)
– central lubrication system for main and eccentric bearings with automatic oil pressure system and minimum grease lubrication
– use of carbon fibre reinforced composites and light alloy
– pneumatic fast clamping systems for needle boards

Dedusting systems for the needling zone offered by DILO and FEHRER keep the bed plate clean and extend cleaning intervals. High operating efficiency requires at the same time high machine efficiency and minimum maintenance.

6.1.3
**Surface structuring**

Needle felts normally have a smooth surface. Due to the penetration of the barbed needles in the felt, the mostly horizontally oriented fibres or fibre portions are vertically arranged as tufts in the felt. Depending on penetration depth and stitching direction of the needles there are fibre sections or loops at the stitching channel's bottom, which are not corrected when the needles are removed. These fibre loops are bent over during the subsequent transport and bound into the felt by the following needle insertion. One can distinguish between the needle's entry and exit side of the needle felt. Structured needle felts, also called nonwoven pile fabric, on the other hand, are characterized by a distinguished surface texture on their surface through loops or velvet surface. Until now, products of this kind have been produced by weaving, knitting and tufting. Structured needle felts can be produced more economically and have proved successful in floor coverings and as moulded parts for automotive interior linings. The structuring of pre-needled felts is done by speciality needles, e.g. fork needles, which push fibre piles (a kind of loop) out of the preconsolidated felt and thus the top layer forms a loop structure. In order to achieve this structure, it is necessary that the fibre loops remain in their vertical position which means that they are not brought down. This is made possible by the use of a lamella table instead of the perforated bed plate. The distance between the lamellas determines the needling gauge and thus the number of loop rows per inch. Due to efficiency reasons several needles are arranged behind each other. Fig. 6-15 shows the creation of fibre piles between the lamellas and the needle arrangement for forming rib and velour structures.

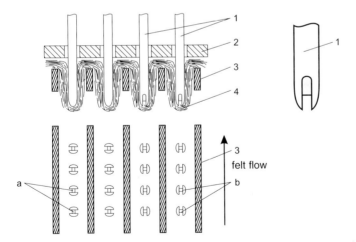

**Fig. 6-15**  Forms of fiber piles for the pole structure and fork needle position for rib and velour qualities
1 fork needle, 2 stripper plate, 3 pair of lamellas, 4 fiber pile
Positioning of fork needle: *a* rib quality, *b* velour quality

These machines are marketed under the description DI-LOOP (DILO) and NL 11-SE (Fehrer).

The fibre piles must be formed very closely to the surface. The number of fibres per loop is determined by the fibre fineness and length as well as the needle design. The loop's height is determined by the penetration depth. Penetration depths ranging from 6 to 10 mm are common. The loops' density depends on the needle rows arranged one behind the other, on the lamella table gauge and on the advance per stroke. When very dense structures surfaces are to be needled, strong fibre piles must be built resulting in high needling forces. The machine load is then multiplied several times. Modern needle design, fibre lengths ranging between 60 and 75 mm and the use of suitable fibre finish aid loop creation and minimize fibre damage.

Patterning of the needle felt's structure is possible when the needles in the needle board are arranged according to the pattern. By lifting and lowering the lamella table the fibre piles are built according to the pattern. Varying the penetration and transportation time of the felt and the number of repeats the patterning variety in regard to the repeat length is increased at a constant needle arrangement.

Messrs. DILO supply needle looms for structuring and patterning with stroke frequencies of up to 2000 $\text{min}^{-1}$ and throughput speeds of more than 20 m/min. The lowering and lifting of the lamella table for patterning is done hydraulically. Apart from the hydraulic patterning also a so-called advance patterning is possible. This leads to relief patterns (3D patterns).

Dr. Ernst Fehrer AG has solved the problem of structuring needle felts by means of a dynamically digitally controlled patterning. This means the high-frequency stroke movement is overlaid by a movement to change the penetration depth assuring that the lamella table stays in a fixed position [14].

The transportation system is firmly connected to the machine frame. When the needles' penetration depth is changed, it stays in its position [15].

The described processes, even if one includes the DI-LOFT machine [16] developed by DILO, allow the formation of loops in rows. The study done by Lünenschloß and Gupta [17] show the limits in increasing pile densities of these techniques.

A new process using a mobile brush conveyor instead of the stationary lamella table allows a random arrangement of loops. This so-called DI-LOUR II machine, developed by DILO, achieves a classic velour with high pile density. When moulding the felts, as it is common when used as automotive interior linings, this random velour has definite advantages.

The needling principle is shown in Fig. 6-16. This process led also to the development of new needle types, so-called crown needles and fine fork needles (cf. Fig. 6-23). The fibre tufts are very small but there are more of them and they are distributed over the whole surface. This is the reason why this machine is being built with two needle boards and a high number of needles. The fibre tufts are formed in the brush apron and simultaneously transported until the structured felt is pulled out of the brush apron by a pair of delivery rollers.

Only a short time after the introduction of this technique, further work on an increase in pile density and on needle felted velour patterning was conducted.

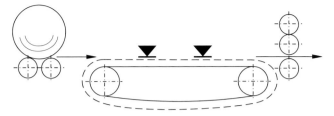

**Fig. 6-16** Needling principle of the DI-LOUR machine, O. Dilo Maschinenfabrik KG

The following here are two examples to explain the machine technical solutions of OSKAR DILO Maschinenfabrik KG, Eberbach:

The machine DI-LOUR DS is a structuring and patterning machine which uses two needling heads with a common brush conveyor and possibly an additional yarn tacker (cf. Fig. 6-17). The following production modes are possible:

1. Production direction from left to right:
   A plain pre-needled felt roll is continuously pre-structured. In front of the second needling head a pattern of different yarns or other textile pieces are being needled on the base felt by the yarn tacker. Colour effects show up on the surface.

2. Production direction from right to left:
   The velour base layer is built by two needle boards with a high stitching density. The introduction of a second, possibly coloured, needle felted substrate results in a very high pile density. In addition, by lifting and lowering the software-driven brush apron, fibre tufts of a different colour can be created in the base velour. The needle board must be equipped according to the desired pattern.

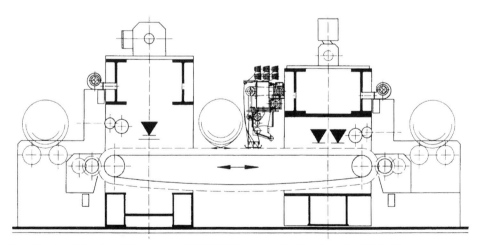

**Fig. 6-17** Principle of needling machine DI-LOUR DS, system O. Dilo Maschinenfabrik KG

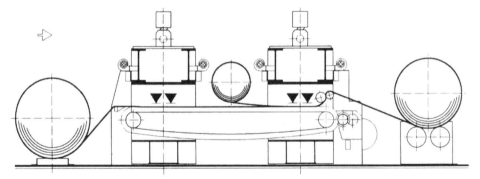

**Fig. 6-18**   Principle of needling machine DI-LOUR IV, system O. Dilo Maschinenfabrik KG

The DI-LOUR IV needle loom (Fig. 6-18) has also two needling units, however, with two needle boards each in normal or wide execution. Both units work on a common brush conveyor. The first felt layer is being needled very intensely by the first needling unit. The felt which is now formed to fibre piles stays in the brush. The second fed felt forms the back layer when passing the second needling zone and is tightly bonded to the first layer resulting in more fibre piles. This allows the production of velour with high pile density and stability. The velour has a tight surface which has a velvet-like character after shearing with a good hand. The capacity of the machine is characterized by over 30,000 needles per meter of working width, stroke frequencies exceeding 2,000 min$^{-1}$ and production speeds of up to 15 m/min.

It is also possible to work with one felt layer only and to use both needling units to increase production.

### 6.1.4
**Paper machine clothing**

The production of needled paper machine clothing has lead to new process developments in order to fulfil the requirements. Endless needle felts serving as press or dry clothing in 10–12 m working width must have the following characteristics: regular pressure distribution over the whole working width, minimum flow resistance for a good drainage when pressing the pulp, high strength – especially in longitudinal direction – and an even surface smoothness. They are reinforced with a carrier of flat or tubular fabric. The weft threads of the fabric in the needle felted tube reinforce it in longitudinal direction. The production of needled paper machine clothing is realized in two steps: pre-needling which means the production of a needle felt, and finish-needling on an endless carrier fabric. Normally, during pre-needling the felt is being produced on an installation consisting of pre-card, intermediate cross-lapper, main card and horizontal cross-lapper in the required working width. The fibres within the felt's plane therefore are predominately cross-oriented. An important factor during needling is to keep the controlled draft between infeed and delivery at a minimum. The present pre-needle looms deliver the pre-needled felt after the delivery rollers via a slide to the winder which can be equipped with a weight balancing system.

Due to the required high strength characteristics in machine direction of the press clothing, it is often advantageous to arrange the fibres within the needle felt in longitudinal direction. This requires that the longitudinal fibre orientation of the web has to be preserved in the needle felt. The felt is being built by doubling the fibre webs in longitudinal direction. Fig. 6-19 shows the base principle of the BELTEX process of DILO. With this method, the fibre web from the card is divided in 4 layers, building a felt by doubling in the longitudinal direction and then needled from both sides by two needling heads. The needle felt is processed in circumferential direction via two rollers having a fixed distance and needled spirally laterally offset into a needle felted tube. High precision for a precise coincidence of the layers without overlapping is a prerequisite for an even distribution of the mass in longitudinal and cross direction. By a continuous feeding of the felt and its entanglement in already needled felt layers, the layers are entangled with each other and intensely consolidated. The needle felted tubes have a high regularity and a high felt density. At the same time it is possible to feed and entangle reinforcing monofilaments. Very advantageous to press clothing is that these felt tubes do not have splices. Furthermore, possible scrim damages during finish-needling are minimized as the stitching density can be reduced.

The finish-needling of paper machine clothing of up to 16 m width requires a special engineering solution. The carrier fabric is covered on both sides with pre-needled needle felts. The upper and the lower needle felted layer can be differently composed according to the functions which have to be fulfilled, which means they

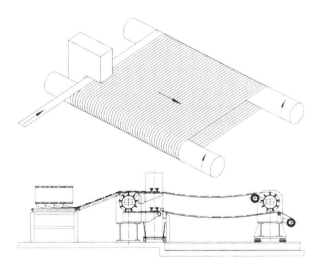

**Fig. 6-19** Needling installation for needling papermachine felts according to the BELTEX system [16]

may have different surface smoothness, water permeability characteristics, high dimensional stability and high strength.

Large needle looms can have up to eight needling zones (needle boards), needling either individually or in pairs in a needling unit the needle felt-carrier fabric-compound from outside to inside and from inside to outside either simultaneously or successively.

A needling installation according to Fig. 6-20 consists of the needle loom itself and a movable tension carriage which enables the tensioning of the carrier fabric and felt transport during finish-needling. One side of the machine has to be opened for mounting the carrier fabric (in tubular form) and for delivering the needle felt. Several machine suppliers offer different solutions for this operation. In any case, the middle part of the needle loom has to be cantilevered to deliver the tube and to mount the new carrier. The weight of this machine part can be more than 100 t depending on its working width. The needling units can separately be switched on or off and thus each needle beam can be brought out of operation. According to the production program, installations of this kind are automatically controlled. A necessary intense needling of the felt-carrier combination can lead to damage of the carrier fabric and thus to a loss in strength. Lünenschloß and Gupta [18] have studied the influence of needling conditions on the damage to the carrier fabric. In general, one can say the following:

The degree of damage increases with the density of the fabric. Finer fibres and higher area masses minimize the damages to the carrier fabric and increase its remaining strength. At a very high consolidation of the needle felt, the tear strength of the scrim can be so low that the strength of the needle felt also decreases. The damage to the fabric heavily depends on the kind of needle, the operational angle of the needle barbs to the monofilaments and the actual penetration depth. Remedial action may be taken by the use of the HYPERPUNCH system developed by DILO which enables the needles to follow the felt during needling.

The requirements for paper machine clothing depend on the requirements of the paper machine and may vary from machine to machine. The various

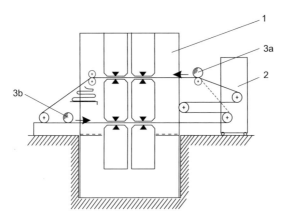

Fig. 6-20 Installation for finish-needling of papermachine felts 1 needle loom with 8 needle boards, 2 movable tension carriage, 3 pre-needled felt a outer layer, b inner layer

technological possibilities when finish-needling these felts allow fulfilment special requirements for different uses.

## 6.1.5
### Needle characteristics

The felting needles are the tool to reorientate and to form fibre or filament tufts in the felt's vertical plane. In this way, due to the compression friction, the consolidation of the felt is achieved. Fineness, design and execution of the needles determine the degree of consolidation. This degree heavily depends on the number of fibres which are reoriented at each penetration of the needles without damage and how many penetrations per unit of area are made. Fig. 6-21 shows the principle drawing of a needle.

Important segments of the needle are the crank, shank, intermediate blade, blade, the barbs and the point. The crank angle to the blade is 0° on standard needles. Thus the barbs are arranged with angles of 60°, 180° and 300° on the blade. Other crank angles are possible. Needles are still named and sold according to gauge. The gauge-system is based on chosen coefficients of measure. Different coefficients of measure are chosen for the shank, the intermediate blade and the blade. For the blade the coefficient of measure refers to its diameter before conversion to a triangle with three identical sides. The crank and the shank are used

single reduced          double reduced

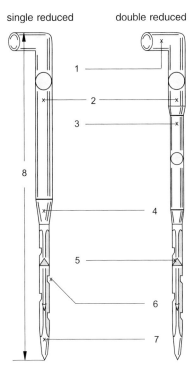

**Fig. 6-21** Felting needle designations
1 crank, 2 shank, 3 reduced shank, 4 shoulder,
5 working blade, 6 barb, 7 point, 8 needle length

**Table 6-2** Dimension of barb needles

| Gauge gg | Blade (inch/mm) | Reduced blade (inch/mm) | Working blade (inch/mm) | Needle length (inch/mm) |
|---|---|---|---|---|
| 12 | 0.105/2.67 | | | |
| 14 | 0.080/2.03 | 0.077/1.95 | 0.077/1.95 | |
| 15 | 0.072/1.83 | 0.069/1.75 | 0.069/1.75 | |
| 16 | 0.064/1.63 | 0.059/1.50 | 0.059/1.50 | |
| 18 | | 0.047/1.20 | 0.047/1.20 | |
| 20 | | | 0.037/0.95 | |
| 22 | | | 0.035/0.90 | |
| 25 | | | 0.031/0.80 | |
| 30 | | | 0.028/0.70 | |
| 32 | | | 0.026/0.65 | |
| 34 | | | 0.024/0.60 | |
| 36 | | | 0.022/0.55 | |
| 38 | | | 0.020/0.50 | |
| 40 | | | 0.018/0.45 | |
| 42 | | | 0.016/0.40 | |
| | | | | 4.0/103.4 [1] |
| | | | | 3.5/90.7 [1] |
| | | | | 3.0/78.0 [1] |

[1] Needle length in inch is from crank inside to needle's point.
   Needle length in mm is indicated from crank outside to point.

to fix the needles in the needle board. At the same time also the needle size determines the needle's characteristic when penetrating the felt. The blade with the barbs is the working part. The point is the element which first penetrates the felt and suppresses the fibres to enable the barbs to seize the fibres and transport them. In general one can say that for the processing of finer fibres finer needles are used. The most important needle dimensions are mentioned in Table 6-2.

The development of the needle felting technology has also lead to different needle developments of the needle manufacturers. Groz-Beckert Nadelfabriken, Albstadt, [19] and Singer Spezialnadelfabrik, Würselen [20] recommend suitable needles for the most different applications [21].

The barbs are arranged on the three edges of the triangular working blade. In general, there are two or three at each edge. The barbs can be chiselled or die pressed or rounded into the working blade. The fibre transport at the needle's penetration into the needle felt is mainly determined by the barb depth and the barb length. The angle determines the barb's ability to hold the fibres. With die pressed needles the edges are rounded, which positively influences the needling of synthetic fibres. The barbs are arranged on the edges but avoiding two barbs are at the same level. The distance between the barbs can be chosen from big to small. It is e.g. 6.3 mm at regular barb needles (RB) and 3.3 mm at close barb ones (CB). The distance between the barbs and the number of barbs on each edge determine the working blade's length which on the other hand influences the nee-

barb dimension

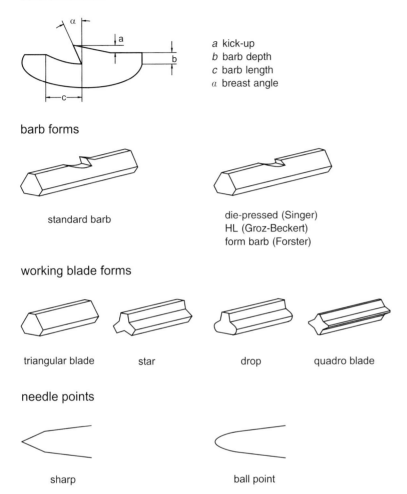

a  kick-up
b  barb depth
c  barb length
α  breast angle

barb forms

standard barb

die-pressed (Singer)
HL (Groz-Beckert)
form barb (Forster)

working blade forms

triangular blade          star                 drop              quadro blade

needle points

sharp                          ball point

**Fig. 6-22**  Barb and working blade forms

dle stability. The needle point can either be sharp or rounded. The point design essentially influences the penetration characteristics of the working blade into the felt (Fig. 6-22).

Apart from the standard executions of the needles there are a number of variation possibilities which can be adopted to the fibres to be processed, the needling conditions and to the final product. There are some examples listed hereafter:

**Barb dimensions**
Barb depth and length generally depend on the needle fineness. The barb length may be reduced by chiselling, which favours the transport characteristics of the

fibres. Die pressed or rounded barbs favour fibre transport at a long lifetime of the needle.

### Working blade

In the case of standard needles, the working blade is triangular from shoulder to point. This ensures that the resistance against deflection is the same in all directions. Star shape, square or drop shape blades allow a change in the kind of barbs and their distribution.

For needling paper machine clothing or filter media with carrier fabric, needles with a drop shape working blade and barbs on only one edge have proven themselves. They minimize damage to the carrier fabric. Barb distances and depths can be reduced distributing the barbs very densely on the edges. The barbs can also be distributed on only one edge or on two edges.

A conical working blade with barbs which become finer at the point reduces the needle penetration force and lends more stability to the needle. The deeper the needle penetrates into the felt, the more fibres are being transported by the barb. This reduces the danger of needle breakage. This needle, however, leaves heavy markings on the material surface.

### Point

In general, the distance between needle point and first barb is 6.4 mm. It may, however, be smaller (3.2 mm). The point itself can be sharp or rounded. In the meantime there are some hundred different kinds of needles from each needle manufacturer.

For surface consolidation of structured needle felts there are in essence two different needle types – fork and crown needles. The first kind is mainly used for coarser fibres. The fork needle point is shaped like a fork (Fig. 6-23), while with

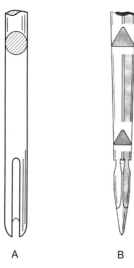

A              B

**Fig. 6-23** Structuring needles.
*A fork needle, B crown needle*

the crown needle three barbs are arranged at almost the same height. Width and depth of the fork determine the amount of fibre which is transformed into a loop.

The finenesses are also expressed in the gauge system. The position of the structuring elements to the needle crank is essential for the formation of rib or velour surfaces on needle felts (Fig. 6-15).

## 6.1.6
### Influence of needling conditions on the needle felt's characteristics

The characteristics of the needle felts depend on the fibres used, the technological conditions of web formation and above all on the consolidation. The knowledge of existing correlations is the basis for application-oriented improvements of product and process technology. Generally important characteristics of needle felts are the degree of felt compression, the strength-elongation ratio and the permeability characteristics.

For the influence of fibre characteristics apart from fibre strength and elongation, the geometrical dimensions – fibre fineness, length, crimp, diameter shape – and surface quality have to be considered. Lünenschloß found out in his studies [22–25] that longer fibre lengths result in higher strength, higher felt density and less air permeability. Hearle and Sultan [26] explain the influence of the fibre length on the strength by the fact that when using longer fibres the movement of the fibres into the vertical plane during transformation is minimized. According to Kosova and Krčma the optimum fibre length for needle felts is in the range of 50–80 mm [27, 28]. Finer fibres lead to smaller felt thickness and to lower air permeability [22–25]. The needling of finer fibres requires inevitably also the use of finer needles to achieve sufficient strength characteristics. Lünenschloß [22] also concludes that a higher crimp results in a higher tear resistance and elongation and a better dimensional stability of the needle felts.

The characteristics and the structure of needle felts also depends on the web structure and the area mass. Machine oriented web results in a high strength in the longitudinal direction and predominantly cross oriented webs result in a high strength in cross direction. Kosova [27] found a linear correlation between web area mass and strength of the needle felt. According to Lünenschloß [23], the web area mass has a great influence on air permeability.

The process values which are active during needling, i.e. kind of needle, stitching density, penetration depth and draft have influence on the needle felt characteristics. Beyond a certain felt density further needling can lead to felt destruction which is called deneedling.

Böttcher [29] has studied the structure formation of felts during needling to find conclusions on the permeabilities of needle felts which are essential for technical uses. According to him a needle felt consists of fibre web parts in which the fibres are arranged mostly parallel to the fibre surface. They form pores. At the penetration points of the needles, fibres or fibre parts are vertically reoriented to fibre plugs. The vertically parallel fibres form the channels with the included hollow spaces. Important values for the formation of pores and channels are:

| | |
|---|---|
| Fibre diameter: | derived value of fibre density and fineness |
| Amount of fibres per area: | derivable value of web mass, fibre length and fineness |
| Needle diameter: | determines the space required by the barbs for the fibre transport. It depends on the fibre fineness, i.e. coarser fibres mean coarser needles (smaller gauge number) and vice versa |
| Penetration depth: | penetration depth of the needle point out of the felt underside and the number of barbs penetrating the felt cross section |
| Stitching density: | important value for the number of penetrations (hollow spaces) in the needle felt and their size |

Theoretical and practical tests have shown that the amount of fibre sections arranged in a fibre tuft depend on the barb's size, the number of barbs and on the fibre diameter. The theoretically calculated amount depending on the barb's area capable of transporting fibres is double the amount of fibres which is actually transported. There is no linear correlation between the fibre sections arranged in a fibre plug and the number of penetrating barbs.

For practical estimation of the amount of fibre reoriented in the vertical plain at each needle penetration, Böttcher gives the following equation (equation 7):

$$a_{F,PF} = \frac{A_{WH}}{A_F} \cdot 0.67 \cdot K_{WH} \tag{7}$$

$a_{F,PF}$ = amount of fibres in the fibre plug
$A_{WH}$ = barb area capable of transporting fibres in mm$^2$
$A_F$ = area of fibre cross section in mm$^2$
0.67 = ratio factor of the sum circle to triangle
$K_{WH}$ = correction factor of the fibre amount in the plug depending on the number of penetrating barbs

Böttcher [30] found in microscopic examinations that the area ratio of the fibre plugs in the needle felt are in the range of 2–12%. The fibre length of the plug is 6–20% and the fibres are more densely packed in the fibre plug than in the needle felt.

The fibre movement in the felt during needling results in an area change which has impact on the area mass of the needle felt see equation (3), page 274). It depends on the web mass, stitching density, penetration depth and draft. It is generally assumed that the draft during needling has to be minimized as it leads to an uncontrollable dimensional change in the felt and thus to an irregularity.

In general, one can say:

$$m_{NV} = f(m_{vl} Ed, Et)$$

$m_{NV}$ = area mass needle felt
$m_{vl}$ = web mass
Ed = stitching density
Et = penetration depth

The area mass of the needle felt generally increases with the web mass. The impacts of stitching density and penetration depth are more complex. The area mass of a needle felt reduces during needling as in general the change in machine direction is many times bigger than one in cross direction. It may increase if the dimensional change percentage in cross direction is bigger than the one in machine direction. The following equation (8) by Voigtländer [31] is valid for the area mass change dependent on stitching density and penetration depth:

$$\Delta m_{NV} = a_1 \cdot Ed^{c_1 \cdot Et} \tag{8}$$

$\Delta m_{NV}$ = change in area mass
$a_1, c_1$ = constant

The same correlation exists between needle felt density, stitching density and penetration depth.

$$\rho_v = a_2 \cdot Ed^{c_2 \cdot Et} \tag{9}$$

$\rho_v$ = needle felt density
$a_2, c_2$ = constant

Further parameters, e.g. batt feeding system, stripper plate design and stitching plate, number of needles, dwell time of needles in the felt, delivery speed, fibres position in the web, fibre fineness and fibre length, length of the needling zone, one-sided or double-sided needling, also influence draft and thus the dimensional change. For this reason, it is very difficult to predict draft.

The type of needle, especially the design of the working blade, has great influence of the web consolidation and the web density [32]. As the needle felt density rises with increasing stitching density, the felt thickness reduces.

The web consolidation after needling determines the strength and stretch characteristics of the needle felt. With a cross laid batt, i.e. the fibres have a predominantly cross oriented position in the web, the cross strength will be greater than the strength in machine direction. The longitudinal change of the needle felt during needling means more or less a fibre reorientation in machine direction and thus also a change of the strength ratio of machine direction to cross direction. Even without fibre reorientation or at a minor reorientation the strength increase in machine direction is bigger than in cross direction.

The highest tensile strength ratio $F_{LQ}$ is the quotient of longitudinal highest traction $F_{HL}$ to cross highest tensile $F_{HQ}$. A value equivalent to one means the

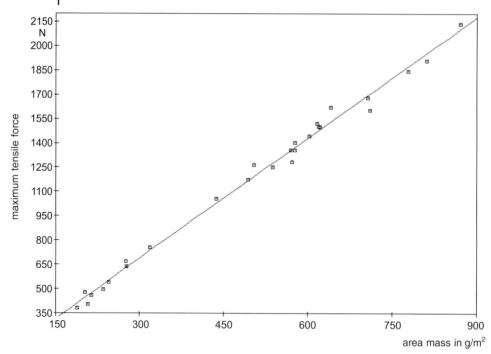

**Fig. 6-24** Medium highest tensile $F_H$ related to the area mass $m_{NV}$ of the needle felt with PP 7,0 dtex/90 mm

strength is the same in both directions. The medium highest tensile strength $F_H$ is the mean of the highest tensile in machine and cross directions. It is directly proportional to the area mass and the thickness of the needle felt with the same needling conditions (Figs. 6-24 and 6-25).

The same dependencies are valid for spunbondeds which consist of endless filaments in the applications they are used e.g. for geotextiles and roofing material. Numerous technical applications require a homogeneous highest tensile strength ratio $F_{LQ}$. It has been proven that increasing stitching density, penetration depth and draft reduce this ratio [33] during the production of needle felts.

If the functional correlation is known, the process parameters can be chosen to achieve a highest tensile ratio equivalent to 1:1. It is possible in principle to needle a felt with a certain draft between infeed and delivery. This means, however, an uncontrolled fibre movement and thus an irregularity of the needle felt. Furthermore, there is a high needle strain. Dilo J.P. [34] has studied the drafting at the example of needle felts for technical applications. This process consists of the steps pre-needling, drafting to reorientate the fibres and finish-needling. The fibre plugs which are built during pre-needling are used as joints for the fibre rotation from the cross direction in machine direction in the following drafting process. A drafting unit with adjustable, controllable draft has been developed.

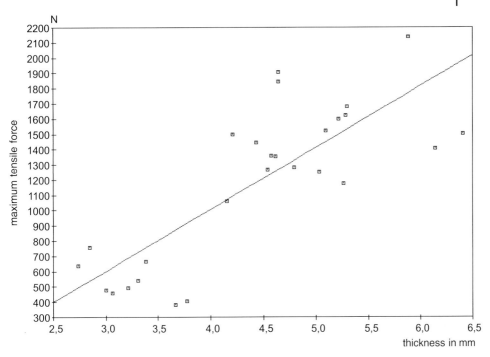

**Fig. 6-25** Medium highest tensile $F_H$ related to the thickness $d_v$ of the needle felt with PP 7,0 dtex/90 mm

The following finish-needling provides the drafted felt with its strength at a highest tensile ratio of about 1. The drafting unit VE (Fig. 6-26) works as four-zone drafting unit according to the rope-friction principle.

Contrary to already known drafting units, where the minimum length of the drafting zones is determined by the diameter of the rollers, the length of the drafting zone in this unit can be adjusted to zero and to the fibre length.

The total drafting factor S is the product of the single drafting factors as according to equation (10)

$$S = s_1 \cdot s_2 \cdot s_3 \cdot s_4$$
$$S = \frac{v_1}{v_0} \cdot \frac{v_2}{v_1} \cdot \frac{v_3}{v_2} \cdot \frac{v_4}{v_3} = \frac{v_4}{v_0} \tag{10}$$

$v_0$   = felt speed at the infeed  
$v_1$–$v_3$ = felt speed at the 1st to 3rd roller  
$v_4$   = felt speed at the exit

Each two of the four drafting zone lengths can be adjusted by vertically moving the intermediate rollers (2nd and 4th roller). Dilo [34] shows that the strengths in machine and cross direction come closer together with a controlled fibre reorienta-

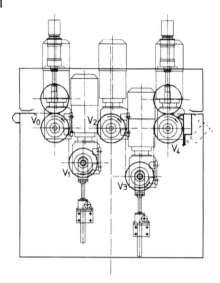

**Fig. 6-26** Drafting unit VE of O. Dilo Maschinenfabrik KG

tion and a secured high felt regularity. The results shown in Table 6-3 confirm that at the same needling conditions for a geotextile the highest felt density, a homogeneous strength ratio and the highest bursting strength have been achieved at a drafting factor S of 25%.

Further increase of the drafting factor leads to a reduction of thickness and density. A highest tensile strength ratio of less than one means for example also a reduction of the bursting strength, which is an important characteristic for geotextiles.

Gupta [35] has studied the strength characteristics of carrier reinforced needle felts. By entangling carrier structures as e.g. scrims or hosiery goods, the dimensional stability of needle felts can be improved essentially. Reinforced needle felts are used as paper machine clothing and as dry and wet filter media. The reinforcement improves strength characteristics and minimizes elongation. Needling, however, may also damage the carrier fabric. The selection of the needle type,

**Table 6-3** Strength characteristics related to drafting factor S for PP-F 7.0 dtex/90 mm, area mass 400 g/m$^2$, stitching density 400 stitches/cm$^2$

| Drafting factor S (%) | Felt density $\rho_v$ (g/cm$^3$) | Highest tensile strength in machine direction $F_{HL}$ (N) | Highest tensile strength in cross direction $F_{HQ}$ (N) | Burst strength $F_D$ (kN) |
|---|---|---|---|---|
| 0 | 0.126 | 1352 | 1550 | 4.012 |
| 25 | 0.137 | 1519 | 1425 | 4.200 |
| 50 | 0.126 | 1371 | 1009 | 3.200 |

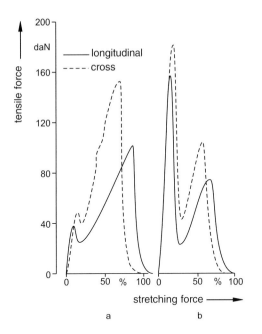

**Fig. 6-27** Characteristic tensile strength – elongation diagram of two reinforced needle felts in machine and cross direction [34]: a) low carrier strength, b) high carrier strength

penetration depth and stitching density influences the degree of damage to the scrim. Fig. 6-27 shows two important strength-elongation curves of reinforced needle felts.

The strength of the carrier fabric of curve a is less than the one of curve b. In general, the warp threads of the scrim are more heavily damaged than the filling yarns. The stitching angle of the needle barbs to the yarn is important (Fig. 6-28).

It becomes clear that a triangular needle is more likely to damage the yarns more heavily than a needle with one edge. Apart from many other factors the damage of the carrier fabric depends on the yarn density in the scrim, on the fibre twist and on the yarn tension when the combination is being needled. A minimum damage of the carrier fabric can be achieved if yarn and scrim construction as well as needling conditions are well tuned to each other to give a maximum retained strength of the scrim.

The permeability characteristics, e.g. air and water permeability of needle felts, are important characteristics for classifying needle felts, press clothing for paper machines and filter media. Voigtländer [36, 37] studied filter characteristics based on a simple model for a cross laid fibre felt and has submitted proposals for the construction of needle filter needle felts. Important characteristics are the pore radius and the air permeability. The pore radius for an ideal felt structure depends on the fibre density, fibre fineness and on the needle felt density

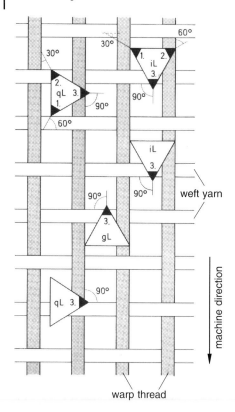

**Fig. 6-28** Effective stitching angle of a (standard) needle with one edge and a (standard) needle with three edges for the different needle positions to the machine direction; position of the 3rd edge to the machine direction [34].
iL=in machine direction,
qL=cross to machine direction,
gL=contrary to machine direction

$$r_{p_{id}} = 2r_F \left( \sqrt{\frac{\rho_F}{2 \cdot \rho_v}} - 0.798 \right) \tag{11}$$

$r_{Pid}$ = ideal pore radius in μm
$r_F$  = fibre radius in μm
$\rho_F$  = fibre felt density in g/cm$^3$
$\rho_v$  = needle felt density in g/cm$^3$

The air permeability is directly proportional to the pore radius and the pressure differential and indirectly proportional to the flow pressure. Pore radius and air permeability depend on ideal and real conditions from the "filter technical needle felt characteristic" KF. This is defined by Voigtländer as follows:

$$K_F \equiv r_F \left( \sqrt{\frac{\rho_F}{2 \cdot \rho_v}} - 0.798 \right) \equiv \frac{r_{p_{id}}}{2} \tag{12}$$

When studying the correlation between air and water permeability in the horizontal and vertical directions of needle felts, Böttcher [30] found out that fibre diameter, web mass and stitching density are important limiting parameters on permeability characteristics. Fibres with a greater diameter form greater pores (equation 11). At an increasing web mass there are more fibres per area unit which influences the needle felt density at a constant thickness. At an increasing stitching density more fibre plugs are created which can lead to a decreasing permeability of the needle felt to a horizontal flow. The following linear correlation exist between air and water permeability:

In vertical direction:

$$\psi = 0.12 \times L_{d,v} + 0.177 \qquad (13)$$

$\psi$ = permittivity $s^{-1}$
$L_{d,v}$ = air permeability, vertical $m^3/m^2$ min

In horizontal direction:

$$\theta = 0.034 \times L_{d,h} + 1.174 \qquad (14)$$

$\theta$ = transmissivity in $cm^2/s$
$L_{d,h}$ = air permeability, horizontal $m^3/m^2$ min

Fig. 6-29 shows the correlation for the ideal pore radius and the needle felt density according to equation (11) for polypropylene fibres [33].

With increasing felt density, the pore radius decreases. Finer fibres result in a smaller pore radius at a constant needle felt density, which is also due to the use of fine needles. As Fig. 6-30 shows, the specific inner surface of a needle felt, which means the total fibre surface, 1 $cm^3$ of felt has a linear dependence on felt density.

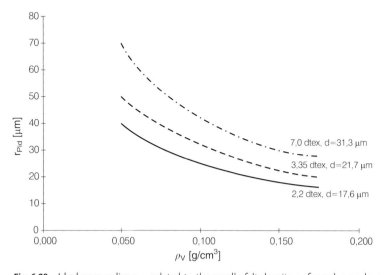

**Fig. 6-29** Ideal pore radius $r_{Pid}$ related to the needle felt density $\rho_v$ for polypropylene fibres of different fineness

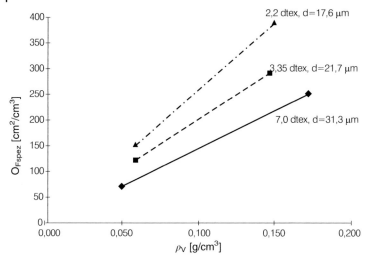

**Fig. 6-30** Inner specific surface of the needle felt $O_{Fspez}$ related to the needle felt's density $\rho_V$ for polypropylene fibres of different fineness

The influence of the fibre fineness becomes visible. This explains the influence of the specific inner surface on the permeability characteristics. Concerning filter media, this means the screening efficiency of a needle felt improves when finer fibres are used and when the needling parameters are chosen to achieve a high needle felt density.

## 6.2
### Loop formation processes
J. Schreiber, A. Wegner, W. Zäh

There is a variety of processes used to bond webs through loop formation by means of threads or fibres. Following [38], the main groups are warp-knitting, stitch-bonding (which is based on warp-knitting) and weft-knitting (Fig. 6-31).

Stitch-bonding, plants and products manufactured this way are called MALIMO in Germany, ARACHNE in the Czech Republic and ATSCHW in Russia.

Historically seen, warp-knitting and weft-knitting are much older than stitch-bonding [39, 40]. The first two mentioned above mainly focus on a mechanical way of making threads into fabrics. Stitch-bonding means to process warp yarns or webs of fibres/elementary fibres by means of loop formation. Within the recent 40 years, stitch-bonding has become industrially important in the context of nonwovens which are made by loop formation.

Basically, warp-knitting means to intermesh the threads of one or several thread systems with one another in a loop-like way so as to make them into a fabric [41]. To achieve this, a variety of types of structure may be used. As to the main direc-

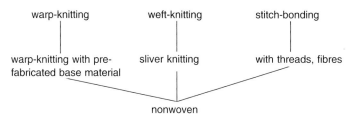

**Fig. 6-31** Loop-formation technologies

tion, the threads in the fabric run longitudinally or, respectively, warpwisely, which explains the name warp-knitting. Warp-knitting processes all execute the steps to form the loops simultaneously at each needle [39].

Weft-knitting means the loops are formed at the needles one after another [39]. As to its main direction, the thread in the fabric runs crosswisely [42].

Stitch-bonding equipment allows to produce textile fabrics either to the stitch-bonding or to the web-knitting processes [43]. The first one creates loops of threads, the latter loops of fibres. Stitch-bonding comprises two processes [38], which are, on the one hand, sewing (i.e. the stitching and bonding of areas, e.g. warp yarns and webs) and, on the other hand, knitting (i.e. the simultaneous formation of loops of threads or, respectively, fibres).

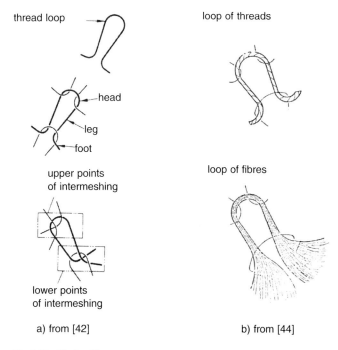

a) from [42]     b) from [44]

**Fig. 6-32** Kinds of loops

The web-processing technologies are distinguished by the stitch-bonding processes Maliwatt and Malivlies and pile web-knitting (Voltex). In addition, there are the innovative processes of Kunit, Multiknit and Kunit layer-bonding.

Stitch-bonding Maliwatt means bonding the fibre web into a loop system of threads. Malivlies and pile web-knitting mean the fibres of the web are used to form the loops. According to DIN 61211 [43], the names "Vlies-Nähgewirke" (stitch-bonded nonwoven) or "Maschen-Vliesgewirke" (stitch-bonded nonwoven of loops) are equally used.

Fig. 6-32 shows elements of a loop of both a thread loop and a fibre loop. A loop is characterized by four intermeshing points. The loops being formed of threads, the thread runs continuously from the loop head via the loop foot on to the next loop. The loops being formed of fibres, there is no visible continuity between the neighbouring loops.

### 6.2.1
### System of processes

Fig. 6-33 gives a systematic overview of processes using **webs**. Focusing on the standard formulations, we have tried to systematically sort the terms generally found [43, 45]. Loop formation is a process which is also well suitable to manufacture nonwoven composites (see Section 6.6).

The brand names known with the different processes have also become common for the equipment that is used and the products manufactured.

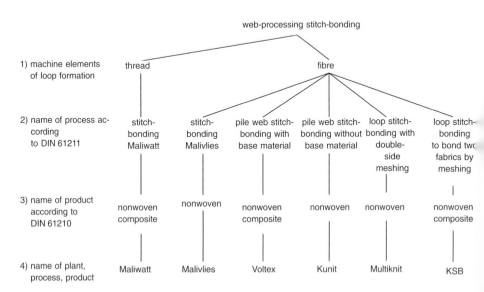

**Fig. 6-33**   System of loop-formation processes to bond nonwovens

## 6.2.1.1 **Stitch-bonding processes**

Stitch-bonding Maliwatt means bonding fibre webs or spunbonded webs, prefer-ably transverse fibre webs, using threads to achieve the bonding. The fibres are bonded into the loops but do not contribute to loop formation. The basic types of structure are pillar-stitch and tricot-stitch (Fig. 6-34). It is also possible to use two systems of threads so both the types of structure can be simultaneously applied. Further types of structure are cord-stitch, velvet-stitch, atlas and inlay.

The loop formation process based on threads is shown in Fig. 6-35. The web-bonding process with its main elements compound needle, closing wire, com-pound needle hook and guide can be seen in process steps a to g.

Fig. 6-36 shows the schematic diagram of the equipment required to produce a stitch-bonded nonwoven Maliwatt. The web formation plant consisting of carding machine and cross lapper is directly connected with the stitch-bonding machine. Fig. 6-37 shows a web bonded by means of loop-formation with threads. This pro-cess makes it possible to produce a very wide variety of nonwoven composites. Their characteristics [46] may range from light to heavy, from softly flowing to compact. Their thickness and strength may vary very much.

The web stitch-bonding process can be carried out both continuously and, if preferable, discontinuously by means of web-formation plants [47, 48]. In addi-tion, a variety of materials such as pre-fabricated fabrics or granulates can be in-corporated into the nonwoven composite.

Further process variants are equipment-related solutions [49] such as

– Maliwatt/G plant, type 14022 with chopper, to process textile glass
– Maliwatt P4 plant, type 14022, to increase crosswise strength by means of parallel weft insertion
– Maliwatt-Intor plant, type 14012, to use short-taped fibres from used materials and spinning mill waste
– tailor-made equipment, for instance to manufacture sandwich textiles on a Maliwatt/S plant

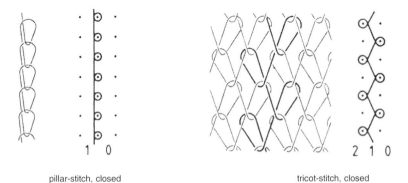

pillar-stitch, closed        tricot-stitch, closed

**Fig. 6-34** Pillar-stitch and tricot-stitch types of structure

a) starting position

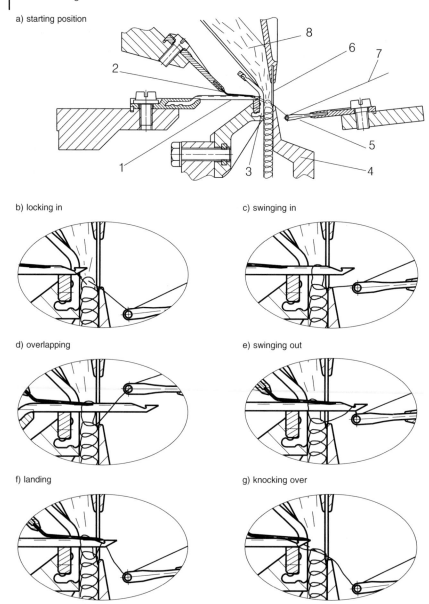

b) locking in

c) swinging in

d) overlapping

e) swinging out

f) landing

g) knocking over

**Fig. 6-35** Stitch-bonding point and loop-formation cycle of a Maliwatt stitch-bonding machine
*1* compound needle, *2* closing wire, *3* knocking-over sinker, *4* support rail, *5* guide, *6* counter
retaining pin, *7* warp yarn, *8* fibre web

**Fig. 6-36** Manufacture of the nonwoven composite Maliwatt
*1* Maliwatt machine, *2* warp beam creel, *3* knitting threads, *4* lap layer, *5* web, *6* stitch-bonded material, *7* pad lapper

The degree of bonding achieved with web stitch-bonded materials depends on the number of loops of threads per unit area. This number is made up of the density of loop wales (number of wales per length unit) and course density (number of courses per length unit). Wales in the nonwoven run in machine-direction, courses run crosswisely.

The density of wales is determined by machine gauge, e.g. the number of knitting elements per 25 cm of working width. As for stitch-bonding, it is currently possible to place a maximum of 22 loop wales within 25 cm of working width. The corresponding machine gauges are 7 F, 10 F, 14 F a.s.o.

Course density is determined by stitch length. Equipment available allows stitch lengths of 0.5 to 5 mm. Machine gauge and stitch length are process values de-

a)        b)

**Fig. 6.37** Nonwoven composite Maliwatt: a) loop surface, b) reverse surface

pending on the product to be made. The basic tools applied in the stitch-bonding process are compound needle, closing wire and knocking-over sinker (Fig. 6-35). Compound needles, for instance, may be coarse, medium, fine or very fine. For better stability, it is common practice to combine two different needle gauges, such as fine/very fine. The length of the compound needle corresponds with lower gauge whereas its width corresponds with higher gauge. Except for the basic knitting tools, further types of tools [50] are available, which are used in the different variants of the stitch-bonding process.

Due to the technological process, stitch-bonding also requires guides (which guide the knitting threads), a support rail and counter retaining pins. The latter serve to fix the fibre web to be needled (see Fig. 6-35 a). Corresponding with the machine gauge in question, the single knitting tools are sealed in needle leads 25 mm wide. They are fixed to needle bars. Via connecting rods, both compound needle and closing wire bar are connected with the driving cams, the knocking-over sinker being inflexible. If any of the knitting elements suffers damage, the needle lead in question is exchanged.

Webs bonded by means of this process show fibres which, generally, show crosswise orientation. It is possible to process any sort of fibre material that can be made into fibre webs [38]. For process-related and technological reasons, fibres of greater length, e.g. wool-type fibres, are preferable. Suitable threads are yarns, twisted yarns, crimp-free or textured filament or film yarns.

Maliwatt stitch-bonding machines are equipped with one or two guide bars. Lapping is achieved by means of two movements, swinging and shogging. While swinging is generally effected by means of a rotary cam and a crank drive, shogging also allows the use of a cam disc. The application of two guide bars makes it possible to use one for pillar stitch, the other for tricot-stitch.

The use of a cam disc opens the way into patterning repeats of 4 or 8 loop courses in the nonwoven. Except for the basic structures of pillar stitch and tricot-stitch, it is then possible to make cord-stitch, velvet-stitch, atlas and inlay. In combination with particular threading-in such as filet-stitch, openwork structures can be manufactured (see Fig. 6-38 a).

Using pile sinkers instead of needle leads (see Fig. 6-35, position 4), web stitch-bonded materials with pile can be created, which allow colour patterns (see Fig. 6-38 b) or embossed patterns.

A large variety of patterned stitch-bonded materials are achievable [51, 52] by means of
– change to bonding
– change to the initial position of the cam discs (turning the cam discs round their axis)
– change to the direction of rotation of the cam discs (turning around the cam discs)
– change to the initial position of the guide bars
– change to the thread insertion and
– use of coloured warps of coloured yarns

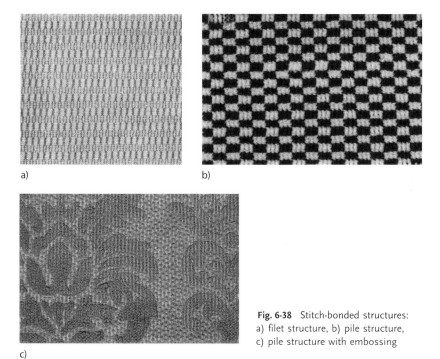

a)

b)

c)

**Fig. 6-38** Stitch-bonded structures:
a) filet structure, b) pile structure,
c) pile structure with embossing

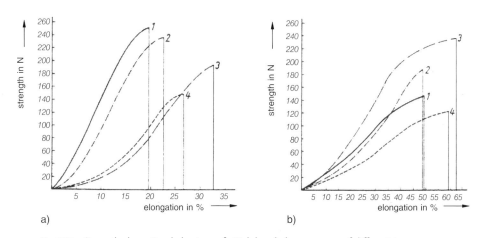

a)

b)

**Fig. 6-39** Strength-elongation behaviour of stitch-bonded nonwovens of different types
of structure: a) longitudinal direction, b) crosswise direction
*1* pillar-stitch, closed; *2* pillar-stitch, open; *3* velvet-stitch, open; *4* tricot-stitch, closed;
length of stitch: 1.02 mm; mass per unit area: 125 g/m$^2$

**Table 6-4** Technical data of a Maliwatt-type stitch-bonding machine Malimo [49]

| Model no. | 14022a) | 14022b) | 14022c) | 14022b)-P | 14023a) | 14023c) | 14021a) | 14023a) |
|---|---|---|---|---|---|---|---|---|
| Max. working width (mm) | 1,700 | 2,800 | 2,800 | 2,800 | 4,100 | 4,100 | 5,100 | 6,100 |
| No. of guide bars | 1 | 1 | 2 | 1 | 1 | 2 | 1 | 1 |
| Machine gauge F (needles/25 mm) | 3.5/7/10/14/18/22 | 3.5/7/14/18/22 | 7/10/14/18 | 7/10/14/18 | 3.5/7/10/14/18/22 | 7/10/14/18 | as N | as N |
| Length of stitch (mm) | 0.5–5.0 | 0.5–5.0 | 0.5–5.0 | 0.5–5.0 | 0.5–5.0 | 0.5–5.0 | 0.5–5.0 | 0.5–5.0 |
| Max. rpm | 2,500 | 2,500 | 1,600 | 1,200 | 2,200 | 1,400 | 3,600a) | 3,600a) |
| Mass per unit area of the stitch-bonded material | —— 30–1,500 to 2,500 with particular materials (glass) —— | | | | | | | |
| Type of structure | tricot-stitch, pillar-stitch | tricot-stitch, pillar-stitch | tricot-stitch, pillar-stitch, cord-stitch, atlas, inlay, combinations | tricot-stitch, pillar-stitch, cord-stitch, atlas, inlay combinations | tricot-stitch, pillar-stitch | tricot-stitch, as 14022c) | tricot-stitch, pillar-stitch | tricot-stitch, pillar-stitch |

Supplementary equipment with all models

**Table 6-5** Technical data of typical applications of a Maliwatt stitch-bonding machine Malimo

| Article | Type of ma-chine | Gauge (F) | Length of stitch (mm) | Sewing thread (tex) | Web material | Mass (g/m²) | rpm (U/ min) | Perfor-mance eff. ca. (m/h) |
|---|---|---|---|---|---|---|---|---|
| Furnishing fabric | 14 021 | 14 | 1.6 | PET/8.4 | PET/CV | 150 | 1,500 | 115 |
| Blanket (emergency equipment) | 14 021 | 7 | 2.5 | L1 PET/25 L2 PET/25 | CV/Bw | 280 | 1,000 | 100 |
| Base material to be coated | 14 022 | 22 | 1.0 | PA/4.4 | CV | 140 | 2,000 | 96 |
| Carpet secondary backing | 14 021 | 14 | 2.0 | PET/15 | PP | 200 | 1,500 | 150 |
| Glass cotton wool | 14 022 | 3.5 | 5.0 | PET/25 | glass fibres | 1600 | 700 | 175 |
| Wound dressing | 14 022 | 14 | 3.5 | elast. thread material | nonwoven CV | 60 | 2,500 | 400 |
| Shoe lining | 14 022 | 18 | 1.5 | PET/7.6 | PET | 110 | 2,500 | 190 |
| Lining material | 14 022 | 18 | 3.0 | PET/7.6 | PP-spun-bonded nonwoven | 60 | 2,500 | 380 |

Loop formation causes the incorporation E of the knitting thread [53] according to the equation (15)

$$E = \frac{l_f - l_w}{l_f} = 1 - \frac{l_w}{l_f} \qquad (15)$$

$l_f$ = initial length of thread
$l_w$ = length of thread in the warp-knitted material

This should be taken into consideration when calculating the material inventory. The ratio E depends on the value of shogging of the type of structure used as well as on length of stitch, mass per unit area, thread tension and, if available, the height of pile sinkers.

Things are simple using one guide bar. Here, incorporation becomes smaller with growing length of stitch and thread tension whereas it becomes larger with growing mass per unit area and height of pile sinker [38]. Due to the larger shogging of the thread, tricot-stitch shows a larger incorporation than pillar stitch.

With two guide bars, tendencies are the same, however, the ratio thread tension to thread tension should not be forgotten [54]. How far the web is tied up also depends on length of stitch, mass per unit area and thread tension (to a smaller degree).

A number of technical data of the Maliwatt type of a stitch-bonding machine Malimo can be found in Table 6-4. For some common practical applications see Table 6-5.

Comprehensive knowledge of the properties of stitch-bonded materials was published by Scholtis a.o. [54, 55] as well as by Böttcher [56].

Strength and elongation properties in longitudinal direction are mainly influenced by the properties of the knitting threads. The incorporation of the fibres into the loops of the knitting threads is essential for the strength in width-wise direction. It depends, for instance, on length of stitch, length of fibre, kind of fibre material. Using two knitting thread systems, a variety of types of structure (showing large shogging such as cord-stitch, velvet-stitch, inlay) may influence strength in width-wise direction essentially. Fig. 6-39 describes how the strength and elongation properties of stitch-bonded materials depend on the types of structure used.

### 6.2.1.2 Stitch-bonding process Malivlies

Contrary to the stitch-bonding process Maliwatt, no threads are used to form the loops but a web whose fibres are, preferably, crosswisely oriented. In this process, the compound needle hook takes up fibres from the web layer facing the laying-in sinker and draws them through the web layer facing the knocking-over sinker (Fig. 6-40). This means part of the fibres are meshed, while another part of them are only incorporated in the loops. One fibre may be both used for loop formation and incorporated in another loop. The fibre length reserves needed to sink the loops are available through the fibres of the unbonded web being flexible, through their crimp and through the lapping angle resulting from the formation of the web. This is why the rule of sinking is not usually neglected.

Comparing Figs. 6-40 and 6-35, it becomes obvious that the laying-in sinker is in operation instead of the guide. It is positioned inflexibly and serves to fix the fibre web.

Fig. 6-41 shows a Malivlies-type stitch-bonding machine Malimo, model 14021, with a working width of 5,600 mm, inclusive of carding machine and cross lapper.

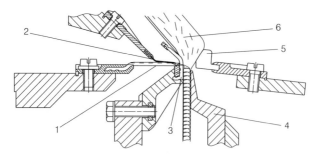

**Fig. 6-40** Stitch-bonding point of a loop web-knitting machine
*1* compound needle, *2* closing wire, *3* knocking-over sinker, *4* support rail,
*5* laying-in sinker, *6* fibre web

**Fig. 6-41** Malivlies-type stitch-bonding machine Malimo, model 14021. Nominal width: 5,600 [49]

As with the stitch-bonding process Maliwatt, machine gauge and length of stitch are used to control bonding intensity. In addition, the degree of bonding may be influenced by a number of changes to the compound needle: to the slot width, to the distance of the laying-in sinker and to the position of the closing point as seen from the knocking-over sinker. Thus, the number of fibres contributing to loop formation can be controlled.

In the loop web-knitting process, long fibres are preferable. The longer the fibres, the more they contribute to loop formation or respectively, the more they are tied up in the loops. Due to their low flexibility, brittle fibres such as glass or carbon are less suitable to be meshed. There are only two ways to achieve patterned effects: the use of webs of layers (e.g. of different colour) and/or the use of flexible laying-in sinkers.

Contrary to the stitch-bonding process Maliwatt, loop formation with fibres results in a higher degree of the web being tied up. This depends on the degree of meshing. A

**Table 6-6** Technical data of a Malivlies-type machine [49]

| Model no. | 14022 | 14022 | 14021 | 14021 | 14021 |
|---|---|---|---|---|---|
| Nominal width (mm) | 1,600 | 2,400 | 3,600 | 4,400 | 5,600 |
| Max. working width (mm) | 1,700 | 2,800 | 4,100 | 4,900 | 6,100 |
| Length of stitch (mm) | 0.5–5.0 | 0.5–5.0 | 0.5–5.0 | 0.5–5.0 | 0.5–5.0 |
| Machine gauge F (needles /25 mm) | | | 3.5 to 22 | | |
| Max. rpm | 2500 | 2500 | 1650 | 1650 | 1650 |
| Performance | 75–675 | 75–675 | 40–405 | 40–405 | 40–405 |
| Mass per unit area of the stitch-bonded nonwoven (g/m$^2$) | 128–1,148 | 210–1,890 | 164–1,600 | 196–1,985 | 244-2,470 |
| Additional equipment (g/m$^2$) | 120–1,200 | 120–1,200 | 120–1,200 | 120–1,200 | 120–1,200 |

Machine gauge, length of stitch, number of revolutions per minute, mass per unit area and web material depend on one another.

**Table 6-7** Technological data of typical applications of a Malivlies-type stitch-bonding machine [49]

| Article | Model of Machine | Fineness (F) | Length of stitch (mm) | Web material | Mass (g/m²) | Revolutions per minute (rpm) | Performance eff. ca. (m/h) |
|---|---|---|---|---|---|---|---|
| Car headlining | 14021 | 18 | 1.8 | PET/CV | 180 | 1,400 | 136 |
| Base material to be coated | 14021 | 18 | 2.0 | PET | 160 | 1,400 | 150 |
| Cushioning | 14021 | 7 | 3.5 | reclaimed fibres | 1,200 | 800 | 145 |
| Polishing cloth | 14022 | 22 | 1.1 | CF/PET | 140 | 2,200 | 130 |
| Packing material | 14022 | 10 | 2.0 | reclaimed fibres | 260 | 1,200 | 135 |
| Geotextile (greening mat) | 14021 | 7 | 2.25 | reclaimed fibres | 300 | 800 | 92 |
| Laminated backing | 14021 | 18 | 1.0 | PET | 150 | 1,400 | 76 |
| Lining material | 14022 | 18 | 2.0 | PAC | 100 | 2,500 | 270 |

high value of the ratio E may result in layer separation in the fibre web fed as that part of the layer of web whose fibres are used to form the web is better tied up. Tables 6-6 and 6-7 contain machine-related technical data and examples of typical applications.

Due to their nonwoven structure, loop web-knitted materials usually show higher strength in crosswise direction than in lengthwise direction. They are characterized by high voluminosity, softness, good absorbent behaviour, good elasticity and air-permeability. Comprehensive investigation into the strength and elongation behaviour depending on the properties of the fibre material and the process conditions was carried out by Scholtis, Ploch, Böttcher a.o. [55, 57–59].

With machine-knitted webs, the moment of the closing wire closing the compound needle hook (closing point) strongly influences mass unevenness. Table 6-42 shows that, depending on the distance of the closing point from the front edge of the knocking-over sinker, parts of web layers of varying thickness are incorporated into loop formation.

It has been found [57] that mass unevenness is at a minimum if the closing point is about one third of the total web thickness away from the front edge of the knocking-over sinker. That is the point at which the full capacity of the compound needle hook to load fibres is reached. The closing point being further away from the front edge of the knocking-over sinker, both thickness and mass unevenness of the web will result in either more or less fibres being incorporated. This causes locally varying degrees of web incorporation, which adds to mass fluctuation.

Generally, the rule of loop sinking used in loop formation with thread-like materials [39] also applies to loop formation with fibres [60]. It is the angle of fibre position $a_F$ (angle between compound needle and fibre direction) which determines whether the compound needle hook can take a fibre, or whether it is taken

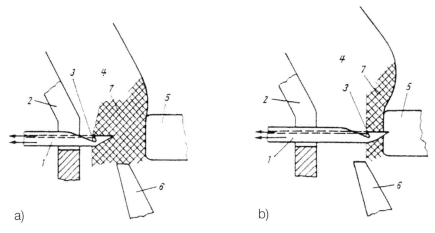

**Fig. 6-42** Point of the closing wire closing the compound needle hook [57]
a) closing point close to knocking-over sinker, b) closing point close to laying-in sinker
*1* compound needle, *2* knocking-over sinker, *3* closing point, *4* web, *5* laying-in sinker,
*6* support rail, *7* layer of web from which fibres are available for meshing

simultaneously by several compound needle hooks, or whether it cannot be taken at all. Ideal conditions given (straight fibres throughout the web) and using a "fine" compound needle, Mägel [61] found that, in case the rule of loop sinking is neglected, damage to the fibres is very likely to occur below an angle of fibre position $a_F < 10$. At about $a_F = 35°$, nonwovens are achieved of approximately equal lengthwise and crosswise strength. The angle of fibre position $a_F$ becoming larger, meshing becomes more rare. From $a_F = 69°$, no fibres can be taken. Looking at minimum damage to the fibres caused by meshing, Mägel [61] calculated an optimal angle of fibre position $a_F$ between $10°$ and $35°$, which relates to ideally straight fibres. He also states that in industry, angles of fibre position $a_F$ between $0.5°$ and $12°$ are common. Nevertheless, there is no damage to the fibres owing to the fact the fibres are not ideally straight, they show a crimp, and as the web is voluminous, there are sufficient reserves available to sink the half-stitches.

Comprehensive research has been carried out to improve the properties of loop web-knitted materials, in particular, their strength and elongation behaviour. In order to characterize the geometric arrangement of the fibres in the loop web-knitted material, Scholtis [62] speaks of the degree of meshing $\eta_m$. The degree of meshing determines the ratio the sums of the loop-likely arranged fibres $\Sigma l_m$ to total fibre length $\Sigma l_f$

$$\eta_m = \frac{\sum l_m}{\sum l_f} \qquad (16)$$

To determine the length of the loop-likely arranged fibres $l_m$, the Munden equation [63] is used. A corrective value is added of 1/3 D, which takes into consideration the thickness of the web and follows the equation

$$\overline{l_m} = \sqrt{\frac{K}{R \cdot S}} + \frac{1}{3}D \tag{17}$$

K = constant
R = course density (in 1/mm)
S = wale density (in 1/mm)
D = thickness of web

Fig. 6-43 gives a geometric overview.

In order to receive the sum of the lengths of the loop-likely integrated fibres, it is necessary to determine the number of fibres per loop. Using this value, the sum of the loop-likely arranged fibres can be calculated, as related to a single loop:

$$\sum l_{m(M)} = \overline{F_{(M)}} \cdot \overline{l_m} = \overline{F_{(M)}} \left( \sqrt{\frac{K}{R \cdot S}} \right) + \frac{1}{3}D \tag{18}$$

$\overline{F_{(M)}}$ is the arithmetic average of the number of fibres per loop determined.

The total fibre length of a web-knitted loop can be calculated to the equation (19)

$$\sum l_{f(M)} = \sum l_{f(A)} \cdot R^{-1} \cdot S^{-1} = \frac{m_A}{T_t} \cdot R^{-1} \cdot S^{-1} \tag{19}$$

$m_A$ = mass per unit area in g/m$^2$
$T_t$ = fibre fineness

The degree of meshing can be received to the equation (16):

$$\eta_m = \frac{l_{m(M)}}{l_{f(M)}} = \frac{\overline{F_{(M)}} \left( \sqrt{\frac{K}{R \cdot S}} + \frac{1}{3}D \right)}{l_{f(A)} \cdot R^{-1} \cdot S^{-1}} \tag{20}$$

In the framework of investigation into modelling the bonding process with loop web-knitted materials, Offermann, Mägel, Ponnahannadige and Jenschke [64–66]

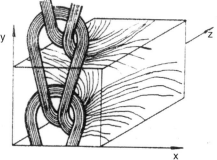

**Fig. 6-43** Geometric presentation of loop-likely arranged fibres $l_m$

determined the length of the fibre loop, bearing in mind the geometric proportions at the stitch-bonding point of the stitch-bonding machine.

Following them, the loop length $l_m$ can be calculated to equation (21):

$$l_m = 2(l_{Apl} + l_H) + t + \frac{3}{2}\left(\frac{t}{2} - \frac{B_{SN}}{2} - \frac{B_{Apl}}{2}\right) \tag{21}$$

$l_{Apl}$ = distance between front edge of knocking-over sinker and back dead centre
   of the compound needle point (in mm)
$l_H$   = length of compound needle hook (in mm)
$B_{SN}$ = width of compound needle in the space of fibre take-in (in mm)
$B_{Apl}$ = width of knocking-over sinker
$x$   = distance between compound needle and knocking-over sinker
$t$   = $2x + B_{SN} + B_{Apl}$

Combining this value with the operation height $h_W$ of the compound needle hook
and the angle of fibre position $a_F$, it is possible to determine fibre incorporation
in the web.

Bearing in mind both fibre material and process variables, the know-how available of the meshing of webs allows to theoretically determine the points where
the fibres are tied up in the loop web-knitted material. This may take us to conclusions concerning the ratio of lengthwise to crosswise strength.

Meshing a web, the material undergoes high mechanical stress, which may lead
to damage to both the fibre material and the knitting tools. The mechanical stress
results from the frictional forces incurred in piercing the fed-in materials (fibre
web or base material) on the one hand and, on the other hand, from forming the
threads or fibres into tuck loops and meshes. Investigation to determine the piercing and tying-up forces was carried out by Ploch [67–69], Zschunke [70] and Nede-

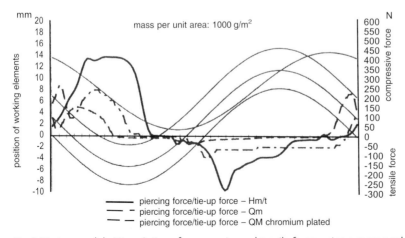

**Fig. 6-44** Loop web-knitting: Action of compressive and tensile forces using a compound needle

wa [71]. With regard to a particular web, Schmalz [72] looked into the compressive and tensile forces incurred in a loop formation cycle (see Fig. 6-44). Except for the compressive and tensile forces mentioned above, it is interesting to watch when they occur in comparison with needle motion. In parts of the process, immense forces act on the knitting elements and the fibre materials. In the example given, a machine gauge of 7 F was used, the compressive forces acting on a compound needle being 64.3 N, the tensile forces 35.7 N.

### 6.2.1.3 Pile stitch-bonding process with base material

Contrary to the stitch-bonding and loop web-knitting processes, a fibre web is tied up in a base material which shows fibres preferably oriented in processing direction, e.g. a card web [73, 74]. Fig. 6-45 shows both the process and the knitting tools.

Using a brush (5), a fibre web (8) ranging between 10 g/m² and 80 g/m² is pressed into the compound needle hook (1).

The fibres taken by the brush are drawn through the base material (7) and formed into loops on the surface facing the knocking-over sinkers (3). Depending on the height of the pile sinkers used, pile folds of corresponding height are created. The ratio velocity of web-feed to loop formation (which can be set) results in the pile being compacted (from 1:4 up to 1:10), the mass of the pile layer ranging from 100 to 800 g/m². Cloth, knitted materials, nonwovens, films etc. can be fed in as base materials. It is important they can be pierced by the compound needles without much damage being done to the materials. The total mass of the pile web-knitted material Voltex, consequently, results from the sum of the masses of the base material and the pile web layer used. The continuous plant consists of a carding machine for pile formation, the pile connection support, a stitch-bonding machine type Voltex with feed system for the base material and a roll stand (Fig. 6-46).

The brush used to compact the web is operated at constant height of oscillation. It does not influence the height of the pile created. Pile height is exclusively deter-

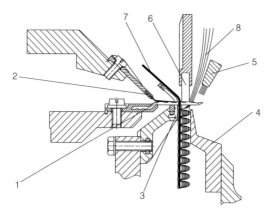

**Fig. 6-45** Stitch-bonding point in pile web-knitting with base material:
*1 compound needle, 2 closing wire, 3 knocking-over sinker, 4 support rail, 5 brush, 6 pile sinker, 7 base material, 8 fibre web*

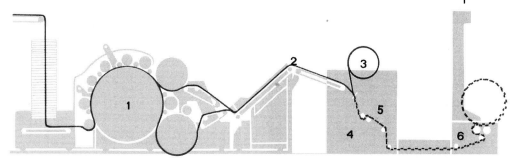

**Fig. 6-46** Continuous plant for the manufacture of pile web-knitted products with base material: *1* webber, *2* pile connection support, *3* base material, *4* stitch-bonding machine, *5* pile web-knitted material, *6* roll stand

mined by the height of the pile sinker. In order to achieve large heights of pile and to tie the fibres up properly, it is necessary to use long fibres. Fibre length should be above 60 mm.

The essential technical data describing a Voltex-type stitch-bonding machine Malimo are contained in Table 6-8.

Processing blends of fibres of a variety of kinds and finenesses results in the pile web structure varying from coarse to fine.

Both course and wale density are essential factors influencing pile density. They strongly determine voluminosity, thermal insulation, touch and look. Strength and elongation depend on the base material used [76].

**Table 6-8** Technical data of stitch-bonding machines Malimo, types Voltex, Kunit and Multiknit [75, 80]

| Technical data | Voltex | Kunit | Multiknit |
|---|---|---|---|
| Nominal width (mm) | 1,600, 2,400 | 1,600, 2,400 | 1,600, 2,400 |
| Max. working width (mm) | 1,700, 2,500 | 1,700, 2,500 (2,800 special design) | 1,700, 2,500 |
| Machine gauge F (needles/25 mm) | 10, 14 | 3.5 7 10 12 14 18 22 | 3.5 7 10 12 14 18 22 |
| Height of pile (mm) | 2 3 4 5 6 7 9 11 (13 15 18 20 23 with 10 F) | 1–17 | |
| Rotations per minute (rpm) | from 500 to 1,200 [1] | up to 1,200 [1] | up to 1,200 |
| Range of stitch length (mm) | 0.7–5.0 | 1.0–5.0 | 1.0–5.0 |
| Mass per unit area (g/m$^2$) | 300–800 [1] | to 700 | 120–800 one-layer goods; 150–1,500 multi-layer goods |

1) Subject to materials used, article specifications and nominal width.

**Fig. 6-47**   Blanket "Molly"

Finishing the raw web-knitted material by means of raising, cropping or tumbling, an even raised pile is achievable in heights ranging from 2 to 17 mm. Such pile web-knitted fabrics with base material are suitable for the manufacture of blankets (Fig. 6-47), shoe lining, soft toy material and lining for winter garments.

### 6.2.1.4   Loop stitch-bonding without base material

To create stitch-bonded materials with pile structure, the novel Kunit process [77] uses fibre webs whose fibres preferably show lengthwise orientation. Those parts of fibre that create the pile preferably take a vertical orientation in the nonwoven. Unlike other stitch-bonding processes, the compound needle folds the fibre web fed in so the brush can press it into the compound needle hook (Fig. 6-48). The compound needle is not pointed since there is no need to pierce the web or the base material. In addition, this helps to sink smaller loops.

Strong loops are created although there are no retractive forces tending to retract surplus loop length [78].

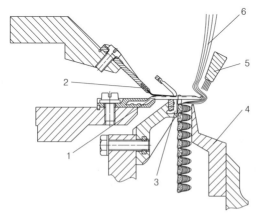

**Fig. 6-48**   Pile stitch-bonding process without base material. Stitch-bonding point:
*1* compound needle, *2* closing wire, *3* knocking-over sinker, *4* support rail, *5* brush, *6* pile

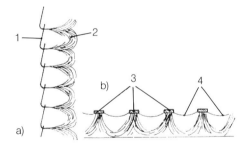

**Fig. 6-49** Pile stitch-bonded material Kunit: a) longitudinal section, b) cross-section.
*1,3* wales, *2* layer of pile folds, *4* crosslinking of wales

The Kunit process does not use a pile sinker, which means the process and the stitch-bonding point of the machine become simpler. Contrary to the Voltex process, pile height is set through the height of oscillation of the brush. Today's machines allow to continuously set it within a range of 8 to 69 mm [79]. The same as with the Voltex process, the card web is compacted (from 1:4 to 1:10). The mass per unit area of the nonwoven Kunit ranges from 100 to 800 g/m². Comparing the Voltex nonwoven composite with the Kunit nonwoven, there is a difference in structure, which also becomes visible. With the Voltex nonwoven composite, the pile tuck courses are, optically, oriented in lengthwise direction since the pile sinkers determine loop orientation. With the Kunit nonwoven, the pile tuck courses show crosswise orientation. The Kunit nonwoven shows a layer of fibre pile, the other surface being a layer of loops (Fig. 6-49).

Analogously to the manufacture of pile stitch-bonded materials with base material, the Kunit-type stitch-bonding machine continuously works with a carding machine. Both wale and course density strongly influence the degree of bonding.

Machine performance is contained in Table 6-8. Reducing both the masses of the brush and of the machine components that drive the brush, an increase in the number of revolutions per minute up to 2,000 rpm has been achieved [79].

Due to the nonwoven structure of the materials created, a variety of products can be developed including upholstery materials and laminated composites for the automotive and furniture-producing industries, insulation and filtration materials as well as packing materials.

### 6.2.1.5  Loop stitch-bonding with double-sided meshing

In this process [81], the pile fibres of nonwovens or other textile fabrics with pile loops or pile clusters (such as Kunit or Voltex nonwovens, high-pile knitwear etc.) are meshed. The process (Fig. 6-50) allows the manufacture of one- as well as multi-layer nonwoven composites.

Using the basic variant, a fabric is fed in the stitch-bonding point in such a way that the compound needles can penetrate the pile clusters, take the fibres and make them into a layer of loops. The Multiknit process in its basic variant results in a three-dimensional stitch-bonded material which is meshed on both its face and back (Fig. 6-51). The layers of loops are connected by one and the same fibres.

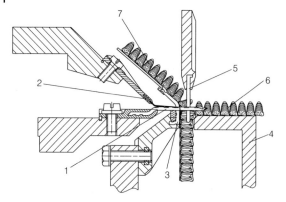

**Fig. 6-50** Stitch-bonding point of a Kunit-type stitch-bonding machine Malimo
*1* compound needle, *2* closing wire, *3* knocking-over sinker,
*4* support rail, *5* pile sinker, *6* Kunit nonwoven, *7* Kunit nonwoven
or textile material with pile structure

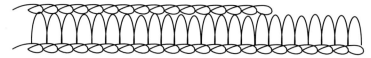

**Fig. 6.51** Basic variant of a Multiknit nonwoven meshed on both its face and back

Another process variant allows to mesh two pile nonwovens so as to make a nonwoven composite (Fig. 6-52).

Inserting nonwovens, other fabrics or scatterable media, nonwoven composites can be made of a diversity of properties [78, 82, 83].

The pile web layer can be discontinuously meshed using a winding-off stand and a stitch-bonding machine. However, Fig. 6-53 gives an example of the continuous manufacture of a Multiknit nonwoven composite.

Whether to use continuous or discontinuous manufacture depends on what is technically feasible and economical as well as on the design of the nonwoven composite.

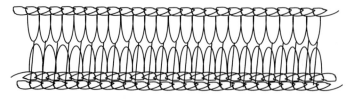

**Fig. 6-52** Meshing a two-layer Multiknit nonwoven composite (basic variant)

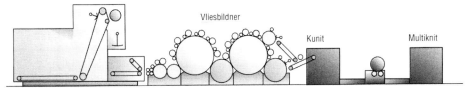

**Fig. 6-53** Continuous plant for the manufacture of Multiknit nonwoven composite-carding machine, stitch-bonding machines Kunit and Multiknit

To mesh the pile layers as evenly as possible, the same machine gauge and stitch length should be used as with e.g. the Kunit nonwoven. For particular effects, other settings are possible. Machine performance in the continuous process is determined by the Kunit machine (Table 6-8).

Working discontinuously and making one-layer Multiknit products, the Multiknit machine can work at up to 2,000 rpm. With multi-layer products, the number of revolutions per minute should be adapted to the piercing forces.

In the Multiknit process, it is possible to produce nonwovens for upholstery to substitute PUR foam in the automotive and furniture-making industries as well as filter materials, heat-retardant materials, sound-absorbing materials, garment lining, incontinence articles, anti-bedsore fleeces and filling materials for moulded parts reinforced with synthetic resin.

#### 6.2.1.6 Loop stitch-bonding to bond two fabrics by meshing

In this process [84] known as the Kunit layer-bonding process (KSB process), two fabrics with pile layers are bonded. The compound needles take the pile fibres of both fabrics to make them, in the middle between the two fabrics, into a third layer of loops (Fig. 6-54).

This allows to bond different prefabricated nonwovens with pile structure so as to make composites of a wide variety of properties [78, 83]. A particular advantage

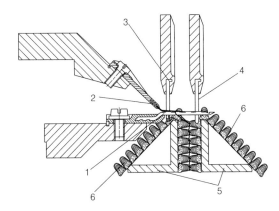

**Fig. 6-54** Stitch-bonding point in the KSB process
1 compound needle, 2 closing wire, 3 knocking-over sinker, 4 counter retaining pin, 5 support rails, 6 Kunit nonwoven

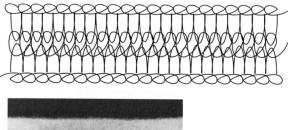

**Fig. 6-55** Nonwoven composite produced to the KSB process

offered by KSB nonwoven composites is their initial volumes are maintained, the nonwoven components remain in their pure original condition. Manufacture requires rolling feed from two large batch rolls so the process is as much as continuous. The degree of bonding is influenced via wale density and course density.

The working tools required in addition to the basic stitch-bonding tools are two support rails and a counter retaining element. Due to the kind of process, the knocker-off sinker needs to be designed open to the bottom. The fibre materials used to form the loops need to allow the conversion via small radii.

A Kunit layer-bonded nonwoven shows a layer of loops on both surfaces, the loops being taken from the materials fed in. In the interior of the nonwoven cross-section, the KSB process creates a third layer of loops bonding the two non-wovens used (Fig. 6-55).

Nonwoven composites made to the Kunit layer-bonding process show, due to their voluminosity, good insulation properties and, as a consequence of their structure, excellent rebound, which can even be enhanced by means of thermal treatment.

### 6.2.2
### Warp knitting

Using warp-knitting machines to bond webs/nonwovens, the fibres are the same as in the stitch-bonding process (Maliwatt), tied up in the loops of the warp-knitted systems of threads. However, they do not create any loops. In addition to the fibre webs, further systems of threads can be tied up [85]:

– bonding through warp yarns by means of one or more guide bars
– bonding through warp yarns and weft insertion
– bonding through warp yarns, weft insertion and pillar inlay
– bonding through warp yarns and pillar inlay
– bonding through warp yarns, weft insertion, pillar inlay as well as yarns
  at angles of +45° and –45° or other angles

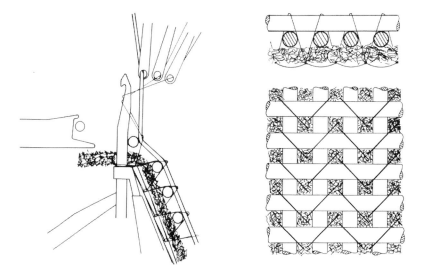

**Fig. 6-56**  Stitch-bonding of a raschel machine RS 3 MUS-V for the reinforcement of nonwoven material and pictogram of a bi-axially reinforced nonwoven [86]

This allows a wide range of products of highly diversified properties. The thread composites inserted focus on higher strength. To meet the requirements in question, it is possible to influence the force absorption capacity. This is achieved adjusting thread arrangement, thread thickness and thread density in accordance with force direction and value. To give an example, they can be tailored to the course of the lines of force in building components.

In Germany, warp-knitting automats and raschel machines [86, 87] for the manufacture of the variants of fabric-formation described are made by Karl Mayer Textilmaschinenfabrik GmbH, Oberhausen and Liba Maschinenfabrik GmbH, Naila (given in alphabetic order).

All manufacturers use the same warp-knitting principle. However, the knitting tools and the ways of how they are positioned in the machines differ slightly. Generalizingly, Fig. 6-56 gives the schematic picture of the stitch-bonding point of a raschel machine.

## 6.2.3
### Knitting

Using circular knitting machines [88], slivers of fibres are fed into the point of operation of the knitting machine, then disentangled by means of mini-cards and fed into the hooks of the latch needles (Fig. 6-57). By means of blast-air nozzles, the fibre ends of the fibre clusters taken over by the latch needles are oriented towards the centre of the needle cylinder. A ground yarn is also processed and, together with the fibres, inserted in the loops.

More advanced mini-cards are driven by means of step motors, which are controlled via microprocessors in connection with the proper software. The needles

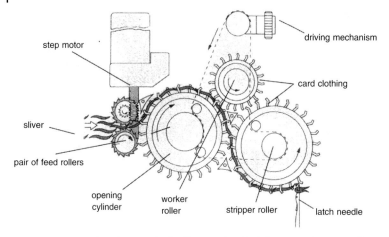

**Fig. 6-57** Mini-card to disentangle the fibres and to feed them into the point of knitting of a Terrot MPK3/MP3 high-pile circular knitting machine [88]

used are chosen electronically and both the quantities of fibres taken in and the colours are influenced. The range of products available this way contains outwear as well as lining pile fabrics for jackets and coats, soft-toy material, upholstery material and floor covering. The widest imaginable range of patterns can be achieved including animal-hair imitation resembling the natural look. For this purpose, fibres of a variety of lengths, titers and cross-sections should be used and the patterning technology should be varied properly. Examples of technical application are filter materials and paint rollers.

One novel development is known as loop technology. In this process, the free ends of the tied-up fibres are blown back to the needle and, together with the ground yarn, made into pile loops, which allows for more patterns and more fields of application.

**6.3**
**Hydroentanglement process**
U. Münstermann, W. Möschler, A. Watzl

The mechanical process of entangling fibre webs and spunbonds by means of water jets has gained importance in industry for about 30 years.

Since the first laboratory tests around 1960 and the patents from that period [89–95] up to the latest figures of installed line capacities and production quantities in 1998 [96, 97] there has been a dynamic development. Especially during the last 10–15 years it has been characterized by an intensive development of process operations and technology as well as product innovations [98–126].

6.3.1
**Process development**

Webs made of fibres, filaments or layers of different fibre structures are bonded by the action of a system of fluid jets or currents with a required minimum energy in that fibres or fibre parts are seized and reoriented by the striking jets or currents and are entangled, intertwined or even knotted with other fibre elements.

Structure and physical properties such as force-elongation behaviour of hydroentangled nonwovens depend on

- fibre properties, web structure, fibre arrangement inside the web
- type and characteristics of the fluid jets or currents, their liquid or even gaseous state of aggregation, jet hole shapes, their linear or sheet-type arrangement and energy quantities
- process-related technical and technological parameters, e.g. structure of web supporting medium for web transportation in the entangling area, striking angle of fluid jets and web speed

This results in the diversity and complexity of the possible influencing factors which have to be taken into account for process technology. The chronological development can be divided into four sections as shown in Table 6-9. A survey of the worldwide production of hydroentangled nonwovens and installations is provided by Table 6-10.

**Physical fundamentals**
Application of fluid jets or currents onto webs by the present state of hydroentanglement technology is done exclusively by means of nozzle elements having a cylindrical opening at the jet exit point and thus providing the jets with a column or truncated cone shape. The use of slot-type nozzles [93] mentioned in literature will be disregarded here.

Bernoulli's simplified equation for a steady flow

$$v_w = a \cdot \sqrt{\frac{2 \cdot \Delta p}{\rho}} \tag{22}$$

shows the relation between the speed $v_w$ of the jet at the exit point, the pressure differential $\Delta p$ between nozzle element and surroundings and the density $\rho$ of the medium (for water as preferred working substance it is $1000 \, kg/m^3$). Value a accounts for friction and other factors which reduce speed $v_w$ under real conditions.

The jet speed is important for the hydroentanglement effect and consequently also for the bonding effect. It also influences the required energy. The specific energy demand $E_s$ in kWh/kg for bonding of 1 kg of dry fibres in web form according to equation (23) can be calculated as follows [127]:

**Table 6-9** Process development of hydroentangled nonwovens

| Period | Technical information | Typical characteristics |
|---|---|---|
| Up to 1969 | – Patterning and after-treatment of nonwovens [89]<br>– First web bonding [94] Jet arrangement analogous with needling machine (hydrolooming) [95]<br>– Term of "high-speed currents" [128]<br>– In 1968 hydroentanglement of filaments into nonwovens | – Water, steam and air jets of low energy for patterning [129] and web bonding with additional binder bonding (spray-bonding)<br>– Spray application nozzles with cone-shaped jets<br>– Development by US companies (Chicopee, DuPont, Johnson and Johnson) |
| 1969–1976 | – High-energy processes [89, 112] with columnar water jets<br>– Basic patents by DuPont de Nemours (USA) [131–133] for patterned and plain nonwovens<br>– Terms of "spunlaced nonwovens" and "hydroentanglement"<br>– In 1973 launching of "Sontara" [96] | – Use of high-energy high-pressure water jets<br>– Studies into the influence of impulse and energy onto the bonding effect for various web masses<br>– Influence of nozzle pressure, nozzle cross-section, web speed<br>– Testing of various fibre types |
| 1976–1987 | – 23 May 1976: Release of DuPont key patents, beginning of new development period [89, 101]<br>– Process subdivision [89, 112] high-energy processes $\geq 1.3$ kW/kg, $\geq 10$ MPa medium-energy processes 0.4–0.8 kWh/kg, 5–9 MPa low-energy processes $\sim 0.1$ kWh/kg, 2–5 MPa<br>– 20 installations worldwide with a total capacity in 1987 of 45,000 t [112]<br>– Maximum in 1987 [89] 3.6 m working width 90 m/min web speed 9,000 t line capacity for 75 g/m$^2$ web | – Differentiated process methods with regard to water jet pressures and energy quantities<br>– Out of patents hardly any information on process and line technology<br>– No line offers from machinery suppliers |
| After 1987 | – In 1987 line offers [89, 96, 112] by Valmet Honeycomb and ICBT Perfojet (F)<br>– In 1995/96 offers by: Fleissner GmbH (D) and Courtaulds Engineering (GB)<br>– In Asia [96] by Mitsubishi Engineering (J), Taiwan Spunlace Group<br>– In 1997 worldwide production of $\sim 150,000$ t of hydroentangled nonwovens, about 100 lines<br>– Hydroentangled nonwovens for technical applications, wiping cloths, medical and sanitary purposes, interlinings, coating substrates, household applications | – Offer of complete lines for hydroentangled nonwovens<br>– Improvement of energy efficiency<br>– Performance of AquaJet line by Fleissner (1999) pressures up to 60 MPa speeds up to 300 m/min, for wet-laid webs and spunbonds up to 500 m/min working widths up to 5.4 m web masses of 15–600 g/m$^2$ |

**Table 6-10** Survey of suppliers, lines and production rates for hydroentangled nonwovens

| Continent/country | Number in 1998 | | Production rate in tons in 1997 |
| --- | --- | --- | --- |
| | Suppliers | Lines | |
| Europe | | | |
| – in 1995 | 15 | 22 | |
| – in 1997/98 | 33 | 42 | 40,000 |
| Asia (except Japan) | 16 | 20 | 10,000 |
| Japan | 18 | 23 | 18,000 |
| USA | 5 | 11 ⎱ | 80,000 |
| Latin America | 2 | 3 ⎰ | |

$$E_s = \sum_1^n \frac{\dot{V} \cdot p \cdot 10^{-3}}{m_v \cdot v_v \cdot AB \cdot 60} \tag{23}$$

$\dot{V}$ = flow rate per jet head (in m$^3$/s)
p  = overpressure in jet head (in N/m$^2$)
$m_v$ = mass per surface unit (in kg/m$^2$)
$v_v$  = web speed in (m/min)
AB = working width (in m)
n  = number of jet heads

Optimum process control and bonding effect among other things depend on the flow rate and the product of mass per surface unit and speed of web under the action of the water jets.

In the printed patent specification [134] the following equation (24) is given for the product of work E of the water jet acting on the fibre web and the impact force I in J · N/kg under concrete conditions

$$E \cdot I = K \cdot p^{2.5} \cdot d_D^4 \cdot n_D / m_v \cdot v_v \tag{24}$$

where K is a process-specific constant and $d_D$ is the nozzle diameter, $n_D$ = number of holes per unit of length, $m_v$ = web mass and $v_v$ = web speed.

For bonding the required energy $E_S$ applied per mass unit of web according to [135] is described as follows:

$$E_S = K \cdot C \cdot p^{1.5} \cdot d_D^2 \cdot n_D / m_v \cdot v_v \tag{25}$$

C is the so-called "nozzle discharge coefficient" and depends on pressure p. It is 0.77 for 2.1 MPa and 0.62 for 12.6 MPa. Taking into account the findings gained from equations (22) to (25), the state of development is characterized by the following quantities:

- high number of nozzles per m of working width with hole diameters between 0.08 mm and 0.15 mm
- pressure generators creating p up to 60 MPa in the nozzle elements
- web speeds of more than 300 m/min and web masses of up to 400 g/m² as can be found in the latest line quotations [124–127, 136]

Hydroentanglement is therefore adapted to the efficiency of the dry and wet forming systems, particularly in the low mass areas up to 100 g/m², for continuous process control [137]. The expenditure of energy for bonding of 1 kg of fibre web has been reduced by half within six years [96]. It sank from 1.10 kWh/kg to 0.50 kWh/kg.

In addition to an evaluation of the processing power, a specific energy coefficient $E_F$ can be determined when taking into account the achievable strength level of hydroentangled nonwovens [138]:

$$E_F = E_S/F_{VL} \tag{26}$$

Quantity $F_{VL}$ is the force in N per g/m² taken up by a hydroentangled nonwoven for a given mass per surface unit at a given tensile stress. The optimum range is reached for a minimum specific energy coefficient $E_F$ [139]. Outside this range a reduction of elongation sensitivity or an increase of maximum tensile stress can only be achieved with a superproportionally increasing expenditure of energy. The strength of hydroentangled nonwovens decreases with further increasing expenditure of energy due to structural damage [140–142].

Quantity $E_F$ also points out the differences in the degree of entanglement for various fibre types (polymer, fineness, length, crimp) and web types.

### Entanglement Process

The processes occurring when high-energy water jets hit the fibres in the web are hard to be observed directly.

The dwelling period of a web section in the direct action range of a water jet with a diameter of 0.1 mm at a web speed $v_v$ of 100 m/min lies in an order of magnitude of $10^{-4}$ s. The jet speed $v_w$ according to equation (22) is 10 to 100 m/s. The fibres have a cross-section ranging from 10 to 500 µm².

Taking into account the findings from the fibre movement analysis in the OE spinning process [143] with comparable time/space conditions, examinations were made using short-time cinematography. Starting at nozzle pressures of 2 MPa already, the water jets entrain air between nozzle exit and web [111]. With the air quantity contained in the web, which corresponds to 95–98% of the web volume, aerosols are formed with the water drops forming when the jets hit the web. The influencing zone during the web's passage through the row of jets according to [111, 144] is a short distance.

In four phases

– fibres are parallelized and reoriented preferably in longitudinal direction
– fibres are curved downward due to the beginning densification of the web and pulled towards the jet line
– the original structure of the web layers is condensed by the factor 10 to 20 when passing the jets, and deflected horizontally
– the web layers are soaked from the bottom side only after they have passed the jets which neutralizes part of the web densification

Figure 6-58 illustrates this process.

The impact forces occurring during the web's passage under the water jet can be measured [145]. One of the decisive factors for the fibre entanglement effect is the size of web movement $v_v$ in relation to these forces.

When $v_v$ is zero during a machine standstill, fibres are displaced instead of entangled which causes web perforation and beginning fibre destruction. This results in the known cutting method by means of water jets [146, 147]. High impact forces combined with a low web speed $v_v$ and low web mass $m_v$ causes fibre reorientation, preferably in vertical direction. In this connection it is undesirable that fibres should penetrate the web carrier material and bond with it. This process is desirable only when several webs or other layers are to be combined in a compound material [148].

The development of highly efficient filtration systems [105, 106] made it possible to hydroentangle natural fibre webs as well. Today the splittability of modern man-made fibres such as Lyocell and bicomponent types is deliberately used for the development of new hydroentangled nonwoven types, e.g. for filters [104, 116, 127].

Optimum bonding of the web is shown in the structure of a hydroentangled nonwoven which is characterized by arrangements of parallel bundles of fibre sections which in turn are wrapped by entangled fibre sections [149]. Fig. 6-59 illustrates such a structure.

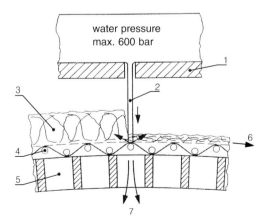

**Fig. 6-58** Principle of hydro-entanglement
1 jet head, 2 water jet,
3 unbonded web, 4 web support,
5 drum, 6 nonwoven, 7 air and water return

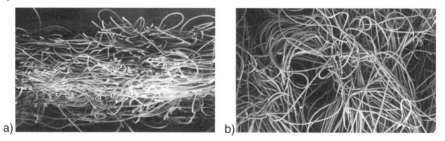

a)　　　　　　　　　　　　　　b)

**Fig. 6-59**　Structure of a hydroentangled nonwoven: a) cross-section b) top view

Due to an increasing orientation of the fibres in longitudinal direction during their passage underneath several rows of water jets and simultaneous reduction of the web cross-section towards the jet action, it is normally necessary to increase the applied pressures in the subsequent nozzle elements. The structure of the presented web is also decisive. Webs with random or cross-fibre structure are favourable to the process of fibre entanglement and the strengths to be achieved.

For an optimum bonding effect, a web must pass a rather large number of subsequent nozzle rows. The fibre entanglement process first concentrates on the layers of both web sides close to the surface. The fibres on the side facing the arriving jets are influenced differently from the fibre layer at the bottom which lies on the permeable web support.

Changing the operating side after each row of nozzles, if possible, will improve the degree of bonding [109, 114, 119, 136, 141, 144] and with increasing pressures will gradually also result in bonding of the inner layers of thick webs and consequently in a homogenization of structure. It is also possible to produce hydroentangled nonwovens of various structures (Fig. 6-60) or with two different surfaces or with varying density through the web cross-section. The structure of a non-

**Fig. 6-60**　Hydroentangled nonwovens with various structures according to [114]

woven is determined mainly by the structure of the supporting wire mesh which also allows structured hydroentangled nonwovens to be produced.

## Hydroentangled nonwovens

Nonwovens produced by the hydroentanglement process are characterized by the following properties:

- They are normally free from binding agents, obtrusive impurities and other foreign matters (with the exception of a deliberate addition of agents for defined requirements and applications) and are therefore especially suitable for medical and sanitary products requiring high hygienic standards.
- They are soft, drapeable and absorbent with high absorption and retention values (expressed by specific quantities such a % or g/g of nonwoven) and can be lintfree and abrasion-resistant so that polishing, wiping and cleanings cloths even for clean-room requirements represent a wide range of applications.

In the current technical stage of development, they comprise a wide range of mass per surface unit (20 to >400 g/m²) and a wide spectrum of fibre composi-

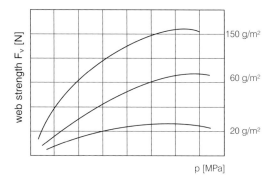

**Fig. 6-61**  Web strength in relation to water jet pressure and various area weights

**Table 6-11**  Maximum tensile force of a polyester hydroentangled nonwoven (100 g/m²) as a function of pressure – sum of 3 to 9 jet heads [149]

| Pressure (MPa) | Maximum tensile force (N/5 cm) | | MD:CD ratio |
| --- | --- | --- | --- |
| | machine | cross | |
| 4 | 2.2 | 1.8 | 1.22 |
| 9 | 16.7 | 11.3 | 1.48 |
| 20 | 46.5 | 29.8 | 1.56 |
| 31 | 102.4 | 64.0 | 1.69 |
| 44 | 127.0 | 73.2 | 1.73 |
| 58 | 150.1 | 91.1 | 1.65 |
| 74 | 145.3 | 77.7 | 1.87 |

tions with regard to polymer and geometry. The degree of bonding depends on pressure as shown in Fig. 6-61.

Table 6-11 illustrates how the nonwovens strength values in longitudinal and cross direction change with increasing pressure and thus influence the MD:CD ratio. The force/elongation behaviour of the MD:CD ratio can be influenced in a defined manner by thermal after-treatment of the nonwoven and can be adapted to the technical demands made on the nonwoven.

## 6.3.2
### Fibre and process influences

Table 6-12 from [111] provides a survey of the essential factors of the hydroentanglement process.

**Table 6-12** Influencing factors in the hydroentanglement process

| *Fibre influences* | *Process influences* |
|---|---|
| Polymer | Web structure |
| • density | • mass per surface unit |
| • force/elongation behaviour | • fibre arrangement |
| • structure, e.g. fibrillation ability | |
| • stiffness, flectional behaviour | Speed |
| • friction | |
| • fibre modification | Web carrying element (support) – structure |
| | • weave |
| Geometry | • opening shape and size |
| • length | • wire gauge etc. |
| • fineness | |
| • texture (crimp characteristics) | Nozzle/water jet |
| • cross-section shape | • cross-section shape |
| • surface structure | • diameter |
| | • quantity |
| Others | • arrangement, distribution |
| • finish, wettability etc. | • pressure/energy |
| | • influencing mode, e.g. angle, processing of web surfaces (1 or 2) |

**Fibre influences**
Although basically all natural and synthetic polymer types can be processed [115, 116], some have shown to be especially suitable:

– cellulosic fibres (cotton, viscose, Lyocell) [93, 99, 106–109, 113]
– polyester fibres [101, 109, 111, 138, 144]

The above represent the main share of fibres used so far, often as blends.

The qualification criterion stressed in favour of cellulosic fibres is their hydrophilicity [139]. One also has to mention, however, their relatively high polymer density in comparison with PA6 or PP because aramides and phenolic fibres are also successfully used for hydroentanglement [89, 150].

With regard to fibre fineness it can be said: the finer the fibre, the better the hydroentanglement effect under identical conditions [151]. Fibres of 4 to 6 dtex are recommended as upper limit [115]. Above that, hydroentanglement is hampered by increasing stiffness and smaller specific surface [107, 113, 138, 144], while conversely fibrillation and splitting of originally coarser fibres increase the bonding effect [89, 99, 104, 107, 127].

There is no limitation of fibre lengths. Short fibres, processed into webs after wet laying process or air-laying process and usually combined with layers of staple fibres or filaments, are also increasingly used [89, 96, 99, 101, 126, 152]. One example are hydroentangled nonwovens of 50% polyester fibres and 50% fluff with masses per surface unit of 60 to 80 $g/m^2$.

In the usual range of fibre lengths between 20 and 60 mm, longer fibres act to increase the strength [111, 113, 144, 149] because structural formation of parallel fibre bundles is supported. Fibre producers make allowance more and more for these demands with the development of special fibre types [113]. This also includes suitable fibre finishes and crimping parameters. When applying pressures of more than 20 MPa, spunbonds also can be successfully hydroentangled [153]. Spunbond nonwovens are characterized by an almost linear force/elongation behaviour.

**Process influences**

The web formation method is doubly important:

- The preferential direction of fibre orientation also decides the entanglement effect.

- The web speed $v_v$ is an important parameter for economy calculations for hydroentanglement processes [89, 96, 112].

When hydroentangling filament webs, wet-laid and air-laid webs, the second factor is not critical [154]. Processing of staple fibres, on the other hand, requires a web formation technology which is co-ordinated with range of application and product quality, with both random-laid and cross-plaited webs being found in practical operation. Owing to the lower production rates, several web formation units – such as cards – are often required for one bonding line [89, 96, 136].

Apart from the fibre throughput of a web formation unit, the mass per surface unit $m_v$, the total pressure in the form of energy input and the water quantity of all jets as well as the concrete aim of web bonding are decisive for the height of $v_v$ [93, 108, 109, 116, 136, 138, 154]. For optimum realization of each process, empirical methods are still dominating.

The design of the web supporting element to a large degree determines the structure of the hydroentangled nonwovens.

The basic rule is: For patterned nonwovens – mostly in the shape of regular apertures – coarse sieves or roller combinations with perforated drums are used which often are provided with profiles, thick wires and large open areas. The supporting elements for hydroentangled nonwovens with closed surface, however, are fine-wire sieves with a high mesh number. Patent literature contains many references to that effect [158].

The possibility of exchanging the supporting element within a short period when changing to another product is one of the basic demands made on modern bonding lines.

Technology with regard to the construction of jet heads, jet heads and the design of the individual nozzles themselves (cross-section, hole quality) has particularly progressed during the last decade. Being the core of the bonding process, this is what is responsible for the improvement of the energy balance.

The following parameters are widely used [89, 109, 112, 124, 127, 138]:

- Cylindrical nozzle cross-sections with determined ratios of length and diameter, the latter ranging from 0.08 to 0.15 mm.

- Arrangement of nozzles in parallel, often staggered lines (by 90° in web running direction), distance between nozzle holes less than 1 mm and distance of outlet openings from web surface in the centimetre range.

Other technical solutions have been described mostly in patent literature, covering among other things slot-type nozzles [93], traversing jet heads, funnel-shaped nozzle cross-sections [159].

To maintain constant conditions for water jet generation and effect on the presented webs, the following influencing variables are important criteria characterizing the state of the art:

– capacity of high-pressure pump(s) (piston or centrifugal pumps)
– removal of water/air mixture from the webs (suction system)
– type of filter system for fresh water and circulating water
– service life of nozzle elements, cleaning devices, possibility of replacements without machine stop

### 6.3.3
### Bonding lines

Since 1996 machine engineering developments for hydroentanglement have been offered by three companies:

- ICBT Perfojet in co-operation with Valmet-Honeycomb [96, 120, 125, 136], and M & J Fibretech [155] for airlay processes

- Fleissner GmbH & Co. in co-operation with Dan-Webforming A/S (airlay processes) [96, 124, 126, 127, 137, 141, 142, 152, 154–156]

- CEL International (formerly Courtaulds Engineering) [96, 121, 155]

According to [96] the JETLACE 2000 line conception of ICBT Perfojet has been realized with about 20 production lines worldwide by 1998. The modified AIRlace

**Fig. 6-62**  Section of a JETLACE 2000 line of ICBT Perfojet

2000 combines spunlacing technology with an airlay process [136]. Fig. 6-62 [157] shows part of a JETLACE 2000.

The machine configuration comprises a number of drums for alternating bonding of web sides. Each bonding stage consists of one or two jet heads [136]. For a working width of 3.5 m and a useful width of 3.3 m, injectors with pressures of up to 40 MPa are used. Hydraulic locking technology enables a quick change of nozzle elements [155].

Further technical details are: one pump for each injector, pressure pulsation damper, special treatment of pistons to prevent vibrations, fully automatic filter system, computer-controlled process control system.

Compared with the state of the art in 1995, the energy requirements were reduced by 25% and the strength of hydroentangled nonwovens made of PET and CV was increased by 40% for a comparable web mass.

Table 6-13 contains data for characterization of line capacity [136]. Fresh water requirements are 3 to 5 m³/h.

With its AquaJet Spunlace System, Fleissner GmbH & Co. has developed into market leader during the past years with 16 lines [155] installed so far (Fig. 6-63).

**Table 6-13**  Energy requirements $E_S$ for web speed $v_v$ for various web compositions for JETLACE 2000

| Fibre type | mv (in g/m²) | $E_S$ (in kWh/kg) for $v_v$ (m/min) | |
|---|---|---|---|
| | | 50 | 250 |
| PET | 35 | 0.20 | 0.18 |
| | 60 | 0.29 | 0.24 |
| CV 70%/PET 30% | 60 | 0.22 | 0.17 |

**Fig. 6-63** Three-stage Fleissner AquaJet Spunlace line, with drum dryer and automatic winder

The use of this system has been extended since 1997 with the inclusion of the Airlay process by Dan-Webforming [126, 152].

The "AquaJet" system is characterized by the following features:

- Working widths up to 4.2 m; pressures up to 60 MPa for web speeds up to 300 m/min depending on web characteristics and bonding target (jet heads are designed for pressures up to 60 MPa for webs of 600 g/m$^2$ and more [126, 137]).

- One to five stage bonding process, line versions depending on intended use [124, 127, 154], usually with web treatment from both sides [141, 142].

- Preliminary treatment of incoming webs with a compacting belt, resulting in compaction and deaeration without shifting of fibre layers [124, 154], with optional subsequent preliminary moistening and preliminary bonding by means of water jets of low pressure.

- Combination of preliminary bonding drum and jet heads in a first bonding stage; easily replaceable jet heads with one or two rows of holes each (hole diameter 0.10 to 0.15 mm, hole spacing 0.6 mm), self-sealing effect; drum shells and spunlace wire mesh on top can also be easily replaced; the drum is supported by the suction cylinder which continuously removes the water/air mixture [124, 126, 152, 154].

- Further bonding stages comprise drums or belts and separate high-pressure pumps for individual pressure application and energy input; apertures of drum and supporting wire mesh are adapted to the concrete treatment target (bonding, patterning).

- Water/air separator, reversible filter elements (belt, cartridges) or flotation sand filters for processing of cotton and fluff; final dewatering before drying, high-capacity through-air dryers with up to 92% of open area and finishing system complete the line.

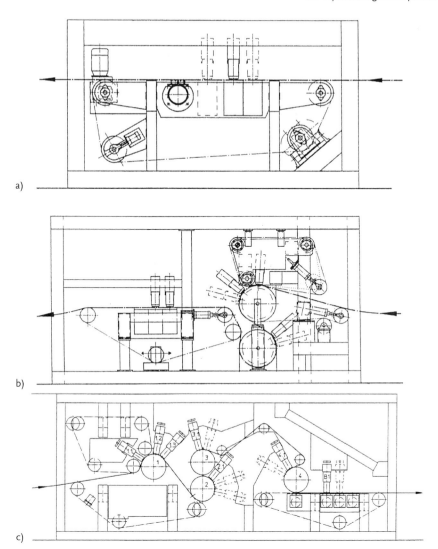

**Fig. 6-64** Principles of AquaJet processing technology of Fleissner GmbH & Co:
a) single-stage line, b) three-stage line, c) five-stage line

Fig. 6-64 illustrates function principles of AquaJet lines. The single-stage line (Fig. 6-64) is used for bonding of a single web side. It is also suitable for bonding two components into a nonwoven compound. In the two-stage line, the web is bonded from both sides. It is recommended for use with light-weight webs. The same line can be used after web bonding from both sides to provide the nonwoven with a surface finish or perforation in a third stage (Fig. 6-64). The five-stage AquaJet with repeated changes of the treatment side was developed for heavy

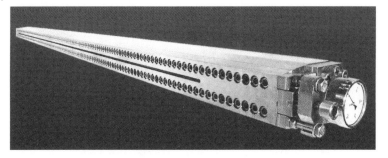

**Fig. 6-65**   Head with 4.20 m working width of the Fleissner AquaJet

webs from 150 g/m$^2$ to more than 600 g/m$^2$ (see Fig. 6-64). The belt dewatering system incorporated into each line minimizes the total energy requirements for bonding and drying of nonwovens.

One of the major components of the AquaJet lines is the jet head (see Fig. 6-65). The jet heads of Fleissner GmbH & Co. are protected by patents and have a maximum working width of 4.20 m.

Energy demand $E_S$ is

– 0.2–0.4 kWh/kg for customary web bonding, and
– up to 0.8 kWh/kg [124, 125] for high-quality products such as layer substrates.

The various versions of the Fleissner Spunlace System comprise:

• **AquaJet type:** hydroentanglement of webs from 10 to 600 g/m$^2$ at 300 m/min max. and a max. pressure of 600 bar for closed and textured nonwovens up to 5 m of working width.

• **AquaSplit type:** processing of micro-segment split fibres [127] with high impact forces (jet speeds of up to 300 m/s) in three or five bonding stages with high split rates.

• **AquaPulp type:** production of airlaid/carded composites [152] from two (fluff on fibre web) or three (web – fluff – web) layers.

• **AquaSpun type:** spunbond hydroentanglement system for max. 600 m/min and max. 5 m of working width.

• **AquaTex type:** hydroentanglement system for textiles using the BBA-Interspun$^{TM}$ technology for improving textiles with regard to density, abrasion resistance, affinity for dyestuffs, textile hand and aspect.

The "Hydrolace" line of Courtaulds Engineering is built on the twin-drum principle [121]. Version H.350 is intended for bonding of heavy webs up to >400 g/m$^2$. A second version (H.352) is recommended for lighter hydroentanglement nonwovens [155].

Characteristic features: Few injectors of high jet energy, jet stability due to improved nozzle geometry, special nozzle material for high service life with possible pressures of up to 60 MPa. The equipment also comprises a filtration system and the automatic process control system. The two bonding drums are arranged vertically [116].

According to one patent [160] constructions with six and more bonding rollers/ jet heads are also possible.

## 6.4
## Thermal processes
A. WATZL

Thermal bonding processes according to ISO/DIS 11 224 (see Fig. 6-1) comprise hot-air treatment, calendering and welding of nonwovens. Chemical bonding by means of binder dispersions for curing and cross-linking of binder molecules as well as drying of the impregnated webs requires the use of suitable processing equipment operating with hot air. It will therefore be appropriate to discuss also nonwovens drying processes at this point.

### 6.4.1
### Drying

Drying of impregnated webs or nonwovens is the removal of moisture and possibly other volatile components by the use of thermal energy. Chemical bonding of nonwovens with binder dispersions can make it necessary in connection with the drying process that coagulation of the dispersion must take place at 40–70 °C and curing of binder molecules at temperatures of 130–160 °C.

During the heating process, the water contained in the nonwoven must be heated to evaporation temperature. The wet bulb temperature for evaporation of the water lies between 70 and 80 °C. As long as evaporation takes place, the nonwoven temperature remains lower than the temperature of the hot air used for drying. With rising hot air temperature, the temperature difference is increased which causes a higher heat transfer and a higher drying speed.

Drying may lead to binder migration which is the result of binder particles being carried to the web surface by the steam. This means an unwanted irregular binder distribution across the web cross-section which in turn can result in delamination.

For selection of a suitable drying process the following has to be observed:

– type of binder and binder quantity in the nonwoven
– air permeability of the nonwoven and
– production speed

Possible drying methods are:

– convection drying
– contact drying
– radiation drying

### 6.4.1.1 **Convection drying**

Convection drying [161] is one of the most frequently used processes. It can be applied when air can flow through the web. Fig. 6-66 illustrates the air permeability ranges for various nonwoven product groups depending on their mass per surface unit.

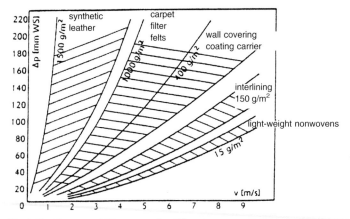

**Fig. 6-66**  Air permeability ranges for different nonwoven product groups

Convection drying is characterized by the rules of heat transfer for

– transmission of heat into the nonwoven to be dried and
– mass transfer for moving liquid and steam from inside the nonwoven towards the surface and towards the drying air

Hence the drying process can be regarded as a two-stage process.

In the first stage, water is evaporated at the surface of the nonwoven. Heat transfer takes place from the hot air through the boundary layer to the web surface and by transportation of the formed steam through the boundary layer towards the air. During this process, only the boundary layer offers resistance.

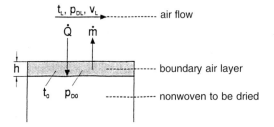

Heat flow $\quad \dot{Q} = A \cdot a \cdot (t_L - t_O)$ $\hfill$ (27)

Mass flow $\quad \dot{m} = \dfrac{A \cdot \beta}{R_D \cdot T} (p_{DO} - p_{DL})$ $\hfill$ (28)

$$d = \frac{\lambda_L}{h} \hfill (29)$$

$$\beta = \frac{D}{h} \hfill (30)$$

| | | |
|---|---|---|
| A | = surface involved in transfer | m$^2$ |
| $a$ | = heat transfer coefficient | W/m$^2 \cdot$ K |
| $\beta$ | = mass transfer coefficient | m/s |
| $t_L$ | = drying air temperature | °C |
| $t_O$ | = nonwoven surface temperature | °C |
| $v_L$ | = air speed | m/min |
| $p_{DO}$ | = steam partial pressure at surface | N/m$^2$ |
| $p_{DL}$ | = steam partial pressure of drying air | N/m$^2$ |
| $R_D$ | = gas constant of steam | J/kg °C |
| T | = temperature of boundary layer | °K |
| $\lambda_L$ | = thermal conductivity in boundary layer | W/mK |
| D | = diffusion coefficient | m$^2$/s |

In this process section, the drying speed can be increased by increasing the temperature and steam partial pressure difference between nonwoven surface and air. The heat and mass transfer values increase when the speed of the drying air rises. This reduces the boundary layer thickness and consequently its resistance. An increased surface also results in an increase of the drying speed.

In the second process section, the evaporation location is moved to the interior of the nonwoven. As a consequence, both heat and mass transportation has to cover longer distances through nonwoven layers already dried. For this process the following relations apply:

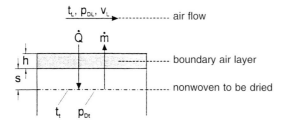

Heat flow $\quad \dot{Q} = A \cdot K \cdot (t_L - t_t)$ $\hfill$ (31)

Mass flow $\quad \dot{m} = \dfrac{A \cdot \sigma_D}{R_D \cdot T} (p_{Dt} - p_{DL})$ $\hfill$ (32)

$$K = \frac{1}{\frac{1}{a} + \frac{s}{\lambda_s}} \tag{33}$$

$$\sigma_D = \frac{1}{\frac{1}{\beta} + \frac{\mu_D \cdot s}{D}} \tag{34}$$

| | | |
|---|---|---|
| $K$ | = heat transfer coefficient | $(W/m^2 \cdot K)$ |
| $t_t$ | = temperature at drying boundary | $(°C)$ |
| $\sigma_D$ | = steam pressure permeability coefficient | $(m/s)$ |
| $p_{Dt}$ | = steam partial pressure at drying boundary | $(N/m^2)$ |
| $\mu_D$ | = diffusion resistance coefficient | |

In this section, both the external diffusion conditions which are influenced by temperature and air speed and the diffusion properties of the nonwoven are important. The more the drying location is moved inside the nonwoven, the more the influence of heat and mass transfer coefficients is reduced and consequently also the influence of air speed on drying speed. For final drying of thick nonwovens of high density the air speed is of low importance only. The drying period is then determined mainly by the temperature being part of the steam partial pressure difference.

When drying by addition of heat, the air guiding method is important. In mixed air drying, the air is constantly circulated over the nonwoven. An exchange of exhaust air by fresh air takes place to a limited degree. This allows the flow speed at the nonwoven to be maintained on a high level. With regard to heat economy, this means a favourable increase of heat transfer coefficients and uniform drying over the dryer length. Ventilation systems are operating either with co-current, counter-current or cross current. The counter-current system in comparison with the co-current system offers the advantage of drying to a low final moisture content without damaging the nonwoven.

Fleissner supplies the following ventilation systems:

- through-air ventilation: perforated drum dryer, belt dryer
- air jetting: belt dryer
- parallel ventilation: hotflue dwelling section

For convection drying the counter-current principle is applied.

When the materials to be dried are nonwovens, paper or tissue of a high initial moisture content and when high production speeds shall be achieved, a high-performance through-air dryer with perforated drum is required. When the drying temperature layout is already at its maximum, there is only the possibility of increasing the circulating air quantity. These dryers operate with high-speed special fans that create high vacuums inside the drum and thus force sufficient air quantities through the web. Infinitely variable fan motors allow the capacity to be adapted to the product to be dried. However, this also requires high motor capacities to be installed for the fans. Depending on web type, temperature, moisture

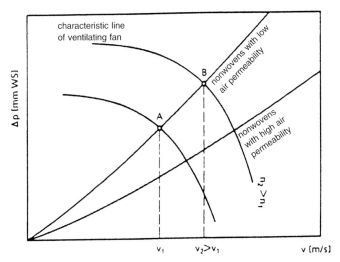

**Fig. 6-67** Working diagram of a through-air dryer

content as well as air quantity and differential pressure, specific drying rates between 15 and 300 kg water/m²/h and more can be reached.

Fig. 6-67 shows the correlation between low pressure and the desired air flow speed as a function of the web's air permeability.

At the intersecting point of material characteristic and fan characteristic the respective operating point is given for various speeds.

The illustration shows the possibility of increasing the capacity of a high-performance dryer by changing the fan speed. With increasing fan speed and resulting higher air speed, however, the fan power consumption rises as well.

The size of the fan is designed such that its working point lies within the optimum efficiency range. As in single-drum lines the web passes through all conditions from wet to dry on the same drum, the co-ordinates of the given working point are average values with regard to Δp and air speed. For highly air-permeable nonwovens (dry-laid, wet-laid, spunbond) the cost-efficient perforated drum dryer or the web-type drum dryer (perforated shell with mounted axial webs) can be used. The stability of the perforated drums is fully sufficient for the customary production speeds of 300–400 m/min.

The perforated drum dryer offers the great advantage of a direct combination of perforated drum and fan in one chamber with the resulting favourable flow efficiencies (Fig. 6-68).

The free open area is 48% to max. 76% for perforated drum dryers and up to 95% for the web-type drum.

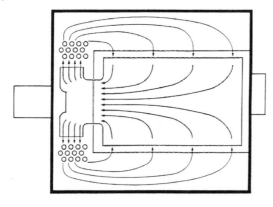

**Fig. 6-68** Cross-section of a perforated drum

Advantages of the perforated drum construction:

- low flow losses
- low heat consumption
- efficient heating (direct gas, direct oil system)
- fan capacity infinitely variable

Fleissner perforated drum dryers can be supplied with diameters of 1400 mm, 1880 mm, 2600 mm and 3500 mm and working widths between 400 and 7000 mm. They can be realized as single-drum or multi-drum dryers with additional equipment for various nonwovens qualities. Fig. 6-69 illustrates possible executions.

The following advantages and special features can be listed for the perforated drum dryer:

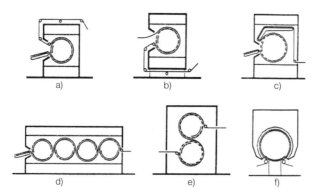

**Fig. 6-69** Various executions of perforated drum dryers:
a) single-drum dryer; web guided above the dryer, b) single-drum dryer; web guided outside the dryer, c) single-drum dryer with dwelling belt for binder curing, d) four-drum dryer with alternating air flow through the web, e) two-drum dryer in vertical design, f) perforated drum dryer of omega design

- By installing the large radial fan inside the drying chamber, flow losses in the system are kept to a minimum. This reduces the power consumption of the fan drive.

- Fleissner multi-drum dryers turn the web during transfer from one drum to the next allowing the air to penetrate the material from both sides; this results in continuous and uniform drying. Each drum in vertical execution is wrapped by 3/4. The space requirement for this dryer is especially low due to the compact design.

- Standard perforation sizes are 3 mm, 5 mm, 8 mm in diameter and also have square apertures with different open areas.

- Wire mesh covers of various mesh sizes on the perforated drums or on the web-type drums prevent perforation marks on delicate material.

- Drying of all types of webs is possible, from extremely light webs of 12 g/m$^2$ to very heavy air-permeable webs and felts up to 3000 g/m$^2$ and more.

- Speeds of 300 m/min and more are possible.

- A temperature accuracy of $\pm 1.5\,^\circ$C even with temperatures of 250 $^\circ$C is guaranteed.

- A double perforated jacket in the drum guarantees an even air flow across the drum width in case of large working widths.

- The heat contained in the material leaving the dryer is returned to the dryer with the fresh air sucked in. Cooling of the material is therefore an integral part of the system.

- For varying material widths, a working width cover is provided inside the drum to prevent energy losses.

- The heating type is determined by the required drying temperatures. The perforated drum dryer can be heated with steam, thermal oil, by direct gas heating, with hot water or electric energy. Direct mixing of drying air and combustion gases in the direct gas heating system allow to achieve a firing efficiency of 100%. This is an about 30% higher efficiency than for other heating systems and consequently results in a saving of combustible of about 30%.

The following are examples for use of perforated drum dryers for drying of nonwovens:

- single-drum lines for impregnated sanitary nonwovens and interlinings
- multi-drum lines for impregnated needled nonwovens for floor coverings in the automobile sector, for filters, wiping cloths, bitumen and synthetic leather coating substrates (Fig. 6-70)
- high-performance perforated drum dryers for wet-laid nonwovens and spunbonds

**Fig. 6-70** Perforated drum through-air dryer in linear arrangement

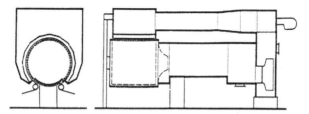

**Fig. 6-71** Through-air dryer with high-performance drum, Fleissner system

High-performance through-air dryers are used when large drying capacities at high speeds are required (Figs. 6-71 and 6-72).

The drum construction consists of a new modular system with bolted drum jacket.

The dryer is characterized by:
– drum diameters of up to 5,400 mm and more
– working widths of up to 10,000 mm
– large open area of about 96%
– high rigidity of drum resulting in high truth of running
– high production speeds of up to 3,000 m/min

Enabling:
– flow-through of large quantities of air at high air speeds and minimum pressure loss and correspondingly high energy transfer
– high specific water evaporation rates of up to 550 kg/m$^2 \cdot$ h
– high vacuum in the drum of more than 1,000 mm water column which makes the drum suitable for drying of denser paper qualities and tissue and nonwovens of low air permeability
– uniform air distribution across the working width
– uniform temperature distribution of up to 1 °C across the working width

**Fig. 6-72** High-performance perforated drum dryer for light nonwovens

This flow-through principle opens up new application areas:
– drying of hydroentangled nonwovens
– drying and bonding of impregnated nonwovens
– drying of air-permeable papers, filter papers, tissue, towel and toilet papers, for industrial and consumer use (Fig. 6-73)
– use as vacuum dewatering rollers and transfer rollers

The through-air ventilation system can also be applied to belt drying lines when – depending on the type of nonwoven – the web must be supported on a horizontally guided belt at low pressures of more than 300 Pa. Fig. 6-74 shows the cross-sections of belt dryers distinguished mainly by the design of the transport element and the air guiding system. The web is supported on the belt and penetrated by drying air flowing from top to bottom.

In the perforated belt dryer, the air is sucked off by a fan underneath the belt, passed over heating elements, distributed uniformly across the width by means of nozzle boxes and returned to the material. The same thing happens in the plate

**Fig. 6-73** HighTech through-air dryer

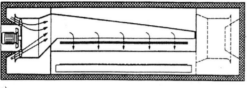

a)

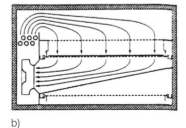

b)

**Fig. 6-74** Cross-sections of belt dryers with through-air ventilation: a) perforated belt dryer, b) plate belt dryer

belt dryer where a fan blows the air heated by heating elements into a chamber where a uniform pressure is given, resulting in uniform air distribution and uniform air flow across the width through the material.

The nonwoven is fed into the dryer and delivered outside the dryer by the transport element. The material width cannot be controlled or only by means of tensioning chains or hooks of a stenter. In the air-jetting system, the surfaces of a nonwoven are subjected to a vertical air jet. The vertical flow component on the surface approaches zero and the air jet is deflected by 90° so that an air flow parallel to the surface is produced forming a laminar boundary layer. As opposed to the through-air principle, there is only a reduced hot air flow in the air-jetting system which penetrates the interior of thick nonwovens. Despite high air speeds, the drying speed of the air-jetting system is more than 10 times smaller than that of the perforated drum dryer. The heat transfer coefficients of both systems differ as follows:

Through-air system:     $a$ 290–470 W/m$^2$ K
Air-jetting system:     $a$ 175–190 W/m$^2$ K

The air-jetting system is used, for example, when the nonwovens to be treated are of very low air permeability or when the nonwoven must be dried very gently (Fig. 6-75).

The most common dryer with air-jetting system is the tenter frame as shown in Fig. 6-75 a. The material is held at both edges by means of needles or tenter hooks and transported through the dryer. This enables controlling of material width. The use of additional equipment allows to set an advance speed in order to achieve the requested mass per surface unit.

Hence, the transport element for the material consists of a chain taking up the material outside the drying zone, passing through the drying zone and the cooling section together with the material and then releasing the material again outside the drying/cooling section.

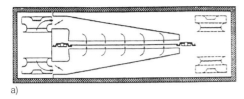

a)

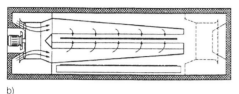

b)

**Fig. 6-75** Cross-sections of belt dryers with air-jetting system: a) tenter frame, b) nozzle belt dryer

The air guiding system is characterized in that the drying air is forced by radial or axial fans through nozzle boxes with round, oval or slot-type nozzles so that it hits the material vertically from both sides, is deflected towards the side and then returned to the fans. In the process the air is heated up again by heating elements. The entire dryer is divided into several zones and provided with an exhaust air duct in each zone through which part of the steam-loaded circulating air is discharged. An identical amount of fresh air must be supplied to the system and heated. As the air leaving the nozzles with speeds of up to 40 m/s is slowed down and deflected when hitting the material, this is a mixture of air-jetting and parallel ventilation. Owing to the design of the transport element and the type of the air guiding system, this dryer is not suitable for loose material, for light tension-sensitive material and for substrates having little strength and needing to be bonded first.

Another machine with air-jetting system is the nozzle belt dryer which is very similar in design to the tenter frame (Fig. 6-75 b). In the place of a chain transport element, it comprises a steel or textile fabric conveyor belt which transports the material. Here also the transport element leaves the drying section for taking up and releasing the material. The material is kept on the belt by means of separate settings for top and bottom air. A defined width control, however, is not possible. As regards the type of ventilation system, there is hardly any difference compared with the tenter frame. The air-jetting system is also used for one-sided air-jetting, for multi-deck dryers and for separate air guiding systems.

Other developments aim to achieve a carrying effect by a corresponding design of nozzle system and air guidance to keep harmful tensions from the material. Usually this requires the air speeds and consequently the specific output to be reduced. These machines are designated as suspension dryers, suspension jet dryers, air cushion frames, carrying jet dryers etc.

#### 6.4.1.2 **Contact drying**

In contact drying the nonwoven to be dried receives heat exclusively by contact with the heated rollers. The steam diffuses from the warmer roller surface through the nonwoven towards the colder side. As a result of the forming liquid gradient, moisture is sucked back towards the roller surface by capillary forces. The drying speed remains constant as long as the balance between liquid evaporation and liquid transport towards the warm contact roller is given. The drying speed is reduced when the drying barrier is moved from the heating surface towards the nonwoven surface and the heat transport distance is increased. Overheating of the nonwoven is avoided by guiding the nonwoven over several heated contact cylinders so that drying takes place from both sides. A comparison of drying rates for convection and contact drying yields considerable differences for the heat transfer coefficient $a$:

$$\frac{a_{\text{convection}}}{a_{\text{contact}}} \approx 4 - 10$$

Contact drying (Fig. 6-76) is used when certain calibrating or ironing effects shall be achieved. When guiding the nonwoven over several rollers, high processing tensions occur.

The cylinders are combined 6 to 12 at a time in frames. Customary cylinder diameters range from 570 to 800 mm. Depending on the drying rate, several frames can be installed one after the other. The cylinders are arranged staggered in two lines, either vertically or horizontally.

The drying cylinders can be heated with saturated steam, hot water, thermal oil or gas. The intensity of water evaporation in the nonwoven to be dried depends very much on structure and thickness of the nonwoven. For thin nonwovens, where the surface with regard to the volume is very big, a higher drying rate is reached. The drying speed sinks rapidly with increasing thickness. The steam must be removed from the surface to avoid saturation of the ambient air and thus a standstill of the drying process. The drying rate of contact drying can therefore be considerably increased by air-jetting. The contact dryer can be used as predryer in combination with a perforated drum dryer. This combination to a great extent avoids the hard touch caused by mere contact drying.

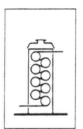 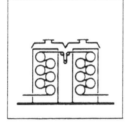

**Fig. 6-76** Cylinder dryer with vertical roller arrangement

### 6.4.1.3  **Radiation drying**

Energy in radiation drying is transferred to the nonwoven without any intermediate carrier by special heating elements, the radiators. With correspondingly high radiation temperatures, higher heat quantities can be transferred then by convection drying. According to the Stefan-Boltzmann law, the heat quantity depends on the fourth power of the radiation temperature. Heat transport is influenced by the radiation wave length and by absorption in the material. Water has a distinctive absorption maximum (Fig. 6-77) in a wave length range from 1.8 to 3.5 µm. Suitable radiators are ceramic IR dark radiators, medium-wave IR radiators from quartz glass tube or flat metal foil radiators. Radiation dryers are preferably used during the first drying phase only because the nonwoven might be damaged at temperatures of more than 95 °C. They are used for pre-drying of impregnated webs or for coagulation of impregnated webs. IR radiation energy is also characterized by high operating cost.

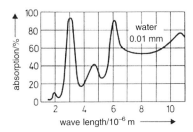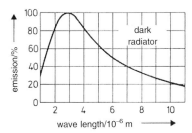

**Fig. 6-77**  Radiation

### 6.4.2
#### Hot-air bonding

Hot-air bonding – also called fusion bonding – is a dry bonding process requiring the presence of thermoplastic fibres. With the development of synthetic thermoplastic fibres, fusion bonding processes have gained in importance. The development of processes and equipment made it possible to use this bonding method for a variety of nonwovens with masses per surface unit from 20 to 4000 g/m$^2$ and thicknesses of up to 200 mm. A comparison with chemical bonding methods yields the following advantages: soft and textile-like nonwovens, no use of binders and consequently high economic efficiency at low machinery expenditure. Uniform and thorough bonding of thick webs is possible. When using thermoplastic fibres of the same polymer type, complete recycling is possible.

### 6.4.2.1  **Basics**

The thermal properties of synthetic fibres, apart from textile/mechanical properties, are also important for thermal web bonding. Polymer selection, structure and manufacturing process influence their thermal behaviour, especially their thermoplasticity during heating and cooling.

**Table 6-14** Selection of thermal fibre characteristics (from [166])

| Fibre | Transformation temperature (°C) | Melting temperature (°C) | Decomposition temperature (°C) | Softening range (°C) | Specific heat $(J \cdot g^{-1} \cdot K^{-1})$ | Thermal conductivity $(W \cdot m^{-1} \cdot K^{-1})$ |
|---|---|---|---|---|---|---|
| Viscose | | | 175–205 | | 1.35–1.5 | 0.3–0.6 |
| Polyamide 6 | 40–60 | 215–220 | 310–380 | 170–200 | 1.5–1.9 | 0.29 |
| Polyamide 6.6 | 45–65 | 255–260 | 310–380 | 220–235 | 1.5–1.9 | 0.25 |
| Aramide (Nomex) | ∼260 | | ∼370 | | 1.2 | 0.13 |
| Polyester | 70–80–(100) | 250–260 | 283–306 | 230–250 | 1.1–1.4 | 0.2–0.3 |
| Polyethylene | –(70–100) | 124–138 | | 105–120 | 1.4–1.9 | 0.35 |
| Polypropylene | –(12–20) | 175 | 328–410 | 150–155 | 1.6–2.0 | 0.1–0.3 |
| Polyvinylalcohol | 75–90 | | 230–238 | 200 | | |
| Ceramic | | 1815 | | | 0.8–1.0 | |
| E-glass | 850/960 | 1300–1500 | | 675–850 | 0.7–0.8 | 3.56 |

Part I, Sections 1.2.2 and 3.3, describes in detail manufacturing processes and properties of synthetic fibres, also used as binder fibres. In addition, Table 6-14 comprises a number of thermal characteristics of fibres which are important for fusion bonding.

The transformation temperature frequently, also in the preceding paragraphs, is referred to as vitrification temperature or as vitrification point. It should be noted that with a drop in temperature – "transformation temperature" – the movement of large molecule segments in the non-crystalline ranges is "vitrified". Above the "vitrification temperature" polymer chain segments start to move due to Brownian movement which results in a rubber-like condition of the fibres.

The binder fibres required for thermal bonding of nonwovens can be homofibres or bicomponent fibres of various cross-sectional structures. Common cross-section shapes of bicomponent fibres are core/sheath, side-by-side and fibres with island structure or orange cross-section. Polymer blends or polymer modifications allow to lower the melting range and minimize shrinkage. These fibres have a high melting speed. They can lose their fibre shape in melted condition and deposit on fibre crossings as beads. Bonding between fibres during thermofusion can occur by cohesion or by adhesion. Bonding is cohesive when intermolecular interaction takes place between fibres of the same polymer. A practical example is bonding between bicomponent fibres with polyester sheath and undrawn polyester fibres. The undrawn fibre surface becomes soft at transition temperature (glass temperature), the fibre turns sticky and adhesive. This process is irreversible. Once crystallization is complete, the bond is thermostable. Adhesive bonding at the cross-over points between matrix and bonding fibres takes place when the bonding fibres have turned plastic at a certain temperature.

The properties of a thermally bonded nonwoven depend on how big the share of bonding fibres in the web is. The selection of fibre polymer, fibre fineness and fibre arrangement in the web also, apart from the nonwoven mass, determines thickness, bulkiness and strength/deformation behaviour. Thermofusion allows to produce high-bulk nonwovens, so-called "high-loft" or "fibrefill" nonwovens which are required, for example, as fibre mats for foamed material substitution in the upholstery industry, in vehicle seats and in the filter industry. There are various definitions for characterizing high-loft nonwovens. Saindon [167] has appropriately deduced a variable C determining the number of fibres passing through 1 mm$^2$ of a vertical web cross-section:

$$C = 9000 \cdot \frac{\rho_V}{T_d} = \frac{1.27 \cdot \rho_V}{\rho_F \cdot d_F^2} \tag{35}$$

$C$ = fibre surface (1/mm$^2$)
$\rho_V$ = web density (g/cm$^3$)
$T_d$ = fibre fineness (den)
$\rho_F$ = fibre density (g/cm$^3$)
$d_F$ = fibre diameter (mm)

The percentage of free area in cross-section $F_F$ is

$$F_F = 1 - \frac{\rho_V}{\rho_F} \cdot 100\% \tag{36}$$

### 6.4.2.2 Process technology

Heating of the web to the required melting temperature and subsequent thermal bonding between the fibres can be achieved by application of the through-air principle or the air-jetting method using hot air.

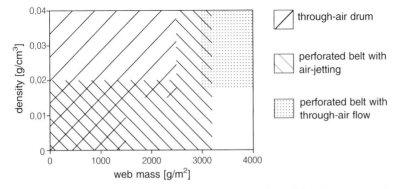

Fig. 6-78 Application ranges of through-air drum and perforated belt for hot-air bonding

Fleissner supplies installations with HighTech drums, perforated drums and perforated belt lines for this purpose. The choice of method depends on the demanded properties of the final product, in particular its web mass and its web density (Fig. 6-78).

It is important that the web be heated quickly to the melting temperature of the melt/bonding fibres. Once the temperature is reached, the air flow is reduced so that the volume of the web is maintained. Afterwards the web is cooled to minimize possible shrinkage of the bonding fibres and thus a reduction of web thickness. The heating and cooling process of a web is as shown in Fig. 6-79.

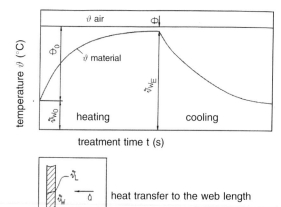

Fig. 6-79  Heating and cooling processes during thermofusion with hot air

The following equation applies to the stationary process:

$$d\dot{Q} = c_{pm} \cdot \dot{m} \, d\vartheta_w = a \cdot dF \cdot (\vartheta_L - \vartheta_w) \tag{37}$$

$\dot{Q}$ = thermal energy      (kcal/h)
$c_{pm}$ = specific heat      (kcal/kg °C)
$\dot{m}$ = m · B · v · 60      (kg/h)
     m web mass      (g/m²)
     B web width      (m)
     v speed      (m/min)
$\vartheta_w$ = material temperature      (°C)
$\vartheta_L$ = air temperature      (°C)
$a$    = heat transfer coefficient      (kcal/m² · h °C)
$F$    = surface      (m²)
$dF$ = B · v · 60 · dt
$\Theta$    = temperature difference = $\vartheta_L - \vartheta_w$

Solving and transposing of the differential equation results in the following web temperature:

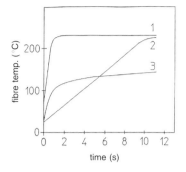

**Fig. 6-80** Comparison of heating periods for various technologies
1 Fleissner through-air principle, 2 air-jetting principle, 3 contact heating

$$\vartheta_W = \vartheta_L - \frac{\vartheta_L - \vartheta_{Wo}}{e^{\frac{a \cdot t}{m \cdot c_{pm} \cdot 3600}}} \qquad (38)$$

While the specific heat $c_{pm}$ is a fibre-specific value, the heat transfer coefficient $a$ is also influenced by process-specific conditions. The heat transfer coefficient $a$ can be calculated by heating trials. When it is known for a certain process, then time t can be determined after which the web has reached the desired temperature. Fig. 6-80 shows the qualitative relationship between fibre temperature and treatment period for various process solutions.

When heating the web to the melt temperature of the bonding fibres, heat transfer takes place by a combination of heat conduction and convection both in the through-air process and the air-jetting system. In the through-air system hot air flows through the web.

The air-jetting system is suitable both for light-weight and for heavy nonwovens. The web is submitted to hot air jets from both sides and thus a convective heat transfer is achieved. While part of the air quantity is sucked off after flowing through the web, another part of the air rebounds from the web surface. Depending on the web structure, a certain ratio of air volume passing through to air volume returned is obtained.

The difference between these two heating systems is also shown by the different heat transfer coefficients. Generally, the following equation applies:

$$\frac{a_{through\text{-}air}}{a_{air\text{-}jetting}} \sim 3:1$$

The higher the web mass and density of the nonwoven, the smaller the difference between the $a$ values. Thermobonding by means of contact heating is out for economic reasons due to excessive heating periods – especially for thick bulky nonwovens.

Hot air temperature and air speed influence the strength of the nonwoven achieved with the thermobonding method. The necessary hot air temperature depends on the melt temperature of the bonding fibres. A temperature increase allows to reduce the treatment period and consequently increase the production speed while the strength of the web remains the same (Fig. 6-81). Excessive treat-

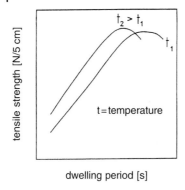

tensile strength [N/5 cm]

$t_2 > t_1$

$t_1$

t = temperature

dwelling period [s]

**Fig. 6-81** Strength of nonwoven as a function of dwelling period and bonding temperature

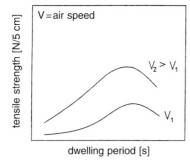

tensile strength [N/5 cm]

V = air speed

$V_2 > V_1$

$V_1$

dwelling period [s]

**Fig. 6-82** Strength of nonwoven as a function of dwelling period and hot air speed

ment periods at elevated temperatures can result in a decrease of the nonwoven strength due to structural changes in the fibres. A higher hot air speed with otherwise identical conditions achieves an increase in strength when this results in a higher number of bonding points (Fig. 6-82). The air speed can only be increased to the extent no reduction of thickness and consequently of the nonwoven volume is caused.

The use of pressure sieves for web guidance during heating with hot air is recommendable when the web tends to shrink at the fibre crossing points during thermal bonding. The result is an irregular web surface. Web guidance must be maintained until the nonwoven has cooled off. In addition, a pressure sieve will cause a higher energy consumption.

It is particularly favourable for the bonding process when a calibrating unit with defined gap setting is used at the discharge end of the line. By adjusting the gap setting to the final thickness of the nonwoven, the degree of bonding, volume and nonwoven thickness as well as surface smoothness of the nonwoven can be influenced. The use of a cooling drum with through-air flow has proved successful for reducing the cooling period of the bonded nonwoven. This avoids possible residual shrinkage of the nonwoven and reduces its tension sensitivity due to speed differences that might occur between the individual transport elements. The cooling air sucked through the web is returned to the circulation system. At the same time, finish vapours and other volatile matters are sucked off as well.

### 6.4.2.3 Line technology

For technological solutions for thermofusion with through-air or air-jetting systems on drum or belt lines (Fig. 6-83) Section 6.4.1 should be compared where perforated drum and belt dryers for drying of nonwovens are treated.

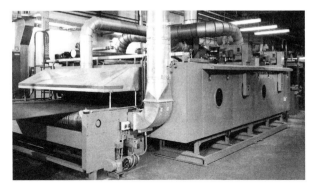

**Fig. 6-83**  Thermofusion line with wire mesh belt

There are many correspondences under the aspect of line technology. Lines with through-air drums can be used both for drying of impregnated webs and for thermofusion. Fig. 6-84 shows a thermofusion line with revolving pressure belts.

At high production speeds or high temperature and flow uniformities, the use of HighTech through-air drums offers advantages over perforated drums when nonwovens of low permeability are to be bonded. Belt lines with air-jetting system are useful when a high nonwoven bulk shall be maintained. Also air-jetting and through-air systems can be combined in a belt line.

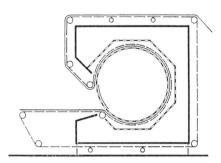

**Fig. 6-84**  Thermofusion line with revolving pressure belts

Fig. 6-85 is a survey of possibilities for cooling and calibrating of nonwovens with perforated drums and belt lines.

The advanced double-belt ROTOSWING developed by Fleissner (Fig. 6-86) provides a hot air bonding line where the top belt can be adjusted with regard to the bottom belt. Infinitely variable adjustment of top belt and nozzle height of the top

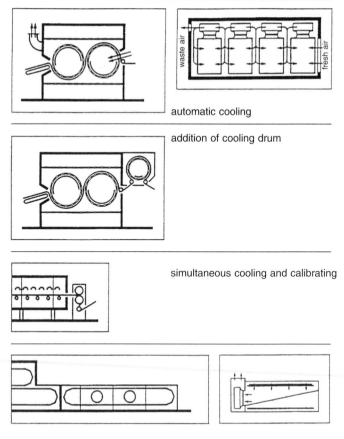

automatic cooling

addition of cooling drum

simultaneous cooling and calibrating

separate cooling section with air suction (also possible with air-jetting)

**Fig. 6-85**   Examples of web cooling by through-air flow or air-jetting

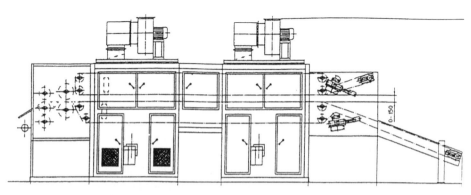

**Fig. 6-86**   Double-belt ROTOSWING line

nozzles allows optimum calibration allowing for thickness and density of the non-woven.

This line can be used to bond, for example, cotton web from 20–4,000 g/m² with a thickness from 1–200 mm and more. A high uniformity of temperature and air flow across the width ensure a consistent bonding effect.

## 6.4.3
## Heatsetting

Heatsetting is a process achieving high dimensional stability of nonwovens made of or with thermoplastic fibres at elevated temperatures. The reason lies in the thermally caused molecular mobility of the fibres. Dimensional stability can be characterized by stability of shape and resistance to shrinkage. During web formation and processing, the fibres are submitted to tension and are stretched to a greater or lesser degree. The web structure therefore contains latent tensions which can only be reduced when the caloric content is such that sufficient molecular mobility occurs. In this case, tension reduction causes shrinkage of the web if the web is heatset freely on a drum or on a conveyor, or shrinkage is avoided when the web width is fixed.

Heatsetting can be achieved by means of hot air, saturated steam or hot water. The fibres are supplied with a sufficient amount of heat energy so that the inter-molecular and intramolecular high tensions caused by deformation are reduced and newly formed on a minimum energy level during subsequent cooling. The heatsetting effect not only depends on the temperature, but also on dwelling period, tension and cooling speed. The maximum heatsetting temperatures for hot-air heatsetting of the known man-made fibres are as follows:

| | |
|---|---|
| PP | 150 °C |
| PA6 | 190 °C |
| PA6.6 | 225 °C |
| PAC | 220 °C |
| PET | 230 °C |

These temperatures approach the softening area and lie about 20–40 °C below the melting point. PET fibres are best heatset with hot air. In case of PA, saturated steam heatsetting is more effective than hot-air heatsetting. The heatsetting process also influences order and orientation of the macromolecules. This is connected with restructuring of the semi-crystalline fibre substance. Namely an increase of crystallization can be observed as a result of heatsetting together with relaxation of the random part caused especially by shrinkage during heatsetting. The heatsetting conditions temperature and time can be defined as a measure for the possibility of free shrinkage, e.g. advance in machine direction on the tenter or at the drum intake. Normally, in case of highly air-permeable nonwovens, the material is heated to heatsetting temperature within 1–2 seconds (see also Fig. 6-80) so that the heatsetting period is sufficient when using through-air drums and does not influence the degree of contraction. The structural changes reach a state of equilibrium after a short time already.

As regards the influence of the heatsetting temperature, the following has to be noted: the higher the heatsetting temperature, the smaller the contraction forces, i.e. the smaller the shrinkage of the re-heated material. It is a well-known fact that the degree of crystallization rises with increasing temperature, reducing the areas of less orientation. These areas are responsible to a considerable extent for the degree of shrinkage. Therefore it can be said that the smaller their share, the higher the dimensional stability and vice versa.

During the heatsetting process the material keeps shrinking under the influence of the heat until the full length of the specified advance has shrunk. From this point on the subsequent heatsetting takes place under tension. Hence the higher the advance, the lower the tensions.

Heatsetting of nonwovens is done
– for dyestuff fixation
– for reduction of tensions in the web
– to obtain dimensional stability
– to avoid width shrinkage during subsequent finishing processes
  such as coating etc.

The Fleissner through-air principle allows very quick heating of the web to heatsetting temperature; this means a very compact machine with all the advantages of low energy consumption. The required shock-like cooling is realized on the following cooling drum. Installations with one or more drums are used for heatsetting (Fig. 6-87).

Many nonwovens finishing processes require the material width to be controlled and possible shrinkage to be avoided throughout the entire treatment period. Such lines are required, for example, for heatsetting of PET webs for coating substrates, for heatsetting and relaxation of spunbonds etc.

Apart from conventional horizontal stenters with air-jetting system, mainly Fleissner circular stenters and Fleissner single-drum dryers with needle strips are used for this purpose (Fig. 6-88). While the circular stenter is used mostly where different material widths require to change the stentering width within wide

**Fig. 6-87** Thermobonding line with perforated drum

**Fig. 6-88** 7 m wide heatsetting line

limits, the dryer with needle strips offers great advantages where only few material widths have to be processed.

Another possibility to heatset webs while at the same time avoiding web shrinkage is offered by the Fleissner single-drum line with revolving pressure belt (see also Fig. 6-84). The web is guided between perforated drum and a wire mesh belt revolving around the drum. The wire mesh belt lies on the drum with a certain initial tension which avoids shrinkage of the web. At the same time, the fibres are heatset and the web strength is increased. While still kept in place, the web is cooled by fresh air sucked through. Spunbonds can be heatset in the same manner as fibre webs.

### 6.4.4
### Thermal calender bonding

During thermal calender bonding a web made of thermoplastic fibres or filaments is bonded in the roller gap of a calender by the influence of temperature, pressure and time. This process method is called thermobonding and has gained particular importance – especially for bonding of light-weight webs made of polypropylene – during the past years. The use of a heated engraved roller and a smooth roller causes cohesive bonding of the web at the positive engraving points. The fibre or filament structure between these bonding points shall be maintained. The fibres are plastified and bonded with each other at the bonding points. The strength of the nonwoven is mainly determined by the fibre properties, by number and form of bonding points and by the process conditions of the thermobonding process. This process offers advantages especially for bonding of light-weight nonwovens by

– low energy consumption
– high production speeds and
– maintaining of textile properties as e.g. hand and porosity between the bonding points

### 6.4.4.1 **Process technology**

The combination of web formation and thermobonding process feeds the formed web to the calender where it is condensed in the roller gap between two driven and heated rollers and melted at the contact points within a very short period. The resulting bonded nonwoven is cut to the desired final width and wound up. The degree of web bonding is influenced by:

– the fibre properties such as fineness, length, crimp, elasticoviscous behaviour under thermodynamic conditions and their softening and melting temperatures
– the mass per surface unit of the web to be bonded
– roller diameter, line pressure in the roller gap
– temperature
– web speed
– type and share of engraving on the roller surface

It is the listed visible fibre properties in particular that determine the number of fibres in the web and the web structure for a constant mass per surface unit. Finer fibres make it possible to produce lighter webs. The lowest mass per surface unit of nonwovens today ranges from 5 to 10 g/m$^2$. Table 6-15 lists the roller temperatures for melting of fibres for various fibre polymers [168]. Bonding of a web in the roller gap is shown in Fig. 6-89.

The web enters the calender at a speed $v_{VE}$. Its web thickness $d_{VE}$ depends on the mass per surface unit and on web density. When getting into contact with the heated rollers, the web is condensed and at the same time heated. Heat transfer

**Table 6-15** Roller temperature ranges for various fibre material to [168]

| Fibre material | Temperature range (°C) |
|---|---|
| Low-pressure polyethylene | 126–135 |
| Polypropylene | 140–170 |
| Polyamide 6 | 170–225 |
| Polyamide 6.6 | 220–260 |
| Polyester | 230–260 |

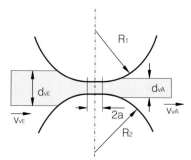

**Fig. 6-89** Geometrical conditions during thermo-bonding process
$R_1$, $R_2$ radius of calender roller; $d_{VE}$ web thickness at intake; $d_{VA}$ web thickness at outlet; $v_{VE}$ web speed at intake; $v_{VA}$ web speed at outlet; $2a$ roller flattening

to the web by means of conductivity takes place during contact time $t_c$ between the rollers and the web. The applied loading force P results in a pressure surface of width 2a and length 1. The occurring roller flattening can be calculated with the Hertzian equation (39) [169].

$$a = 1.52 \sqrt{\frac{P \cdot R}{E_{St} \cdot L}} \tag{39}$$

$$\text{with} \quad \frac{1}{R} = \frac{1}{R_1} + \frac{1}{R_2}$$

This equation (39) only applies for the assumption of mere elastic bodies and that flattening with regard to the body dimensions is low.

R = roller radius (in mm)
P = loading force (in kp)
$E_{St}$ = modulus of elasticity for steel (in kp/mm$^2$)
L = web width (in mm)
a = flattening (in mm)

Contact time $t_c$, i.e. the dwelling period of the web in the roller gap, is calculated for $v_v$ = web speed as follows:

$$t_c = \frac{2a}{v_v} \quad (v_v \text{ in m/s}) \tag{40}$$

Heating of the web must be done within the contact period in such a way that formation of bonding points takes place with simultaneous application of pressure. When the web leaves the roller gap, it can be observed that the web thickness increases. Fig. 6-90 makes it clear that the contact period decreases with increasing web speed. It can be increased by an increasing linear load and/or a bigger roller diameter.

This results in the possibility to increase the web speed and consequently the line capacity for a constant contact period by varying the linear load.

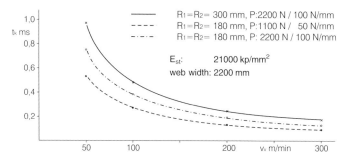

Fig. 6-90  Contact period inside roller gap as a function of web speed

Studies by Bechter and others [170] show that, depending on the temperature, a maximum occurs for the average maximum tensile stress at various line pressures. When the fibre-specific temperature is exceeded, a reduction of strength occurs. Temperature data always refer to the temperature at the roller surface. It must be selected so high only that bonding between the fibre components results in optimum strength and that the fibre structure in the bonding area is maintained. Wei and others [171] have found that the thickness of webs made of hardly oriented fibres remains constant within a wide temperature range. Webs made of highly oriented fibres show a decrease of thickness with increasing temperature which is due to hot fibre shrinkage. Part of this process is deformation heat which according to Warner [172] is partially consumed again by melting of the polymer.

Pressure is another important variable. A pressure increase results in an increase of the polymer melting temperature.

It is the pressure's task

– to improve heat transfer from the rollers to the web and
– to create bonds between the fibres melted at the surface

The pressure for thermal web bonding is generally indicated as line pressure. It should correctly refer to the web width and not to the roller width. Pressure generation has to be considered observing contact geometry and thermodynamic conditions during web deformation. As engraved rollers are generally used, still elastic and plastic deformation conditions of the web overlap in the area of the engraving points. Klöcker-Stelter [173] has dealt with process modelling through web behaviour in the gap in thermal calender bonding and based on theoretical and

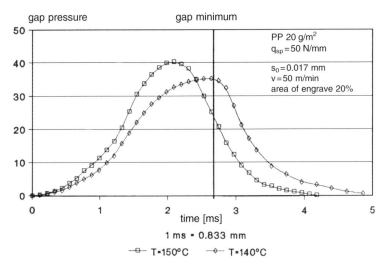

**Fig. 6-91** Pressure at engraving point with temperature variation for polypropylene web [172]

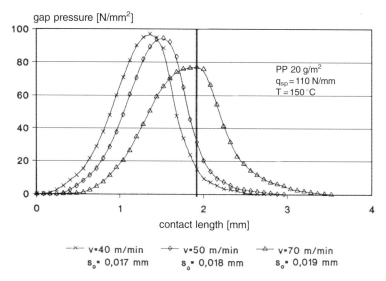

gap pressure [N/mm$^2$]

PP 20 g/m$^2$
$q_{sp} = 110$ N/mm
$T = 150\,^{\circ}$C

contact length [mm]

$\times$ v·40 m/min
$s_o$· 0,017 mm

$\diamond$ v·50 m/min
$s_o$· 0,018 mm

$\triangle$ v·70 m/min
$s_o$· 0,019 mm

**Fig. 6-92** Influence of speed on gap pressure for polypropylene web [173]

experimental studies has come to the following results: While the maximum gap pressure coincides with the gap minimum when using the Hertzian equation, the pressure maximum occurs before the gap minimum according to Fig. 6-91.

In this connection it has to be observed when using an engraved roller that the gap pressure is transferred to the web by the engraving only. The pressure acting on the engraving point therefore depends on the engraving pitch. Compared with a smooth roller it has to be increased by this amount. The gap load must be increased as a function of engraving pitch, surface mass of the web and contact length. The roller temperature for larger web surface masses should be increased such that the pressure maximum has the same distance from the gap minimum. The pressure maximum is reduced by the amount of the increasing contact length. An identical pressure effect is also achieved when the gap load is increased. According to Fig. 6-92 the pressure maximum is moved towards the gap minimum when the speed is increased.

Good strength properties of thermobonded nonwovens are achieved at critical gap pressure when softening of the fibres is caused. For this purpose, the softening temperature of the fibres must be reached before the gap minimum so that the maximum bonding pressure can act on the web before the narrowest point of the gap.

### 6.4.4.2 Line technology

Web bonding by means of calenders is directly coupled with web formation from fibres or filaments. The unbonded web is fed to the calender by transport conveyors and wound on large batches after thermal bonding. Rising line capacities and working widths of more than 5 m determine the development of line technology with regard to

- web guidance inside calender
- roller construction with guaranteed constant linear gap load over the working width
- roller temperature with temperature constancy of ± 1 °C over the width
- design of roller engraving according to the demands made on the nonwoven

Generally two-roller or three-roller calenders are used for the thermobonding process. The two-roller calender, for example, operates with an engraved heated steel roller and a smooth heated floating roller [174].

Three-roller calenders permit roller combinations with one or two engraved rollers [175, 176]. This allows to do web bonding with two different engravings, different line loads and/or with web smoothing. Fig. 6-93 shows possible web paths in a three-roller calender.

The required line loads can range from 15 to 200 N/mm and must remain constant over the entire width. This asks for suitable measures for elastic line compensation of a roller pair (see also Section 6.5.3.2). Hydraulically controlled rollers [175, 176] represent the most advanced solution. On the one hand, pressure control allows the demanded line loads to be realized, while on the other hand the roller can be simultaneously heated by the pressure medium oil and the required thermal energy input in the roller gap can be maintained highly constant. The principle of construction is based on the fact that inside a hollow drilled roller an oil pressure is generated which is directly opposed to the line load. This is the principle of operation of the THERMO Hydrein roller of Kleinewefers with temperatures up to 250 °C and 600 m/min and the heated S-type rollers 170 and 250 of Küsters Maschinenfabrik [177]. Calender bonding of webs with smooth heated rollers results in thorough thermal bonding and provides the nonwoven with a foil-type character. The type of roller engraving and the number of bonding points not only influences the bonding effect, but also the softness of the nonwoven [178]. Engravings can have round, square, rectangular, oval and rhombic raised patterns. To maintain the textile character of a nonwoven, the share of bonded surface in the total nonwoven surface should be as small as possible. It should not exceed an amount of 20–30%. The surface mass of the nonwoven determines the depth of the engraving. Steep faces of an engraving are

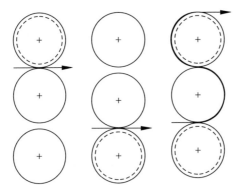

**Fig. 6-93** Possible web paths through the three-roller calender

favourable for locating the bonding point and for high softness and flexibility of the nonwoven.

### 6.4.5
### Ultrasound bonding

So far nonwovens have been bonded in individual spots in the way of a quilting seam by means of ultrasound. The basis of this method is electrical energy which is converted into mechanical vibration energy. The following applies:

$E = A \cdot t$      electrical energy input
               $A$ = work
               $t$ = time

$A = F \cdot s$
               $F$ = force
               $s$ = vibration amplitude

The bonding horn operates with a vibration amplitude of, for example, 100 μm. At these vibration points the fibres are softened and welded with each other at their crossing points in the web.

The use of ultrasound technology for web bonding by means of a calender requires the vibration amplitude to be constant over the entire web width. High precision of the rollers and stable mounting of the ultrasound heads allow to maintain a constant gap. The ultrasound calender developed by Küsters Maschinenfabrik GmbH & Co. KG and Herrmann Ultraschalltechnik GmbH fulfils these conditions at a speed of 100 m/min and working widths of more than 2,000 mm [177].

### 6.5
### Chemical methods
P. EHRLER, W. SCHILDE

### 6.5.1
### Introduction

Methods for the adhesive and cohesive bonding of webs will be described in this section. Many of these methods are neither web-specific nor specific to any form of bonding, i.e. they are also suitable for other textile substrates, and they are not only used for the actual bonding, but also for subsequent finishing treatments, e.g. coating. Overlapping with the subject matter of Chapters 7 and 8 is thus unavoidable. In addition it should be noted that chemical bonding does not have to be the sole method of bonding a nonwoven, but may be an extra method of bonding in association with mechanical (pre-) bonding.

As a rule the chemical bonding of webs comprises at least two stages: firstly the application of a substance, followed by the triggering-off of the bonding by heat treatment. The present chapter will be dealing with methods of application. For methods of heat treatment please refer to Section 6.4.

**Adhesion and cohesion**

A cohesive bond forms between two identical web fibres without any bonding agents. A cohesively-bonded nonwoven is therefore free of binders and consists of fibres having identical properties. On the other hand an adhesively-bonded non-woven contains binders (crosslinked and coagulated binder fluid or solidified binder fibre droplets) which bond the matrix fibres to one another; in addition it may consist of different fibre types. Nonwovens which contain binder fibres have both adhesive and cohesive bonds. Up to the early eighties the majority of chemically-bonded nonwovens were adhesively bonded with binder fluids. Their importance has shrunk since then in favour of principles which are reported below.

In adhesive bonding the bonds are based on intermolecular forces [179, 180] which occur in the boundary layer between the fibre and the binder layer. This two-body system reaches maximum adhesion and, at the same time, elastic extensibility, when the two components in the boundary layer are brought to within a short distance of one another, but which, viewed in the molecular dimension, still do not touch one another (Fig. 6-94). With the cohesive bonding of fibres from identical polymers there is no boundary layer.

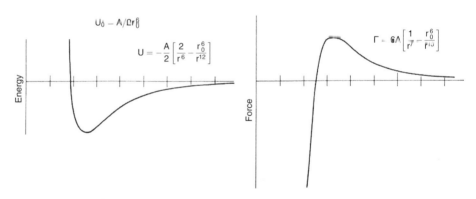

**Fig. 6-94** Distance-dependent curve for the intermolecular force F and the energy level U in the boundary layer of two adhesively-bonded molecules according to [179]
A = constant, r = distance between molecules, $r_0$ = distance in equilibrium condition

**The boundary layer of an adhesive bond**

A continuous boundary layer can only be produced on a fibre surface which is fully wetted by the binder sites which contain air bubbles are a clear indication of inadequate wetting. The fibre wettability must therefore be adjusted in such a way using a processing chemical, and in conjunction with the binder viscosity, that the binder is concentrated at the fibre intersections and contact points (Chapter 3). Contrary to widespread opinion, this applies not only to binder fluids, but also to the melts of binder fibres and powders.

As a rule, synthetic fibre lubricants (Section 2.5) assist wetting because these are mostly hydrophilic systems. Stenemur [181] confirmed practical experience via laboratory investigations. How great the effect of the spin finish can be becomes clear from the great web strength differences which emerge typically from a comparison of several different types of fibres and hence several spin finishes. If the boundary layer which is responsible for the adhesion is to be optimized still further, the reactive groups (radicals) in the binder and the fibre have to be maximized (Chapter 3).

**Bonding site morphology and arrangement**
The bonding site morphology (Fig. 6-95) is considerably influenced by the binder properties (Chapter 3) and by the application technology. This includes:

– binder fluid coagulate: point to laminar bonding
– binder fluid film: excellent laminar bonding
– powder melts: excellent laminar bonding
– binder fibres, molten or softened: predominantly point bonding

In the spatial distribution of bonding sites a distinction can be drawn between the following:

• *Total bonding:* a homogeneous distribution of bonding sites over the surface and thickness of the nonwoven fabric: typical of bath application and the homogeneous addition of binder fibres.

• *Surface bonding:* the bonding sites, distributed uniformly in the surface, are concentrated on one side of the nonwoven fabric: typical of spray or squeegee application methods.

• *Partial bonding:* the surface of the nonwoven, mostly in the form of patterns, is bonded locally: typical of application on one side by the print bonding technique, in embossing calendering.

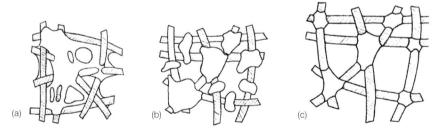

(a)    (b)    (c)

**Fig. 6-95** Bonding site formation [229]
a) area enveloping of fibre intersection points; b) small-area and punctiform enveloping of fibre intersection points; c) punctiform enveloping of fibre intersection points

• *Gradual bonding*: the bonding site concentration varies continually over the thickness of the nonwoven fabric: typical of binders which have migrated in bath application and in application by the slop-pad method.

## 6.5.2
## Methods for cohesive bonding [182]

Cohesive bonding offers the opportunity to produce nonwoven fabrics from homogeneous fibre without the use of binders, which is a crucial structural advantage for many technical applications. With dyed nonwovens the homogeneous composition encourages uniformity of shade.

Webs which have been cohesively bonded over their total area are characterized by high levels of stiffness and dimensional stability. The stiffness can be reduced at the expense of dimensional stability if the web is bonded only locally, in the form of small zones, e.g. in the case of embossed nonwoven fabrics.

### 6.5.2.1 Cohesive bonding through thermoplasticity

Embossing calendering and thermofusion are currently the best-known methods of consolidation which utilize cohesive bonding. Its use is based on the thermoplasticity of fibres which can be bonded to one another in the plasticized state under pressure (Section 6.4).

However, thermofusion is also used to consolidate webs, the fibres of which are used as binder fibres for adhesive bonding. For this reason reference should be made to it at this point. The special fibres used have been known for approximately three decades. On one hand they are classical bicomponent fibres (heterofil fibres) which consist of two polymers having different melting points (Section 1.2), and on the other they are undrawn polyester fibres (Chapter 3).

With skin/core fibres the low-melting polymer is in the fibre skin which becomes soft during heat treatment, while the high-melting polymer remains sound. A nonwoven which consists exclusively of such fibres contains cohesive and adhesive bonds.

Only cohesive bonds are produced during the consolidation of undrawn polyester fibres if the web, consisting of these fibres, is heated to the glass transition temperature which is lower than 100 °C. The fibre surface becames sticky, so that fibres can be locally bonded together under pressure. The softening/plasticizing is accompanied by irreversible changes in crystallinity, as a result of which the, bonding is stable even at high temperatures. Such nonwovens are used for special forms of electric insulation and in filters.

### 6.5.2.2 Cohesive bonding in the case of a dissolved fibre surface

The classical example of this type of bonding is paper. Reactive groups are located on the surfaces of moist pulp fibres which can form covalent bonds (hydrogen bonds) between adjacent fibres. These hydrogen bonds have no role to play in cohesively-bonded webs; cellulose pulp can only be used in adhesive-bonded nonwovens (dry-laid).

The solvent method has attained a certain level of importance for nonwoven fabrics: The fibre surface is temporarily dissolved with a fibre-specific solvent, but the solvent does not remain in the nonwoven, but volatilizes or is washed out. Basically speaking, any fibre can be bonded by this method. In addition to methodological problems, however, massive ecological problems restrict its application. Three versions have attained practical significance:

The surfaces of viscose rayon fibres are dissolved at ambient temperature by the 'Eisenhut method' [183], using 7% caustic soda solution. Following the removal of excess caustic soda solution, the fibres are bonded at the points of contact. This method, which is simple in principle and is used for the production of special nonwovens, is very demanding on fibre quality and the process.

The critical 'Bondolane process' [184], which was used in the seventies, was based on cyclic tetramethylene sulphone, a thermally activatable organic solvent. Acrylic, modacrylic, acetate, triacetate and chlorofibres were sprayed with this solvent with add-ons of 10–15% in front of the carding machine. The web containing the solvent then underwent a heat treatment, during which the volatilizing solvent dissolved the fibre surfaces. The web was then bonded in this state under pressure. These nonwovens are characterized by a high level of dimensional stability.

Acetate fibres, which are available in the form of multifilament webs, are treated with acetone. The solvent volatilizes during subsequent heat treatment and the fibres are bonded locally, forming a stable shaped body. These are suitable for cigarette filters and felt tips.

## 6.5.3
### Adhesive bonding methods [185]

The binders (Chapter 3) are fed to the web by different add-on methods. The choice of method depends on the properties which the nonwoven is required to have and on the type and consistency of the binder:

- excess application of the binder fluid (bath application)
- binder fluid or foam add-on from one side: spraying, 'print bonding', spreader roller, squeegee
- mechanical feeding of solid binders: scattering of thermoplastic powder, blending-in of binder fibres

Heat, and in certain cases also pressure (Section 6.4) must be applied to achieve bonding after the binder has been added. In the case of hinder fluids (Fig. 6-96) they are used for coagulation, and drying and, with a suitable formulation, for crosslinking; with solid binders they are used for melting and/or plasticizing.

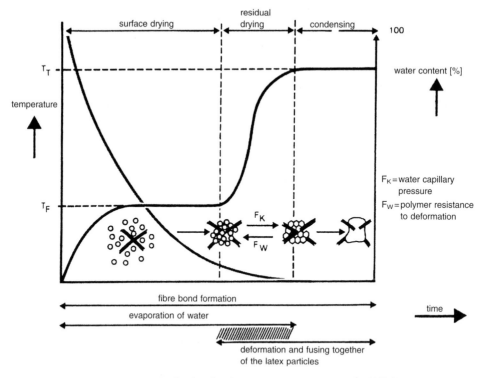

**Fig. 6-96** Heat treatment of webs after the application of binder fluids [219]. Web temperature and binder droplet water content plotted against time

### 6.5.3.1 Application of liquid binder in excess

The so-called application from the bath covers both technological stages: liquor entry in the web and the removal of excess liquor by expression or suction.

**Liquor application**

The liquor is either in a trough located in front of the padding mangle or directly in the padder nip. The excess liquor is squeezed out by the padding mangle rollers. This padding should be regarded as a universal method and is suitable for medium-to heavyweight webs.

The nip method of application (Fig. 6-97) is characterized by the following advantages over the trough method of application:

– simple method principle
– minimized liquor volume
– no separate liquor return
– easy cleaning
– low residual liquor volume

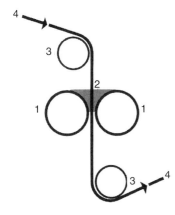

**Fig. 6-97** Principle of the padder nip method [219]: *1* horizontal arrangement of nip rollers; *2* binder fluid liquor in nip; *3* guide rollers; *4* web

Moreover, the nip method severely taxes the wettability of the fibres because the web has to be fully impregnated with the binder fluid within a very short time – of the order of 0.1 s. The wetting rate limits the web weight which can be processed.

The trough method is characterized by a wide range of applications, especially as it is possible to improve the wetting conditions by using the trough dimensions and an intermediate squeezing station (Fig. 6-98). In addition it evens out the binder fluid distribution and improves liquor absorption. The disadvantage of large quantities of liquor, large quantities of residual liquor and contamination with dissolved fibrous impurities, can be offset using liquid-level displacement.

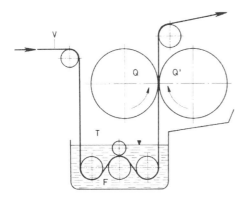

**Fig. 6-98** Trough method of applying binder fluid [185]. Improved liquor pick-up as a result of intermediate squeezing station in the trough
F = liquor, Q = padding rollers, T = trough, V = web

As a rule binder fluids cannot be applied directly to lightweight or non-consolidated webs because of the risk of distortion. In such cases, if no thermal bonding is possible, the web is transported through the bath on a perforated belt up to the squeezing unit. The belts in such 'saturators' need to be cleaned constantly.

**Additive application technique (wet-on-wet application)**

The so-called wet-on-wet application technique reduces multistage methods by dispensing with an intermediate drying stage; a first wet treatment with the application of a first effective substance and an expression process is followed directly by the second wet treatment. This method helps to save energy, but has a reputation for delivering moderately reproducible results because the add-on during the second stage depends appreciably on the expression during the first stage. In addition the residual liquor from the first stage may accumulate uncontrollably in the storage container of the second stage. Rieker, Braun [187] see the solution to the problem in an efficient on-line testing technique.

The 'Flexinip' technique from Kuesters and the 'Optimax' technique from Menzel have been developed as special wet-on-wet methods. They are mainly used in textile finishing, but the principle is also suitable for the application of binders:

- A small trough is designated the 'Flexinip' unit (Fig. 6-99) and the substrate to be processed which comes from a high-production nip unit runs through it from top to bottom for additive application [188]. The trough exit is designed as a second nip unit. The wet web coming from a padder passes through the small trough and picks up extra liquor. The lower pair of nip rollers then holds back the excess liquor.

- The 'Optimax' unit consists of a two-stage nip mechanism which the substrate to be treated runs through from bottom to top. The nip of this first squeeze unit acts as the second application stage. Schlicht [189] emphasizes the very rapid liquor pick-up by the hydroextracted and deaerated substrate in this nip.

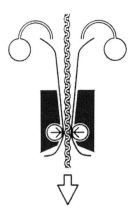

**Fig. 6-99** 'Flexinip' method for wet-on-wet application [188]

### 6.5.3.2 Removal of excess liquor

Two basic operations have proved their suitability for this task: expression and suction.

The padding mangle, which is widely used in finishing, is suitable for expression/squeezing. This padder is fitted with two or three rollers. Moreover, the concept 'squeezing' gives an inadequate description of this function. The squeezing process

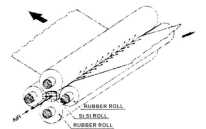

RUBBER ROLL
St.St.ROLL
RUBBER ROLL
AIR

**Fig. 6-100** Four-roller method ('Suprajet') for mechanical and pneumatic squeezing [190]

affects not only the amount of liquor removed from the web, but also the binder distribution in the nonwoven fabric and its residual bulk. An elevated nip pressure evens out the binder distribution over the thickness of the nonwoven fabric and reduces the drying costs, but also reduces the bulk of the nonwoven fabric. A three-roller arrangement further increases these effects. Nikko Seisakusho even uses a four-roller nip unit in which not only the nip pressure, but also extra pneumatic pressure is created by means of an integral airtight zone [190] (Fig. 6-100).

A space is created by sealing the roller arrangement at the sides; this can be placed under pneumatic (pressure and assists) mechanical hydroextraction.

Removing excess liquor by suction (instead of squeezing) is used when the web consists of pressure-sensitive fibres, e.g. glass fibres, or when the web bulk must not be impaired. It takes place at the slit of a suction box which the impregnated web passes. The vacuum at the slot must be at least 0.5 bar; it must be constant over the whole working width. The binder which has been sucked away is separated from the air and fed back to the applicator. With the suction method one generally has to reckon with a concentration gradient in the binder distribution, the binder concentration is higher on the web side facing the slot.

With porous, preconsolidated webs excess liquor can also be removed by blowing through. This makes it possible to make considerable improvements to the residual porosity, and hence to the air permeability of a nonwoven fabric.

### Measures for achieving uniform expression

At high nip pressures: typical line forces (force/roller length) are in the range 1,000 to 2,000 N/cm, the nip rollers inevitably sag. As a result the local pressure and nip effect vary over the web width. In the case of such important properties as strength, bending rigidity, porosity and absorbency this irregularity creates a troublesome width-dependent profile. The pressure along the roller gap must therefore be evened out. Various solutions have been developed to this problem.

With constant squeezing conditions during the whole time the machine is in use, the problem can be solved using a special barrel-shaped roller. The diameter variation is in the millimetre range and the roller clearance assumes a straight-line form as a result of sagging.

With variable squeezing conditions – the usual case in web bonding – the above roller system fails to have an effect. Therefore the appropriate measures are used to correct the shape of at least one of the rollers: the counterroller or the roller acted on by pressure (typical diameter variation 0.1 to 1 mm). The construction principle of such devices is based on replacing the usual cylindrical roller by a system consisting of a stiff roller core and a flexibly deformable jacket. Thus other types of roller geometry can also be used in addition to the barrel shape, corresponding to the parabolic shape of the bent roller.

The different mechanisms vary primarily with respect to the way in which the roller jacket is deformed; mechanically, hydraulically or by pneumatic means [191]. The tendency is for the design of such devices to become more complex, as the requirements relating to regularity grow and the working width and web. For example, the 'Artos Variflex' padder, the 'flexible pressure roller' from Benninger and a more recent development from Suchy [192] are based on the mechanical principle. Well-known hydraulic systems include the 'floating roller' (S roller) from Kuesters with hydrostatic pressure [193] and the 'Nipco roller' with hydrodynamic pressure, preferably found in calenders. The Bicoflex roller [194] (Fig. 6-101) operates on the pneumatic principle. Further developments are awaited.

Air cushions located on the roller core deform the flexible roller jacket in the appropriate zones and the deformation is adjustable. The roller clearance can thus be adjusted appropriately.

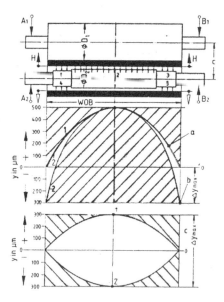

**Fig. 6-101** Pneumatic deflection compensation using two 'Bicoflex' rollers [194]

### 6.5.3.3 Metered binder application to one side
The methods discussed below relate to the metered application of binder fluids.

#### Using doctor blades (coating)
Binder fluids can be applied to one side of a web in the form of paste or foam using a doctor blade or squeegee. The doctor blade may consist of a section with

a knife-shaped edge, or it may be a doctor roller which is positioned at the periph-
ery of larger rollers as a small-diameter stripper roller. A layer of binder fluid of
adjustable thickness is located on the running web, corresponding to the distance
between the lower edge of the doctor blade and the web plane. Excess binder fluid
accumulates at the fixed doctor blade in the form of a characteristic rotating
'bulge'.

The depth of penetration of the binder fluid into the web is decisively influenced by the 'abutment which supports the web vis-à-vis the doctor blade (Fig.
6-102):

- the web is not supported in the case of an air doctor blade: the doctor operates
  against the air. The minimum layer thickness which can be attained is greater
  than in the case of the arrangements which are described below; the depth of
  penetration is minimal.

- With the rubber sheet doctor the web lies on a continuous rubber belt which,
  because of its elasticity and special geometry, can be regarded as a soft abutment. The depth of penetration is greater.

- With the table doctor, a level 'hard' table forms the abutment so that, with this
  doctor position, still thinner layers and greater depths of penetration can be
  achieved.

- With the roller-supported doctor a large-diameter roller forms the abutment. As
  a result of convex roller curvature there is minimum contact length between
  the doctor and the web and maximum pressure per unit area. Accordingly, this

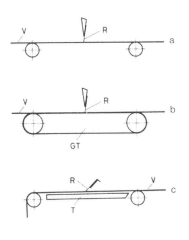
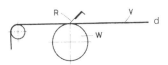

**Fig. 6-102** Squeegee/doctor blade operating
principle [185]:
a) air doctor; b) rubber belt; c) table doctor;
d) roller-supported doctor
GT=rubber sheet, rotating; R=doctor/squeegee;
T=feed table; V=web; W=roller

arrangement is characterized by the minimum achievable layer thickness and the maximum depth of penetration.

The layer thickness and depth of penetration of the binder fluid depend on a large number of variables [195.] The theoretical aspects of these complex relationships are examined by Jin et al. [196].

The wettability of the web by the pasty binder fluid has a considerable effect on the quality of the applied layer. With moderate wettability air bubbles are formed which get into the rotating bulge at the edge of the doctor blade and reappear at some time as bubbles which cause trouble in the applied layer. The doctor blade or squeegee technology is intended for the post-bonding of pre-consolidated webs. Preconsolidation is unavoidable because the web has to be fed through under tension during this process, which has special importance for foam application and the back reinforcement of substrates.

### Total-area application via roller surfaces (slop-padding)

When the running web comes into contact with the binder-fluid covered surface of a rotating roller, the kiss roller, binder fluid can be transferred over the whole surface of the web. The fluid transfer is regulated by wetting effects. The kiss roller picks up the liquor either directly the liquor storage tank or indirectly in accordance with the 'reverse-roll coating' procedure, via an intermediate roller rotating in the liquor. It operates without a counterroller. The add-on is controlled by the angle of wrap and the fabric speed.

The slop-pad method is frequently used for the back reinforcement of heavyweight needlefelt fabrics (floorcoverings, wallcoverings, shoes), and also for filters.

### Small-area application using engraved cylinders

With this method an engraved cylinder transports the binder fluid in dot form to the web (repeat pattern or random point pattern (computer point)). A counterroller is needed because a specific pressure is required for the binder fluid transfer which is controlled by the wettability.

In the case of a high-pressure engraved cylinder the raised areas transfer the viscous binder fluid to the web; with a low pressure cylinder the deep-lying areas perform this function. The low-pressure makes great demands on the wetting behaviour of the web, and the binder fluid must have excellent thixotropic viscosity, as the small depressions must be completely emptied at each passage. High production rates can be reached if the web wettability is adequate. Prewetting of the web reduces the risk of fibres sticking to the cylinder.

Engraved cylinders are also suitable for the application of hot melts if the cylinder and the storage tank are heated (Fig. 6-103) [197]. Great demands are made on the thermal stability of the hot melt in the case of this melt printing method.

The small-area binder patterns, covering 15–80% of the web area, result in nonwovens with low bending rigidity, a soft handle and high water absorbency. The strength which can be achieved is low so that this method is often used for further bonding. According to Welter [197], typical end-uses include interlinings,

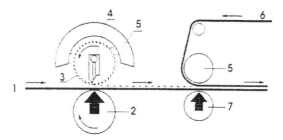

**Fig. 6-103** Engraved-cylinder application system with rotogravure
roller for hot-melt adhesives ('melt-print') [197].
*1* web, *2* counterroller, *3* engraved cylinder, heated; *4* hotmelt
adhesive liquid, *5* storage tank, heated, *6* substrate, *7* pressure roller

composites (for shoes, pursemaking fabrics, car interiors), insulating materials,
filter laminates and packaging materials.

### Small-area application via rotary screens

With this method a rotary screen which is normally used in screen printing is
used for the application of small-area patterns instead of the engraved cylinder.
The binder fluid, either in paste or foam form, flows to the inside of the rotary
screen via a printing line, from where a doctor blade forwards it on to the web via
perforations (Fig. 6-104). The paste, which has a special formulation, must not
'pull any threads'. One well-known version of these rotary screens is the 'Zimmer
magnetic roller' system [198] with a magnetically-fixed squeegee.

Welter [197] describes a method which is known as the 'hot-melt screen print',
whereby hot-melt adhesives can be applied via rotary screens. The screen and the
mechanism supplying the hot-melt adhesive are heated to a working temperature
of a maximum of 180 °C. Welter [197] expects numerous uses in view of the low
pressures needed. Instead of that Endress [199] recommends the application of

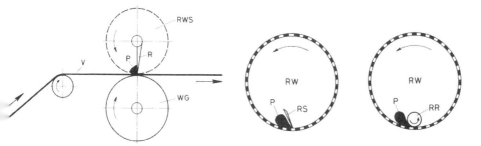

**Fig. 6-104** Rotary screen for the application of binder fluids in paste form [185]
*Left:* Web path with counterroller; *Right:* Paste distribution via normal or roller squeegee
RR = roller squeegee; RS = normal squeegee; RW = 'screen roller' in the form of a rotary screen;
RWS = 'screen roller' in the form of a rotary screen/sieve; V = web; WG = counterroller

hot melts using a nozzle with a wide slot which is subdivided into small, individu-ally-controlled segments. This should even give add-ons of the order of $1 \text{ g/m}^2$.

### Multifunctional machines

The development of chemical bonding indicates that the nonwoven fabrics manu-facturer must use more and more binding methods for different applications. This trend runs counter to the usual demand for low-cost production conditions. As a solution to this problem, Welberg [200] recommends a multifunctional ma-chine suitable for several application techniques which has a modular construc-tion and combines various squeegee versions and rotary screen technology with a padding mangle.

### Spraying

The binder fluid, which, in the form of small drops, is sprayed on to the top face of the web which is laid on a perforated belt. The droplets penetrate into the inte-rior of the web to a certain extent. Suction applied to the lower face of the web as-sists this migration. In certain cases both sides of the web are sprayed one after the other. The spray method is particularly useful for the bonding of aerodynami-cally laid webs because the individual airborne fibres can be contacted with the binder droplets on the way to the sieve drum. In addition the tile-shaped fibre ar-rangement facilitates the binder migration into the interior of the web.

Compared to the bath method of application, in spray bonding one generally has to reckon with lower strength and a variable binder content over the web thickness. The side of the nonwoven to which the spray is applied always has the greater concentration.

The fabric weight range which is suitable for spraying is from 15 to $1,500 \text{ g/m}^2$ [201], the binder content is between 10 and $30 \text{ g/m}^2$. The main end-uses for spray-bonded nonwovens are in filters, stuffing, upholstery, insulating material and wadding. In the eighties and nineties the spraying method lost a considerable amount of ground to foam technology; as before, the main end-uses are in fine-count quilting, carrier substrates for use in emerizing and dry-laid pulp-fibre non-wovens [202].

The droplets come in the form of a spray jet of limited width (conical jet, flat jet). Several atomizing units have to be arranged next to one another for this rea-son. Web widths of up to 12 m can be achieved [201]. To prevent stripy add-on dif-ferences in the area where the jets overlap, they traverse in those cases where it is necessary. At high web speeds the traverse mechanism by an elliptical web guide ('oval runner').

The atomizing units are housed in a spray cabin. Excess spray has to be re-moved from this by suction and suitable methods (e.g. baffle-type separators, band filters or air scrubbers) have to be implemented to make it environmentally acceptable [201]. Steps are being taken towards the recycling of binders. The con-siderable soiling of the conveyor belt and the atomizing units necessitates permanent and thorough cleaning. Special wide-mesh perforated belts make cleaning easier.

The droplets are produced in atomizing units, using shear forces. The mean droplet size is in the range 20–200 µm, depending on the atomizing method. The normal methods are classical air atomizing and a decompression method – known as 'airless' atomizing:

- In air atomizing use is made of two-media nozzles (Fig. 6-105), where air atomizes the liquor and transports the droplets to the web. The mean droplet size is more than 50 µm as a rule. Because of the high air outlet velocity and the associated risk of the web being 'blasted', a distance greater than 0.5 m must be maintained between the two-media nozzle and the web.

- In 'airless' atomizing, pressure instead of air is used for atomizing. The liquor is initially compressed by a piston pump (30 to 300 bar) [203]; during the subsequent decompression in a special nozzle it is broken down into small droplets (mean size: smaller than 50 µm) which leave the nozzle as an aerosol almost without any pressure. The small droplets are transported to the web directly and loss-free, so that a particularly uniform binder distribution is achieved. In addition the binder fluid for this method must be specially formulated.

The piezoelectric technique is no alternative to the atomizing methods, because not every binder fluid can be atomized ultrasonically. On the other hand a two-stage air atomizing technique, as used in medical technology for inhalation aerosols, is universally applicable [204].

Goossens [205] reports on a new spray bonding method, the hot-melt spray technique.

External mixing system

Internal mixing system

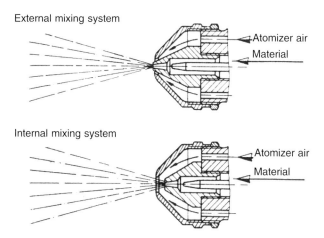

**Fig. 6-105** Principle of a two-media nozzle for the atomizing of liquids using air [203]

### 6.5.3.4 **Application of foamed binder fluids**

**Foam production**

From the physical/chemical point of view the foam produced from binder fluids has a complex structure [206–209]. Expenditure increase in the case of foams which contain a filler (e.g. chalk) [210].

The foam is produced mechanically in a foam mixer which resembles a high-speed stirrer which basically consists of a stator and a rotor [202, 207, 208, 211–215]. One alternative, foam production using a static mixer [202], has not made a breakthrough so far. Liquid dispersal and the mixing-in of air at high shear stresses are effected via pins or perforated plates. The pressure in the mixer, which results from the flow stress yield resistance of the foam at the mixer outlet, is between 3 and 6 bar. As the crucial process parameter, it is primarily influenced by the rotor speed, the air volume and the air pressure, so that it is advantageous to have a control system [211].

Basically speaking, the same binder fluids can be used for the foam bonding of webs as for bonding with binder fluid liquors. Foaming agents and foam stabilizers are used to give them foamability [206, 207, 213, 216]. Concentrations of the order of 10 g/l are needed for this.

Foam is characterized by its mass and stability. The foam mass is in the range 30 to 300 g/l. The foam stability is characterized by the disintegration rate of the bubbles and hence affects the processing behaviour. Unstable foams give rise to regular binder distribution [216]. With stabilized foams a foam-like, porous film which protects the surface is formed on the nonwoven fabric. Other characteristics are discussed by Reinert and Kuthe [207], Fiebig [206], Kroezen et al. [208, 209] and Engelsen et al. [210]. Isarin et al. [217] describe a method which can be used to determine the bubble size distribution of generated foams online.

**Foam application**

Originally the padding mangle, squeegee and rotary screen techniques were used for the application of foam. The 'Variopress' unit (Fig. 6-106) has established itself as a means for supplying a rotary screen with foam (as an alternative to the roller squeegee) [213, 218, 219]. It can also be used separately as a foam applicator. Foam

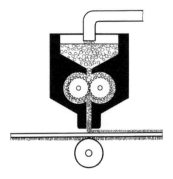

**Fig. 6-106** *Variopress equipment for the metering of foam [219]*

having a specific flowability is conveyed by a pair of grooved rollers either directly on to the web or into the interior of a rotary screen. Special methods of application were added at the beginning of the eighties [207, 212–214, 216, 218, 220]:

- Roller application method ('Kuesters roller') (Fig. 6-107)
  A layer of foam is transferred via a squeegee roller to a carrier roller which then comes into contact with the web. Two identical units apply foam to both sides of the web.

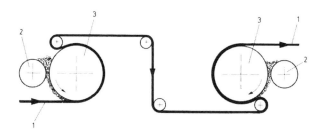

**Fig. 6-107** Application of foam to both sides using the 'Janus' unit from Kuesters [214]
*1* goods, *2* squeegee/doctor, *3* contact/carrier roller

- 'Vaku-Foam' from Monforts (rubber blanket) (Fig. 6-108)
  Foam is applied to the web using an air-impermeable rubber blanket backcloth. The vacuum created by a vacuum sieve drum ensures forced wetting of the web. The forced wetting produced by this method facilitates the reproducibility of foam application.

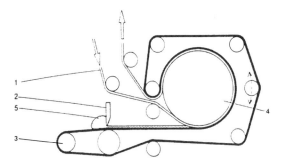

**Fig. 6-108** 'Vaku-Foam' applicator from Monforts
*1* web, *2* metering squeegee, *3* air-impermeable back grey, *4* vacuum sieve drum, *5* foam

- Flow coating process from Gaston County
  The foam, which is metered from a nozzle, is left to its own devices; it penetrates into the web depending on the wettability of the web and various foam properties.

**Characteristics of foam bonding**

Foamed binder fluids contain less water than the corresponding binder fluid liquors. In the case of foam the solids content is approximately 40 to 50%, with impregnating liquors it is around 15%. This results in lower drying and hence energy costs. Reinert and Koethe [207] doubted whether, when everything was taken
into consideration regarding foam generation, there was actually any energy saving. Despite this foam technology gained very greatly in importance in the eighties at the expense of the impregnating technique for web bonding because it was
realized that there were many other advantages [202, 207, 212–216]:

– higher production rates
– a wide range of add-ons, right down to minimum add-on
– more uniform binder distribution over the surface
– less materials to dispose of
– less risk of migration during drying
– greater strength at reduced bending rigidity as a result of point bonding
– higher nonwoven air permeability and greater nonwoven bulk
– improved textile drape

Of course, these advantages were known a long time ago; wadding used in making-up and upholstery was bonded with foamed 'Darmstadt hide glue' as early as
the thirties [216].

The basic disadvantages are [207]:

– irregular binder distribution over the web thickness at high fabric weights
– pronounced processing behaviour dependence on the web wettability
– the extra cost of foam generation
– reproducibility problems on account of variable foamability and foam stability

In the meantime foam technology has become established as the standard method for web bonding. It is suitable for webs in the range 15 to 2,000 g/m$^2$ at
speeds of up to 500 m/min and working widths of up to 6 m [202].

### 6.5.3.5 Application of powder

Bonding powders (see Section 2.3) are preferably used in the finishing of nonwoven fabrics (Part III: 'Finishing of nonwovens'). However, they have also become
established in the meantime as an alternative to binder fluids or binder fibres in
the bonding of webs, although they were originally regarded as expensive bonding
agents. Moreover, this cost argument was never convincing, since reprocessed cotton nonwovens bonded with phenol resin powder have been used for two decades
for mouldable nonwovens in car interiors [221]. In the eighties the method was
used in the fabrication of body parts for the Trabant car in East Germany. In the
nineties there was a steady increase in the importance of powders for web bonding because of the following clear advantages [197]: high production, contactless
application, no residue, low energy costs.

In the meantime the following have also become typical end-use sectors: interlinings, composite fabrics for shoes, pursemaking fabrics and floorcoverings, nonwovens from recycled materials. In the meantime, therefore, a wide range of powder types has become available in addition to the phenolic powders with a typical particle size approaching 100 µm [222]: polyethylene, nylon copolymer, polyester copolymer. (note: phenolic resin powders as thermosetting plastics behave like thermoplastics during processing, because they become soft.)

The powder is either mixed in with the web as early as the production stage or spread over the web surface later on. On aerodynamic web-laying systems the mixing is usually carried out in front of the sieve drum on which the web is formed. This permits a largely homogeneous distribution of the powder over the web thickness. This is not possible in the case of card webs. For this reason Fleissner has developed a spreading/steaming method which is suitable for heavy nylon card webs. Before the web reaches the sieve drum dryer the powder is scattered over it and it is then steamed. Only then is bonding carried out in the dryer. The steam treatment prevents the sieve drum from becoming contaminated by powder particles and by fibres sticking to it [223]. A further possibility is to mix fibres and powder in advance (Schott [226]). Thus two different materials, e.g. short fibres and bonding powder, can be spread together using a 'twin spreading unit'.

In the metering of powder spreading units have made a breakthrough (see Fig. 6-109), although the electrostatic metering units which are normally used during powder coating are also suitable (also [221]). The spreading units basically consist of a metering roller with axial grooves and a brush roller which brushes the powder out of the grooves and on to the web surface [224]. The powders used for this must have high fluidity and a high bulk density; any caking which may occur during transport or storage [197] must be completely eliminated.

A great deal of expense is incurred to maintain uniform powder distribution when powder is subsequently metered to the web surface [223]. In addition the

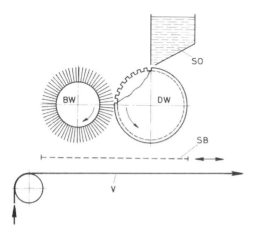

**Fig. 6-109** Powder spreading unit [185]
$BW$=brush roller; $DW$=grooved roller; $SB$=vibrating perforated belt; $SO$=powder feed tank; $V$=web

web tension must be strictly controlled because it affects the depth of penetration of the powder particles.

Dominik [225] describes a spreading system which can scatter particles in the size range between 50 μm and 20 μm in patterns. A monitoring unit supplies a rotary screen with particles; these pass from the screen on to the web as a result of a vacuum.

In the nineties we became aware of the 'double-point coating technique', a method which had been used for a long time for interlining fabrics. It combines powder technology with rotary screen technology [226, 227]. Great-store is set by this method, because it facilitates the controlled distribution of powder particles over the surface of a substrate: a rotary screen is initially used to apply a crosslinkable binder fluid in the form of dots of paste to the substrate surface, after which excess powder is scattered over it. Powder particles adhere to the sticky dots of paste, the rest of the powder is removed by suction. The dots of paste are 'sintered' during a subsequent heat treatment process. The bondable thermoplastic powder particles remain on the surface of the substrate.

**Environmental aspects**

Numerous environmental aspects have to be taken into consideration in the chemical bonding of nonwoven fabrics. They relate primarily to procedures for dealing with exhaust air, solid waste and effluent which come into being, for example, when cleaning contaminated machine parts and disposing of residual liquor. The method used for bonding the web and the selection of agents and auxiliaries are crucially influenced by such aspects. Fischer [228] summarizes these aspects in an impressive catalogue of control measures ('environmental panorama'), although it is confined to the processing and use of binders:

- 7th amendment to the EU guideline on 'Classification, packaging and characterization of dangerous substances'
- EU guidelines Nos. 92/39/EWG and 128/90/EWG relating to the legal authorization of monomers with respect to foodstuffs and relating to 'Materials and objects which are meant to come into contact with foodstuffs'
- DIN EN 71 on the safety of toys
- Appendix 38: Textile production and textile finishing; 38th textile wastewater management regulation; 1984 ff
- law on effluent clarification; 1992
- maximum workplace concentration and toxicity indices
- 4th regulation in the Federal emission protection law; 1991 addition
- Regulation on air pollution
- Circulation efficiency regulation
- Requirements relating to the burning of residues and solid waste
- Requirements relating to degradability
- Formaldehyde release
- Various quality certificates for characterizing the environmental compatibility of textiles

**6.6**
**Composite materials**
P. BÖTTCHER

Both the quality and quantity of technical textiles have been quickly developing. As a result, a much wider range of textile composites is available for many different uses. Textiles of any kind and made by means of whatever manufacturing process are combined with one another or with non-textile materials. This mainly aims at better processability (with stress-bearing composites), multi-functionality (with large area composites meant to transport or filter out liquids) as well as a storage function (with composites showing a non-textile functional element which is centrically inserted).

6.6.1
**Nonwoven Composites**

Nonwoven composites are textile fabrics of webs or, respectively, bonded webs which are combined with other textile elements, their web or their felt character being pre-dominant [230]. According to [230], these web-type composites are divided into the groups below, following their manufacturing processes:

• Nonwoven composites made up of several layers
These are composites made in one separate process from at least one pre-fabricated or pre-bonded web and at least one more pre-fabricated textile fabric (e.g. web, nonwoven, cloth, knitted fabric, net, thread composite). In this process, the structure of layers can be maintained. Bonding the layers to one another may, for example, be achieved by means of needle-punching, stitch-bonding, high-frequency welding, ultrasound or laminating. The structure of layers can also be cancelled. Usually, such nonwoven composites are named thread-reinforced nonwovens, cloth-reinforced nonwovens, paper machine felt.

• Nonwoven composites reinforced by means of thread loops
These are nonwoven composites made in one process and consisting of a pre-fabricated fibre gauze, a web, a pre-fabricated bonded web and a high number of threads which penetrate the fabrics and bond them. To give one example, pile-knitted nonwovens such as Voltex belong here.

More advanced technologies make new processes available to bond the webs which can also be used to manufacture nonwoven composites. Such processes are, for instance, entangling, meshing in the web layer, cover-seaming, thermofusion and thermal calendering.

Today, it does not appear to be important whether or not such composites maintain their web character. Looking at things from the product, two aspects are essential:

– one or more components of the web-type composite consist of a web
– bonding is achieved by means of a web-bonding process

### 6.6.1.1 **Variants of processes**

Speaking of the variants of web-bonding, there is a wide variety of processes to make the composite, ever new developments coming up in this field [231, 232]. Figs. 6-110 and 6-111 provide a survey of both the mechanical and chemical ways of bonding. In addition, they give the specialist further information.

While in Fig. 6-110 the thermal bonding processes are sorted to physical principles, Fig. 6-111 follows chemical principles. In most cases, chemical bonding is based on adhesive effects.

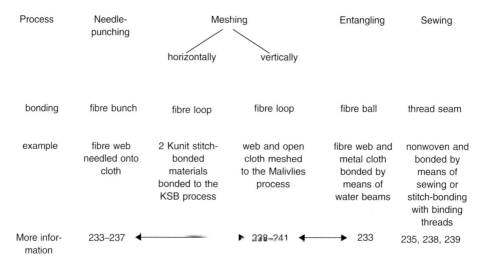

**Fig. 6-110** Variants to mechanically bond nonwoven composites

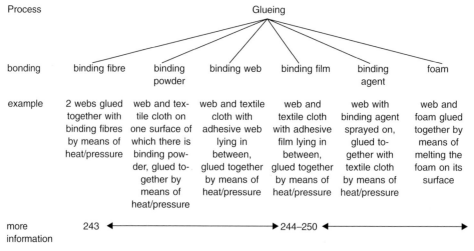

**Fig. 6-111** Variants to chemically bond nonwoven composites

### 6.6.1.2 Bonding by means of needle-punching

The manufacture of composites by means of needle-punching is an efficient process. With regard to the thickness, number and kind of the composite components, there are few limits. It is also common practice to centrically deposit non-textile functional components [233, 239]. The following materials are generally used in industry:

– dried herbs used for medical purposes
– water-storing materials such as super-absorbers (incontinence products, nappies for babies)
– grass seed, fertilizers (for greening purposes or application in erosion protection mats)
– highly swellable bentonites (sealing sheets used in the construction of waste dumps) and in
– hydraulic engineering
– sand, gravel or other mineral materials (loading covering mats or filter mats in hydraulic engineering)

The majority of materials used for insertion range from 30 to 5,000 g/m$^2$ and require, depending on their functions, nonwoven components ranging from soft to highly solid. These materials are distinguished

• by their make-up:
  – suitable to be tipped
  – suitable to be laid on one another
  – suitable to be stuffed
  – suitable to be taken down

• by their shape as:
  – grained
  – fibre-type
  – powdery
  – wide
  – thread-type

• by their specific properties as:
  – meltable
  – swellable
  – heat-proof
  – resistant to rotting
  – rotting
  – partly rotting

The materials mentioned above are also applicable in stitch-bonded composites produced by means of fibre or thread loops (see Section 6.6.1.3).

Needle-punching requires parts of fibres from one or more of the layers to be bonded. As to setting the needle-punching parameters, this experience is available [251]:

- Needle gauge: Needle gauge should – similar to web-bonding – be set to the fibre diameter of the fibre parts to be transported by the barbs. For practical reasons, a needle somewhat larger (lower gauge number) is advisable.

- Depth of stitch: Stitches should be deep enough to allow the fibre parts of the one component of the composite to enter into the cross section of the other. However, no fibre parts must, as a consequence of stitches being too deep, appear on the surface of the lower component of the composite.

- Density of stitches: Tearing resistance is influenced by the density of stitches rather than the depth of stitches.

One interesting variant of composite manufacture by means of needle-punching is known as 'fibre sewing'. In this process, the barbed needles are arranged in the needle board in a tight, row-like way to create a seam-like connection. The surface structure is similar to a "hill" with the part not needled, and similar to a "valley" with the needled seam-like trace. This allows for particular functional effects, such as water storage (with mats used for greening or embankment protection) or proper support for a layer of humus soil.

As the vertically arranged fibre parts enter into one of the composite layers either on one particular side or reciprocally, the materials are blended. Given common needle-punching parameters and measuring by the total fibre length which is theoretically available, a share of no more than 2 to 10% is transported into the other component of the composite. However, this effect is worth being mentioned when talking about the properties and functional advantage of a particular nonwoven composite.

Furthermore, it is important to know that needle-punching as a mechanical bonding process generally results in the composite being compressed and its thickness being reduced.

As for the manufacture of nonwoven composites with stitch-bonded nonwovens, let us look at the following example [252]: Bonding two stitch-bonded nonwovens by means of needle-punching, the fibre bunches created by the fibres and fibre parts of the nonwoven stitched first are entered into the cross-section of the second nonwoven, which causes the bonding. Logically, one important prerequisite is the availability of as many as possible flexible, horizontally arranged fibres or parts of fibres on the surface where the barbed needles stitch in. Table 6-16 shows to what extent a variety of stitch-bonded nonwovens are generally suitable.

The vertical fibre bunches resulting from the process described serve to improve elasticity when surface pressure is applied. Preferably, these vertical fibre bunches should be packed as densely as possible and as equally long as possible. With regard to the choice of needle, this is best to achieve by means of needles known as points. These are needles showing only one barb on each edge of the working shaft, all three barbs being placed at the same distance from the needle point.

Tests to vary needle gauge and stitch density have resulted in the following:

- Needle-punching as a bonding process causes a reduction in thickness and a rise in compactness. This effect decreases with growing needle gauge and increases with growing stitch density.

- Additional thermal bonding with binding fibres results in a rise in tearing resistance and elasticity.
- As to tearing resistance, the current limit value relating to textile composites is at least 12 N/5 cm. Among other industrial applications, it is important in the automotive industry (interior lining). This value is reliably surpassed only by variants of above 150 stitches/cm$^2$ or, respectively, 75 stitches/cm$^2$ with additional thermal bonding.
- Fibre bunches needle-punched of Maliwatt nonwovens show higher tearing resistance than those of Malivlies materials.
- Fibre bunches of needle-punched Maliwatt nonwovens show slightly higher tearing resistance than comparable ones of pile-knitted nonwovens Kunit.

**Table 6-16** Suitability of a variety of stitch-bonded nonwovens for bonding by means of needle-punching

| Stitch-bonded nonwoven | | Suitability for needle-punching | |
| --- | --- | --- | --- |
| Type | Position to stitch-in surface | Good | Poor |
| Maliwatt | Same | Many flexible fibres lying crosswisely | Thread of the loop is destroyed |
| Malivlies | Loop side | Loop side remains | Few flexible fibres lying crosswisely, loop side disappears |
| | Fibre side | Many flexible fibres lying crosswisely | |
| Kunit | Loop side | Few flexible fibre parts of the fibre loop lying crosswisely | |
| | Fibre side | | No fibre parts lying crosswisely |

### 6.6.1.3  Bonding by means of stitch-bonding

Section 6.2 will discuss a variant of how, according to the layer-bonding process, two nonwovens and/or cloths or knitted fabrics can be horizontally bonded with a vertical fibre surface by means of fibre loops. This variant is of considerable technological and product-related interest.

Stitch-bonding with its process variants Malivlies and Maliwatt (see Section 6.2) allows the use of extensive composite components such as cloths, spunbonded nonwovens, films a.o. with both the nonwoven bonding process and as a separate process. The binding elements are fibre loops (Malivlies) or thread loops (Maliwatt). By the example of the Maliwatt process, Fig. 6-112 shows the different ways to combine materials available with this bonding technology.

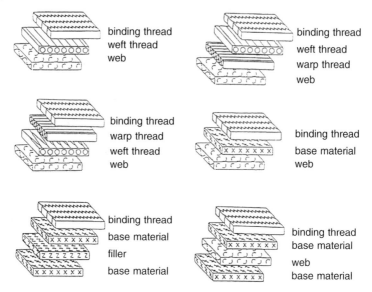

**Fig. 6-112**   Nonwoven composite structures to the Maliwatt process [238]

#### 6.6.1.4   **Bonding by means of entangling**

The novel bonding technology using high-pressure water beams (see Section 6.3) is also suitable to effectively manufacture particular nonwoven composites [243]. This concerns a composite of a voluminous stitch-bonded nonwoven and a thin fibre web, the latter being compacted and, at the same time, bonded with the stitch-bonded nonwoven by means of water beams.

To manufacture this composite material, 6 jet bars were used, the pressure exercised by the water beams rising from bar to bar. It is important the water beam pressure of the first jet bar does not surpass 0.2–0.5 MPa. Hydroentangling is followed by drying and setting. The composites of stitch-bonded nonwoven and spunlaid fabric (Fig. 6-113) made to this process show high strength and voluminosity, a fine-pore web layer on one or two surfaces and, due to their good sliding properties, excellent processability. The bonding between the stitch-bonded nonwoven and the hydroentangled nonwoven is evenly strong.

**Fig. 6-113**   Composite of stitch-bonded nonwoven and hydroentangled nonwoven

Composites of stitch-bonded nonwovens and spunlaid fabric provide an interesting range of alternative products which are found, for instance, in

– depth filters for use in both wet and dry filtration
– insulation material
– upholstery material
– absorbent materials for technical, medical and hygienic uses

Composites of stitch-bonded nonwovens and spunlaid fabrics show a number of important advantages. They are strong, voluminous and, as they also slide well, easy to process. Their edges are strong. They offer the possibility to combine a number of functions, e.g. the good attenuation effect provided by stitch-bonded nonwovens with the flame-retarding character and the heat-resistance of the web layer as well as high storage capacity with the capacity to separate smallest particles when used as a filter medium, which is combined with conductivity respectively antistatic behaviour and high moisture capacity.

### 6.6.1.5 Bonding by means of glueing

On the one hand, the manufacture of nonwoven composites by means of glueing is a traditional bonding technology, analogous to textile composites of textile fabrics of cloth, loop materials etc. On the other hand, it is economical and also eco-friendly, as long as solvents are not used, emissions and non-recyclable waste are avoided.

Thermoplastic adherends are known as adhesive webs and adhesive films, which bond large-area nonwovens with one another or with textile or non-textile materials in an economical and eco-friendly way. Such glueing materials are available in all kinds of polymers in a large range of mass per unit areas from 12 to 200 g/m$^2$. Their melting points vary as much (Table 6-17).

Their advantages are:
– economical use
– eco-friendliness (with both processing and manufacture)
– cleanness
– easy processability
– excellent adhesiveness
– dry and solvent-free processing
– clean and space-saving when stored
– soft, penetrable glueing with webs and split films

Table 6-18 generally compares the characteristics of adherends with other adhesive systems. Table 6-19 provides a survey of adhesive webs.

Using adherends, process variables as seen from the product are the polymer and the melting point. Process variables as seen from the plant are temperature, pressure and dwell time. In addition and seen from the polymer, variations are possible with respect to the time needed to allow crystallization and with respect to melting viscosity.

**Table 6-17** Range of melting temperatures of thermoplastic adherends

| Kind | Polymer | Range of melting temperatures (°C) |
|---|---|---|
| Glueing web | Copolyamide | 80–135 |
| | Copolyester | 90–135 |
| | Polyolefin | 110–170 |
| Glueing film | Copolyamide | 75–190 |
| | Copolyester | 65–135 |
| | Polyurethane | 65–135 |
| | Polyethylene | 60–130 |
| | Polypropylene | 135–150 |

**Table 6-18** Variety of glueing systems compared (completed to [230])

| Process | Powder point | Paste point | Adhesive web | Adhesive film | Flame lamination | Spraying | Sheeting die |
|---|---|---|---|---|---|---|---|
| Polymer | PA/PET | PA/PET | PA/PET/ PO[1] | PU/PET/ PO[1]/PA | PU | PA/PET/ PU | PA/PET/ PU |
| Minimal weight $(g/m^2)$ | 5 | 10 | 12 | 15 | 20 | 1 | 1 |
| Infrared radiator | yes | yes | yes | yes | no | no | no |
| Melting system | yes | yes | yes | yes | yes | no | no |
| Emission | little | little | no | no | yes | little | little |
| Plant cleaning | yes | yes | no | no | yes | yes | yes |

1) Polyolefin

As to machine equipment, all heat and pressure-exercising systems are applicable, such as flat-bed laminating plants, presses, calenders, felt calenders etc. Adhesive webs provide high bonding strength with all textile structures, such as cloths, knitted fabrics, nonwoven materials and with many non-textile components as, for instance, films of PU, PET, PE, PVC, PO, leather, wood, paper, metal, glass, soft foam, hard foam of polyurethane, polyether, glass materials, ceramic materials, aramid materials and of carbon fibres.

One novel development is seen in an effective mechanical combination of these adherends with a nonwoven. This results in a web bonded by means of binding threads, in whose manufacture a light-weight adhesive web is directly tied up on its surface.

Another process that has a long tradition is the use of melting glueing materials. To give an example, adhesive powder may be put on a component to be glued. It is exposed to heat so the glue can melt and then, the second component is glued onto it, either using the whole of its area or only spots of it. As with all bonding technologies, the single components must meet requirements depending on the process.

**Table 6-19** Spunlab adhesive webs

| Type | Polymer | Range of melting points (°C) | Thermal resistivity (°C) | Plasticizer-proof | CR[1] | WP[2] (°C) | Steam reactive | HF-weldable |
|---|---|---|---|---|---|---|---|---|
| PA 1001 | Copolyamide | 97–110 | 90 | on condition | yes | 40 | yes | yes |
| PA 1008 | Copolyamide | 100–115 | 90 | on condition | yes | 40 | yes | yes |
| PA 1300 | Copolyamide | 120–135 | 100 | no | yes | 60 | yes | yes |
| PA 1541 | Copolyamide | 87–100 | 60 | on condition | yes | 60 | yes | yes |
| PA 1545 | Copolyamide | 90–105 | 70 | on condition | yes | 60 | yes | yes |
| PE 2900 | Copolyester | 115–125 | 100 | yes | no | 30 | no | yes |
| PE 2942 | Copolyester | 120–135 | 90 | yes | no | 60 | no | yes |
| EV 3007 | Polyolefin | 110–125 | 80 | no | yes | 30 | no | no |
| LD 4000 | High pressure polyethylene | 100–125 | 100 | no | no | 30 | no | no |
| PP 5000 | Polypropylene | 165–170 | 130 | no | no | 25 | no | no |
| VI 6010 | Terpolymer | 105–115 | 110 | yes | yes | 60 | yes | yes |
| SL 7005 | Elastomer | 109–170 | 140 | yes | yes | 60 | no | yes |

1) CR = dry-cleaning proof.
2) WP = washable.

In this case, these are:

– sufficient thermal resistivity to survive the process temperature of the melting glue
– a sufficiently large surface so the melting adhesive powder cannot fall through or get lost in the structure
– a flat surface so the dispersed powder cannot shift on the substrate

The powdery melting adhesive may be applied by means of a dispersing device or a system using engraved rolls. The latter is also capable to apply liquid melting adhesive in a punctual way.

The melting adhesives preferably offered are powders, granulates or films made from co-polyamides. In addition, co-polyester, ethylene-vinyl-acetate polymers and polyethylene are found. They are used to laminate thermally bonded materials with membrane films so as to make lining materials that can breathe, or to laminate resin-reinforced mats of glass fibre with decorative wall tiles, or to laminate thermally bonded materials with polyurethane foam serving to line the interior of motor vehicles.

Except for the melting adhesives mentioned, plastisols, contact adhesives, pressure sensitive adhesives as well as solvent adhesives or, respectively, dispersion adhesives are known.

## 6.6.2
### Nonwovens suitable to make composite materials

Composite nonwovens were first known as glass-fibre reinforced nonwoven composites. Today, it is more preferable to use high-performance fibre materials so as to make high-quality articles such as thread composites designed for the performance wanted, wovens, loop materials, stitch-bonded materials [231, 232, 234, 254–256].

The following is known with regard to quality and finishing:

The surface of the nonwovens being exposed to radiation, abrasion, chemical impact, rock fall and the like, surface stability is essential with glass-fibre reinforced composite materials. If the surface is left unfinished, the reinforcing textiles may become externally visible, some of the reinforcing fibres looking out. These will function as wicks for the attacking medium. As a consequence, all the laminate will suffer.

The most economical and reliable way to finish the surface is provided by surface nonwovens based on glass fibres and chemical fibres. Surface nonwovens consist of hydrophobic fibres which are bonded by means of binder agents that are well compatible with the plastic matrix. However, a number of requirements should be met:

• The bearing reinforcement needs to be completely covered up (even image of nonwoven).
• The nonwoven cover needs to be fine and dense as well as absorbent. This reduces shrinkage and balances out the proportion of resin contained.

- The fibres need to be strong so as to reduce abrasion and crack formation.
- The fibres need to be resistant to yellowing so transparent parts are not affected by light.
- Surface nonwovens should stick well on the plastic matrix.

Choosing the most suitable processes of web formation and the most preferable fibres and bonding technologies, surface nonwovens can be made which meet the requirements of further processing and are sufficiently resistant to media attacking the parts in question.

## References to Chapter 6

[1]  Jacobasch HJ (1984) Oberflächenchemie faserbildender Polymerer, Akademie Verlag Berlin
[2]  ISO/DIS 11224 Textiles–Nonwovens–Webformation and bonding–Vocabulary, 07.10.1996
[3]  Computersoftware CORA (Computer organized Random Needle Arrangement) der O. Dilo Maschinenfabrik KG, Prospektinformation 1998
[4]  Nencini A (1997) On-screen manipulation of needle felt structures, Nonwovens Report International 319: 26–27
[5]  Jakob H (1993) Neue Entwicklungen in der aerodynamischen Vliesbildung und der Vernadelungstechnologie, Melliand Textilber 74: 291–293
[6]  Sievers K (1969) Der heutige Stand der Nadelfilz-Technik im Hinblick auf Verfahren und Maschinenkonstruktionen, Melliand Textilber 50: 151–156
[7]  Zocher J (1967) Verfahren mit gebogenen und geraden Filznadeln, US-Pat. 3 340 586
[8]  Dilo JP (1997) Hyperpunch, DE 196 15 697 A1
      Dilo JP (1998) Hyperpunch, US 5 732 453
      Dilo JP (1996) Hyperpunch, IT MI96A 001877, FR 96.10796 AT A 1603/96
[9]  Heinze EF (1968) Neue Märkte für Vliesstoffe, Probleme-Produkte-Praktiken, Z Ges Textilind 70: 417–422, 544–548
[10]  Dilo R (1975) Vorrichtung zum kontinuierlichen Herstellen von rohrförmigen Nadelvliesstoffen, DBP-No. 2552243
[11]  Kunath P (1998) Rontex S 2000, eine neue Nadelmaschine für Schlauchfilze, Taschenbuch für die Textilindustrie Berlin: Schiele & Schön
[12]  Seyam AM, Mohamed A, Kim H (1998) Signal Analysis of Dynamic Forces experienced by individual Needles in High Speed Needle Punching, Text Res J: 296–301
[13]  Tänzler W (1973) Der Einfluß von Avivagen auf die mechanische Vorverfestigung (Vernadelung) von Faservliesen und ihre Bedeutung für Loop- und Velours-Endprodukte, text praxis internat: 45–48
[14]  Kapeller M (1990) Musterung von Nadelvliesen, Chemiefasern Text Ind, Ind Textilien 40/92: 7/8
[15]  Strukturier-Nadelfilzmaschinen (1998) Prosp der Dr. Ernst Fehrer AG, Linz
[16]  Dilo H (1976) Di-Loft-Verfahren, Melliand Textilber 57: 642
[17]  Lünenschloß J, Gupta VP, Berns K (1979) Nadeleinstichkräfte bei der Herstellung strukturierter Nadelvliesstoffe, Textilbetrieb 97, 7: 29–32
[18]  Lünenschloß J, Gupta VP, Berns K (1977) Die Abhängigkeit des Vernadelungsvorganges und der Eigenschaften von Nadelfilzen mit eingenadelten Trägergeweben von den Herstellungsbedingungen, Textilbetrieb 10: 32–35, 11: 23–28, 12: 47–54
[19]  Prospektinformationen der Groz-Beckert Nadelfabriken KG, Albstadt (1998)
[20]  Prospektinformationen der Singer Spezialnadelfabrik GmbH, Würselen (1998)
[21]  Böttcher P, Kittelmann W (1998) Trends der mechanischen Vliesverfestigung, ITB Vliesstoffe-Techn Textilien 4: 8–16

[22]  Lünenschloß J (1972) Einfluß der Faserlänge, Faserfeinheit, Kräuselung und Mattierung auf den Vernadelungsablauf und die Nadelfilzeigenschaften, Melliand Textilber: 144–151

[23]  Lünenschloß J, Janitza J (1971) Die Untersuchungen des Vernadelungsvorganges bei der Nadelfilzherstellung und Eigenschaften des Nadelvlieses, Z Ges Textilind: 142–147, 208–214, 270–273

[24]  Lünenschloß J, Kampen W (1975) Ergebnisse einiger Untersuchungen der Nadeltechnik mit unterschiedlicher Nadeleinstichgeometrie, Melliand Textilber: 285–288

[25]  Lünenschloß J (1973) Die mechanische Verfestigung von Vliesen aus Chemiefasern durch Vernadeln, einige neuere Erkenntnisse, Melliand Textilber: 1163–1171

[26]  Hearle JW, Sultan MAJ (1968) Journ of the Textile Institute, 59: 137–147

[27]  Kosova RA (1972) Textilnaja Promyschlennosti: 53–55

[28]  Krčma R (1976) Neue Erkenntnisse über die Struktur, Parameter und Eigenschaften von Vliesstoffen, Textiltechn 26, 6: 347–351

[29]  Böttcher P (1993) Strukturuntersuchungen an Nadelvliesstoffen bezüglich ihrer Durchlässigkeit und Ableitungen auf die Konstruktion von Nadelvliesstoffen für technische Einsatzgebiete, AiF-Forschungsvorh No 261 D

[30]  Böttcher P (1993) Zusammenhänge zwischen Luft- und Wasserdurchlässigkeit von Nadelvliesstoffen, ITB Vliesstoffe-Techn Textilien 4: 15–16

[31]  Voigtländer G (1992) Untersuchungen zum Nadelprozeß, Melliand Textilber 5: 391–396

[32]  Kittelmann W, Brodtka M (1998) On-line-Messung der Qualität in der Nadelvliesproduktion, ITB Vliesstoffe-Techn Textilien 2: 42–46

[33]  Kittelmann W (1995) Erarbeitung funktioneller Abhängigkeiten zwischen on-line-meßbarer Vliesstoffdichte, qualitätsrelevanten Produkteigenschaften und Prozeßvariablen beim Vernadeln von Vliesen als Basis zur Qualitätssicherung, AiF-Forschungsvorh No 9729 B

[34]  Dilo JP (1985) Vernadelungstechnologie für Geovliese, textil praxis internat 4: 370, 375–380

[35]  Gupta VP (1982) In: „Vliesstoffe" by Lünenschloß/Albrecht, Georg Thieme Verlag, Stuttgart/New York: 149–160

[36]  Voigtländer G (1991) Filtervliesstoffe: I Vliesstoffmodell als Basis zur Berechnung filtertechnischer Parameter, text praxis internat 4: 304–309

[37]  Voigtländer G (1992) Filtervliesstoffe: II Modellgestützte Konstruktion von Filtervliesstoffen, text praxis internat 1: 24–26

[38]  Ploch S, Böttcher P, Scharch D (1978) Malimo-Nähwirktechnologien, VEB Fachbuchverlag, Leipzig

[39]  Offermann P, Tausch-Marton H (1978) Grundlagen der Maschenwarentechnologie, VEB Fachbuchverlag, Leipzig

[40]  Lünenschloß J, Albrecht W (1982) Vliesstoffe, Georg Thieme Verlag, Stuttgart/New York

[41]  DIN 62050 (1990) Gewirke und Gestricke, Part 2: Kettengewirke

[42]  DIN 62050 (1990) Gewirke und Gestricke, Part 1: Gestricke und Kuliergewirke

[43]  DIN 61211 (1976) Auf Nähwirkanlagen hergestellte textile Flächengebilde

[44]  Ploch S (1989) Vlies-Nähwirken, Part 2, Allgemeiner Vliesstoff-Report 7: 46–47

[45]  DIN 61210 (1982) Vliese, verfestigte Vliese und Vliesverbundstoffe auf Basis textiler Fasern, Ausg 01/82

[46]  Prospekt Nähwirkmaschine Malimo, Typ Maliwatt; Karl Mayer Malimo Textilmaschinenfabr GmbH Chemnitz (1999)

[47]  Schreiber J, Kolinsky O, Kopal J (1995) Vliesbildung und Verfestigung trennen? Vliesstoff Nonwoven Internat 11–12: 346–347

[48]  Scholtis W (1984) Getrennte Vliesbildung und -verarbeitung – eine Variante bei der Herstellung von Vlies-Faden-Nähgewirken, Textiltechnik 34, 11: 614–617

[49]  Technik und Einsatzmöglichkeiten der Typen Maliwatt und Malivlies, Prospekt Nähwirkmaschine Malimo; Karl Mayer Malimo Textilmaschinenfabr GmbH Chemnitz (1999)

[50]    Lieferprogramm Wirkwerkzeuge, Karl Mayer Malimo Textilmaschinenfabr GmbH Chemnitz

[51]    Schreiber J (1983) Erzeugnisausmusterung Part 1 bis 6, F/E-Ber des ehemaligen Forschungsinstituts f Textiltechnologie Chemnitz

[52]    Scholtis W (1978) Mustermöglichkeiten mit der Nähwirkmaschine Malimo, Typ Maliwatt, Modell 14012, Textiltechnik 28, 11: 699–703

[53]    DIN 53852 Bestimmung von Garnlängenverhältnissen in Geweben und Maschenwaren

[54]    Scholtis W (1976) Einfluß der Bindung auf die Eigenschaften von Vlies-Faden-Nähgewirken, Textiltechn 26, 12: 768–773

[55]    Scholtis W, Ploch S, Böttcher P (1971) Maliwatt- und Malivlies-Nähwirkverfahren – einige neue technologische Erkenntnisse, Deutsche Textiltechnik 21, 8: 513–515

[56]    Böttcher P (1978) Eigenschaften von Nähgewirken und Möglichkeiten ihrer zielgerichteten Beeinflussung, Textiltechnik 28, 7: 424-428

[57]    Böttcher P, Hunger M (1978) Untersuchungen zur Verbesserung der Festigkeit von leichten Vlies-Nähgewirken, Textiltechnik 28, 1: 56–61

[58]    Assmann B (1987) Ausgewählte Zielgrößen der Malivlies-Erzeugnisse in Abhängigkeit wesentlicher Einflußgrößen, Dipl Arb TU Chemnitz, Sekt Textil- u. Ledertechnik, TMT

[59]    Schmalz E (1998) Technologische Untersuchungen zum Angleichen des Längs- und Querfestigkeitsverhaltens von Vlies-Nähgewirken „Malivlies" für technische Anwendungsbereiche, F/E-Ber Sächs Textilforschungsinst eV

[60]    Offermann P, Mägel M (1985) Modellierung des Verfestigungsprozesses beim Nähwirkverfahren Malimo[1]), Typ Malivlies, Part 2, Textiltechnik 35, 6: 308–310

[61]    Mägel M (1983) Experimentelle und theoretische Untersuchungen zur mechanischen Verfestigung von Vlies-Nähgewirken, Dissertation TU Dresden, Fakultät f Maschinenwesen

[62]    Scholtis W (1975) Bindungen von Vlies-Nähgewirken, Textiltechnik 25, 12: 764–769

[63]    Munden DL (1959) The geometry and dimensional properties of plain knitted fabrics, J Text Inst 50, 7: 448–471

[64]    Offermann P, Jenschke D (1982) Theoretische Strukturmodellierung von Vlies-Nähgewirken, Textiltechnik 32, 3: 187–191

[65]    Offermann P, Mägel M, Ponnahannadige K (1982) Modellierung des Vliesverfestigungsprozesses beim Vlies-Nähwirken, Textiltechnik 32, 4: 247–250

[66]    Offermann P, Mägel M (1985) Modellierung des Verfestigungsprozesses beim Nähwirkverfahren Malimo[1]), Typ Malivlies, Part 1, Textiltechnik 35, 5: 265–267

[67]    Ploch S (1976) Ermittlung von Durchstich- und Einbindekräften bei Nähwirkmaschinen Malimo, Typ Voltex, Textiltechnik 26, 12: 773–778

[68]    Ploch S (1978) Höhe der Durchstichkräfte bei Nähwirkmaschinen Malimo, Typ Voltex, Textiltechnik 28, 7: 418–424

[69]    Ploch S (1979) Einbindekräfte bei Nähwirkmaschinen Malimo, Typ Voltex, Textiltechnik 29, 4: 230–235

[70]    Zschunke H (1973) Durchstich- und Einbindekräfte bei Malivlies, Dissertation

[71]    Nedewa (1973) Durchstich- und Einbindekräfte bei Maliwatt, Dissertation

[72]    Schmalz E (1997) Verarbeitung von einheimischen nachwachsenden Rohstoffen auf Nähwirkmaschinen, Vortr 5, Chemnitzer Textilmaschinentagung

[73]    Ploch S, Scholtis W, Zschunke H, Verfahren zur Herstellung eines Textilstoffes und Vorrichtung an Nähwirkmaschinen zur Durchführung des Verfahrens, Patent DD 39819, IPK D 04 B,

[74]    Ploch S, Scholtis W, Zschunke H, Verfahren und Vorrichtung an Nähwirkmaschinen zur Herstellung eines Textilstoffes, Patent DD 39820, IPK D 04 B

[75]    Techn Daten der Nähwirkmaschine Voltex, Prospekt Karl Mayer Malimo Textilmaschinenfabr GmbH Chemnitz

[76]    Ploch S (1993) Vliesgewirke – Struktur und Eigenschaften, Melliand Textilber 5: 390–393

[77]   Ploch S, Zschunke H, Schreiber J, Dietrich K-H, John M, Rödel J, Verfahren zur Herstellung eines Vliesgewirkes, Patent DD 282 585, IKP D 04 B 21/14

[78]   Ploch S (1995) Kunit – Multiknit – KSB, Neue Varianten der Vlies-Wirktechnik, Melliand Textilber 6: 404–408

[79]   Erth H, Fuchs H (1999) Verfahrenstechnische Optimierung der Nähwirkstelle beim Kunit-Verfahren, Vortr 7, Chemnitzer Textilmaschinentagung

[80]   Kunit – ein Vlies-Wirk-Verfahren zur Nähwirktechnik Malimo, Prospekt Karl Mayer Malimo Textilmaschinenfabr GmbH Chemnitz

[81]   Heilmann H, Ploch S, Roth G, Stein E, Vogel W, Zeisberg P, Zschunke H, Vliesstoff sowie Verfahren und Vorrichtung zu dessen Herstellung, Patent DE 42 20 338 A1, IPK D 04 H1/70

[82]   Tröger J (1995) Neues Verfahren zur Herstellung von Abstands-Vliesgewirken und Verbundstoffen, ITB Vliesstoffe-Techn Textilien 2: 32–37

[83]   Schreiber J (1997) Textiltechnologische Untersuchungen zum Kunit-Schicht-Bindeverfahren, Melliand Textilber 11–12: 824–826

[84]   Ploch S, Zeisberg P, Verfahren und Vorrichtung zum mechanischen Verbinden von voluminösen Flächengebilden, Patent DD 44 03 473, IPK D 04 1/45

[85]   Böhm C (1995) Vliesverfestigung auf maschenbildenden Maschinen, Techn Textilien/Vliesstoffe, Maschenindustrie 45, 5: 391–393

[86]   Prospekt Hochleistungs-Raschelmaschine für Vliesstoffverstärkung RS 2V und RS 3 MSU-V, Karl Mayer Textilmaschinenfabrik GmbH Obertshausen (1999)

[87]   Prospekt WARP KNITTING MACHINES, LIBA Maschinenfabrik GmbH Naila (1999)

[88]   Kunde K (1999) Modernste Highpile-Technik, Vortr 7, Chemnitzer Textilmaschinentagung

[89]   White C, Zevnik F (1989) Nonwoven technology. An appraisal of hydroentanglement technology, EDANA Nonwovens Symposium Papers: 37–67

[90]   Holliday Th (1991) Spunlace patents and processes, TAPPI Nonwovens Conf Papers: 305–310

[91]   Fluid entanglement principles and systems, Nonwovens Ind 12: 32–33 (1992)

[92]   Moore GK, White CF (1988) Aspects of hydroentanglement, Pira-Conference papers: 1–7

[93]   Chang WJ (1993) Effect of processing parameters on the structure and strength property of Hydroentangled Nonwovens, Ind & Technical Textiles Conf Papers, Univ Huddersfield: 1–10

[94]   Patentschrift US 28 62 251 vom 02.12.58 und (1962) Patentschrift US 30 33 721 vom 08.05.62 (Chicopee Manufacturing Corp) (1958, 1962)

[95]   Patentschriften US 32 14 819 v 02.11.65 (Du Pont de Nemours) (1965)

[96]   Coppin Ph (1998) Spunlacing worldwide …Insight 1998, Section II: 1–18

[97]   Koslowski H-J (1998) Vliesstoffmärkte in Zahlen, 13. Hofer Vliesstoffseminar: 1–2

[98]   White C (1990) Hydroentanglement technology. A review and projection of markets, Nonwovens Ind 10: 37, 39–41

[99]   Hardy C (1991) Fine dtex cellulosic fibres offer significant advantages, Nonwovens Rep Int Yearbook: 50–52

[100]  Wilharm M (1991) Spunlaced nonwovens: Trends, uses and water treatment, Nonwovens Ind 12: 48–49

[101]  White CF (1990) Hydroentanglement technology applied to wet formed and other precursor webs, TAPPI Nonwovens Conf Papers: 177–187

[102]  Vuillaume AM (1991) Manufacturing of spunlace nonwovens, TAPPI Nonwovens Conf Papers: 311–314

[103]  Rueher JA (1991) Neuere Entwicklungen in der Wasserstrahlverfestigung, 6. Hofer Vliesstoffseminar: 35–38

[104]  Ohmura K (1991) Japanese microfiber nonwovens, Nonwovens Ind 5: 16–18

[105]  Guss DB, Hebert RP (1990, 1992) Recycling water in high pressure needle showers, TAPPI Nonwovens Conf Papers: 5–6 and avr 2: 24, 27, 28

[106] Cotton used in spunlaced wipes, Nonwoven Rep Int 263: 7–8 (1993)

[107] Woodings CR (1993) High pressure hydroentanglement of cellulosic fibres, INDEX'93 Congress Papers Section 5c, VI: 1–17

[108] Woodings CR (1989) The hydroentanglement of a range of staple fibres. Fiber developments/Polymer based Nonwovens Conf, Sect IV: 1–15

[109] Vuillaume AM (1989) Latest advances in the hydroentanglement technology, Fiber developments/Polymer based Nonwovens Conf, Sect IX: 1–14

[110] Möschler W (1993) Möglichkeiten der Einflußnahme auf Struktur und Eigenschaften von Wirbelvliesstoffen, 8. Hofer Vliesstoffseminar: 1–12

[111] Möschler W, Meyer A, Brodtka M (1995) Faserstoff- und Prozeßeinflüsse auf die Eigenschaften von Wirbelvliesstoffen, ITB Vliesstoffe-Techn Textilien 26, 28: 30–31

[112] Turi M (1988) The outlook for spunlaced nonwovens, Nonwovens Ind 11: 30–32, 34, 36

[113] Bartholomew A, Abercrombie A (1989) The use of rayon in water-entangled nonwovens, EDANA Nonwovens Symp: 69–93

[114] Hendler J, Brodtka M (1988) Vliesverfestigung durch Faserstoffverwirbelung – Norafin, Textiltechnik 38: 430–434

[115] Hendler J (1990) Wirbelvliesstoffe, 5, Hofer Vliesstoffseminar: 1–17

[116] Ward DT (1997) Wasserstrahlverwirbelung für schwere Vliesstoffe, ITB Vliesstoffe-Techn Textilien 1: 24, 28, 29

[117] Forsten HH (1985) Neue Anwendungsbereiche für die Spunlaced-Technologie, Chemiefasern 3: 203

[118] Crouse JL et al. (1990) Pesticide barrier performance of selected nonwoven fabrics, Text Res J 3: 137–142

[119] Mehr als tausend Walzen, avr 3: 296, 298, 299 (1990)

[120] Jetlace 2000 – Eine neue Generation von Spunlace-Maschinen, ITB Vliesstoffe-Techn Textilien 1: 92 (1996)

[121] Courtaulds Engineering debuting "Hydrolace" technology, Nonwovens Ind 1: 76 (1996)

[122] Widen CB (1991) Forming fabrics for spunlace applications, TAPPI Nonwovens Conf: 315–321

[123] Specklin P (1987) Liage des nontissés par jets de fluide, Ind Textile 1176 4: 383–384

[124] Watzl A (1997) Fleissner-Aquajet Spunlace System, XXV Intern Nonwovens Colloquium Brno: 1–9

[125] Völker Kl (1998) Wasserstrahlverfestigung bei Geschwindigkeiten von über 200 m/min mit JETlace 2000, 13. Hofer Vliesstoffseminar: 1–7

[126] Combinations of hydroentangling and airlaid, Nonwovens Technology Conf Aarhus (1999)

[127] Watzl A (1998) Mikrofasern – Eine neue Herausforderung für die Spunlace-Technik, 13. Hofer Vliesstoffseminar: 1–2

[128] Patentschrift US 3353225 v 22.11.67 (Du Pont de Nemours) (1967)

[129] Patentschrift US 3391048 v 02.07.68 (Eastman Kodak Co.) (1968)

[130] Patentschrift US 3240657 v 15.03.66 (Johnson & Johnson) (1966)

[131] Patentschrift US 3485706 v 23.12.69 (Du Pont de Nemours) (1969)

[132] Patentschrift US 3486168 v 23.12.69 (Du Pont de Nemours) (1969)

[133] Patentschrift US 3493462 v 03.02.70 (Du Pont de Nemours) (1970)

[134] Patentschrift EP 0108621 A2 v 16.05.84 (Du Pont de Nemours) (1984)

[135] Patentschrift EP 0446432 A1 v 18.09.91 (Internat Paper Co) (1991)

[136] Noëlle F (1999) Wasserstrahlverfestigung bei Geschwindigkeiten von über 200 m/min, Techn Textilien 2: 63–64, 66

[137] Fleissner (1999) beginnt eine neue Ära der Spunlace-Technik; AquaJet Spunlace-System für Geschwindigkeiten bis 600 m/min und Drücke bis 600 bar, Pressemitteilung

[138] Möschler W (1997) Technologische Einflußfaktoren auf den Verwirbelungseffekt bei der Wasserstrahlverfestigung von Feinfaservliesen, XXV Internat Nonwovens Colloquium Brno: 1–7

[139]  Münstermann U (1996) AquaJet for Lyocell fiber nonwovens, Chemical Fibers Internat 46

[140]  Watzl A (1998) Fleissner-Aquajet Spunlace System, Firmeninformation

[141]  Watzl A (1997) Energieverbrauch beim Wasservernadeln mit dem Fleissner-Aquajet Spunlace System, mittex 2: 8–9

[142]  Watzl A (1997) Multi-step route along the waterway, Nonwovens Rep Int 5: 16–17

[143]  Ülkü S, Acar M et al. (1993) Fiber alignment and straightening in opening for open-end spinning, Text Res J 6: 309–312

[144]  Möschler W (1995) Some recent results in the fields of hydroentanglement of fibrous webs and of their thermal after-treatment, EDANA's Internat Nonwovens Symp Bologna, Book of Papers XIII–1–23

[145]  Leihkauf B (1992) Bestimmung der Wasserstrahlprallkraft als Voraussetzung für die Optimierung der Spunlaced-Technologie, Diplomarb TU Chemnitz

[146]  Stöferle F et al. (1974) Trennen mit Hochgeschwindigkeitsstrahlen, Werkstatt u Betrieb 107: 257–259

[147]  Lutze H et al. (1992) Trennen mit Flüssigkeitsstrahl, Wissenschaftl Schriftenreihe TU Chemnitz H 2

[148]  Patentschrift DE 27 31 291 v 19.01.78 (Mitsubishi Rayon) (1978)

[149]  Der Einfluß der Fasergeometrie und -physik auf das Verhalten von Fasern bei der Vliesverfestigung mittels Wasserstrahlverwirbelung, AiF-Projekt No. 9256 B, STFI Chemnitz, Abschlußbericht (1995)

[150]  Brodtka M, Meyer A, Möschler W (1995) Verarbeitungs- und Produkteigenschaften von Wirbelvliesstoffen aus Spezialfasern, Techn Textilien 3: 13–16

[151]  Meyer A, Möschler W (1996) Technologieaspekte beim Herstellen von Wirbelvliesstoffen aus Fein- und Splittfasern, 11. Hofer Vliesstoffseminar: 1–4

[152]  Watzl A, Spunlacing von airlaid/carded composites, Fleissner GmbH, Mitteilung File KS 16737

[153]  Münstermann U (1997) Imagine the future of viscose technology, Lenzinger Berichte 76/97

[154]  Münstermann U (1996) Das Fleissner Aquajet-Spunlace-System, 11. Hofer Vliesstoffseminar

[155]  Perfojet's promise to keep increasing the pressure, Nonwovens Rep Int 3: 22, 24, 26, 28 (1999)

[156]  Münstermann U (1997) Anwendungen des Aquajet Spunlace Systems, 12. Hofer Vliesstoffseminar

[157]  Spunlace lines Jetlace 2000, Airlace 2000, Firmenprospekt ICBT Perfojet (1999)

[158]  Patentschriften EP 01 30 070 A2, EP 02 10 777 A2, EP 02 23 614 A2, EP 02 15 684 A2, EP 04 46 432 A1, EP 06 36 727 A1, WO 92/088 32, EP 02 23 965 A2, EP 01 08 621 A2, EP 01 47 904 A2, DE 31 32 792 A1, US 57 36 219

[159]  Patentschriften EP 01 32 028 A2, EP 04 00 249 A1, DE 33 02 709 A1, US 41 52 480, WO 97/29 234

[160]  Patentschrift WO 97/29234 (Courtaulds Engineering)

[161]  Watzl A (1989) Erweiterte Anwendung des Durchströmprinzips für die Trocknung von Nonwovens und Papier, Melliand Textilber 70: 911–920

[162]  Watzl A (1997) Thermofusion, Thermobonding und Thermofixierung für Nonwovens, Firmeninformation der Fleissner GmbH & Co

[163]  Schmidt F (1998) Gegenüberstellung des Durchbelüftungs- und des Düsenbelüftungsprinzips (energetische Betrachtungen, Möglichkeiten der Temperaturführung bei zonenweiser Aufteilung)

[164]  Watzl A (1987) Anlagen für die Herstellung von Vliesstoffen für die Automobilindustrie, text praxis internat 11: 1344–1346, 1351–1354

[165] Watzl A (1995) Verbesserung der Eigenschaften von Geotextilien durch Thermoverfestigung und Thermofixierung von Vliesstoffen, 23 Internat Kolloquium über Vliesstoffe Brno/CS

[166] Bobeth W et al. (1993) Textile Faserstoffe, Beschaffenheit und Eigenschaften, Springer-Verlag Berlin – Heidelberg

[167] Saindon R (1992) High-Loft Nonwovens, Inda Journ Vol 1, 1

[168] Loy W (1989) Taschenbuch für die Textilindustrie, Fachverl Schiel & Schön GmbH Berlin

[169] Sass F, Bouché Ch (1955) Dubbels Taschenbuch für den Maschinenbau I: 404

[170] Bechter D, Kurz G, Maag E, Schütz J (1991) Über die thermische Verfestigung von Vliesstoffen 1. Mittlg: Der Einfluß der Kalanderbedingungen auf die Festigkeit von Polypropylenvliesstoffen, text praxis internat 11: 1236–1240

[171] Wei KY, Vigo TL, Goswani BC (1985) Structure-property relationship of thermally bonded polypropylene nonwovens, Journ of Appl Polymer Science, Abschn 30: 1523–1534

[172] Warner SB (1989) Thermal bonding of polypropylene fibres, Text Res Journ 3: 151–153

[173] Klöcker-Stelter St (1992) Entwicklung eines Prozeßmodells zum Verhalten textiler Gebilde im Spalt biegekompensierter Walzenkalander am Beispiel der Vliesverfestigung, Dissertation an der Uni Bremen, FB Produktionstechnik

[174] Prömpler S (1998) Kalander-Technologie zur Herstellung von Vliesstoffen für technische Verwendungszwecke, Taschenbuch für die Textilind, Fachverlag Schiel & Schön GmbH Berlin

[175] Firmeninformation der Küsters-Maschinenfabrik GmbH & Co KG Krefeld (1999)

[176] KTM Kleinwerfers Textilmaschinene GmbH, Krefeld (1999)

[177] Ward DT (1999) Vorteile der Heißverfestigung, Vliesstoffe, Techn Textilien 3: 58–59

[178] Schilde W, Wolf K (1995) Wissenschaftliche Untersuchungen zur Modellierung des Bindemechanismus von thermisch gebundenen Faservliesstoffen unter Berücksichtigung der Fasereigenschaften und des Radialdruckes, AiF-Forschungsvorh 9412 B STFI eV Chemnitz und ITB der TU Dresden

[179] Patrick RL (1967) Treatise on adhesion and adhesives, Vol 1, M. Dekker, New York

[180] Bikerman J (1968) The science of adhesive joints, 2nd edition, Academic Press New York/London

[181] Stenemur B (1988) Einfluß von Präparationsmitteln wie Faseravivagen auf die Vliesbindung, zitiert in Nonwoven-Verfestigungsprinzipien in der Gegenüberstellung, AVR-Allgem Vliesstoffrep 16, 9: 185, 187

[182] Ehrler P (1982) Kohäsive Verfestigung von Vliesen, in: Lünenschloß J, Albrecht W (eds.) „Vliesstoffe", Georg Thieme Verlag Stuttgart/New York

[183] Jörder H (1977) Textilien auf Vliesbasis (Nonwoven) Keppler-Verlag Heusenstamm

[184] van Dorp T (1973) Anlösen von Synthesefasern – ein Weg zu self-bonding-Vliesstoffen, Reutlinger Kolloquium „Bindung von Vliesstoffen"

[185] Ehrler P, Vialon R (1982) Bindemittelapplizierung bei adhäsiver Verfestigung in Lünenschloß J, Albrecht W "Vliesstoffe", Georg Thieme Verlag Stuttgart/New York

[186] Clauss B (1993) Induced electron beam polymerization in the web binder hardening, INDEX 1993 Geneva

[187] Rieker J, Braun W (1985) Verfahrenstechnische Grundlagen der Naß-in-Naß-Auftragstechnik, Melliand Textilber 66, 7: 514–520

[188] Heintz P (1993) Verfahren und Optimierung beim Additionsauftrag, Internat Textile Bull 39, 2: 50–51

[189] Schlicht W (1993) Praxiserfahrungen mit dem Universalauftragswerk Optimax, text praxis internat 48, 2: 148–150

[190] Abquetschen und Entwässern, Internat Text Bull Veredlung 41, 1: 48–49 (1995)

[191] Eltz H-U von der, Olpeter G, Walbrecht H (1984) Das kontinuierliche Färben von Polyester-Cellulosefaser-Mischungen, text praxis internat 39, 7: 689–692

[192] Pretschner R (1997) Foulard und Hochleistungsquetschwerk mit neuem Walzenanpreß-system, Melliand TB 78, 5: 340–341

[193] Itgenshorst D (1989) Entwicklung und Einsatz der Schwimmenden Walze, text praxis internat 2: 140–142

[194] Meisen K (1986) Neue Durchbiegungs-Ausgleichwalze und ihre Regelmöglichkeiten, Melliand Textilber 67, 3: 184-188

[195] Rouette H-K (1995) Lexikon für Textilveredlung, Laumann Verlag, Dülmen

[196] Jin KK, Byong R (1993) Theoretical study of knife coating process: The effect of the contact angle of the knife on the knife coating, Plastics, Rubber and Composites, Processing and Application 20, 2: 101–105

[197] Welter C (1995) Maschinen zur Herstellung textiler Verbundwerkstoffe mittels Hotmeltklebern, Vortr 7 Internat Techtextil-Symp Frankfurt, Techn Textilien 41 (1998) 1: 27–30

[198] Hartmann H (1994) Maschinen für die Veredlung von Vliesstoffen, Blick hinter die Kulissen, Vliesstoff-Nonwoven Internat 1–2: 30–31

[199] Endress KM (1995) Das Nordson Porous Coat System, eine umweltfreundliche Methode zur Laminierung und Beschichtung flächiger Textilien, 7. Internat Techtextil-Symp Frankfurt

[200] Welberg AJG (1997) Modulare Beschichtungstechnologie: Die Flexibilität der Zukunft, 12. Hofer Vliesstoff Seminar

[201] Maßschneiderei im Maschinenbau, AVR-Allgem Vliesstoffrep 16, 6: 350–352 (1988)

[202] Fleissner (1998) Fleissner Kontinue Ausrüstungsanlagen für Nonwovens, Firmenschrift

[203] Unger J (1986) Moderne Sprühsysteme zur Oberflächenverfestigung, Textilbetrieb 6: 66–77

[204] Ehrler P, Institut für Textil- und Verfahrenstechnik Denkendorf (1998) Interne Untersuchungen zur Nutzung der Aerosoltechnik in der Plasmatechnik

[205] Goossens F (1992) International developments in fabric finishings improvement of surface structure of filter media by bonding, laminating, coating and impregnating, 2. Internat High-Performance Fabrics Conf Boston

[206] Fiebig D (1982) Chemische und physikalische Grundlagen von Schäumen, Textil Praxis 37, 4: 392–398

[207] Reinert F , Kothe W (1982) Minimalauftrag mit Schaum – die neue Technologie für die Ausrüstung? Melliand Textilber 63, 2: 138–144

[208] Kroezen ABJ, Groot-Wassink J (1986) Foam generation in rotor-stator-mixers, J Soc Dyers Colorists 102, 12: 397–402

[209] Kroezen ABJ, Groot-Wassink J (1987) Bubble size distribution and energy dissipation in foam mixers, J Soc Dyers Colorists 103, 11: 386–394

[210] Engelsen CW, Gooijer H, Groot-Wassink J, Warmoskerken MMCG (1996) Foaming of suspensions, Vortr 17. Kongr d Internat Föderat d Vereine d Textilchemiker u Coloristen Vienna

[211] Campen (1982) Computergesteuerter Mixer für Schaumfärbung und Schaumausrüstung, Chemiefasern 32/84, 12: 903

[212] Stakelbeck HP (1982) Kritische Betrachtungen zur Schaumveredlungstechnik, textilveredlung 17, 1: 22–27

[213] Tetzlaff N (1982) Stand der Technik in der Schaumausrüstung und Schaumbeschichtung, Chemiefasern 32/84 12: 896–901

[214] van Wersch K (1982) Schaumauftrag – maschinentechnische Realisierung, Melliand Textilber 63, 2: 147–152

[215] Watzl A (1990) Anlagen zur Herstellung von PES-Vliesstoffen für Dachbahneinlagen: Spinnvliese – Stapelfaservliese – Sandwichbahnen, text praxis internat 5/6: 522–531

[216] Stepanek W (1989) Vliesstoffverfestigung mit verschäumten Acrylharzdispersionen, Taschenbuch f d Textilind: 213–220

[217] Isarin JC, Kaasjager ADJ, Holweg RBM (1995) Bubble size distribution during the application of foam to fabrics and its effects on product quality, Text Res J 65 (1995) 2: 61–69

[218] Schaum-Auftragungs-Geräte, Melliand Textilber 63, 2: 137 (1982)

[219] Polymer-Latex GmbH, Acrylpolymer-Dispersionen. Chemie und Anwendung, Techn Inform 45764 Marl

[220] ITS-Charts: Beschichtungsmaschinen, Internat Text Bull Vliesstoffe-Techn Textilien 43, 1: 16–19 (1997)

[221] Eisele D (1988) Vom Faservlies zum technischen Vliesstoff, Chemiefasern 38/90, 8: 684–687

[222] Kuznezova EI, Gorcakova VM (1994) Einlagevliesstoffe mit pulverförmigen Bindemitteln (russ.). Textilnaja Promyslennost 54, 1: 24–25

[223] Vliesverfestigung mit Bindemittelpulvern auf dem Siebtrommeltrockner, Chemiefasern 39/91, 7/8: 867–868 (1989)

[224] Schott & Meissner, Blaufelden: Pulver-Streuaggregat; Twin-Streuaggregat, Firmenschrift

[225] Dominik M (1999) Streuen und Thermofixieren, KU Kunststoffe 89, 8: 64–67

[226] Technisch hochstehende Pulverbeschichtung: Der Doppelpunkt-Beschichtungsprozeß, TechTex Forum 4: 108 (1998)

[227] Pössnecker G (1999) Klebstoffe für Einlagen, Internat Text Bull 45, 2: 16–19

[228] Fischer K (1992) Binder und Binder-Additive – Variationsbreite unter Umweltgesichtspunkten, 7. Hofer Vliesstoff-Seminar, textil praxis internat (1993) 9: 709–714

[229] Krcma R (1970) Handbuch der Textilverbundstoffe (non-wovens), Deutscher Fachverlag

[230] DIN 61 210 (1992) Vliese, verfestigte Vliese und Vliesverbundstoffe auf Basis textiler Fasern, Ausgabe 01/92

[231] Loy W (1994) Textile Verbundstoffe – eine Übersicht, Techn Textilien 5: T 46–T 49

[232] Böttcher P (1991) Faserverbundwerkstoffe – ein wachsender Markt für technische Textilien, ITB Nonwovens 2: 10–21

[233] Böttcher P (1986) Vernadelung von mittig gefüllten Vliesstoffen, Techn Text 4: 103–105

[234] Schramm (1993) Möglichkeiten und Grenzen fadenverstärkter Vliesstoffe, Taschenb für die Textilind 1993: 253

[235] Zeisberg P (1993) Mechanische Verfestigungstechnologien und Composites aus Vliesen, Fäden und Flächen, Allg Vliesstoffrep 21, 3: 22

[236] Mieck K-P a.o. (1995) Vernadelte Hybridvliese, textile Halbzeuge für die Faserverbundherstellung; Techtextil-Symp 1995, Vortr 314

[237] Böttcher P (1996) Vliesstoffe für Verbundstrukturen, Allg Vliesstoffrep 1: 26–32

[238] Plesken P (1996) Malimo-Textilien für Verbundwerkstoffe, 35. Chemiefasertagung Dornbirn

[239] Zeisberg P (1996) Mechanisch verfestigte Verbundtextilien mit Medieneinlagerung, Kettenwirk-Praxis 2: 67–70

[240] Grenzendörfer D (1993) Vliesverarbeitende Nähwirkverfahren – Varianten und technische Möglichkeiten, Melliand Textilber 74: 300

[241] Ploch S (1993) Vielfältige Möglichkeiten der Nähwirktechnik Malimo, Melliand Textilber 74: 1245

[242] Schmalz E (1998) Produktinnovationen mit Nähwirk/Spunlaced-Verbunden für technische Anwendungen; Techn Textilien 9: 148–151

[243] Böttcher P (1980) Zusatzverfestigung von Vliesstoffen durch Bindefasern;, Textiltechnik 4: 242–245

[244] Böttcher P (1998) Trends der thermischen Verfestigung, Vliesstoffe-Techn Textilien 4: 20

[245] Schmidt G (1992) Kaschierung von Autopolsterstoffen mit Faservliesen, Melliand Textilber 73: 479

[246] Welter C (1995) Die wirtschaftliche und umweltverträgliche Herstellung von Verbundtextilien, Techtextil-Symp 1995, Vortrag 221

[247] Welter C (1998) Maschinen zur Herstellung textiler Verbundwerkstoffe mittels Hotmeltklebern; Techn Textilien 2: 27–30

[248] Steenblock R (1992) Schmelzkleber: neue Problemlösungen bei der Verbundherstellung, CTI 42/94: T 101

[249] Jahn H (1993) Verbunde mit Vliesstoffen unter Verwendung von Schmelzklebern, Melliand Textilber 74: 297

[250] Kauderer H-J (1993) Verfahrenstechniken der kontinuierlichen Textillaminierung, tpi 48: 895

[251] Untersuchungen zum Einsatz der Nadeltechnik als mechanische Veredlung von Nähwirkvliesstoffen; Forschungsber STFI eV Chemnitz, BMWI 77/96 (1996)

[252] Untersuchungen zur Erhöhung der Druckelastizität von Nähwirkvliesstoffen, Forschungsber STFI e V Chemnitz, BMWI 134/95 (1995)

[253] Menges G (1990) Faserverbundwerkstoffe mit Kunststoffmatrix: Stand und Aussichten, CTI 40/92: T 23

[254] Hinz B (1991) Aufgabe und Funktion der textilen Verstärkung in hochbelasteten Faserverbundwerkstoffen, CTI 41/93: T 34

[255] Wulfhorst B, Büsgen A, Weber M (1990) Dreidimensionale textile Halbzeuge zur rationellen Fertigung von Bauteilen aus Faserverbundwerkstoffen, Melliand Textilber (1990): 672

# Part III
# Finishing nonwovens

The same as wovens or knits, nonwovens are textile fabrics. One might think finishing them and equipping them for particular purposes could be done as with those, using the same equipment, chemicals and processes. This is true in many cases, however, there are numerous examples of nonwovens requiring treatment different from wovens or knits. The following aims to discuss such differences where commentary seems necessary.

Both textile and nonwoven finishing are widely diversified. This is why systematic approach is helpful. It has turned out first efforts to distinguish by dry and wet processes are not very practical. Frequently, finishing can be successfully executed by means of both dry and wet processes. It is more sensible to distinguish by mechanical and chemical processes even if the latter are hardly possible without any mechanical components, which is machines.

Thinking of costs and ecology, it seems clear that finishing, wherever possible, will be a dry process whether to save energy or to avoid harm to the environment. This takes us round to a challenge that has become of greatest interest within no more than the last twenty years. For this reason, ecological questions are discussed in a particular paragraph at the end of this Chapter.

# 7
# Mechanical finishing
K.-H. STUKENBROCK

## 7.1
## Shrinking

### 7.1.1
### Creation and removal of distortions

Like woven and knitted fabrics, nonwovens are also subjected to stresses during production, mainly in the longitudinal direction, and therefore distortions are not uncommon.

In the case of wet bonding with chemical binders (adhesive bonding), a drying process is necessary in any case and it is therefore relatively easy, with the right drying guide mechanism and suitable machines, to remove existing distortions by means of relaxation.

After mechanical bonding, for example by needling webs, a special shrinkage process may be needed to improve the dimensional stability in later use.

### 7.1.2
### Deliberate shrinkage

In many cases, the compression that inevitably occurs during shrinkage is utilized to achieve a higher weight per unit area, higher density, more bulk, higher strength or better cleavage properties (see Section 7.5). This is particularly common during the production of base materials for synthetic leather.

The shrinkage process is carried out wet or dry depending on the type of fibres used. Dry shrinkage by means of a heat treatment is only suitable for nonwovens made from 100% or primarily synthetic fibres and is particularly effectively when the fibre blend contains shrinkage fibres. They are fed through screen dryers, perforated drum dryers or short-loop dryers with rotating rods, whereby the nonwoven is fed into the heating zone with overfeed, that is it is delivered faster than it is drawn off.

Wet shrinkage is indicated whenever the nonwoven contains significant amounts of natural fibres, for example. The nonwoven is passed through a bath of hot water to trigger the shrinkage and is dried without tension after squeezing or hydrosuction. There are also particular synthetic fibres that respond to both wet shrinkage and dry heat shrinkage.

A variation of the wet shrinkage method that helps to save on drying energy is steam shrinkage.

By needling together shrinkable and non-shrinkable webs, decorative, relief structures can be obtained after shrinkage that are suitable for wall coverings or structured carpets, for example [3] [1].

## 7.2
## Compacting and creping

Some nonwovens initially lack the hand and drape properties required by the manufacturer and the consumer. They seem papery. Many attempts have there-fore been made to alter this papery character and to give the material more vol-ume and softness. This is achieved to a large extent using purely mechanical pro-cesses among which, for example, compacting and creping are the most impor-tant and the most technically advanced. These finishing methods are carried out by the Clupak and Micrex microcrepe processes.

## 7.2.1
## Compacting – the Clupak process

The Clupak process, invented by the American Sanford L. Cluett, is similar to the sanforizing process known in the textile industry for many years. It was used for the first time in the paper industry in 1957 and was carried over from there to wet-laid nonwovens. The equipment (Fig. 7-1) consists of a continuous rubber belt approximately 25 mm thick with a fabric lining, which abuts a heated, chrome-plated and polished drying cylinder in the working region and to which pressure

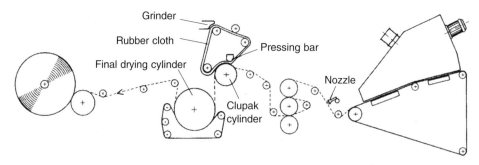

**Fig. 7-1** Diagram of the Clupak process (according to [4])

1) References for Chapter 7 see page 457

is applied at the first point of contact with the cylinder by a (in contrast to sanforizing) non-rotating clamping bar.

This causes the rubber blanket to be compacted lengthwise, which is transmitted to the nonwoven web between it and the cylinder, resulting in compacting and crimping of the fibres in the longitudinal direction. This compacting – provided the nonwoven is fed in to the gap between the cloth and the cylinder in a moist, that is plasticized state – is set during the drying process, the resulting cushion of steam enhancing the relative motion between the nonwoven and the cylinder.

The effect of the Clupak process depends on various factors. Hydrophilic fibres, such as cellulose and viscose, are more suitable than hydrophobic fibres. Polyolefin fibres are less suitable not only because of their lack of water absorption, but also due to their heat sensitivity. With regard to fibre orientation, longitudinally oriented webs give a more distinct effect than cross-laid or random-laid webs. A high moisture content of 20% enhances the plasticity of the fibres and therefore the compacting, while a high binder content (>50%) reduces it almost completely. Thermoplastic binders increase compacting, but also the adhesion of the web to the hot cylinder. Elastomeric binders cause practically no adhesion, but neutralize the compression effect to a certain extent due to their high elasticity.

## 7.2.2
### Creping – the Micrex microcrepe process

In the Micrex process, the nonwoven web is compacted to such an extent that it is evident in a visible creping and a measurable increase in the elongation and weight per unit area. The surface per unit area is increased and the flexibility is improved even more than with the Clupak process.

The equipment for the Micrex process consists of: a rotating carrier roller with spiral grooves running along the surface; one fixed and one elastic projecting guide plate, the latter mounted on the cylinder, between which the nonwoven web is fed in; and a blade-like compacting device at an acute angle to the roller surface.

The web is compacted in the first gap, can then lift off the cylinder in the relaxation zone and is compacted a second time in front of the blade and finally removed from the cylinder.

By adjusting the processing conditions, a fine or coarse crepe can be produced, without significantly reducing the strength, despite production speeds of 150–250 m/min. This is because, unlike the Clupak process, this process is carried out dry and at a considerably lower temperature.

This method is suitable for creping longitudinally oriented carded webs, wet or dry-laid random fibre nonwovens, spunbonded nonwovens, spunlaced products and papers with a weight per unit area of between 10–150 g/m$^2$. The degree of creping can be established simply by looking at the crepe folds, but more precisely by measuring the weight per unit area, which always increases significantly (increases of 50% are possible) and the elongation.

## 7.3
## Glazing, calendering, pressing

As in the textile industry, various opportunities exist in nonwovens finishing for calenders and presses. They are generally used for surface finishing, the most important processes being glazing and patterning. Discontinuous presses are mainly used as shaping presses for webs that have been cut to length.

In continuous calendering, the web or nonwoven is fed between one or more pairs of rollers under pressure, whereby steel rollers are normally heated.

### 7.3.1
### Glazing or rolling calenders [5]

Sometimes it is possible to give the nonwoven a smooth surface during web formation and bonding. This is mainly the case after wet bonding, when contact drying is carried out on polished, often Teflon-coated cylinder drying machines.

The well-known effect in textile finishing of finish breaking has gained little importance the finishing of nonwovens up to now.

However, as a pretreatment for subsequent coating, for synthetic leather, for example, calendering is essential. Also, for evening out and reducing the pore size of heavy nonwoven filter fabrics, a calendering treatment is indispensable. It greatly improves the filter performance and removal of dust.

"Drips" in spunbonded nonwovens can be broken by a combination of two steel rollers in the calender.

### 7.3.2
### Embossing or goffering calenders

Embossing or goffering calenders have many uses. A significant one, for example, is the reinforcement of webs made from natural and synthetic fibres that serves both as a bonding and a finishing process. By dampening slightly and hot goffering, longitudinally oriented webs made of cotton or viscose fibres with a weight per unit area of approximately 10–30 g/m$^2$, for example, can be stiffened and reinforced to such an extent that they are suitable for use as milk filters and are not flushed through a wire strainer. Similar embossing methods are also important for paper nonwovens for use as hygiene wipes, household cloths and napkins.

Also of interest is the hot embossing of synthetic fibre webs, which allows them to be strengthened significantly even when the fibres are only longitudinally oriented due to the fusing of fibres in the embossed areas. Grid, line or spot designs can be used. The temperature of the heated cylinders is normally 20–30 °C above the melting point of the fibre and the nip roll pressure is 20–50 dN/cm, depending on the volume and synthetic fibre content of the web. This type of reinforcement is particularly suitable for webs with a weight per unit area of up to 50 g/m$^2$. The hand is determined, on one hand, by the fibre orientation and, on the other, by the engraved pattern. The highest strengths are achieved for cross-laid webs by spot embossing and for longitudinally oriented webs by line embossing. Conversely, the hand is more supple the fewer fibres are fused together within a certain area.

Impressive visual effects can be achieved with ciré embossing using engravings with different heights, which not only give a very three-dimensional surface structure, but also allow a dull lustre or shine to be produced according to the hatching. These finishing methods are of importance primarily for tapestries and decorative nonwovens.

The embossed effect becomes even more three-dimensional when it is simultaneously combined with dyeing of the recessed or raised areas.

For example, the "grain" on artificial or synthetic leather is created in this way, whether it is based on polyvinyl chloride or polyurethane. Whether this grain is produced directly using embossing calenders or indirectly, as in transfer processes, by embossing the release paper, depends on the technology used and also on the chemical composition of the coating. The large number of possibilities means that there are almost no limits to the imagination. As well as leather grain effects, woven structures, plaster, brush stroke, cord or braid patterns, also well known are mock tile engravings such as those on cushioned vinyl floor coverings.

Another important end-use for heated calenders is the production of laminates and sandwich products. Here, thermoplastic fibres, scrims or films are placed between two non-plastic webs and used to bond the outer web layers by means of heat and pressure. These laminates are used as tablecloths, upholstery and cushion covers in railway carriages and aeroplanes.

### 7.3.3
### Felt calenders, transfer calenders

Felt calenders also belong, in name at least, to the category of calenders. Their use is described in Section 8.3.4 "Transfer printing".

### 7.3.4
### Rotary pressing machines

The oldest form of surface finishing of nonwovens by pressing is the smoothing of wool felts, especially felts for collar linings, on rotary pressing machines, which not only gives a smooth surface, but at the same time also improved strength and greater lustre.

### 7.3.5
### Moulding, stamping

Moulding and shaping are an important finishing process, particularly for nonwovens in the automotive sector (see Section 15.7). Thus, for example, using phenolic resins (Novolac plus hardening agent), pre-bonded nonwovens made from secondary fibres, preferably cotton waste, are manufactured by moulding and shaping under pressure and heat for headlining, rear shelves, dashboards, door linings, wheel case covers, boot linings and entire floor coverings (see Section 15.7). Here, use is made of the chemical conversion (three-dimensional crosslinking) of thermoplastics into thermosetting plastics. If a decorative shell (headlining fabric, rear shelf) is simultaneously laminated to it (see Section 8.6) and the shaping press is also used for blanking, then the nonwoven is transformed into a laminate or composite. As well as adhesively prebonded natural secondary nonwovens, nonwovens made from thermoplastic fibres (for example, polypropylene, polyester) can also be moulded and laminated.

If a knitted fabric made from polyester yarns is used for the headlining and a copolyester is used as the laminating adhesive, the result is a composite material that is 100% recyclable, as required by the car manufacturers [6].

## 7.4
## Perforating, slitting, breaking

Despite all the technical advances in web formation and bonding technology, some nonwovens, especially for the clothing sector, may have a hand that is too stiff. This is linked to the fact that the individual fibres cannot move as freely in relation to one another as the yarns in a woven or knitted fabric. There has been no lack of attempts to improve the draping property of nonwovens. This is achieved by perforating and slitting.

### 7.4.1
### Perforating

In a process developed by the company Artos, hot needles are used to perforate a chemically bonded web. This not only creates holes, but also fixes the openings due to the crosslinking and condensation of the binder.

The Hungarian firm Temaforg uses a similar method to perforate webs made from reclaimed synthetic fibres and finely pulverized film waste and simultaneously fuse around the stamped holes to produce nonwovens for the construction and insulation sector with good strength properties and adequate suppleness.

### 7.4.2
### Slitting

Slitting has become an important method, originally developed to improve the softness and drape of films. It was first applied to nonwoven interlinings by the company Breveteam and was then carried over to fusible adhesive interlinings in particular.

The optimum cutting lengths and distances between the slits that give the maximum softness and drape without impairing the strength too much can be established for many nonwoven interlinings by means of systematic investigations. The effect of slitting, whereby the greatest flexibility is at right angles to the slitting direction, is supported by the application of hot-sealing adhesive in broken stripes. If, in addition to slitting, but at right angles to it, the nonwoven is stretched and – in the case of thermoplastic nonwovens – set, it is possible to obtain three-dimensional effects, such as those needed for bra pads.

Slitting is carried out using a roller fitted with small blades in a staggered arrangement 1.7 mm apart, for example, which make slits up to 6.5 mm long. Instead of small blades, circular knives with an intermittent cutting edge arranged on a cylinder with spacers can also be used.

With the Xironet process [7] and the Smith-Nephew process, polyethylene or polyamide films shaped by slitting or embossing and stretching, can be used successfully as an air-permeable bonding layer for laminating (see Section 8.6.2) nonwovens.

## 7.4.3
## Breaking

The oldest method for enhancing the softness, suppleness and drape of a textile by purely mechanical means is by breaking. The nonwoven to be treated is fed under slight tension through breaking calenders or button breakers, such as those commonly used in the textile industry.

## 7.5
## Splitting, emerizing, suede finishing, shearing, raising

## 7.5.1
## Splitting

In the leather industry it is common to split thick skins, especially cow hides, several times parallel to the grain side in order to obtain a larger, if also thinner area of leather from a single skin. The value of the split pieces decreases from the grain to the flesh side. As well as the larger area and greater yield, this also produces leather that is too thin, which is used in large quantities for making bags and as lining leather and is preferred over thinner skins of small animals due to its size and low price.

As nonwovens are often used instead of leather, here too the splitting method is used on thick, dense, high strength synthetic fibre needle-punched nonwovens to produce thin, supple, leather-like fabrics that can be used as belts, shoe linings and – as a coating substrate – also as shoe uppers and outer materials for bags.

Splitting is carried out on "splitting machines" in which a continuous rotation hoop knife is guided precisely into the gap between two carrier rollers. The distance between the rollers is determined on one hand by the thickness of the nonwoven and on the other by the required "split". Needle-punched nonwovens of 500–1,000 g/m$^2$ that have been bonded with elastomers are suitable for splitting.

With working widths of up to 2 m, split nonwovens can be "cut" down to 150 g/m² and can be processed as roll goods with defined dimensions more rationally than natural split leather with its varying sizes.

### 7.5.2
### Emerizing, suede finishing

After splitting, either roll calendering (see Section 7.3.1) or embossing (see Section 7.3.2) are carried out or, to make the surface even, emerizing and polishing, giving the nonwoven a velours or suede quality – so-called suede finishing.

Several consecutive machines or passages are normally used, first to emerize the surface coarsely and then to polish it increasingly finely, producing an almost velvet character. After polishing, the fabric is brushed, beaten or vacuum cleaned to remove the grinding dust. Top quality emerized and suede finished products stand out due to their soft hand, elegant drape and velvety surface (so-called peach hand).

### 7.5.3
### Shearing, raising

In the past, needle-punched nonwovens for floor coverings normally had a smooth face layer as a result of multiple needling.

With the introduction of structuring needles and the use of particularly coarse fibres, nonwoven surfaces with loop pile or velours structures can now be produced. They give a high-quality carpet-like effect and are used as floor coverings in cars. Raising and shearing increase the bulk and improve the look and feel of these nonwovens [8, 60].

### 7.6
### Singeing

Also included under mechanical finishing is *singeing*, which is used for burning off protruding fibres from needle-punched nonwovens for use as filters. The process is the same as conventional singeing used in the textile industry and is carried out, for example, on so-called gassing machines on which the fabric is passed over an open gas flame. This gives the nonwoven a smoother surface, making it easier to clean dust or dry filters, for example, in the cement industry or other types of filtration involving powdery materials.

## 7.7
## Sewing, quilting and welding

As has been referred to above, webs and nonwovens are often joined together to give composite materials. There are various possible ways of doing this. In addition to the conventional bonding techniques of sewing and quilting are welding processes such as thermal impulse, ultrasonic and high-frequency welding.

### 7.7.1
### Cover-seaming and quilting

These finishing methods are so well known that a more detailed description is unnecessary.

### 7.7.2
### Ultrasonic welding

Webs and nonwovens made from thermoplastic fibres or nonwovens bonded using thermoplastic binders are suitable for ultrasonic welding. In this process, the frequency of an alternating current is increased from 50 Hz up to 20 kHz by an oscillator and this energy is converted into a corresponding mechanical vibration by means of an electromechanical converter. It is transferred to a hammer (sonotrode), which hits the anvil with a lift of 50–100 μm. The fibres to be bonded, lying between the hammer and anvil, are heated to the extent that they start to melt and bond together. If the work of the sonotrode is transferred to a wheel, then endless "seams" can be produced or the composite can be separated along the fused joint (separation welding) [9].

This technique is important for tea bags, sanitary towels, panty liners, diapers and incontinence products.

### 7.7.3
### High-frequency welding

Whereas with ultrasonic welding, the energy is supplied from outside, with high-frequency welding, the heating occurs from the inside out, as it were. The main current is transformed from 50 Hz to 26.2 MHz in a high-frequency generator. The molecules of the thermoplastic fibres or thermoplastics, which have a pronounced dipole character with a high phase-angle difference, are subject to such rapid pole reversal in the high-frequency field, according to the frequency, that they rub against each other and the thermoplastic spontaneously starts to melt. When the field is switched off, the fused joint formed under pressure cools down again almost as rapidly as it was heated up.

PVC fibres and copolymer fibres made from 85% vinyl chloride and 15% vinyl acetate (MP-Fibre from Wacker) are especially suitable for this technique. They are used for tea bags or for bonding two polyurethane foam sheets by the needle punching technique for upholstering car seats. These are permanently bonded to the exterior layer using the high-frequency welding technique. By arranging the welding electrodes appropriately, ornamental effects can be achieved. Unlike ultrasonic welding, which can be carried out as a continuous process, high-frequency welding can only be carried out in batches, that is discontinuously.

Both welding techniques are also suitable for non-thermoplastic fibre structures as long as these are coated with a sufficient quantity of a thermoplastic binder or with dipolar plastics.

## 7.8
## Other mechanical finishing methods

Purely for the sake of completeness, cutting, stamping, batching, winding and packaging should be mentioned briefly, as in many cases specialist machines had to be developed for the nonwovens industry to suit the particular requirements of these materials. Here, the varying weights per unit area of 15 g/m$^2$ to over 2,000 g/m$^2$ have to be taken into account, as well as the thickness, which can range from fractions of a millimetre to several centimetres.

Centre winders, ascending batch winders and carrier drum rollers are used for batching nonwovens.

Various different possibilities exist for cutting. Rotating cutting blades are well known for continuous roll cutting along the fabric length and also for discontinuous cutting across the fabric. There are also stamping machines for cutting fabrics to length or blanking out shapes, for example, as tiles for floor coverings or moulded articles for cars. Laser technology is also often used for cutting.

Finally, it is worth mentioning the vacuum packaging of bulky filling nonwovens to save space during transportation.

# 8
# Chemical finishing
K.-H. STUKENBROCK

The term "chemical finishing" was chosen because in this section on the finishing of nonwovens, chemical or textile auxiliaries assume a significant role in the processing.

## 8.1
## Washing

The purpose of washing is to remove unwanted substances from the nonwoven by means of a wet process. It requires a suitable washing machine, normally water as the washing medium and sometimes a detergent to enhance the effect.

Washing machines that enable the fabric to be fed in flat at full width are usually used, as any folds or cracks produced only have to be removed again at a cost.

Winch becks are normally only used for washing continuous needle-punched nonwovens for paper machines and they are then finished (dyeing, softening, impregnating) on the same machine before they are dry.

Otherwise, roller vats and perforated cylinder washers with divider inserts are used, the later sometimes being designed as additional rotors (Fig. 8-1) [10].

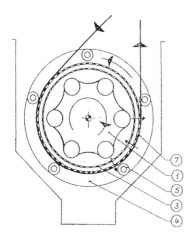

**Fig. 8-1**  Diagram of a vibrating cylinder washing machine (according to [10])
1 rotor and divider insert, 3 and 4 perforated cylinder inside and outside, 5 stainless steel bearing, 7 fabric

In conjunction with spray pipes, they help to save and give extra turbulence to the water, promoting the removal and flushing out of unwanted impurities or chemicals.

These include dyeing assistants, dye thickeners from dyeing and printing, water-soluble, pore-forming substances, for example salts, urea and swelling agents, which are sometimes used in the production of imitation chamois leathers, wiping cloths and the like, printing adhesives and foaming or wetting agents used in foam coating, which would impair the properties of the finished nonwoven.

The addition of a special scouring agent is indicated when a large amount of impurities are present. Here, attention should be paid to the compatibility of the ionic charges, as well as good anti-redeposition power to prevent redeposition, low foaming and, last but not least, to the legally prescribed biodegradability (see Section 8.9).

Some anionic scouring agents also often have a softening effect (superfatting), non-ionic auxiliaries have the advantage of universal compatibility, but "exhaust" the nonwoven more and are more effective in their washing effect in some temperature ranges than others due to their own turbidity point. During washing, as is the case with all wet and drying processes, the nonwoven should be subjected to as little tension as possible. Too much tension in the longitudinal direction is particularly undesirable.

After or at the same time as washing, a softener or other required auxiliary can be applied.

After washing, the nonwoven is normally dried, for which machines normally used for drying binders are available (see Section 6.5.1).

## 8.2
## Dyeing

It is not easy to estimate the proportion of dyed nonwovens compared to undyed or natural coloured ones. Colour is always present, be it single-shade or printing, whenever the nonwoven has a decorative task, for example for wall or floor coverings, as table or bedlinen or as a furnishing fabric. Shirt and blouse interlining nonwovens are also dyed to match the colour of the face fabric, which, especially in workwear (hospital, surgical, clean-room clothing, protective suits and so forth), is often made out of nonwovens.

### 8.2.1
### Stock and spin dyeing

Apart from transfer printing, dyeing and printing are wet processes and are therefore time-, energy- and cost-intensive. Wherever possible, the dyeing of nonwovens is combined with the wet processes necessary for bonding or fibres and filaments that are already dyed are used. Stock dyeing of fibres and spin dyeing of synthetic fibres are of great importance. The fact that cotton coloured (cream) purely by cultivation is now available on the market, once again underlines the high value placed on dyeing.

However, this chapter deals with the subsequent dyeing of already formed webs and nonwovens.

## 8.2.2
## Dyeing and bonding

As long as the webs in question still have to be chemically bonded, it is appropriate to add the dye to the binder bath. Some interesting observations have been made in this regard. Assuming that the binder liquor, and therefore the binder, covers all the fibres in the web evenly, then it would be most practical to colour the binder liquor in the required shade. For this, finely dispersed pigment dyes, for example, could be used. They would stick to the surface of the fibre due to the binders and would have the excellent light fastness properties of pigments in general, but otherwise the same fastness properties as the binders, such as wet and dry rub fastness, fastness to perspiration or ironing fastness to name just a few. However, it is not necessarily desirable for the binder to cover all the fibres in a web evenly. It may be preferable only to apply it to the intersections between the individual fibres to achieve the softest hand possible. If dyeing was attempted simultaneously, this would lead to unlevelness. However, if dyes with good affinity are added to the binder liquor, then even if the binder is distributed unevenly, level dyeing will be achieved with a fastness comparable with that of a thermosol process [11].

If there are sufficiently large quantities and it is a web made from only one type of fibre, then continuous bonding and dyeing is perfectly feasible and is almost without problem. Thus, dyeing can be carried out evenly and deeply and with the fastness properties customary of conventional dyeing, by dyeing cotton and viscose webs with direct dyes, polyamide webs with acid and polyester webs with disperse dyes, for example. The only consideration is that the ionic character and pH value of the binder should be compatible with the dye.

## 8.2.3
## Subsequent dyeing

It becomes more difficult to dye and bond simultaneously in the same bath when, as often happens in practice, the web contains blends of different types of fibre. In this case, but also in many cases when there is only a single fibre raw material, dyeing has to be carried out in a subsequent process. The nonwoven is then considered in almost the same way as a woven or knitted substrate and is dyed using the same processes and dyes as conventional dyeing for the individual types of fibre or their blends.

8.2.4
**Different dyeing methods**

Both discontinuous and continuous dyeing methods are in use and possible. Dyeing is normally carried out in open width, although in special cases nonwovens are successfully dyed on winch becks or even by jet dyeing if the structure, strength and above all the weight per unit area allow this. In general, heavy and bulky nonwovens are dyed continuously because the quantities of them that discontinuous machines, such as jigs or dyeing beams, can lift are too small to operate economically.

Exceptions to this are continuous needle-punched nonwovens for paper machines, which, as was mentioned in Section 8.1, are dyed on winch becks.

Light nonwovens, on the other hand, can be dyed without problem on beam dyeing machines. However, it is necessary to take into account the binder that may be present. Thermally crosslinked binders based on acrylic esters can be dyed as deeply and with the same fastness as polyester or polyamide fibres with most disperse, but also many acid dyes. The dyeing rate is even higher as the macromolecules of the binder are less oriented than those of the fibres and therefore take up the dye molecules more quickly. Vulcanized or vulcanizable elastomers, such as polyisoprene, butadiene acrylonitrile and butadiene styrene latexes have a resist effect on most dyes and only have a poor dye affinity.

In beam dyeing, thermoplasticity, especially of soft acrylic binders, plays an important role. To save time, it would be preferable to dye light nonwovens made from normal polyester fibres, for example, on the beam under high-temperature conditions. However, it has been shown that, at temperatures above 102 °C, certain binders start to stick the separate web layers together, and consequently the beams are not guaranteed to unwind properly. In this case, carrierless dyeable polyester fibres that can be dyed at 90–100 °C may be advantageous.

Nonwovens made from a single fibrous material that are bonded purely mechanically without binders can be dyed by the same methods and on the same machines as corresponding woven or knitted fabrics.

8.2.5
**Cold pad batch dyeing**

An interesting process is the method patented by *Farbwerke Hoechst* for dyeing bonded fibre webs made from polyamide fibres by the batch method. Nonwovens for furnishings and table linen produced by the melt spinning or card/cross layer process and bonded with acrylic esters are cold padded with acid or metal complex dyes with the aid of acids as hydrogen donors and cold wetting agents and migration auxiliaries. They are then batched up and left for 24 hours rotating and covered with polyethylene film, after which they are given a warm rinse, followed by soaping, a further rinse and press finishing. In this way, dyeing can be carried out in all the current fashion shades with sufficient fastness [12].

8.2.6
**Continuous dyeing**

As was mentioned previously, heavy nonwovens are dyed continuously because it is almost impossible to handle them discontinuously due to their weight. Flow coating methods are preferred (for example the Küsters System) with subsequent steaming to fix the dye, as in the conventional pad-steam process. With dye liquor applications of 400–500% of the fabric weight, the danger of frosting on protruding fibres in the steamer can often arise due to condensate formation. In an attempt to combat this, the flow coating liquor is thickened and a dyeing assistant that foams at high temperatures is also used. After steaming, rinsing and washing are carried out as usual.

8.3
**Printing**

The fact that nonwovens are increasingly found in the home textiles sector has led to a growth in the desire for colour by dyeing and also by printing.

8.3.1
**Printing of lightweight nonwovens**

When dealing with lightweight nonwovens of 50–55 g/m$^2$, preferably made from one type of fibre, as is the case, for example, with spunlaid nonwovens, printing is more or less problematic, just as it is with other textile fabrics.

Screen printing and rotary screen printing are the most commonly used methods. As with any other textile substrate, the nonwoven is attached to the printing blanket or backing cloth, printed with the dyes normally used for the fibres, pre-dried, fixed by steaming, washed (see Section 8.1) and press finished.

Pigment printing has special significance, especially for blotch printing, as additional bonding of the nonwoven by the pigment binder can be achieved, particularly in the case of spunlaid nonwovens. Moreover here, drying and steaming are replaced by condensation – in principle nothing more than a very intensive drying process – for the pigment binders. In some circumstances, when working with thickeners with a low solids content, it is possible to dispense with the subsequent washing process. This applies particularly when, as is common practice when printing lightweight nonwovens by the rotary screen printing method, no adhesive is used because the accuracy of individual repeats can be controlled by regulating the tension between the fabric feed and take off to the back drier. However, under no circumstances should adhesives be dispensed with on flat screen printing machines as the nonwoven can lift off and stick underneath the screen, severely disrupting the accuracy of the repeat.

Pigment dyes have the advantage that they can be used regardless of the fibres and the nonwoven binder. They can be used in the same way for all lightweight nonwovens and only vary according to the substrate in their wet, washing and dry rub fastness properties depending on the adhesion with which the pigment binder is anchored onto the fibres. Although printed lightweight nonwovens are often seasonal products, the fastness requirements on printed goods play the same role as with textiles that have to be washed and/or dry cleaned often in use throughout the year.

For the printer, the dye concentration for formulating the printing paste recipe is important when printing nonwovens. The dyeing effect perceived by the human eye is a function of the amount of dye applied per unit area. Lightweight nonwovens for tablecloths have a weight of around 60 g/m$^2$, for example, whereas with needle-punched nonwovens for floor coverings the figure is 600–1,200 g/m$^2$. Therefore, to achieve a given depth of a certain shade, a larger quantity of dye in relation to the fibre weight has to be applied to the lightweight nonwoven than to a heavier needle-punched nonwoven. However, as a lightweight nonwoven can take up less printing paste than a heavy needle-punched nonwoven, the dye concentration of the printing paste must be low for the latter and high for the former.

### 8.3.2
### Printing of heavy nonwovens (needle-punched nonwoven floor coverings)

Another problem area with heavy needle-punched nonwovens is dimensional stability and distortions that make printing more difficult. Whereas with lightweight nonwovens it is possible to attach them before printing, thereby fixing the dimensional stability, although this is technically possible with heavy needle-punched nonwovens, it is not economically viable.

Due to the difficulty in handling heavy rolls of needle-punched nonwovens, a discontinuous process involving, for example, printing, intermediate drying, steaming (fixing), washing off and drying on different machines is not usually used. All operations are carried out continuously in a sequence on a combined processing line. This inevitably produces longitudinal tensions, which can lead to irregularities in the pattern or repeat with multicoloured patterns, and even with single-shade designs if they have regular geometric figuring such as circles or squares [13].

To prevent distortions, it is necessary to needle needle-punched nonwovens for floor coverings onto a dimensionally stable backing fabric, which throws up economic as well as technical problems, or to reinforce them in advance.

It has already been stated in the section on dyeing that nonwoven binders can behave very differently with regard to their dye affinity than fibres. It therefore follows that needle-punched nonwovens for printing should not be bonded in a full bath, but only on the reverse, that is they should be slop padded. It is also well known that, with slop padding, it is not easy to apply the binder to an even depth. It is easy to see that binder that has penetrated too deeply or even gone right through would lead to variations in the printing result.

In addition, a printing paste for rotary printing heavy needle-punched nonwovens has to have completely different flow properties than a paste for lightweight nonwovens. Whereas the latter can be printed at speeds of more than 30 m/min, with heavy needle-punched nonwovens a speed of between 2–5 m/min is standard. This results in different shearing forces and varying detachment of the printing paste from the screen or a different adhesion to the substrate. Finally, with add-ons of two to three kilograms of printing paste per square metre, with a relatively low dye concentration, the "frosting effect" is much greater than when printing a lightweight nonwoven.

### 8.3.3
### Spray printing

These difficulties have led to the rediscovery of spray printing, which is carried out by the ink jet printing principle on fully computer-controlled machines.

The *Millitron* and the *Chromojet methods* have been successful in practice. With both methods, printing pastes (inks) are sprayed onto the substrate through fine nozzles with a frequency of 3–300 kHz, the Millitron method working by the continuous jet system and the Chromojet method by the drop on demand system (Fig. 8-2).

The nozzles are mounted on a fixed or traversing printing bar so that 16 per inch are arranged across the length and the width (in the case of rotary printing, by way of comparison, it is 125 dots per inch).

The main advantage of spray printing is that the printed design with up to twelve colours (even more with Millitron) can be transferred directly on the machine, saving on the preparation of the printing screens. The disadvantage is that the definition of the design is not as clear. After spraying, the nonwoven is normally steamed to fix the dye, excess dye is washed off by spraying and it is dried [14].

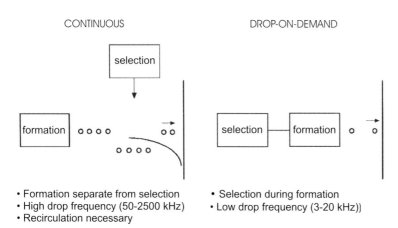

Fig. 8-2  Spray printing system (according to [14])

8.3.4
**Transfer printing**

Also important is the transfer or heat-transfer printing method in which the subli-
mating dye is transferred from a release paper onto the nonwoven under the influence of heat and pressure. Nonwovens made of polyester fibres are especially
suitable for this as they bond very strongly to the dye and give good general fastness properties. While there is no problem transfer printing onto lightweight nonwovens and most binders cope with a temperature of around 200 °C for 30–60 s
without yellowing or sticking, problems once again arise with heavy nonwovens.
These include sufficient diffusion of the dye into the inside of the nonwoven with
differences in shade occurring with combination dyes. Dyes with a small molecule sublimate from 145 °C and therefore penetrate easily and deeply into the nonwoven. Large dye molecules, on the other hand, only sublimate at higher temperatures and therefore later and thus go mainly onto the surface, which means that,
although there is no inaccurate registration, colour changes are likely to occur. In
general, heat-transfer printing is carried out on transfer calenders, based on the
felt calenders mentioned in Section 7.3.3 (Fig. 8-3).

Transfer paper and printing substrate are treated between a heated steel cylinder and a continuous rotating felt. This makes transfer printing the only dry process among the printing techniques mentioned and it is therefore a very environmentally friendly method, apart from the need to dispose of the printed paper
[15].

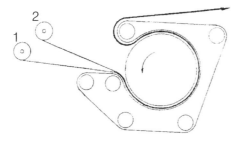

**Fig. 8-3**   Diagram of a transfer calender
(according to [15])
*1* nonwoven, *2* transfer print paper

8.4
**Finishing, softening, special effects**

When processing adhesive bonded nonwovens, the aim is normally to achieve all
the desired properties for the end product via the binder liquor. However, this is
not always possible, particularly when extra effects are required in addition to simply bonding the fibres. A subsequent treatment is usually unavoidable on mechanically reinforced nonwovens that are to be given special properties by means
of a finishing process.

8.4.1
## Machine facts and possibilities

Practically the same machines and methods are used for these treatments as for piece finishing, preference being given to non-contact or at least low-contact methods.

First of all, full bath immersion (impregnation) and squeezing on the mangle should be mentioned. To ensure that the nonwoven is evenly saturated with the finishing liquor, longer or shorter immersion times and immersion lengths are used. Guide rollers and divider inserts prevent too much liquor from being prepared, stored or disposed of. Jacketed heating and cooling troughs, together with these divider inserts, ensure an even liquor temperature. After immersion, the nonwoven is squeezed with as much pressure as possible to save on drying energy.

Surface effects are achieved mainly by spraying. A distinction is made between spraying with single-product (airless) and two-product nozzles, their use depending primarily on the amount to be applied and the mechanical stability of the medium to be applied. Two-product nozzles are used for sensitive liquors that are to be applied in relatively small amounts. Airless nozzles are employed with stable finishes that are also to be applied in large quantities.

For extremely small add-ons, rotor humidifiers or the minimum liquor application method similar to slop padding are used in which single or double-sided liquor application is possible.

Foaming and spreading or printing with finely perforated rotary screens are also common application techniques.

8.4.2
## Stiffening finishes

An example of an additional finishing process is stiffening, which cannot be combined with the initial bonding process. Examples here include vertical louvre and roller blinds, which require an additional finishing process. Non-yellowing polyvinyl acetate and acrylic ester dispersions are mainly used for this. If flame resistance is also required, appropriate additives (see Section 8.4.8) should be added.

The diversity of finishing and stiffening effects possible with nonwovens is illustrated by the following examples.

### 8.4.2.1 Finishing of toe and heel caps on footwear

Toe and heel caps for sturdy footwear were and in some cases still are manufactured on the basis of twill fabrics. Nonwovens have the great advantage over woven fabrics of stretching better in all directions and being mouldable without forming any creases.

Nonwovens for toe and heel caps are finished in several operations with thermosetting precondensates, thermoplastic or elastomeric dispersions. The latter serve as so-called resilient elastic toe and heel linings often referred to as having a "ping-pong effect". Thermoplastic dispersions, especially SBR latexes with a high styrene content, are used for heat mouldable caps that can be reactivated with solvents. Precondensates are used in conjunction with elastomeric dispersions to produce so-called vulcanized caps, although these are now increasingly being superseded by the two other types.

For all these finishing processes, the finish add-ons are up to 300% or more of the web weight, which naturally cannot be applied in one operation.

Instead of elastic stiffening with aqueous dispersions and/or latexes and precondensates, powder coating can be advantageous. Scattering with 200–250 g/m² of an ethylene vinyl acetate copolymer powder, which is melted under infrared emitters and then smoothed out between a pair of cooled steel rollers, not only gives the desired resilience but also makes it possible to iron the toe and heel caps into the shoe instead of sticking them in (see also Section 8.5).

### 8.4.2.2 Finishing of nonwovens for roofing felts
Similarly high add-ons are used for finishing nonwovens for roofing felts. A prebonded web is fed through molten bitumen at 180–220 °C and passes between a pair of heated steel rollers with a set clearance, resulting in a bitumen add-on several times the web weight. It is then scattered with coarse or fine grains of sand and cooled down. With this type of finishing, high tear strengths even at 200 °C are equally as important as are good flatness with high tensile loading and, simultaneously, adequate water repellence as in practical use water absorbed by a capillary action would lead to blistering and splitting of the bitumen layer on evaporation.

### 8.4.2.3 Finishing of glass fibre nonwovens for glass fibre composites
Polyester moulded articles produced by laying up glass fibres by hand, e.g. boat hulls, have a rough, sometimes porous surface. To make it smooth, it is covered with a glass nonwoven of approximately 30 g/m² that has been reinforced with around 15% dry add-on of an aqueous dispersion. This finish must be resistant to the monomeric styrenes contained in polyester resin systems. Certain PUR dispersions meet this requirement.

### 8.4.3
### Softening

Softening can refer to the fibre and/or the binder component of the nonwoven. For subsequent plasticization of binders, phthalic, phosphoric, adipic and sebacic acid esters can be used. For softening most fibres, but also for smoothing, cationic fatty amine condensates or quaternary nitrogen compounds are used among others, which cannot be combined with most anionically emulsified dispersions in the binder bath, lead to precipitation or are at least influenced in their effectiveness.

Especially effective is subsequent softening with silicone products which, due to their smoothness, are used for their surface effect on nonwoven interlinings because they enhance the sewing property. It is possible to combine them with the binder bath, but this has certain disadvantages. Firstly, these relatively expensive products are also applied on the inside of the nonwoven where they are of little benefit to the sewing property and secondly, they prevent the binder from anchoring itself firmly to the fibres to some extent, thereby reducing the strength of the nonwoven.

There are different types of silicone softeners. As well as simple, emulsified silicone oils, which sometimes come in the form of microemulsions and can be washed out, there are methyl and amino-modified silicones which crosslink after drying, thereby producing a permanent softening and smoothing effect [16].

Silicone rubbers give polyester fibres permanent bulk and are successfully used for the finishing of "fibrefill" fibres for nonwoven wadding and filling materials.

The diversity of silicone products is underlined by the fact that certain methyl siloxanes are also suitable as water-repellent finishes (see Section 8.4.6).

As far as possible, softeners are used that are biodegradable or that have no biological or chemical need for oxygen and can therefore be fed into the effluent without any concern.

## 8.4.4
### Antistatic finishing

Antistatic finishes are very important for home textiles [17], especially floor covering materials, wall hangings, wallpaper as well as upholstery and mattress covering fabrics.

As far as possible, suitable products, such as those based on phosphoric acid esters or polystyrene malic acid sodium, for example, are added to the binder liquor or subsequently sprayed onto the right side, as is the case with needle-punched nonwovens for floor coverings, for example.

Use is also made of antistatic treatments in the finishing of dust-bonding nonwoven cloths or so-called antistatic cloths. Here, the antistatic agent is often supplemented with hygroscopic substances such as glycerine or polishes, for example paraffin emulsions.

Permanent antistatic properties are achieved by treating with combination products made from epichlorohydrin and EOPEO chains or by inserting copper wires, for example when laminating cut-pile carpets with needle-punched nonwovens for floor coverings (EP 00 3057).

Finely dispersed carbon black also gives a permanent antistatic effect when mixed into the binder liquor for backings. Lightweight nonwovens that have been impregnated, sprayed or coated with more than three percent dry add-on in relation to the nonwoven weight of finely dispersed carbon black or copper particles, when used as wall, ceiling and door coverings, form a kind of Faraday's cage and shield against X-rays or provide protection from bugging by electronic devices.

### 8.4.5
### Soil-repellent finishing

The distinction should be made here between "soil release" and actual "antisoil" finishing.

Soil release finishing refers to a nonwoven treatment intended to facilitate the removal of dirt from textiles. This applies above all to nonwovens that are "durables", that will be washed several times during their use.

Soil release finishes used are, for example, film-forming, low-molecular and therefore slightly swelling polymers, which initially form a kind of protective layer against the dirt, but can be slightly washed off the fibres.

An antisoil or stain resistant effect is provided by fluorocarbon resin chemicals, which usually also have an oil and water-repellent effect.

Polytetrafluoroethylene dispersions are specialist finishes and are applied especially to high-temperature resistant staple fibre nonwovens for use in filtration to make them easier to clean. Similarly, finely dispersed silicon and aluminium hydrosols act as "invisible dirt" to protect the nonwoven from further soiling.

### 8.4.6
### Water-repellent and oil-repellent finishing

The paraffin emulsions containing aluminium and zircon still used for the water-repellent finishing of woven and knitted fabrics are of practically no importance for finishing nonwovens.

Polysiloxane emulsions are used when, in addition to the water repellency, a softer and smoother hand is also required (see Section 8.4.3).

The fluorocarbon resin emulsions already mentioned give a rather dry, dull hand, but in addition to good water repellency, also offer excellent oil repellency and good protection against acids. Both product classes are often processed in conjunction with precondensates, so-called extenders and require sufficiently high drying temperatures to develop their good and permanent properties.

A new class of perfluorated alkyl triethyloxysilanes form a chemical bond with the fibre surface and provide soil, water and oil-repellency in layers just nanometres thick [18].

It goes without saying that any processing or wetting agent residues present should be carefully removed, as detergents seriously impair the effect of silicon products in particular.

### 8.4.7
### Hygiene finishing

In the hygiene and hospital sector, antibacterial, antifungal or fungicide finishes are required, but care must be taken in their use that they do not have any negative effect on the human body (skin inflammation, diffusion effect) [19].

Auxiliaries that impart the desired properties can be applied both before production of the nonwoven onto or into the fibres and subsequently onto the finished nonwoven. Some have a temporary, others a permanent effect.

Also commonly found are nonwoven finishes with collagen (naturin), tea-tree oil, witch-hazel, aloe vera, camomile or beeswax for covering wounds and skin regeneration, sometimes in laminates with aluminium foil.

In addition to the use of these chemicals and remedies, laminates are also well-known in the hospital sector in which microporous membranes are adhered to nonwovens and other textile fabrics to ensure long-lasting protection against germs, but also against blood (see Section 8.6) [20].

8.4.8
**Flame resistant finishing**

Although flame retardant properties are determined primarily by the fibres and binders used, additional effects can be achieved by supplementary finishing. There are a range of nitrogen/phosphorous compounds – the simplest being, for example, diammonium hydrogen phosphate –, which impart good flame resistance to cellulosic fibres. However, this kind of finish can "bloom" and is therefore sometimes combined with hygroscopic auxiliaries, which for their part reduce the flammability of the fibres due to their hydrophilic property. Other aqueous inorganic substances such as aluminium trihydrate often demonstrate good effects. Organic bromine and chlorine derivatives reduce the flammability of synthetic fibres. However, they often have the disadvantage that the gases produced on thermal degradation are more harmful to people than the open flames. With certain products formerly on the market, skin irritations were possible and there was a risk of cancer and therefore it should be weighed up very carefully where and whether flame retardant finishes should be applied [21, 22].

Intumescent products such as intumescent mica and intumescent graphite, which is patent protected for flameproofing textiles (EP 0752 458) offer an environmentally friendly alternative.

8.4.9
**Absorbent and water absorbent finishing**

Water absorbency, which is important for towels, hygiene and medical products, is determined primarily by the fibres and binders used, but can be significantly increased by adding specialist auxiliaries or by subsequent treatment.

The use of "rewetting agents" – not to be confused with wetting agents – promotes the absorption and thus the drying effect of nonwoven towels. Suitable products are, for example, modified betaines or quaternary oleic acid triisopropanol amine, which also impart softening properties to cellulosic fibres. Combining a nonwoven with cellulose wadding also has a hydrophilic effect. An even better effect, with an absorptive capacity a hundred or more times their own weight, is provided by absorbent fillers such as partially crosslinked polyacrylic acid and polyacrylamide derivatives and modified cellulose derivatives in the form of fibres and powders. Another advantage is that less is required in diapers and sanitary towels than cellulose [23].

A lot of experience is needed to fix these auxiliaries – when they are in powder form – sufficiently securely in the fluff core of a diaper as they make up 60–70% of the core or 15–20 g of a complete diaper with a total weight of around 50 g [25, 26].

With water absorbent finishing, care should again be taken to ensure that the finishes are compatible with the skin and are physiologically harmless. This also applies to the hydrophilic finishing of nonwovens as long as they come into contact with the human body.

Less critical in this regard is the use of superabsorbents in technical products such as, for example, for irrigation nonwovens in greenhouses or as a water barrier for electrical cables to be laid in water or in the ground [24].

Food laws have to be taken into account when nonwovens treated with superabsorbents are used to transport fish in ice or for packaging poultry to absorb any water or blood coming out of them [28].

Agriculture laws have to be observed when nonwovens are used as seed tapes for encapsulated seeds in combination with superabsorbent powders and fertilizers to improve their growth.

### 8.4.10 Dust-bonding treatment

Dust-bonding finishes are important primarily in the technical nonwovens sector. The high dust-bonding property of nonwovens, especially those made from microfibres, is achieved in an almost ideal manner by the labyrinthine structure of the fibres. As long as no cleaning as is required, as is the case with dust filters in the cement industry, for example, and the dust particles are to remain bonded onto the nonwoven for as long as possible, the application of special dust-bonding substances is recommended. These can be either hydrophilic or hydrophobic.

An interesting end-use for such finishes are dust-bonding mats found everywhere in the entrances of public buildings. An oil-containing finish on the mats binds dust that is stuck to the soles of shoes and holds it securely like a magnet. If these mats become sufficiently soiled, they are washed or cleaned and refinished with the "dust catcher" for re-use.

Nonwovens cleaning cloths are finished by the same principle for use as dusters. By combining antistatics, hygroscopics and paraffin emulsions as lustring agents in a full bath impregnation process, effects are obtained that are far superior to those of normal napped dusters.

Filters for the passenger area of motor vehicles consist of statically charged nonwovens made from microfibres. They provide protection from pollen and, to a large extent, from diesel exhaust fumes. If pouches containing activated carbon are incorporated, it also becomes possible to neutralize smells.

**8.5**
**Coating**

A fundamental and extremely important finishing method for nonwovens is coating, the aim of which is to bring about a single or double-sided visual, hand or technical alteration in the substrate or the whole composite. The process depends on the substrate, the coating equipment available, the coating material and, not least, the desired effect.

If one considers, first of all, the coated surface, a distinction can be made between coatings that cover the whole surface and those that cover only part of the surface of the nonwoven. As examples, level foam coatings on floor coverings or imitation leather coatings are classed as the first and heat-sealing dot coatings on fusible interlinings as the second.

Coating of the whole surface is achieved by either direct or indirect methods that work with direct contact with the substrate or that allow non-contact application.

**8.5.1**
**Coating methods**

**8.5.1.1 Kiss roll coating**
One of the best known direct application methods is "kiss roll coating". This refers to application by means of a rotating roller, which has the product to be applied on its surface. This kiss roller is either fed directly with the coating liquor, whereby it is immersed directly into the auxiliary or it is covered with the coating liquor via special feed rollers (Fig. 8-4).

The first case is referred to as direct coating, the second, reverse roll coating. The kiss roller rotates either in the running direction of the substrate or in the opposite direction. If they rotate in the same direction, it is conceivable that the peripheral speed of the roller may be greater than the speed of the nonwoven or it may be slower. Even so, a synchronous running speed is ultimately possible.

In practice, it is common to operate the kiss rollers alternatively in both directions. In this way – with the same liquor concentration – the add-on and the penetration

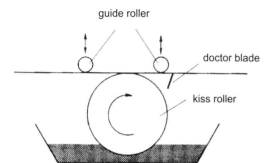

**Fig. 8-4** Diagram of a kiss roll mechanism (according to [1])

depth can be altered substantially. When the kiss roller rotates in the same direction, the liquor normally penetrates more deeply into the substrate than if it rotates in the opposite direction. The add-on is dependent on the speed of rotation. The result can also be varied with the same solid content of the liquor by altering the viscosity. Another variation is possible in so-called hollow application by altering the angle of wrap with which the nonwoven touches the kiss roller.

The application of coating materials by printing rollers can be understood as kiss roll coating the fabric face. It is almost exclusively limited in rotogravure printing to the transfer of so-called top or final coats, also sometimes called varnishing. It is carried out using low-viscosity dispersions or solutions.

### 8.5.1.2 Rotogravure coating

In the last paragraph, the use of so-called stippling rollers for applying a coating material over the whole area was described. However, it is also possible to produce patterned coatings with rotogravure rollers. This is often carried out with heated rollers that are used to apply hot melts, that is, fused thermoplastics. In most cases, the adhesives are intended to form the bond with a second fabric. If it is fed in immediately, it is referred to as laminating (see Section 8.6) and the composite material is called a laminate. However, it is also possible to leave the hot melt as a coating after it has cooled down and to carry out the bonding (laminating) with a second fabric layer – after reactivating the adhesive.

An interesting variation on this method is the powder dot process for coating fusible interlinings (see Section 8.5.2.8).

### 8.5.1.3 Rotary screen coating

For applying hot-sealing coatings to fusible nonwoven interlinings, the so-called paste dot method on rotary screen printing machines has gained far greater importance as it allows greater working widths and higher production rates than the powder dot method.

The process uses both a doctor and magnetic roller squeegee system.

Both aqueous and non-aqueous pastes (plastisols) can be applied by rotary screen coating. Under certain conditions, powder application is also possible if the powders to be applied are sufficiently free flowing and their particle size is in an appropriate ratio to the diameter of the holes in the perforated screen. Here, powders that are as spherical as possible with a particle size of 80–200 μ can be processed successfully. The single inner doctor normally used for pastes is replaced for powder application by a double doctor. As with normal cylinder printing, during rotary screen printing for this purpose, 9–35 mesh or so-called pixel screens are used.

As well as patterned coatings, finishing of the whole surface can also be carried out with rotary screens. For this, particularly fine-meshed screens with more than 60 holes per linear inch are used. Depending on the hole diameter and wall thickness of the screen and the solid content of the coating material, add-ons from a few grams per square metre (minimum add-on) to over 100 $g/m^2$ are possible.

### 8.5.1.4 **Knife or doctor coating**

The traditional method of coating involves spreading by means of doctor blade mechanisms. The different types are knife-over-roll, blanket and air knife coating (Fig. 8-5).

It should be noted that, during knife or doctor coating, the coating substrate should be kept under longitudinal tension and therefore only nonwovens with sufficient dimensional stability can be knife coated. This includes heavy and densely needle-punched nonwovens, strong lightweight nonwovens such as, for example, longitudinally oriented nonwovens produced by the Eisenhut process, or wet-laid glass fibre nonwovens.

Normally, high viscosity, aqueous pastes, solutions or plastisols are knife coated, but powdery coating materials can be applied by the doctor method with surprising evenness. Decisive factors here include the type of knife coating process chosen, the positioning of the knife in relation to the substrate and finally its shape.

Particularly popular is the knife coating of foamed coating materials because both light and heavy coatings can be applied with minimum tensile loading of the substrate.

Typically, nonwovens can be foam coated with natural or synthetic latex dispersions, for example. Additional foaming agents such as ammonium stearate, potassium oleate or sulphosuccinamate enable the mixture to be foamed to a density of $800-200 \text{ g/dm}^3$, corresponding to 1.2 to 5 times the original volume. Foaming can be continuous or discontinuous. The latter is carried out by means of an eccentric whisk. It is especially effective when it carries out a planetary motion. In continuous foaming, the latex mixture is foamed together with air by special foam mixers between discs set with truncated pyramids that function as stator and rotor. It is possible to add coagulants or sensitizing agents or even dye solutions to the blender. The foam formed is conveyed onto the substrate via a nozzle, which, in the case of wide fabrics, traverses across the web. The foam is then spread by the knife, coagulated or predried under infrared emitters and then vulcanized. With regard to the equipment, a dwell time of 30–120 s in the infrared field with a radiator power of $10 \text{ kW/m}^2$ is standard and for vulcanization 10–15 min at 140–160 °C [29].

The discontinuous production of latex foams by means of blowing agents (yeast and hydrogen peroxide) is no longer of any technical significance.

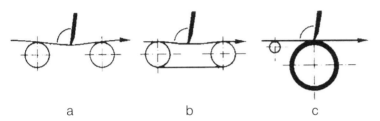

a           b           c

**Fig. 8-5** Diagram of doctor blade mechanisms (according to [1])
a) air knife, b) rubber blanket spreader, c) knife-over-roll

Conversely, with PVC foams it is a different story. Here, chemical foam is as important as foam produced mechanically. Polyvinyl chloride pastes containing blowing agents are, without exception, easier to process. The only prerequisite is that the blowing agent should be mixed in well. However, as polyvinyl chloride pastes are themselves ground via roller frames or stirred at high speeds in dissolvers to distribute the individual components evenly within the mixture, in principle there is no problem mixing in the blowing agent (usually diazo dicarbonamide) as well. Furthermore, stabilizing agents also contained in the paste serve to boost the effect of the foaming agent, making the whole process even simpler and more reliable.

Polyvinyl chloride plastisols are usually spread with the knife-over-roll and subsequently – depending on the composition of the mixture – gelatinized at 135–220 °C. The main end-use sector for coatings are PVC imitation leather, floor coverings and carpet backings, to which we will come back later.

An interesting variation on foam coating is so-called crushed foam. Aqueous plastic dispersions, preferably based on polyacrylates and/or fully crosslinked aliphatic polyurethanes are used with a foaming agent and ammonium stearate as a stabilizer as well as a thickener to adjust the viscosity.

After foaming to a foam volume of 80–400 g/dm$^3$, it is spread onto the substrate in the usual way and predried at 100 °C, without any noticeable loss in volume. This first foam coating is then calendered (crushed) and further foam coating layers can then be added by the same method. After the last application, drying and calendering, crosslinking is carried out at 140–160 °C.

If just one coat is applied, this method can be used to adjust the porosity of needle-punched nonwovens for paper machines or to produce a breathable adhesive layer for flock coating.

In multiple coat processes, for example using a black coloured middle coat, black-out curtains can be produced, which can then be printed or flock coated later for decoration.

The coating of nonwovens with solvent-containing products is relatively rare. An exception is polyurethanes, which are normally dissolved as bicomponent systems from isocyanates and diols in ethyl acetate or applied as single-component systems in dimethyl formamide. Varnishes and top coats with polyurethane varnishes or acrylic ester polymers are often applied as solutions.

### 8.5.1.5 Extrusion coating

In certain cases, coating or laminating of the whole surface (see Section 8.6) are also carried out using thermoplastics by the extrusion method by means of sheeting dies. This type of coating is relevant when impermeable barrier layers are required, for example in the hospital sector (incontinence products, surgical garments) and for packaging that is impervious to gases or smells. Although it only involves one coat, this type of finish is often referred to as a laminate [27].

### 8.5.1.6 **Non-contact coating**

Whereas with direct coating, considerable requirements are placed on the mechanical stability of the substrate, non-contact coating methods can also be carried out on less stable nonwovens and those that do not have a completely smooth, even surface.

One type of non-contact coating is spraying, as has already been mentioned in Section 8.4.1, with the difference that with coating, higher viscosity finishes, usually enriched with dyes and/or extenders, are used.

Especially economical is so-called scatter coating, which is preferred for partial applications, but is also very relevant for coating the whole surface in conjunction with a smoothing process by sintering. Thermoplastic powders containing blowing agents can be applied to very light and voluminous nonwovens particularly efficiently. By means of a subsequent smoothing or embossing process between cooled calender rollers, interesting surface effects are possible. The relevant thermoplastic powders are applied by special scattering machines, electromagnetic scattering channels or roller/brush devices with add-ons of 10–600 g/m$^2$, sintered under infrared emitters and moulded, for example by smoothing or embossing, in the thermoplastic state as they exit the heating zone.

Also included among non-contact coating methods is vapour deposition with sprayable metals in a high vacuum. With the smallest add-ons, coatings that are visually attractive and have a noticeable effect on the properties are achieved. These are sometimes used for decoration, but mostly for reflective safety wear.

### 8.5.1.7 **Release coating**

Release coating processes occupy a middle point between direct and non-contact coating methods. It is the coating of a carrier material with release properties, from which the coating materials can be transferred onto the substrate once they have set sufficiently by surface drying, gelling or the like, to form a coherent film. These kinds of coatings are highly efficient inasmuch as they remain almost entirely on the surface of the substrate with very low add-ons, resulting in a film-like covering, with only an imperceptible effect on the hand (Fig. 8-6).

Typical of this type of coating are self-adhesive coatings (see Section 8.5.2.3) and grained imitation leather coatings (see Section 7.3.2).

An interesting variation is the transfer of metallic coatings (see Section 8.5.1.6) onto nonwovens that are not suitable for direct vapour deposition in a high vacu-

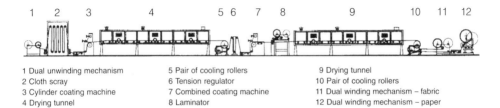

| | |
|---|---|
| 1 Dual unwinding mechanism | 5 Pair of cooling rollers |
| 2 Cloth scray | 6 Tension regulator |
| 3 Cylinder coating machine | 7 Combined coating machine |
| 4 Drying tunnel | 8 Laminator |

| |
|---|
| 9 Drying tunnel |
| 10 Pair of cooling rollers |
| 11 Dual winding mechanism – fabric |
| 12 Dual winding mechanism – paper |

**Fig. 8-6**  Tandem transfer coating line according to [1] (courtesy of Mohr)

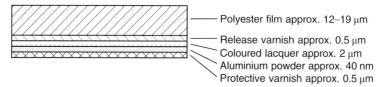

Polyester film approx. 12–19 μm
Release varnish approx. 0.5 μm
Coloured lacquer approx. 2 μm
Aluminium powder approx. 40 nm
Protective varnish approx. 0.5 μm

**Fig. 8-7**  Composition of a transfer film (not to scale) (according to [15])

um. For this, polyester film approximately 15 μm thick is used, which is first coated with a release agent around 0.5 μm thick and, if necessary, also with a coloured lacquer around 2 μm thick. The prepared film is coated with a layer of aluminium around 40 nm thick in a high vacuum by vapour deposition and this is in turn covered with a protective varnish 0.5 μm thick (see Fig. 8-7).

The nonwoven to be metallized is coated with a heat-reactivatable adhesive over the whole surface or printed according to a pattern and an intermediate drying process is then carried out.

On a transfer calender, in a similar way to transfer printing, the metallic layer is transferred from the film to the adhesive, thereby fixing it onto the nonwoven with good resistance to washing and dry cleaning. This finish is also used for decoration purposes and reflective safety wear.

## 8.5.2
## Coating effects

As has already been emphasized several times, the same coating method can produce very different effects, according to the coating material used and depending on the coating substrate. Conversely, the same technical effects can be produced by different methods. Some typical coating effects are therefore described below, although the selection made does not claim to be complete.

### 8.5.2.1  Non-slip finishing

A good example of the same coating effect being achieved by different methods is the non-slip finishing of floor coverings made from needle-punched nonwovens or carpet underlays with latex dispersions. A good effect is achieved, for example, by spraying a lightweight, needle-punched polyamide nonwoven with a butadiene styrene or butadiene acrylonitrile latex. The individual drops of latex anchor themselves onto the surface of the nonwoven and onto the fibres, thereby giving a large number of dull retarding points which, as an underlay, prevent a carpet from sliding on the floor.

The same effect is achieved by kiss roll coating with the same latex. The difference is that a considerably larger quantity has to be applied because some of the slip-resistant coating material sinks into the inside of the web and its surface effect is lost even if, as is normal in kiss roll coating, the coating material has been further thickened. Even more costly for the same effect is knife coating, as here the coating material is applied to the substrate from above and, due to gravity, sinks even deeper into the substrate than with kiss roll coating, which is normally carried out from below. Localized printing, on the other hand, is more economical because the thickened coating material can be placed very precisely on the reverse, for example in the form of small slip-resistant semicircles, without some of the coating material being needlessly sprayed and extracted (spraying loss), as happens with spraying.

### 8.5.2.2 Mouldable coating
A similar situation occurs with the mouldable finishing of carpets for cars. For a long time it was usual to reinforce mouldable needle-punched nonwovens for floor coverings in the traditional way by full bath impregnation and to make it mouldable by scattering on, sintering and smoothing 400–600 $g/m^2$ polyethylene powder. These two processes can be reduced to one if a high-styrene-content butadiene styrene dispersion is kiss roll coated onto the back of the nonwoven with an add-on of approximately 500–600 $g/m^2$ dry add-on and dried below the film forming temperature. When it is reactivated, which is necessary both for the polyethylene coating and also the kiss roll coating described here, the particles in the dispersion, which initially lie separately next to one another, fuse together during the subsequent cold moulding to produce a closed film, giving practically the same dimensional stability as a polyethylene coating.

### 8.5.2.3 Self-adhesive coating
Contact adhesive finishing can also be carried out by several methods.

If only a weak adhesive effect is required, a light spraying with a contact adhesive, which may be in the form of a solution or an aqueous dispersion, is often enough. One potential end-use is nonwovens as linings for trays to prevent the dishes from slipping (use in aeroplanes and other modes of transport where food is served). Another possibility is flow coating, in which a fine curtain of the coating product runs off a sloping surface as a formed film and evenly covers the substrate passing underneath. This method can be used, for example, to finish self-adhesive needle-punched nonwoven floor tiles.

A variation of this is "localized" flow coating, which is used for the same purpose, namely self-adhesive floor tiles. The contact adhesive, here in the form of a dispersion, is expressed through the holes in a horizontal pipe located above the needle-punched nonwoven passing underneath and flows as a fine stream onto the nonwoven, producing a striped coating that is much cheaper compared to coating the whole surface. If aqueous adhesives are used in this process, then the nonwoven has to be dried after coating, but if a hot melt is used as the adhesive, as is the case with the Resimelt process, for example, drying is replaced by cooling, which can be carried out either as air or contact cooling over cooled rollers.

Another method for the striped application of contact adhesive dispersions also found in practice involves knife coating, but instead of a smooth-edged blade, a kind of saw-tooth blade with notches is used.

Especially efficient is release coating, which can be made even more economical for contact adhesive coating when the adhesive applied to a release paper is the same one that would have been necessary anyway to cover the contact adhesive layer. With a smooth and completely level paper surface, the adhesive can be applied with a defined layer thickness and without sinking in. As it leaves the drying zone, the coated paper is calendered together with the nonwoven, causing the adhesive to bond firmly to it so that during later use the release paper can be detached easily. In the transfer process, the adhesive can be formulated as a solution, dispersion or a hot melt adhesive.

Another end-use sector for self-adhesive nonwovens is so-called double-sided adhesive tapes for laying textile floorings, consisting of a nonwoven coated on both sides with contact adhesive, which is rolled up together with a paper coated on both sides with silicon and usually sold as tapes or rolls of different widths.

Large amounts of adhesive are necessary for double-sided adhesive tapes and therefore here too the hot melt process, which is based entirely on contact adhesives, is used.

The basis for contact adhesives consists of polyisoprene or polyisobutylene, polyvinyl acetate plasticized by external or internal plasticization or acrylic esters with relatively high ($C_8$-) alcohols. Non-vulcanized natural rubber, used as a solution or dispersion, produces an excellent adhesive effect, although often with insufficient stability to light and ageing. For self-adhesive hot melt adhesives, EVA resins with a high vinyl acetate content are very popular and have a strong adhesive effect, especially when they are mixed with tackifiers, that is tackifying resins based on hydrocarbon resins, for example.

A particularly important end-use for contact adhesive finishes is sticking plasters and surgical dressings. Naturally, only contact adhesives that are compatible with the skin and physiologically harmless should be used to coat them [32].

### 8.5.2.4 Foam coating

If we look again at the foam coating of needle-punched nonwoven floor coverings, there are certain distinctions with regard to the individual processes and chemicals that should be pointed out in particular.

### Latex foam coating

Continually produced latex foam can be applied in different forms. So-called gelable foam is preferable. The foamed natural or SBR latex mixture is mixed in the blender of the foam mixing machine with a latent acid donor, for example sodium silicofluoride or ammonium acetate. It then undergoes the shock-like infrared gelling process, resulting in a very compact, resilient and abrasion-resistant foam that has a uniform pore structure even at a thickness of several millimetres.

Furthermore this foam is not hydrophilic. Disadvantages of gel foam are the relatively low degree of expansion (max. 75%), the need for sensitizers and the resulting limited pot life.

Non-gel foam has no need for an additional gelling agent, enables degrees of expansion of 150–200%, but the final foam is hydrophilic due to the high "soap" content and therefore requires an additional water repellent in the form of a paraffin emulsion. It has a kind of "skin" on the surface with a very fine pore structure (integral skin foam).

Resin-crosslinked foam occupies the middle point. Here, a carboxylated SBR latex produces a uniformly tough foam backing with excellent resistance to ageing without vulcanizing agents by adding melamine resin precondensates.

Another foam coating process, but with a completely different end result, is the finishing of imitation chamois leathers. The substrate is a carded, cross-laid and lightly pre-needled web made partly from hydrophilic natural or regenerated fibres and partly from highly water-resistant synthetic fibres. The weight per unit area is around $100 \, g/m^2$. The web is full-bath impregnated with a foam mixture and squeezed between two rollers with a set clearance, giving a wet add-on of several hundred percent in relation to the web weight. The impregnating liquor contains a mixture of natural and butadiene styrene latexes. In cases where a specific oil-resistance is required, a nitrile latex is used; as well as suitable vulcanizing agents and anti-agers, a foaming agent, pore-forming substances and often also coagulants, which make the foam coagulate at the applied height when a certain temperature is reached during drying.

These mixtures are foamed to a density of approximately $300 \, g/dm^3$, coated in a full bath – or by knife coating on both sides with intermediate drying –, coagulated, dried and vulcanized. The foaming and pore-forming agent, the coagulants and unused vulcanization agent are then washed off and the web is dried again. Finally, the web is cut lengthwise and widthwise, resulting in a soft, absorbent, wash and boil-resistant, non-fuzzing, wringable chamois leather. The final washing process is often dispensed with and the consumer is advised to do this as a necessary first treatment on the packaging. If the washing-off process is omitted, unwanted stripes occur on the dry leathers.

If certain selected polyurethane dispersions are used instead of the latexes mentioned, the result is an even more leather-like hand and the "squeak" typical of leather when used for wiping.

The absorption can be increased by needling a polyurethane foam several millimetres thick with cellulosic fibres instead of using a normal web substrate and then, as described, impregnating it with the latex foam liquor, squeezing, drying and press finishing. The mechanical strength of a household cloth produced in this way is nevertheless lower than the one described previously.

## Polyvinyl chloride foam coating

Similar variations exist for polyvinyl chloride foam as with latex foam. Regardless of the different production methods (chemical or mechanical foam), it is above all considerably heavier than a latex foam expanded to a similar degree.

Because it is non-aqueous, with the same add-on it dries more quickly (or to be more precise: gels) than latex. This means a higher production speed or shorter driers. On the other hand, polyvinyl chloride foam usually also needs higher temperatures that are not compatible with all fibres. Polypropylene needle-punched nonwovens can also be coated with polyvinyl chloride, although in this case more easily gelling polyvinyl chloride copolymers have to be used, which are more expensive than basic polymers.

PVC-coated nonwovens are often used as imitation leather for bags, but are preferred for floor coverings instead of linoleum.

In the simplest case, needle-punched jute webs were used, which were first of all reinforced in a full bath with heat-resistant elastomer dispersions. Onto these, a relatively highly expanded polyvinyl chloride base coat was knife coated and gelled. They were then calendered and/or embossed and the indentations were filled with a differently coloured polyvinyl chloride paste, wiped off and gelled again. After this they were coated with a colourless unexpanded polyvinyl chloride top layer and finally gelled. An alternative to embossing was printing with polyvinyl chloride dyes, in which certain patterns recurring in the repeat, such as the very popular wooden parquet, were imitated surprisingly well. After printing, it was covered with a particularly abrasion-resistant transparent polyvinyl chloride coating.

This floor covering, which was manufactured easily and reliably on continuous lines, nevertheless had the serious disadvantage of a certain susceptibility to microbes and fungi when it was laid on a floor that was not completely dry due to the jute (or linen waste) substrate. Laying it in damp rooms was inconceivable.

It was therefore a considerable advance when, instead of jute nonwovens, ones made from asbestos were used. However, since it became known that there is risk of lung cancer during the processing of asbestos fibres, work began on the development of more suitable heat-resistant bonded glass fibre nonwovens on which to base a new generation of polyvinyl chloride floor coverings [33].

An interesting variation is cushioned vinyl coverings made up of a combination of a solid polyvinyl chloride bottom layer, a polyvinyl chloride foam layer that has been embossed or moulded by inhibition and printed and a transparent polyvinyl chloride covering layer. The glass fibre web can be used here either as the substrate for the compact base coat or the polyvinyl chloride foam layer.

The decision to use glass fibre nonwovens for this type of floor covering can be explained by a range of marked advantages of this material, which nevertheless place extremely high demands on its manufacture. These include a relatively low weight per unit area with a high density and uniformity, whereby weight fluctuations of only ±5% compared to a given average are permitted. The binder component usually accounts for 20% of the fibre weight. It should also only fluctuate within very narrow limits as otherwise effects on the web strength and the polyvinyl chloride are unavoidable.

A big advantage of this type of reinforced glass fibre web is their high temperature resistance. This is indispensable due to the high temperatures to which they are subjected several times during gelling of the various, usually three, polyvinyl chloride coats and which affect not only the fibres, but also the binders. Another important advantage is their good dimensional stability, which is excellent compared to other nonwoven substrates. Others also include a tensile strength of at least 50 N per 5 cm strip width, an excellent resistance to plasticizers, a minimum amount of volatile components under production conditions of up to 230 °C, and of course good flatness, as polyvinyl chloride floor coverings are manufactured up to a width of 4 m.

The importance of fulfilling all of these criteria is made clear by the fact that, during production of these coverings, all the processes, that is coating several times, printing, embossing and so forth, are carried out continuously on a single line. It only becomes evident in the finished product after around 300 m of running production whether a faulty substrate was supplied at the feed end. The large quantities of polyvinyl chloride floor coverings and especially cushioned vinyl floor coverings produced show that the requirements stipulated have been fulfilled to a very large extent.

**Polyurethane foam coating**

Finally, polyurethane plain or embossed foam should not go unmentioned. Although it involves relatively high chemical costs, it is applied without expensive driers, that is without high energy costs and without long drying tunnels by means of simple machines and produces remarkable technical effects.

While with the Bayer and ICI processes the polyurethane components are foamed "in situ", in the Dow-Chemical process the mixture is foamed mechanically and applied to the nonwoven backing in the same way as a latex foam [34, 35].

### 8.5.2.5 Loose lay coating

Another type of coating for floor coverings is loose lay coating found mainly on square floor tiles. As a heavy-duty backing, it is based on highly expanded finishing pastes. Up to 600% expandable and with a solid content of between 75–80%, latex blends (SBR latex) are suitable. Not quite so expandable, but applied without water and with a higher specific weight are polyvinyl chloride coatings.

Another heavy-duty backing involves formulations based on bitumen and atactic polypropylene. As well as good value for money, they have the advantage that, without requiring a drying or gelling process, they can be poured on as a hot melt and only need time or a suitable machine to cool down before they are ready. As the preparation of the coating material requires special machinery, this process is only carried out by a few manufacturers [45].

### 8.5.2.6 **Microporous coating**

The desire to imitate or – where possible – to improve on a lightweight garment leather or shoe upper leather and to supply it to the garment manufacturer as uniform roll goods instead of pieces of different sizes and thicknesses has led to many different synthetic leathers. The well-known fibrillation of sheath/core fibres will not be discussed here because it has now largely been replaced by the direct production of microfibres of less than 1 dtex.

It is only finishing (coating) processes that are comparable with leather in hand and porosity that are of interest here [36].

It began with the treatment of nonwovens with polyurethane dissolved in dimethyl formamide. After immersion and squeezing, the impregnated nonwoven web was successively diluted by passing it through various water baths until the dimethyl formamide, which can be mixed with water in any ratio, was no longer able to keep the polyurethane in solution. It therefore coagulated in situ, thus forming a film with an immense number of micropores, which were water-vapour permeable but at the same time waterproof. After drying, the treated nonwoven was usually emerized or suede finished. As dimethyl formamide poses a health risk, the whole finishing process has to be carried out on closed machines and all the dimethyl formamide has to be recovered from the water by distillation.

A variation on this process involves pre-emulsifying the dissolved polyurethane in water. The viscose emulsion is knife coated on one side. During drying, the solvent evaporates, causing the polyurethane layer to coagulate as described above. Once the water then evaporates, a polyurethane film with a large number of micropores again remains on the nonwoven. Here too, the solvent has to be recovered by condensation or eliminated by afterburning.

When ionic coagulating polyurethane dispersions are used, there is no need for solvents and finishing can be carried out on conventional coating lines. After impregnating or coating, the applied polyurethane dispersion is coagulated by means of a solution containing acid or opposed ions, the excess water is squeezed or suctioned off and the fabric is deacidified, rinsed and dried. In contrast to the first two processes, in which coagulation proceeds very slowly as the solvent becomes progressively more dilute, in the case of ionic coagulation, it occurs spontaneously, resulting in larger pores, some of which have to be closed again after drying by calendering [37].

### 8.5.2.7 **Drainage coating**

Another type of coating is drainage coating of synthetic lawns, which can be understood as a type of dot coating. It is carried out by means of thick-walled rotary screens with large holes. As the coating material, either thermoplastic powders, polyvinyl chloride plastisols or expanded dispersions or latexes are used. This finishing method is particularly suitable for backcoating outdoor floor tiles that are used to cover terraces, balconies and swimming pool surrounds and guarantee that the covering dries rapidly after becoming wet through.

### 8.5.2.8 Heat-sealing coating

A substantial proportion of nonwovens not designed for disposable products go into the making-up industry as interlining materials. Whereas nonwoven interlinings were originally found only in the ladies' outerwear sector, their properties have now become so diverse and have been improved so much that they have also entered the men's and boys' wear sector with equal status alongside conventional woven interlining materials [38].

**Properties and possible applications**

A large proportion of nonwoven interlinings are used as fusible interlinings. These are interlinings that are covered with a thermoplastic coating and bonded permanently to the face fabric by ironing. This direct bonding to the face fabric has revolutionized garment manufacture and therefore a rather more detailed section should be devoted to fusible interlinings in this chapter.

Interlining materials, the best known being hair linings, were originally only joined to the face fabric at the edges and with a few pick stitches in the middle. They were relatively heavy and were also stiffened, giving the garment its good shape. By bonding the interlining directly onto the shell, a laminate is produced which, in a similar way to multilayer glueing (as in plywood), is much more elastic and dimensionally stable for the same thickness and weight than a corresponding single-layer fabric or a multilayer fabric that is not firmly bonded. In this way the dimensional stability of even light and very lightweight face fabrics can be fixed with fusible interlinings.

The idea of bonding interlinings to the face fabric instead of sewing (blind-stitching) them goes back to the second decade of the last century. However, success was not forthcoming because the adhesives available did not match the requirements of garment manufacturers or consumers. This only changed from the middle of the century, when sufficiently resistant thermoplastics and machines suitable for applying them came onto the market.

**Coating materials** [39, 43]

As well as the actual thermoplasticity, which should be within a reasonable temperature range for all current textile fibres, for coating, adequate fusion is also expected from these products. This means that they should flow far enough during the bonding process, which lasts just a few seconds, to be able to cling sufficiently firmly to the interlining and the face fabric. However, the melt viscosity should not be so low that the adhesive penetrates through the face fabric or back through the interlining material. Requirements include the softest hand possible, good resistance to washing, dry cleaning and ageing and, not least, good adhesion to the extent that the nonwoven interlining would normally split first.

One of the first thermoplastics for this purpose was polyethylene, which fulfils many of the stipulated requirements with good value for money. Better products have been introduced and it has been replaced for certain purposes, but even today, it is still preferred as a heat-sealing coating in the lingerie industry and for fixing small parts. In outerwear and particularly for so-called full front fusing, it has been superseded by ternary copolyamides [40], which offer better adhesion, a softer hand and practically unlimited resistance to dry cleaning, although at a higher price. Polyvinyl chloride copolymers in plastisol form occupy a middle point. With adequate resistance to washing and dry cleaning, they excel above all due to their adhesion to siliconized face fabrics. However, they do not have such a wide fusing plateau as copolyamides and because of the add-on, which is almost double that of copolyamides, they demonstrate an unmistakable tendency to penetrate back through the interlining, especially with lightweight nonwovens.

Other important thermoplastics are copolyesters, polyurethane and finally EVA copolymers and polyvinyl acetate, which are useful because of their low melting point and because of their favourable price for bonding to temperature-sensitive face materials, for example leather in the shoe and bag industry.

**Form of supply and use**

Polyethylene, copolyamide, copolyester, polyurethane, polyvinyl acetate and also EVA resins were originally available as granules and had to be converted into free flowing powder by grinding, in some cases with the addition of liquid nitrogen. These thermoplastics, preferably with a particle size of 100–400 µm, were applied to the nonwoven interlining by means of scattering devices and bonded to the substrate by fusing, usually in infrared heated ovens. In some cases, they were then smoothed by a pair of cooled calender rollers immediately after exiting the sintering zone, fixing them more firmly onto the nonwoven to prevent peeling. The normal add-ons were between 15–30 for polyethylene, 8–15 for copolyamides, copolyesters and polyurethanes and between 30–60 g/m$^2$ for EVA copolymers, or in special cases (see Section 8.4.2.1) up to 250 g/m$^2$.

Polyvinyl chloride copolymers can only be processed when pasted with an equivalent quantity of monomer or polymer plasticizer. They are applied as a plastisol with an add-on of 20–30 g/m$^2$ on rotary screen printing machines through 9–30 mesh or pixel screens and fixed onto the nonwoven by pre-gelling. It became apparent that monomer plasticizers migrated to some extent into the binder of the nonwoven, reducing their strength and leading to inferior sealing. For this reason, only non- or minimally migrating polymer plasticizers are now used in polyvinyl chloride heat-sealing compounds.

The possibility of dot coating with plastisols, which is much more uniform compared to the scatter method, led to investigations into the possibility of applying other thermoplastic powders in dot formation.

For woven and knitted interlining materials, this was achieved using machines known as "powder calenders". On these, the thermoplastic powder, preferably with a particle size of 50–200 µm, is spread into the recesses of a tempered cup roller via a funnel shaped like a doctor blade and transferred under pressure onto the substrate that has been preheated to well over the melting point of the powder. The final bonding of the small heaps of powder applied in this way is carried out by another heater roller or in an infrared field.

This method has not achieved any great significance with regard to nonwovens because the high roller temperature is only compatible with a small proportion of appropriately constructed nonwovens and the production rate of 25–35 m/min is much slower than the printing speeds of rotary screen printing machines.

All the thermoplastics currently in use can be printed with rotary screens if they are processed as very fine powders with a particle size of up to 80 µm into stable aqueous pastes. Here, suitable dispersing agents, protective colloids and thickeners are necessary. Since high viscosity thickeners with a low solids content have been available on the market, it has been possible to produce pastes that can be printed on machines up to 4 m wide at production speeds of over 60 m/min [44].

This "paste dot process" offers another important advantage compared to scatter or powder dot methods. Special plasticizers for copolyamide and copolyester, particularly important as heat-sealing compounds, can be added to the pastes to vary their melting and flow behaviour within certain limits, making it possible to a very large extent to tailor the fusible interlinings to suit the face fabric and the requirements of the processor.

In the "double dot method", pastes of different compositions can be printed one on top of another in a single process with the same screen by means of a specially designed feed system to the screen (Fig. 8-8).

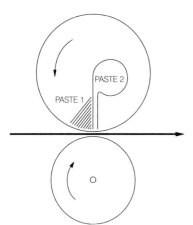

**Fig. 8-8** Principle of double dot coating by the rotary printing method (according to [15])

Another variation on the double dot method consists of printing a base coat of a non-thermoplastic plastic onto the nonwoven by means of a rotary screen. Immediately after this, a thermoplastic powder is scattered on and bonds with the still wet base coat. The unbonded powder is suctioned off and re-used. Base coat and powder are then dried in the usual way and lightly calendered on exiting the machine (see Fig. 8-9).

The advantage of this method is that it is possible to prevent the thermoplastic from penetrating back through the interlining or through the substrate during later sealing. Another advantage is that the quantity of the normally expensive thermoplastics used can be reduced in this way without impairing their adhesive properties, as they cannot sink unused into the substrate, but bond particularly firmly to the face fabric during later sealing.

This coating method rules out any further finishing of the fusible interlining.

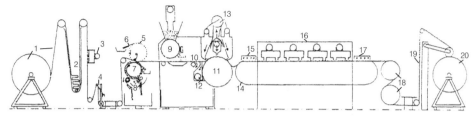

| | |
|---|---|
| 1 Roll of uncoated interlining | 11 Gripping feed roller with slack fabric |
| 2 Cloth scray | forwarding |
| 3 Lint removal device | 12 Suction device |
| 4 Floating regulator | 13 Blowing and suction device |
| 5 Screen, slightly offset to 7 | 14 Conveyor belt |
| 6 Exterior doctor blade | 15 Infrared emitters, drier entrance |
| 7 Rubberized bottom roller | 16 Drier |
| 8 Bottom roller cleaning device | 17 Infrared emitters, drier exit |
| 9 Powder scattering mechanism | 18 Cooling rollers |
| 10 Beater roller | 19 Fabric inspection |
| | 20 Winding of rolls with coated interlining |

**Fig. 8-9**   Diagram of a dot and scatter machine (according to [38])

### Processing of fusible nonwoven interlinings

Coating becomes dry laminating (see Section 8.6.2) when the garment manufacturer irons the nonwoven interlining onto the face fabric by means of heat and pressure. This happens either on discontinuous boards or continuous through-feed presses. As long as the heat-sealing compounds used have a sufficiently high dielectric loss factor, sealing can also be carried out on high-frequency welding presses in which up to ten layers can be welded simultaneously in one operation. However, this process has not achieved any great significance.

There are almost no outerwear garments that do not contain a heat-sealing nonwoven piece somewhere. As well as front fusing, fusible nonwovens are used to reinforce button strips, lapels, slits and pocket flaps. Large quantities of heat-sealing nonwovens are also used in the shoe and bag industry [41].

## 8.6
## Laminating

Laminating is the permanent joining of two or more prefabricated sheet materials. If none of these materials develops adhesive properties under certain conditions, an additional medium is required to bond them together. In general, this type of medium can be termed an adhesive. In practice, the distinction is made between wet and dry laminating. In the first case, the adhesive medium has been temporarily combined with an accompanying substance that serves as a vehicle for the adhesive. With dry laminating, it is possible to dispense with this.

8.6.1
**Wet laminating**

Adhesives used for wet laminating are either dispersed or dissolved in a suitable solvent. There are many adhesives that are suitable for bonding nonwovens to other fabrics or sheet materials. It is not possible to list them all in this section. Particular aspects will therefore be discussed.

The simplest form of wet laminating consists of applying the adhesive to one of the webs to be joined using suitable methods and equipment, adding the second web – normally under gentle or stronger pressure – and drying, hardening or condensing the composite material.

In the case of adhesives dissolved in solvents [46] – these are normally macromolecular, natural or synthetic substances – a significant advantage is the rapid drying of the solvent and therefore a rapid adhesion between the sheet materials. They also have the benefit in many cases of preventing swelling and associated dimensional change phenomena. Disadvantages are the normally low solid content, which has its limitations primarily in achieving a good processing viscosity and the fact that, for environmental reasons, solvents not only have to be evaporated, but also expensively recovered for recycling or afterburning (for heating purposes for drying equipment and/or manufacturing plants).

In the case of aqueous adhesives, the secondary medium is water, which is available cheaply and can be eliminated without polluting the environment. Aqueous adhesives can also be manufactured with a relatively high solid content and easily adjusted to the desired viscosity with appropriate thickeners. It is important that the adhesive remains in the joint and does not penetrate unnecessarily into the webs to be bonded where it is lost to the adhesion process and can lead to unwanted stiffening.

Disadvantages of aqueous adhesives are the difficulty evaporating water and the high condensation temperatures necessary for some dispersion adhesives to achieve adequate fastness properties.

Solvent-containing and aqueous adhesives can only be used when, after joining the webs, the solvent or water can evaporated and escape through at least one of the webs.

Wet adhesives are applied by spraying, kiss roll coating, printing or knife coating. The webs are joined on a squeezer or laminator consisting of two rollers under pressure. Drying is carried out in hot-air driers or sometimes on cylinder drying machines.

8.6.2
**Dry laminating**

For dry laminating, all types of thermoplastics are used in the form of powders, plastisols or hot melt adhesives and are applied to one of the substrates to be joined using appropriate machines. Compared to wet adhesives, they have the advantage of being 100% pure, hardly sinking into the substrate at all and therefore giving the laminate a soft hand. However, they sometimes require special machines for their application and are only suitable for laminates that can tolerate sufficiently high temperatures.

For powders, the aforementioned scattering machines or powder calenders are used, for plastisols, knife coating or printing equipment and for hot melt adhesives or granules, heated spray, flow or roll coating machines. Occasionally, especially for thermoplastic granules, extrusion coating machines with sheeting dies are used [45].

The coated web must be joined to the second web while the adhesive is sill molten, that is tacky. This can happen either immediately after applying and fixing (sintering) the adhesive onto the first web or later, after the coating, which has cooled down and become inactive in the meantime, has been reactivated. The webs are joined under pressure, usually with simultaneous cooling in order to block the thermoplastics.

A variation somewhere between wet and dry laminating is when the laminating adhesive is initially applied wet onto one of the webs (or moulded articles), followed by intermediate drying. It is then joined to the other piece under heat and pressure, for example on shaping presses, as is normally the case with highly moulded headlinings or side panels.

Finally, flame laminating should be mentioned. Here, a thin sheet of polyurethane foam is burnt on the surface by gas burners, becomes molten and is immediately bonded to the nonwoven web with simultaneously cooling. In this way it is possible to produce single- and double-sided laminates in one operation or on one flame-laminating machine. Due to the environmental pollution caused by this process, the machine has to be enclosed and the combustion residues have to be carefully filtered out of the exhaust air.

### 8.6.2.1 Use of adhesive nonwovens

A variation of dry laminating is the use of contact adhesives and adhesive nonwovens.

The latter are manufactured by the spunbonding method from thermoplastics that melt at low temperatures. They are laid between the webs to be laminated and activated by heat and pressure, causing them to bond. This type of laminating can be characterized as particularly clean and environmentally friendly. Thus, with the *Codor process*, for example, two nonwovens or other fabrics can be bonded together quickly and permanently by means of adhesive nonwovens.

Instead of adhesive nonwovens, adhesive films or adhesive nets can also be produced for permanently bonding nonwovens to form laminates.

### 8.6.3
### Examples of laminating

As well as good adhesion, the fastness properties of the composite also play an important role. The main ones worth mentioning are resistance to washing, dry cleaning and ageing, although there are also laminates that only have temporary characteristics. The required fastness properties can be obtained, using polyurethane bicomponent adhesives or crosslinkable aqueous dispersions, for example, and in some cases by using certain thermoplastic powders such as polyamide, co-polyester and other hot melts.

Cynics claim that everything that could not defend itself would be laminated. And although nonwovens were not designed specifically for laminates, there are a whole range of them. One of the most important nonwoven laminates is the bonding of nonwoven interlinings to textiles for use in clothing. There is almost no garment that has not been fused to a nonwoven at some point.

Of similar importance are laminates of nonwovens and membranes, which guarantee an excellent barrier effect against infections and blood in the hospital sector, for example, and especially in the surgical sector.

In the sportswear and leisure sector these laminates are used under the face fabric and provide windproofing and waterproofing at the same time as good breathability [47].

Extremely important are nonwoven laminates in the construction industry. Thus, for example, glass fibre nonwovens are coated with a highly loaded latex compound in a first process in preparation for subsequent foaming with polyurethane foam (EP 015 9514). The laminate is formed in a second process. A system made up of a diisocyanate and a divalent or polyvalent alcohol and other components is applied to the coated nonwoven web using a mixing head and covered with a second web of the same nonwoven. Due to a chemical reaction, a solid polyurethane foam forms between them, the thickness of which can be between two and several centimetres, depending on the amount applied. This composite material forms an excellent insulating panel against heat or cold for use in light construction. The glass fibre nonwoven forms a smooth and decorative surface, which can also be coloured by surface coating with disperse dyes.

If a coating made from oxidizing, elastomeric or elastoplastic bitumen is used as the barrier layer, which is then sanded and/or powdered, then insulating slabs for light construction for flat roofs are obtained by the same process.

Also generally well known are pan scrubs consisting of a composite of scouring webs coated with an abrasive grain and polyurethane foams. The nonwoven used in this composite consists of a carded cross-laid or aerodynamically formed random laid web that is spray bonded. Onto the still wet spray binder, corundum or another suitable abrasive grain is scattered, the web is then dried and the reverse is treated in the same way. The material thus prepared is then passed again through an impregnating bath containing the acrylic resin used for spray bonding, thereby binding the abrasive grain more securely and providing it with the required water-resistance. The web is also simultaneously dyed during this treatment.

Quantitatively very important are nonwoven webs as secondary backings for particularly dimensionally stable floor coverings (tufted and woven carpets), whereby they also improve the resilience and insulating properties. Here, laminating is carried out wet with latex adhesives or dry, using thermoplastic powders.

Also worth mentioning are combinations of nonwovens with paper for wallpapers and with film for decorative and insulation purposes.

Also not to be forgotten are the many laminates used in the car industry, such as headlinings, hat racks, door side panels, linings for A, B and C columns and floor components, the latter also being designed to be sound insulating and vibration damping by means of additional bituminization. Here too, bitumen is used, mixed with atactic polypropylene or SBS block polymers to improve the elasticity and to modify the melting and temperature behaviour. Many of the laminates used in the car industry are produced by the dry or semi-wet laminating method for environmental reasons. Care must be taken that all the nonwovens, textile auxiliaries and binders used in this sector are non-fogging and as odourless as possible.

This list provides examples to indicate the diversity of laminates, without being even remotely exhaustive.

## 8.7
## Flocking

Flocking offers the possibility of adding a third dimension to two-dimensional fabrics.

The flocking of nonwovens does not differ fundamentally from the flocking of other textile substrates and, as with them, is carried out either mechanically or primarily by electrostatic means [48].

Short fibres of less than 1 to a maximum of 20 mm in length, assuming they are not sufficiently electrically conductive already, are coated with an appropriate finish. Zirconium salts combined with softening antistatic agents or tannin are suitable for this and are fixed onto the fibre surface by trivalent metal salts. Here, it is also advantageous to add antistatic softeners.

In the flock beamer, the fibres are charged and fly along the field lines of a strong electrostatic field in the direction of the adhesive-coated substrate where they bore vertically into the adhesive layer, giving it a plush, velvet or velours surface depending on the length and fineness of the fibres. Instead of cut short fibres, fibre dust produced by grinding can also be used, giving the nonwoven surface a peach skin effect similar to emerizing.

Flocking can be carried out over the whole surface or decoratively, according to the application of the adhesive. Similar adhesives are used as for laminating, that is polyvinyl chloride plastisols, polyurethane bicomponent adhesives and all kinds of aqueous dispersion adhesives, those that are thermally reactive or crosslinkable via precondensates giving the greatest fastness properties.

Aqueous adhesives have the advantage of assisting in the generation of the electrical field, having a relatively high solids content and being very variable in their viscosity. This means that, on one hand, they do not offer the flock too much resistance upon impact but, on the other hand, they have sufficiently high viscosity to hold the fibres that have penetrated vertically and not allow them to tilt. The disadvantage of aqueous adhesives is the intensive drying conditions.

The flocking of nonwovens achieved greater importance when pre-foamed instead of compact aqueous dispersions were used as flock adhesives, giving greater softness, flexibility and good breathability. If the foamed flock adhesive is coated onto a release paper – then flocked, dried, crushed and cured – after removing the release paper, the result is, according to the definition, an adhesive-bonded nonwoven that can be treated as one and used for many end-uses [49].

As flocking substrates, nonwovens have the advantage over other textile fabrics of having a uniform, dense surface at extremely low weights. This results in an economical flock adhesive add-on and – linked to their low weight – a high degree of flexibility.

## 8.8
## New processes and products

Finishing processes are the subject of constant research and development. These projects are often carried out with the aim of finding environmentally friendly processes and products.

Although the first applications in the textile sector occurred in the 1960 s and on an industrial scale between 1975 and 1977 [50], research into the treatment of textile substrates with low-pressure plasma [51–54] is experiencing a new, heightened level of interest. Depending on the reactive gas used, it is possible to produce surface activation and enlarging by etching, functionalization by grafting or polymerization of the fibre surface, for example water repellency and oil repellency by using methyl fluoride. Coatings nanometres thick give very good effects [56].

Instead of processing prefabricated polymers, the relevant monomers are applied and fixed to the substrate by radiation crosslinking.

The use of supramolecular chemical components also appears promising. Thus, for example, certain cyclodextrins [57, 58] can be fixed onto fibres like reactive dyes. Cyclodextrins have annular structures with an internal molecular "hollow" in which they can bind (temporarily) the most diverse substances due to the van der Waals forces. Thus, for example, perfumes have been deposited which remain effective over a long period, but they also bind unpleasant smells such as underarm sweat. They can also be used to store biocides or skin-compatible and wound-healing preparations for use in the medical sector. Cyclodextrins are based on natural substances.

With this same prerequisite, research was carried out on chitin and chitosan, the film-forming, wound-healing and complexing properties of which are well known [59].

Attempts, begun a decade ago, to carry out finishing in supercritical carbon dioxide instead of water had the same aim. As is well known, dyeing has already been carried out successfully with selected dyes, as well as water-repellent finishing [61].

It remains a question of time when such developments will be translated from the laboratory stage into full-scale production.

## 8.9
## Ecology and economy

During the course of this section, reference has been made to environmental pollution and environmental protection several times. Since the mid-1960 s, when mountains of foam on German rivers alarmed the media, a large number of laws have been passed relating to the handling and use of dangerous substances and environmental protection of the air, water and soil. In Germany, this is the responsibility of the Federal Ministry for the Environment. This legal framework applies to all the finishes, processes and products discussed in the second part of this chapter.

Problematic substances are the so-called CMR substances, which are carcinogenic, mutagenic or toxic to reproduction. Problematic substances include components which have a toxic effect on watercourses and are not easily biodegradable. These are low-molecular halogenated hydrocarbons, arsenic, lead, cadmium, mercury and its compounds, tri- and tetraorganotin compounds, APEOs, EDTAs and GTPA. The most comprehensive information about individual textile auxiliaries for users and consumers is provided by the safety data sheets according to EC guidelines, which can be obtained from the manufacturers of these products. As well as the composition, these detail the possible dangers and measures for first aid, fire fighting and in case of accidental release and provide advice about handling, storage, limiting exposure and personal protective equipment. Information is given about physical and chemical properties, stability and reactivity and, of course, on the toxicology and ecology, disposal, transport and other regulations [64].

Furthermore in Germany, the trade association TEGEWA has committed itself to voluntarily classifying all textile auxiliaries into three categories "non-effluent-relevant", "effluent-relevant" and "strongly effluent-relevant" [63].

To the extent that nonwoven manufacturers and finishers are becoming or have already become system suppliers, they have a duty to consumers, especially in the car industry, not only to manufacture with as little impact on the environment as possible, but also to design their products in such a way that they are recyclable and form a closed material cycle. Ecological audits and environmental management form the basis of this [62].

All this should contribute to the continuation of the important role played by the finishing of nonwovens in future.


### Epilogue and acknowledgement

Although reliant on many experienced experts in research and development, in academia, science and production, sales and marketing, this chapter has become a subjective description.

Without naming names, I would like to thank most sincerely all those who have assisted in this, as without their advice and help, this chapter could not have been written.

# References to Chapters 7 and 8

[1]   Lünenschloß J, Albrecht W (1982) Vliesstoffe, Georg Thieme Verlag, Stuttgart: 254–282

[2]   Jörder H (1972) Ausrüstungsverfahren für Vliesstoffe und ähnliche nicht gewebte Textilien, Textilveredlung 7: 317–324

[3]   Sievers K (1972) Geschrumpfte Produkte auf Vlies-Basis, Melliand Textilber 53: 156–157

[4]   Schmidt S (1974) Praktische Erfahrungen einer Clupak-Anlage, Kolloquium Veredlung und Verarbeitung von Vliesstoffen und Nadelfilzen, Reutlingen

[5]   Endler H (1975) Der Einsatz von Kalandern bei der Vliesstoffherstellung, Allg Vliesst-Rep 4: 205–208

[6]   Eisele D (1992) Recyclate aus Reißbaumwolle und Phenoplast – ein bewährtes System, Melliand Textilber 73: 873–878

[7]   Colijn JJV (1975) Geschlitzte Vliesstoffe, Text Prax Int 30: 1192–1202

[8]   Montag, Reimann, Dittrich (1997) Vliesstoffe rauhen zur Verbesserung von Optik und Haptik, Melliand Textilber 76: 359–360

[9]   Watzl A (1994) Thermofusion, Thermobonding und Thermofixierung, Melliand Textilber 85: 840–850, 933–940, 1015–1020, 86: 76–78, 170–173, 265–269

[10]  Ott R (1994) Breitwaschanlage für Maschenware und empfindliche Gewebe, Melliand Textilber 75: 997–998

[11]  Weber KA, Langenthal W v (1963) Färben von Vliesstoffen, Bayer Farben-Revue 6: 1–12

[12]  Beiertz H (1970) Farbwerke Hoechst AG, DT 2037: 554

[13]  Homuth H, Weyer HJ (1973) Das Bedrucken von Bodenauslegware, text praxis internat 28: 49–51, 102–104, 156–159, 222–223

[14]  Kool RJM (1995) Die Jetprint-Technologie, Textilveredlung 30, 3/4: 72–78

[15]  Stukenbrock KH (1990) Möglichkeiten des Druckens von Vliesen und Vliesstoffen, Melliand Textilber 71: 303–305

[16]  Hardt P (1990) Umweltfreundliche Textilweichmacher, Melliand Textilber 71: 699–705

[17]  Schmiedgen H (1976) Antistatische Ausrüstung von Vliesstoffen und Nadelfilzen, text praxis internat 31: 890–892

[18]  Anonym (1997) Fluor-Alkylsilan von Hüls, Melliand Textilber 78: 202

[19]  Klesper H (1975) Antimikrobielle Ausrüstung von Vliesstoffen und Nadelfilzen, text praxis internat 10: 1011–1022

[20]  White C (1976) Microbiological contamination, Nonwoven Industry 27, 2: 190–192

[21]  Einsele U (1975) Flammhemmende Ausrüstung von Vliesstoffen und Nadelfilzen, text praxis internat 30: 1672–1687

[22]  Schäfer W (1997) Vliesstoffe und Composites mit feuerblockenden Eigenschaften, ITB 43: 47–49

[23]  Stukenbrock KH, Türk W, Werner G (1984) Möglichkeiten zur Verbesserung des Saugvermögens von Vliesstoffen, Melliand Textilber 65: 173–175

[24]  Kerres B (1996) Superabsorber für wäßrige Flüssigkeiten, Textilveredlung 31: 238–241

[25]  Buchholz FL, Graham AT (1998) Modern Superabsorbent Polymer Technology, Wiley-VCH New York/Chichester: 224–231

[26]  Herrmann E (1998) Permeable Superabsorbents, Function and Importance for the Application in Diapers, Edana Congress 1998, Nonwoven Symposium

[27]  Matson B (1996) Barrier Coating Technologies and their application for Nonwovens, INDEX '96, Congress, Packaging Session

[28]  Lehwald D (1996) Verwendung von Superabsorbern im Verpackungsbereich, INDEX '96, Congress, Packaging Session

[29]  Offermann P, Jansen E (1996) Universal-Beschichtungsanlage für die Textilindustrie, Melliand Textilber 77: 794

[30]  Dugasz J, Szasz A (1991) Metallic coatings on Non-wovens for special purposes, Journal of Coated Fabrics 21, July: 42–52

[31]  Chemical metallization, High Performance Textiles 1993, June: 6–8 (1993)

[32]   Dobmann A, Bamborough DW (1996) Lösungsmittelfreie Haftschmelzklebestoffe für medizinische Anwendungen mit Hautkontakt, INDEX '96, Congress, Rohstoffe Session

[33]   Kooy T (1978) Anforderungen an Glasvliesstoffe für den Fußbodenbereich, Kolloquium, Vliesstoffe für technische und textile Einsatzgebiete, Reutlingen

[34]   Bobe JEA (1976) Polyurethan, die 3. Generation der Teppichbeschichtung, Melliand Textilber 57: 934–939

[35]   Neue Polyurethan-Teppichbeschichtungsanlage, Melliand Textilber 58: 335 (1977)

[36]   Zorn B (1974) Die Beschichtung von Vliesen mit einer mikroporösen Deckschicht, text praxis internat 29: 1706–1712

[37]   Stukenbrock KH (1984) Ionisch koagulierbare Polyurethan-Dispersionen, eine neue Möglichkeit für die Textilveredlung, Melliand Textilber 65: 756–758

[38]   Sroka P (1993) Handbuch der textilen Fixiereinlagen, Hartung-Gore Verlag, Konstanz

[39]   Stukenbrock KH (1977) Heißsiegelbeschichtung, in Handbuch der Textilhilfsmittel (Chwala A, Anger V, Eds.), Verlag Chemie, Weinheim

[40]   Schaaf S (1974) Copolyamidpulver als Textil-Schmelzkleber, Textilveredlung 9: 14–25

[41]   Steukart H (1971) Die Fixiertechnik in der Bekleidungsindustrie, Theorie – Maschinen – Praxis, Textilveredlung 6: 469–482

[42]   Jörder H (1977) Textilien auf Vliesbasis (Nonwovens), P. Keppler-Verlag, Heusenstamm

[43]   Stukenbrock KH (1971) Chemische Grundlagen von Heiss-Siegelklebern für Fixiereinlagen, Textilveredlung 6: 459–468

[44]   Stukenbrock KH (1992) Ein neues Konzept für Heißsiegelpasten, Deutscher Färber-Kalender 96: 188–194

[45]   Lukoschek K (1976) Hot-melt-Rückenbeschichtung und Kaschierung, Melliand Textilber 57: 939–940

[46]   Eisenträger K, Druschke W (1977) Acrylic Adhesives and Sealants in Skeist, I. Handbook of Adhesives, Van Norstrand Reinhold Company, New York

[47]   Hürten (1997) Laminieren von Sympatex Membranen, Adhäsion 41: 34–37

[48]   Maag U (1976) Deflockung von Vliesstoffen, Allg Vliesst Rep 5: 92 96

[49]   Stukenbrock KH (1988) Substratloser Flock, Textilveredlung 23: 58–60

[50]   Rakowski W, Okoniewski M, Bartos K, Zawadzki J (1982) Plasmabehandlung von Textilien – Anwendungsmöglichkeiten und Entwicklungschancen, Melliand Textilber 63: 307–313

[51]   Thomas H, Denda B, Hedler M, Käsemann M, Klein C, Merten T, Höcker H (1998) Textilveredlung mit Niedertemperaturplasmen, Melliand Textilber 79: 350–352

[52]   Bahners T, Ruppert S, Schollmeyer E (1997) Kreatives Gestalten der Oberflächeneigenschaften textiler Flächengebilde durch Plasma-Grafting und -Beschichten, Melliand Textilber 78: 770

[53]   Sigurdsson S (1996) Plasma treatment of Polymers and Nonwovens improving their functional properties, INDEX '96, Congress, R & D Specials Session

[54]   Poll HU, Schreiter S (1997) Industrienahe Plasmabehandlung textiler Bahnware, Melliand Textilber 78: 466–468

[55]   Schollmeyer E et al. (1997) Textilien in 20 Jahren, Textilveredlung 32: 236–237

[56]   Knittel D, Buschmann HJ, Schollmeyer E (1992) Maßgeschneiderte Eigenschaften, B+W/B+M (1992), 12: 34–36, 39–40

[57]   Denter U, Buschmann HJ, Knittel D, Schollmeyer E (1997) Modifizierung von Faseroberflächen durch die permanente Fixierung supramolekularer Komponenten, Teil 2: Cyclodextrine, Angewandte Makromolekulare Chemie 248: 165–188

[58]   Denter U, Buschmann HJ, Knittel D, Schollmeyer E (1997) Verfahrenstechnische Methoden zur permanenten Fixierung von Cyclodextrinderivaten auf textile Oberflächen, Textilveredlung 32, 33–39

[59]   Knittel D, Schollmeyer E (1998) Chitosan und seine Derivate für die Textilveredlung, Textilveredlung 33, 3/4: 67–71

[60]   Eisele D (1990) Nadel-/Polvliesbeläge für den Automobilbau, text praxis internat: 723–727

[61] Knittel D, Saus W, Schollmeyer E (1997) Water-free dyeing of textile accessories using supercritical carbon dioxide, Indian Journal of Fibre & Textile Research 22, Sept.: 184–189

[62] Eisele D (1996) Reißfasergut – Merkmale – Zusammenhänge, Melliand Textilber 77: 199–202

[63] TEGEWA Jahresbericht 1996/97: 34–37

[64] Sicherheitsdatenblatt nach EG-Richtlinie 93/1/12 EWG

**Part IV**
**Peculiarities with regard**
**to ready-making nonwovens**

# 9
# The making up of finished products
H. RÖDEL

## 9.1
## Concepts and definitions

The making up is the final stage within the production process of textiles. The articles produced thus, e.g. clothing, may be delivered directly to the consumer through the traders, or they can be used as semifinished products that are incorporated in further non-textile production and assembly processes, such as the manufacture of textile filters.

The making-up process transforms textile fabrics into ready-made products. This is also underlined by the Latin origin of the word *to confect:* conficere – to make, to complete [1, 2]. However, it is possible to complete the products by adding fabrics, knitwear or even nontextile fabrics, such as foils or membranes, or certain accessories, e.g., buttons, zippers, or Velcro fasteners.

The technology of making up comprises all machines and procedures that are exploited for the industrial production of finished textile goods.

The world's biggest exhibitions of engineering and technology development in the three world economy regions are:

- International Exhibition for Clothing Machines IMB Cologne, Germany
- Bobbin Show, Atlanta, USA
- JIAM, Osaka, Japan

Each exhibition takes place in a several years' rotation. The current exhibition catalogues present the interested with the current market overview of the engineering and technology companies and their supplies.

The technology of making up involves product development, production preparation and production. The production process is subdivided into the preparation of pattern making, pattern making, joining, completing and shipping [3]. The aesthetical design of textile clothing, which can be related to the term fashion, is explicitly not included in these stages of the process.

| Making up process | | |
|---|---|---|
| Process steps | Worksteps | Work function |
| Product development | Aesthetic design | Sketch of fashion, sketch of art |
| | Technical design | Pattern construction<br>Pattern grading<br>Selection of fabrics<br>Selection of accessories<br>Production of examples<br>Product data management |
| Production preparation | Technical production preparation | Pattern design |
| | Organize production preparation | Purchase<br>Planing of capacity<br>Sequence of charges |
| Production | Cutting preparation | Spreading of materials<br>Handling of material defects |
| | Cutting | Cutting of pieces<br>Pick out of pieces |
| | Connecting Assembly | Hemming<br>Sewing<br>Welding<br>Gluing<br>Fusing |
| Press finish and shipping | Press finish | Ironing<br>Packing |
| | Shipment | Storage process<br>Transportation |

**Fig. 9-1** Making up process in process steps and worksteps

## 9.2
## Product development

### 9.2.1
### Product development for garment textiles

The designer drafts the form of the product and chooses the fabrics with regard to patterning, feel, drape and also fibre composition. When he makes his decision the designer hardly takes into account technically definable parameters of the fabric. The product shape results from the pattern design. The basis for the pattern design are product-relevant human physical dimensions and proportions of an *average* human being. Pattern design is two-dimensional [e.g. 4–6]. The construction methods, the tables of body measurements and proportion calculations are specific for the individual country as they result from traditions. Pattern design produces two-dimensional patterns which when combined form the three-dimensional product surface drafted by the designer. The combination of the flexible textile parts results in the desired three-dimensional form of the product. The seam allowances which are necessary for joining the parts are prepared for in the pat-

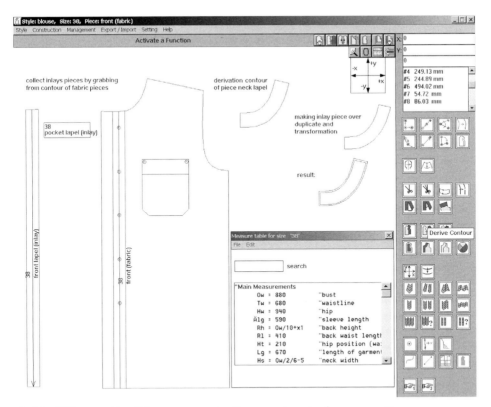

**Fig. 9-2**    CAD-system NovoCut, program construction, example parts of nonwovens [7]

tern design stage. It is common practice to make a specimen to examine the pattern design.

In future a transition from 2-dimensional patterning systems to computer-supported 3-dimensional patterning systems is expected. For clothes closed to the body in the field of sportswear and underwear this already succeeds quite well, whereas the calculation methods for the field of clothing like jackets, suits, skirts and so on are still not perfected [e.g. 8–10].

**Pattern grading** is a means to enlarge and/or reduce the size of the patterns so that the product can be manufactured in all sizes if the order requires so.

The exact knowledge about the internationally common size system is essential. There is an efficient branch specific CAD software being available for pattern grading. Clothing companies that work with the latest technologies use branch-specific CAD software for the design and grading of patterns. This allows a totally computer-aided production preparation process.

Besides a product documentation is created that describes the structure and the composition of products but also the stages of the production which are neces-

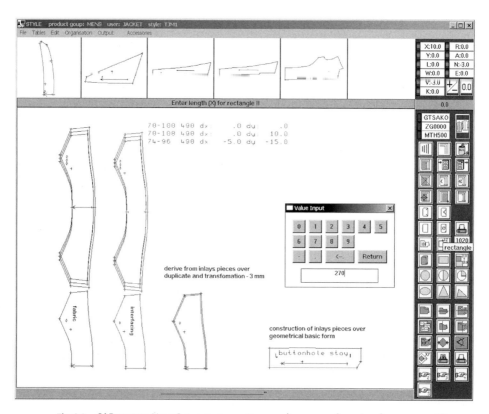

**Fig. 9-3**   CAD-system NovoCut, program pattern grading, example parts of nonwovens [7]

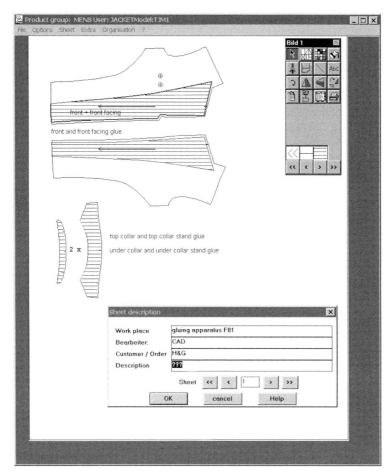

**Fig. 9-4** CAD-system NovoCut, program sewing routine, example parts of nonwovens [7]

sary. In this case CAD systems whose dates are also useable in the software systems for the planning and controlling of the production, for procurement and distribution and also for calculation and costing are suitable, too.

Nonwovens are used as fixation linings in everyday clothing and sportswear. The surface of the nonwovens is processed to make it adhere to the upper material. Cuffs and collars, e.g., are stiffened in this way, the front form of blazers and jackets is stabilised. Depending on the specific product, the nonwovens are patterned from the yard ware. For this reason, the contours of these nonwoven patterns are designed and made together with the patterns for the upper fabric and the lining during the pattern design and grading stages.

Finished nonwoven goods, which can be purchased directly from the manufacturers of nonwovens in specific band forms or as individual parts, do not only support the form stability but also enhance the sewing process if positioned exactly.

Nonwoven fabric assembly units are available on the market, too. Those can be available without patterning and patterning gradation for the production process. The catalogues of the nonwoven fabric manufacturers include sufficient information (e.g. [11]).

Voluminous nonwovens may be inserted between the upper material and the lining of garments as heat insulation. To fix the position of such insulating fleece material stitch welds are often chosen which, however, reduce the insulation effect owing to the thermal bridges that are formed as a result.

## 9.2.2
### Product development for decor fabrics and home textiles

The term decor fabrics and home textiles covers both textiles used for decoration and also linen, towels and other textile commodities. In this sector, nonwovens are above all used as filling materials for heat insulation, as elastic voluminous fillers and as local stiffeners in places of particularly great wear and tear. Simple nonwovens are also processed to make cleaning or sanitary tissue. Further product examples are quilts and upholstery.

Among the manual procedure especially CAD software is suitable for the patterning. The patterning contours that shall be produced are independent of the product designs.

## 9.2.3
### Product development for technical textiles

The requirements made on the service values are defined by the individual technical demands of the planned application and may be sufficiently defined in terms of physics. Such requirements are for example: strength and stretching for static and dynamic loads, impermeability against humidity, water-vapour permeability, translucence, resistance against chemicals, biological effects, light, UV irradiation and/or climate, combustibility, inflammability, ageing stability, maintenance, care and cleaning properties and finally the recyclability of the product at the end of its service time [12]. Numerous national and international standards and regulations may be applied when the product qualities are defined. The precise terms of the admission of construction materials which have to be registered are particularly extensive for technical textiles used in civil engineering.

The geometry of technical textiles may be two or three dimensional. More complex forms with so-called undefined geometries require a more sophisticated pattern construction. If the material width is defined already in the product development stage, savings can be made.

Nonwovens are exploited in many technical textiles. Some of these are ready-made filters, geotextiles, insulating materials, labour protection devices and protective clothing and also sanitary goods. Nonwovens are chosen according to the desired function of the product, and the possible ways of patterning and also of joining the patterned parts as well as seam properties are assessed.

## 9.3
## Production preparation

The delivery orders brought about in the sales negotiations provide the data, such as number of items and delivery times and also the exact product specifications, which are necessary for the preparation of the production process. Moreover, it is essential that product developers provide those preparing the production process with a detailed product description so that they are enabled to compile the production documents and to purchase fabrics, accessories (buttons, zippers, belts, straps, lace, eyes and hooks, labels etc.) and also the sewing yarns in the appropriate quality and quantities.

After the adequate fabrics have been chosen by the product developer and the pattern contours, which are necessary to manufacture the product, have been developed and made available in accordance with size systems as far as the production of garments is concerned, the sufficient quantity of material is optimised by pattern design. While formerly original or reduced patterns were used in marker making, presently CAD systems with interactive or automatic marker programs, e.g. by LECTRA Systems, France, are preferably applied.

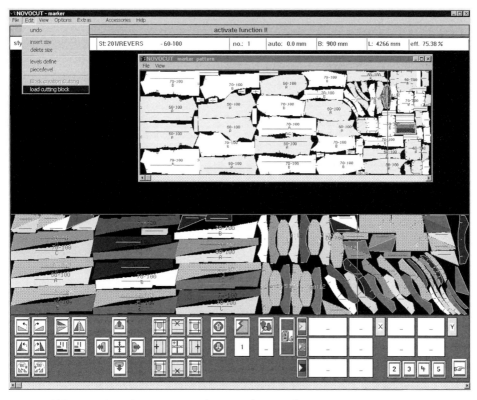

**Fig. 9-5**  CAD-system NovoCut, program marker, example parts of nonwovens [7]

Interactive work means a worker organizes the cutting parts on the screen. Cloth parts which are necessary for the product are taken, in the concerning cloth type, from the computer store. By activating by mouse or light pen they are represented on the cloth surface that is reproduced on a scale. Arrangement principles which were defined before and their tolerances as well as the minimum distances of the cloth parts in the cutting layout conditional on the cutting technique are automatically considered. The worker is permanently informed about the current grade of capacity and the loss part in the cutting layout.

The automatic cutting layout programs usually do not reach such favourable capacity grades but they reduce the deployment.

The techniques and methods of pattern design are mainly aimed at the arrangement principles in the sense of preferred variants for product-typical pattern pieces (marker systematics), the optimum material width for certain products or product sizes, precautions for the later compensation of possible material defects, the fulfilment of production orders in terms of number of items per size and style by so-called size combinations or pattern combinations in the marker layout and also the use of fabrics of varying width by classifying them into width classes. The varying bale lengths may be optimised to minimise remaining unusable material. It is necessary that these theoretical optimisation methods for material use are transferred into the production process in a reasonable time with small personnel resources. The latter limits the application of all these possibilities in practice. It is much more worthwhile to think about the optimum material use by marker design for large numbers than for trendy productions in very small batch sizes.

The successful procurement of all product components and the cutting layout are the basis for the production beginning. The cutting layout information can be used again during the production for the controlling of the laying, the defect treatment and the cutting. The required cutting layout information is written down by a suitable drawing technique on endless paper on a scale of 1:1 with the width of the cloth. So the cutting can be made by manual cutting technique. In the case of the use of CNC-controlled automatic cutting machines special software edits the graphic cutting layout information as way information for the controlling of the cutting medium.

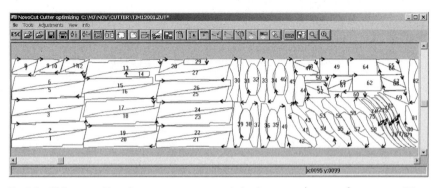

**Fig. 9-6** CAD-system NovoCut, program cutter optimisation, example parts of nonwovens [7]

## 9.4
## Production

### 9.4.1
### Spreading

Spreading the material to form stacks is an auxiliary process because the working speed of the cutting tools is relatively low when they are applied to one-ply cutting, however, they are effective tools to cut multiple-ply pieces. In industrial production, the process of handling material defects is part of the spreading process. The way how plies are spread depends first of all on the type of fabric. Nonwovens can be processed by the efficient zigzag spreading method where the direction may be disregarded. Conversely, face-to-face spreading with the grain makes sure that the direction and the repeat of matched material are chosen correctly.

For the laying a tabletop with the dimension of the cutting layout is necessary. Bolts of cloth can be simply unrolled and manually smoothed and straightened. The manual or machine stripping of cloth from a stationary stored bolt of cloth cannot be recommended especially for nonwoven fabrics with less surface stability because delays and undesirable lasting extension can be caused. On the other hand manually mobile laying machine or electrical laying machines are favourable which make it possible that the cloth taking off happens with less extension, without folds, edge orientated, with defined length and with high productivity.

Tractives in the materials to lay can be avoid by a hollow storage of the bolt of cloth because a girth impulse is created by the circulating conveyor belts. Loading machine for bale of materials in combination with modern storage equipment make a minimum need of staff possible also if there are big bolt of cloth masses.

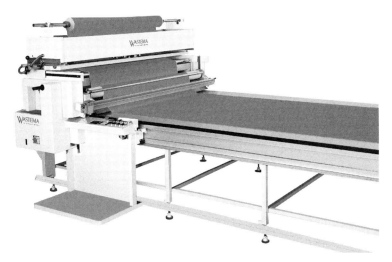

**Fig. 9-7** Spreading machine [13]

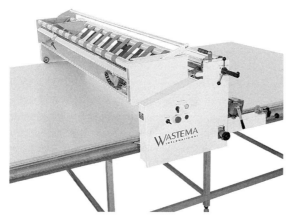

**Fig. 9-8**  Unrolling cradle of spreading machine [13]

For the cloth defect treatment especially two procedures are suitable – the cutting out of defects and the recutting of defect cloth parts. Another technological possibility in the case of less cloth costs is the multiple layer method. For all cloth parts some spare cloth parts are cut by a number of lays (a bit higher than the ordered number). However it can happen that some of the cloth parts are more inquired than others because of defects [3]. If cloth defects are covered by sufficient large cloth parts faultless cloth parts will arise among the defect ones. The defect cloth parts have to be selected before the connecting.

## 9.4.2
## Cutting

The cutting process is divided in the following subprocesses:
– coarse and fine cutting
– marking, inscribing
– removing

Depending on the technological process these process stages are run at the same time. The delivery of all the pattern pieces of a product in stacks which contain the individual pattern pieces in the correct numbers and shapes determines the interface between cutting and assembly. The cutting technology is designed to prevent the cut edges from hardening so that the individual cut pieces may be separated from each other easily. Hardening of cut edges may be caused by heat that is developed when synthetic fibres are processed.

When we want to assess the cutting process it is recommendable to examine how difficult the ready-made task is.

The simplest task in a ready-made process is cutting the yard ware transversely with subsequent seaming of the cutting edges or even of all edges of the cloths. It is somewhat more complicated to fold the pattern piece after the transverse cut-

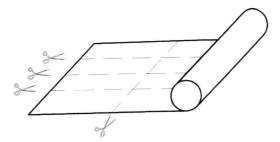

**Fig. 9-9** Princip of automatic cutting system QLA 2000 [14]

ting. Two side seams or one side seam and one base seam will complete the ready-made product – a sack or bag.

The next step of cutting is known from the manufacture of the various types of cleaning tissue. The textile fabrics, which are very efficiently produced in large widths, are cut transversely and also lengthwise. The cutting edges are seamed only if required.

The procedures and machines applied to these elementary assembly tasks are simple. As a rule, mechanical cutting tools, e.g., revolving and spiral blades are used. These transverse or longitudinal cutting machines may be adjusted to the dimensions of the pattern pieces [14].

A task that in the wider sense can be attributed to cutting is the making of cut-edge fabrics of constant width. Cutting tools are rotating blades or laser cutters; the latter additionally strengthen the cut edges of thermoplastic fibres. Hand-operated cutting machines are used to cut any contours, while the exact contour cutting is prepared by cutting the textile stack roughly for easier handling.

Pattern pieces with a constant shape that are needed in the long run may also be cut by die clickers [15, 16]. Parts of the product are cut out of the nonwoven fabric on intermittent assembly lines and stacked. The productivity is very high. If multi-ply material is die-cut, the cut edges of thermoplastic fibres may be hardened, which possibly impedes the further smooth processing of the individual material parts.

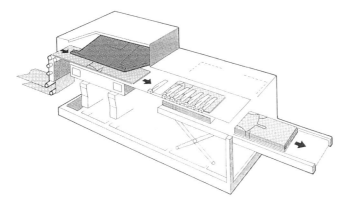

**Fig. 9-10** Die cutting machine, Bierrebi, Italy [15]

### 9.4.2.1 **Conventional cutting techniques**

Hand-operated cutting machines are equipped with various blades that move in a specific way. The necessary relative movement between the cutting machine and the material has to be in correspondence with the cutting contour. Because the textile materials are flexible the cutting machine is most often moved by men along the planned cutting contour. Only in the band-knife machine, the stack is moved according to the machine's operational principle.

**Rotating blade machines** contain a rotating blade and are particularly useful for straight cuts.

**Fig. 9-11**   Rotating blade machine [13]

**Band knife machines** are equipped with an endless band-shaped knife that works at the operating point in a vertical direction at a constant speed. Since the operating point of the band knife is fixed, consequently, the material is moved along the

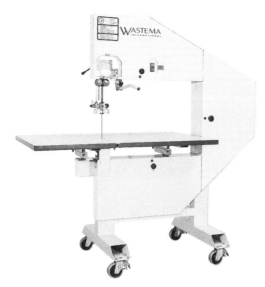

**Fig. 9-12**   Band knife machine [13]

cutting contour. The smoothness of table surfaces may be increased by an air cushion. The clean cut is worth mentioning.

**Vertical-blade machines** or straight knife cloth cutters are characterised by an oscillating straight knife that is moved vertically. The knife is moved through the U-shaped guide. It can be used for any pattern contours. As the vertical-knife machine is guided by a swivel arm, the forward forces are considerably reduced which makes the cutting more exact. Arrangements of this type are known as servo cutters.

**Fig. 9-13**   Vertical blade machine [13]

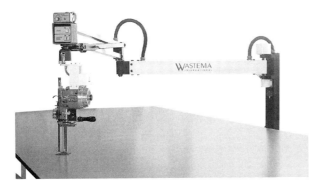

**Fig. 9-14**   Servo-Cutter [13]

### 9.4.2.2  Automatic cutting systems

For more than thirty years computerised cutting together with computer-aided production preparation has been successfully introduced in many businesses as CAD/CAM systems [17, 18].

**Fig. 9-15**   CNC-Cutter TexCut 2050 [13]

Automatic cutting machines guide the cutting medium automatically using the computerised marker layout. They are very efficient for large cutting quantities and elaborate contours. Non-productive set-up and shut-down times should be minimum. Basic requirements are high accuracy through all plies, small distances between pattern parts and little susceptibility to failure. The stitch knife as a special form of the vertical blade is a universal mechanical cutting medium. Due to the hanging bearing the stitch knife is designed without a U-shaped guide. It cuts into the stack at any place so that internal openings of any desired shape may be cut. A bristle mat forms the surface of the table. During cutting the knife penetrates into this mat. The restoring forces per unit area during cutting are taken up by vacuum suction from below a porous table and a thin plastic film that is placed over the layered material and held by this vacuum. The vacuum may reduce the height of the textile stacks to 70 mm which ensures a high productivity. As this capacity may only seldom be exploited in the garment industry

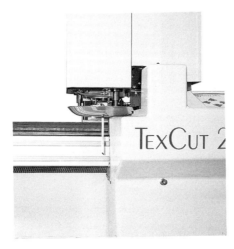

**Fig. 9-16**   Detail of CNC-Cutter, chance position of knife [13]

due to small batch sizes, so-called low or one-ply cutters are also available on the market.

Automated machines for one-ply cutting are presently equipped with a small round knife as the cutting medium and can thus do without the thin plastic film.

**Laser cutting** is useful for very fast, precise cuts and very tight internal openings [19]. The cutting rate may be 100 m/min. The laser beam has a thermal effect so that the cut edges are principally fused. Vapours and smokes have to be discharged effectively. Laser cutters are an economic alternative for the cutting of interlining pieces made of special materials and in small numbers. From the point of view of labour safety, top priority has to be given to the safe control of the laser beam.

There has been done a lot of research work on the plasma jet, however, it is principally not used to cut textiles [3].

The compressed water jet may be used for the cutting of textile fabrics that are made of individual fibres or filaments whose mobility is limited due to coating and rubber films or a composite design. The edges of fleecy fabrics, such as voluminous nonwovens, cannot be cut clean as the fibres may escape the water jet uncut. The energy demand is high, the table must be designed to let the water jet pass through it. Low stack heights may be cut [20, 21].

**Ultrasound** is another cutting medium that is usable for textiles. It can be combined with ultrasonic hardening of the cut edges of thermoplastic fibres, which is particularly efficient. It is advantageous that the costs for producing the cutting energy are reduced. Compared to laser cutting, ultrasonic cutting does not singe or discolour the cut edges [22].

### 9.4.3
### Joining and assembly

The cut parts of the nonwoven fabric may be joined either by sewing as the typical joining technique for textiles but also by welding and sticking [3]. Points, lines and areas are the forms that may be chosen to join fabrics.

Of prior importance are the seam properties that can be obtained with regard to their mechanical strength and tightness against various substances. Every stitch of the seam represents a joining element that determines the overall strength. As a result, the seam is the stronger the more stitches it has (number of stitches per unit length) unless the stitch density causes excessive perforation of the fabric. Moreover, the fabric's strength against slip should be sufficient so that the composite nonwoven structure is not disturbed in case of loads attacking the structure next to the solid seam. Sewing is quite interesting for the processing of nonwovens into quilts, for example, which is done very efficiently directly from the roll in automatic sewing machines.

The stitch type classes that are usually used in the seam technique are internationally standardized. The stitch type classes differ in thread consumption, seam strength, seam elasticity and automation ability. The very strong, less elastic double lockstitch can be carried out more elastically by a zigzag arrangement but it is technologically less automation friendly because of the principle limited thread stock of the bobbin thread.

In the chain stitch elasticity and the ability of being undone are combined. The double chainstitch is characterized by elasticity and sufficient stability. For the cutting edge stabilization the overedge stitch is suitable because threads are conducted around the selvedge. In combination with the double chainstitch the overedge stitch can be carried out parallel on the so-called Safety Sewing Machine. So two seams mean an additional seam safety reserve.

The cover stitch with its high tread consumption is important for the knit goods processing [3, 23].

By sewing also selvedge fixing by sewing round of the cloth parts with overedge stitch can be realized in the pre-production.

In the field of nonwoven fabric processing sewing is interesting for e.g. quilts that can be made directly and with a high productivity in automatic sewing units from the bale of material. Another application of sewing technique in the field of nonwoven fabric processing is padding. Upper clothes and nonwoven inlays are connected by chain stitch in the special form of the blind stitch by more invisible seams distributed over the surface [3, 23].

On account of the geometric design of the sewing technique it is possible to influence essentially the handling of the cloths during the sewing process. So according to the sewing task that shall be realized it is necessary to choose between flat bed machines, cylinder bad machines, post-bed machines and cast-off-arm machines but also other constructions.

The internationally working manufacturers of sewing threads and sewing needles stand by advising the ready-to-wear producers at any time. This concerns both the processing behaviour during the sewing process and the seam characteristics of the product.

**Textile welding** is used to join thermoplastic fibres and thermoplastic coatings of textile fabrics [3, 24, 25]. It is important that the welding seam is adjusted to the textile structure, that it is soft and flexible. Common welding techniques for textiles are high-frequency and ultrasonic welding, hot air welding and hot wedge welding. The most popular product ranges are healthcare garments, protective clothing, sportswear and leisurewear, also with integrated membranes, and landfill sealings. Welding is generally defined as the joining of two or more pieces made of the same material thereby applying heat and pressure. A welding seam is produced without the use of additives. The welding process consists of three individual processes that are carried out partly at the same time at the point of welding:

– heating until melting is initiated
– compression
– cooling down until connection is solid

During use, the transition between the initial textile structure, which is characterised by the fibres and/or the filaments, and the produced textile structure,

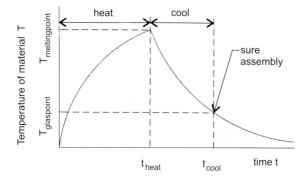

**Fig. 9-17** Welding procedure: temperature of material during the welding process as a function of time.

which is the result of the flow of the thermoplastic polymer material in the welding seam, is particularly sensitive to all types of changing stress. Typical failures are planar defects in this transitional area given that the welding process was carried through properly. Dangerous substances may be a by-product of the welding process. Therefore, these substances should be removed from the work station by suction, as a rule. For safety reasons, the demand in non-destructive test methods of welding seams in textiles is growing; these test tools should be mobile as they are needed on dumpsites during the final assembly.

**High-frequency welding is a technique** in which the textile parts are used as a dielectric between two electrodes and are subjected to a high-frequency alternating electric field, e.g., a field of the industrial frequency of 27.2 MHz [26]. This welding technique is only applicable to fibrous materials that fulfil the following conditions:

– presence of polarisable or polarised macromolecules with comparatively short chains
– good mobility of the molecule segments
– occurrence of inducible dipoles

If these conditions are fulfilled the induced segments start to oscillate as soon as the alternating field is applied. The temperature rises, the secondary valency forces are loosening, and the molecules can interchange their sites. Process parameters are time, pressure and high-frequency voltage. If the available electrodes are long enough, high-frequency welding is suited to produce whole seams in one welding process in a flow-line process. The electrode design may be varied to produce different seams or to trim the edges of the textile pieces.

**Ultrasonic welding** [27, 28] is based on mechanical oscillations of about 30 kHz to generate the necessary heat. This heat is produced by a sonotrode that generates ultrasonic waves inside the material. The molecule segments are excited to oscillate, which produces friction heat in the thermoplastic material. Also this welding technique requires certain properties of the textile fibres: the logarithmic damping decrement factor, the modulus of elasticity and the heat conductivity. As the ultrasonic welding machine is designed analogously to the sewing machine using a sonotrode as the "needle", welding can be carried out as a continuous process [27]. Process parameters are profile and width of the rollers, feed rate and pressure between the sonotrode and the backing strip.

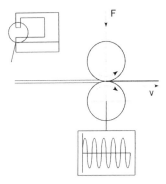

**Fig. 9-18**  Principe of ultrasonic welding

**Hot wedge welding** uses an electrically heated wedge as the heat conductor whose temperature is kept constant by a control unit [3]. The temperature of the wedge is essentially higher than the melting temperature of the fabrics to weld because of efficiency reasons in heat transmission, so the wedge has to be turned out of the working place in cases of process interrupting to avoid damages. The heat emission is directed to the inside of the seam by a hot wedge directly touching the material then immediately joined by being pressed by two rollers.

**Hot air welding** is similar to hot wedge welding, the difference being that rather than a hot wedge touching the material there is only hot air blowing at the material [3]. Welded seams are very often applied to disposable products in the healthcare and labour safety sectors; these products are mainly made of nonwovens or coated nonwovens and are manufactured in large numbers.

**Gluing** is a technique to connect materials; it is particularly important for the processing of nonwovens in the apparel industry. The adhesives of welding tapes are made liquid by hot air stream and immediately pressed to the seam between the two rollers, which produces planar or linear inseparable connections. There is a minimum limit of the width of linear seams.

The adhesive is activated by the influence of heat, pressure and moisture or even by an electric high-frequency field analogously to high-frequency welding. The gluing process is widely used for the joining of the upper material with the interlining material and has proved to be very economical. This type of connection which is preferably applied to flat rectangular structures uses gluing agents or gluing textiles and is also known as the gluing and fixing process. The textile fixation linings, most of which are nonwovens, are treated with adhesives already in the manufacturing process; these adhesives are applied to the nonwoven surface either according to a statistic distribution or in a screen pattern. In his guide to textile fixing pads Sroka [29] presents fixing pads that are relevant for practical use, the technologies for glue application, the various adhesives, the fixing plants for the apparel industry, the possible process parameter variations as well as the consumer requirements and the available test methods.

In the interest of the product quality it is often required that among parts of upper clothes fixed with nonwoven fabrics also the other parts of the upper clothes are exposed to the heat treatment. In that way by the heat energy caused measure changes shall be evenly released in every part of the clothes.

Fixing techniques that are available on the international market are stationary presses, continuous presses and also high-frequency devices.

The manufacturers [e.g. 23] provide the user of the padding they produce with information about the type of adhesive, the glue distribution and also about the relevant process parameters, such as pressure, temperature and residence time.

Kaiser [30] points out the increasing demand of research work in the field of fixation technology in the interest of high-quality products. New materials, higher quality requirements and also new care methods for textiles increasingly cause difficulties that become obvious only after the textiles have been used for a certain time. The upper materials used in the apparel industry become finer and lighter and at the same time more sensitive to temperature and pressure. The materials are treated with heat to activate the latent shrinkage to prevent undesired reduction in size or the formation of wrinkles. This altogether underlines the demand made on precise process parameters and their transfer into practice during the gluing and fixation process to secure product quality during use.

The gluing of textile fabrics along lines is mostly applied in the assembly phase of technical textiles, e.g., for life rafts made of rubber-coated material. This technology is very time-consuming as most of the processes, e.g., application of the adhesives, joining and pressing, require manual labour. The limited open time of the glue, e.g., forbids the use of any dosing and application technique.

There are two basic principles that are decisive for the adhesion of a glue on a surface, these are the mechanical and the chemo-physical (also specific) adhesion. The constellations of the atomic and/or molecular forces, which are responsible for the cohesion within and among the layers, also determine the strength of the adhesive connection. There are both forces in the adhesive film and forces in the boundary layer (adhesive forces) because they are decisive for the adhesion theory.

**Shape-specific adhesive forces** depend on the shape, size, number and uniformness of the capillary tubes that are generated in the surface layers in suitable pretreatment processes. These tubes act as an anchor to affix the adhesive when it hardens. A strong capillary effect may be assumed at least for uncoated textile fabrics because of those parts of the surface which are formed by the individual fibres and/or filaments in a thread structure.

**Material-specific adhesive forces** arise from molecular or atomic interrelations between the bonding partners. The dimension of these adhesive forces depends on the type of fibrous material and the energetic conditions of the surface, its polarity and/or polarisability and also on the type of adhesive and its properties. These forces are independent of the surface properties of the textile parts. The stability of a glued connection decisively depends on the shape and the size of the connection, the mechanical deformation behaviour of the material pieces and also on the inherent stability and the adhesive and deformation properties of the adhesive film. These parameters together determine the diffusion of stress.

### 9.4.4
### Ironing

Ironing is carried out before, during and at the end of the manufacturing process to smooth and shape the textile material and the ready-made products for the production of garments [31]. Pressure, heat and moisture are applied as well as subsequent cooling down and drying to shape garments in three dimensions, to produce permanent creases in textiles, e.g., in the seam area, for darts and pleats, or to smooth and remove undesired alterations of the material that are caused by the production process. Thermal connections between the upper material and the interlining pieces must thereby not be broken, i.e., the ironing temperature must not exceed the fixing temperature.

Various ironing techniques with different manual expenditure and different efficiency are available on the international market:

– electric irons
– steam iron
– iron workplaces with product specific table form
– ironing presses with product specific pressing tool design
– ironing dummies
– trouser pressing
– tunnel finisher

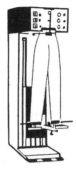

**Fig. 9-19** Examples of ironing techniques [32]

A special attention has to be directed to the correct quantities and parameters of the ironing technique's steam supply, so that the quality of the ironing process will not be damaged.

## 9.5
## Packaging

After cutting, assembly and ironing (for apparel) the product is complete. It is packed, shipped and intermediately stored a several times on its way to the consumer. The customer expects the product to be in a perfect condition. Damages imply not only product faults but also packages that became damaged [33].

The optimum package should, on the one hand, allow an efficient packing process and, on the other, it is expected to protect the product during the circulation process. Essential parameters are the quantity, form and susceptibility of the product. Undesired interactions between package and product must not occur, e.g., corrosion or the transfer of an unpleasant inherent odour of the package.

Home textiles, such as duvet covers, pillowcases, sheets, towels, nappies and blankets are folded, laid into a tubular film made of polypropylene or polyethylene which is sealed by strap and glue, strap and label or by welding. The package can be labelled. It is possible to process 400 to 720 package items per hour [e.g., 34].

The hanging get-up is especially well-suited for upper garments that require much ironing; it is the gentlest way to pack and carry apparel of that type to the customer. To protect the garment on the hanger from dirt it is wrapped in tubular film and sealed. For economic and logistic reasons, the packaging machines for hanging get-ups are integrated in the hanging transport plants of the dispatch bays of the textile manufacturer [35].

Room temperature and relative humidity are closely related with each other and effect numerous wearing or use properties of textiles. Permanent pleats are made under the influence of heat and moisture. The crease resistance can be increased if the textiles are protected from excessive humidity. In particular those creases can be removed easily which were made at low temperatures and little humidity.

## 9.6
## Mechanization and automation

The making-up of textiles is traditionally very labour-intensive.

The continuous CAD/CAM technology of the process preparation and of the CNC cutting is interrupted by the removing of the cloth part layers from the cutting table and by the separation of the cutting waste that have been manual processes so far.

Sewing machines require one worker each, and although great efforts are made toward an optimum design of the work station and the handling the output increase is limited. The sewing process takes up only a very small portion of the whole production process. Substantially more production time is used for getting the textile pieces ready and laying them down again, the geometrical combination of the textile pieces, the removal of broken threads, the cutting-off of sewing threads etc.

Large-scale automation of the making-up process is extremely difficult because of the flexible textile fabrics that are processed and the necessity to design the production plants adjustable to the changing requirements, i.e., varying product designs and dimensions [36]. Solutions for completely automated plants exist for the production of bed linen, towels, quilts and similar products of simple geometric forms. It is essential that these products are manufactured on a very large scale as the flexibility of these complex production plants is rather limited.

The successful solutions start from the fact that the textile material is provided in rolls. The product geometry decides which cutting media or die clickers are chosen for the cutting process. The arrangement of the textile parts which is defined during cutting is maintained during the further processing if possible, the textile pieces are positioned in accordance with the product requirements and are joined using one of the assembly processes. After the products are put down in a defined way their packaging may be part of the automated process.

Despite all engineering finesse of these plants, these solutions are only efficient if the process is run most reliably and standstills occur very seldom.

### References to Chapter 9

[1]   Großes Fremdwörterbuch, 7, VEB Bibliographisches Institut Leipzig (1986)
[2]   Langenscheidts Taschenwörterbuch der lateinischen und deutschen Sprache, 19, Langenscheidt KG Berlin (1957)
[3]   Rödel H (1995) Analyse des Standes der Konfektionstechnik in Praxis und Forschung sowie Beiträge zur Prozeßmodellierung, Habilitation Techn Uni Dresden, Fak Maschinenwesen, Shaker Verlag Aachen 1996
[4]   Jansen J, Rüdiger C (1994) Systemschnitt I: Röcke, Blusen, Kleider, Jacken und Hosen, Fachverlag Schiele & Schön Berlin
[5]   Jansen J, Rüdiger C (1993) Systemschnitt II: Mäntel, Parkas, Bademoden und Kinderbekleidung, Fachverlag Schiele & Schön Berlin
[6]   Körper-Sekundärmaße für die Konstruktion von Herren- und Damen-Oberbekleidung, DIN 61517 (1980)
[7]   CAD cutting line Entwicklungs- und Vertriebs-GmbH Berlin
[8]   Schenk A (1996) Berechnung des Faltenwurfes textiler Flächengebilde, Dissertation Techn Uni Dresden, Fak Maschinenwesen
[9]   Fischer P (1997) Ermittlung mechanischer Kenngrößen textiler Flächen zur Modellierung des Fallverhaltens unter Berücksichtigung konstruktiver, faserstoffbedingter und technologischer Abhängigkeiten, Dissertation Techn Uni Dresden, Fak Maschinenwesen
[10]  Fischer P, Krzywinski S, Rödel H, Schenk A, Ulbricht V (1999) Simulating the Drape Behavior of Fabrics, Text Res Journ 69, 5: 331–334
[11]  Vom Entwurf zur Realisation: Vlieseline ist dabei. HAKA-Zirkel, Firmenschr der Fa. Carl Freudenberg, Einlagestoffe Vlieseline, Weinheim (1993/94)
[12]  Brier C (1988) Grundlagen rechnergestützter Konstruktion und Gestaltung von Steilwand-Campingzelten, Dissertation Techn Uni Dresden, Fak für Maschinenwesen

[13]   KURIS-WASTEMA Vertriebs GmbH Veringenstadt

[14]   (2000) QLA 2000 – Quer- und Längsschneideautomat, Firmenschr Parker Hannifin GmbH EMD HAUSER Offenburg

[15]   Stanzmaschinen-Baureihen LTE und TA, Firmenschriften Fa. Bierrebi, Italien

[16]   Stanzmaschine CA-125, Firmenschr Fa. Llesor, Spanien

[17]   Großkreutz D (1995) Die Gestaltung des Schneidwerkzeuges beim ziehenden Schneiden textiler Flächengebilde, Dissertation Uni Kaiserslautern, Shaker Verlag Aachen, 1996

[18]   Rödel H (1997) Untersuchungen zur Zuschnittautomatisierung in der textilen Konfektion unter besonderer Berücksichtigung der Wirkpaarung von Vertikalmesser und textilem Stoffstapel, Dissertation Techn Uni Dresden, Fak Maschinenwesen

[19]   Bahners Th, Müller B, Schollmeyer E (1992) Der gepulste $CO_2$-Laser – Optimierung des Schneidergebnisses, Bekleidungstechn Schriftenreihe 86, Eigenverlag der Forschungsgemeinschaft Bekleidungsindustrie e. V.

[20]   Henne H, Steinlein J (1978) Schneiden von Bekleidungstextilien mittels Wasserstrahl, Bekleidung u Wäsche, Düsseldorf 30, 21: 1409–1413

[21]   Henne H, Steinlein J (1979) Schneiden von Bekleidungstextilien mittels Wasserstrahl (Nachtrag), Bekleidung u Wäsche, Düsseldorf 31, 3: 141–143

[22]   Randolph V (1992) Wasserstrahl und Ultraschall als Medien beim formgerechten Schneiden, Melliand Textilber 73, 12: 948–949

[23]   Autorenkollektiv (1991) Fachwissen Bekleidung 2, Verlag Europa-Lehrmittel Vollmer GmbH & Co Nourney

[24]   Bäckmann, R (1990) Thermische Konfektionstechnologien für Textilien und Textilverbundstoffe, Chemiefasern Text Ind 40/92, 6: T88–T90

[25]   Bäckmann R (1988) Thermische und adhäsive Verbindungstechniken für technische Textilien, Kettenwirk-Praxis 1: 19–24

[26]   Rische UW (1990) Hochfrequenz-Schweißen in der Kunststoff-Fügetechnik, Eigenverlag der Herfurth GmbH Hamburg

[27]   Rische UW (1982) Ultraschall-Schweißen in der Kunststoff-Fügetechnik, Eigenverlag der Herfurth GmbH Hamburg

[28]   Kontinuierliches Schweißen mit Ultraschall, Techn Textil-Forum, Ober-Mörlen 7, 4: 100 (1994)

[29]   Sroka P, Hefele J, Koenen K (1995) Handbuch der textilen Fixiereinlagen, 3, Hartung-Gorre Verlag Konstanz

[30]   Kaiser A (1995) Optimales Zusammenspiel aller Parameter beachten. Wie man Risiken beim Fixieren vermeidet, Bekleidung/wear Düsseldorf 47, 4: 29–31

[31]   Krowatschek F (1991) Ist der Bügelvorgang ein chaotisches System? Forschung zur Optimierung des Bügelns, Bekleidung u Wäsche/Bekleidung u Maschenware 43, 15: 34–42

[32]   Wir bringen Bekleidung in Form, Firmenschr VEIT GmbH & Co Landsberg/Lech

[33]   Purewdorsh E (1994) Verpackung und Versand von Konfektionserzeugnissen, Diplomarb Techn Uni Dresden, Fak Maschinenwesen, Inst f Textil- u Bekleidungstechnik

[34]   Verpackungsautomaten-Baureihe VA, Firmenschr Texpa-Arbter Saal/Saale (1994)

[35]   Das Technologie- und Logistikzentrum in Landsberg/Lech, Internetpräsentation Schönenberger Systeme GmbH Landsberg/Lech (1999)

[36]   Krowatschek F, Nestler R (1999) Was nun, Konfektion? In: Schierbaum W: Jahrbuch für die Bekleidungsindustrie, Verlag Schiele & Schön Berlin

# Part V
# Characteristics and application of nonwovens

# 10
# Nonwovens for hygiene
W. KITTELMANN

The development of nonwovens for hygiene products is characterized by rising quality demands on the products and an increased growth in volume. The main groups of hygiene products are baby diapers and training pants, incontinence products and feminine hygiene. Hygiene products also include wet wipes and swabs.

Whereas in the past, the aforementioned products were manufactured for multiple use and, especially in the diaper sector, from natural fibres, in the last few decades they have developed into single-use or "disposable" products.

Also included among disposable products are those based on nonwovens for the hospital sector. Due to the danger of cross-infection, for example, surgical gowns are preferably manufactured for a single use.

As well as the physical and chemical properties of nonwovens for medical and hygiene products, in the case of textiles used next to the skin, their thermal-physiological and skin sensory properties – and in some cases their barrier effect – are also important. The products manufactured from or using nonwovens must guarantee as a composite system, that they have a high absorbency and guarantee a quick dispersal and rapid transport of liquids – for example urine – away from the body. There should also be no release of moisture back onto the skin due to the weight of the body. These products are required to be soft, to be a good fit on the parts of the body they are protecting and not put any pressure on the body. This also means taking into account the required comfort and ease of use of the aforementioned hygiene products in the production and choice of the nonwoven.

The growth in hygiene products in the last 20 to 30 years is the result of population growth, particularly in countries with a high living standard, their demographic development and the consequently increased demands for hygiene with a reduction in health risks. The development of new fibres and auxiliaries such as absorbents, films and adhesives, as well as highly productive processing techniques for nonwovens manufacture have led to a situation in which superabsorbent and coverstock nonwovens make up a large share of nonwovens production.

**10.1**

**Diapers**

At the end of the 30 s, cellulose diapers made from creped tissue paper came onto the market. They were followed by two-component diapers in which a non-woven coverstock or film was used as a casing for the cellulose layer. At the start

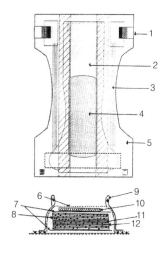

1 Fastening tapes
2 Absorbent tapes
3 Elastic leg cuff
4 Distribution and transport layer
5 Protective film
6 Nonwoven top sheet
7 Wet laid tissue
8 Pulp
9 Leak guard
10 Distribution layer

**Fig. 10-1** All in one diapers [1]

of the 80s "all in one" diapers with superabsorbent polymers were developed. This type of diaper [1] (see Fig. 10-1) has a multi-component structure.

Important elements of this disposable diaper are the nonwoven coverstock, the one- or two-layer nonwoven sheet to take up and distribute the liquid and to draw it into the absorbent core and the backsheet. In addition, elastomeric materials and waterproof elements are incorporated into the diaper. The coverstock is manu-factured from polypropylene as a thermally bonded staple-fibre or spunlaid non-woven. Bicomponent nonwovens are also used. The acquisition and distribution layer can consist of a one- or two-layer nonwoven with fluff or pulp and, if neces-sary, a thermoplastic fibre component for thermal bonding. The absorbent layer has the task of absorbing and storing liquids. This layer undergoes a change from the dry to moist or wet state. The superabsorbent not only stores liquids, but draws moisture out of the damp or even wet fluff/pulp. For this, the gel stability of the superabsorbent must remains intact so that the capillary action of the core matrix is not lost during use [2]. The manufacture and function of superabsor-bents are discussed further in Section 2.4.

The protective film is a multilayer structure based on polyolefins. Microporosity is achieved by adding calcium carbonate.

## 10.2
## Incontinence products

Incontinence products include pants, liners, absorbent pads and waterproof sheets. These products are very important in medical establishments.

Their development began around 20 years ago. A large proportion of these products, particularly for urine and fecal incontinence in adults, are on the market as disposable textiles. Developments are also under way that are concerned with incontinence products, in particular waterproof sheets, as multiple-use products. There are indications of positive developments in pile loop structures for incontinence protection [3].

Knowledge gained from the development of diapers can be carried over to incontinence products. In terms of the degree of protection required, a distinction is made between lighter and heavier incontinence, the latter occurring especially among older people. Of particular importance is that these incontinence pants have a high absorption capacity. The retention of moisture away from the skin is a prerequisite for wear comfort.

## 10.3
## Feminine hygiene

As multiple-use articles, such products have been known since ancient times. At the turn of the 20th century came the development of disposable sanitary towels made from cotton with tubular gauze or net coverings. The development of thermally bonded coverstock made from polypropylene, absorbents and hot melt adhesives was the basis for new combinations enabling soft, thin sanitary towels to be produced. Self-adhesive liners increase the wear comfort. Very thin and absorbent sanitary towels require the use of very light microfilament nonwovens. The absorbent layer must absorb and store the liquid within a small area. Sanitary towels are designed in their structure and shape for effective positioning on the body and are equipped with means of securing them in position.

Also included under feminine hygiene are tampons, which are made from absorbent carded webs that are made into the appropriate shape.

## 10.4
## Quantitative and qualitative trends in nonwovens

As the latest information from EDANA for 1998 [4] shows (Table 10-1), the production of nonwovens for hygiene in Western Europe has increased particularly sharply in the last few years.

It can also be seen that, of the total production of nonwovens in $m^2$, 55.3% are manufactured for hygiene products. Altogether, the USA, Western Europe and Japan account for more than 85% of the demand. The 15 countries of Western

**Table 10-1** Nonwovens production in Western Europe for hygiene products [4]

|  | *Production* | | *Proportion of overall production* | |
|---|---|---|---|---|
|  | *in 10³ t* | *in 10⁶ m²* | *in % of the weight* | *in % of the area* |
| 1993 | 179.8 | 8,003.7 | 31.5 | 52.5 |
| 1998 | 286.9 | 11,788.1 | 34.7 | 55.3 |
| 2000 | 341.4 | 14,452.3 | 33.8 | 57.2 |

Europe are currently showing signs of a decline in the birth rate. Demand is therefore predicted to rise particularly in South America and other countries with a rising birth rate. Table 10-2 gives an overview of nonwoven hygiene products for the afore-mentioned end-uses.

Extrusion nonwovens include spunlaid and meltblown nonwovens, split or embossed films and nonwoven composites made from these structures. Like extrusion nonwovens, thermally bonded staple-fibre nonwovens are also made of polypropylene. As Table 10-2 shows, around 60% of these nonwovens are used in diapers. Production of thermally bonded staple-fibre nonwovens is concentrated in Western Europe. Extrusion nonwovens are likewise important for incontinence products. For feminine hygiene, aerodynamically formed staple-fibre nonwovens are mainly produced.

The qualitative development of extrusion nonwovens based on polypropylene is characterized by the fact that they are made from increasingly fine filaments with a simultaneous reduction in the weight by unit area. By increasing the filament fineness, for example from 2.0 to 1.0 dtex with a weight by unit area of 20 g/m², the fibre length per square metre increases from 100 to 200 km and the nonwoven becomes softer, increasing the wear comfort. With a weight per unit area of 10 g/m², the fibre lengths are halved to 50 and 100 km [5]. A very high degree of surface coverage is achieved in this way with a simultaneous reduction in the weight per unit area. By combining processing techniques and machine technol-

**Table 10-2** Nonwovens production according to manufacturing process for hygiene products in 2000 in Western Europe [4]

|  | *Extrusion nonwovens* | | *Heat-bonded staple-fibre nonwovens* | | *Aerodynamically formed staple-fibre nonwovens* | | *Total* | |
|---|---|---|---|---|---|---|---|---|
|  | *in 10³ t* | *in %* | *in 10³ t* | *in %* | *in 10³ t* | *in %* | *in 10³ t* | *in %* |
| Baby diapers | 87,900 | 61.2 | 98,200 | 74.5 | 19,400 | 29.5 | 205,500 | 60.2 |
| Incontinence | 39,300 | 27.4 | 13,500 | 10.2 | 3,300 | 5.0 | 56,100 | 16.4 |
| Feminine hygiene | 16,400 | 11.4 | 20,200 | 15.3 | 43,100 | 65.5 | 79,700 | 23.4 |
| Total | 143,600 | 100.0 | 131,900 | 100.0 | 65,800 | 100.0 | 341,300 | 100.0 |

ogy for spunlaid (S) and meltblown (M) nonwovens, it is possible to produce composite structures, for example SMS or SMMS, in a continuous process. Based on a fibre diameter of 2–3 µm for the meltblown fibres and 11–15 µm for the filaments, very soft and dense structures are produced [6]. Nonwoven composites with good barrier properties can be produced on high-performance machines with speeds of up to 500 m/min.

More recent developments have concentrated on the production of bicomponent spunlaid nonwovens with a soft hand. The VAPORWEB process from Reifenhäuser Maschinenfabrik GmbH & Co is used to manufacture lightweight, porous spunlaid film structures. After production of a polypropylene spunlaid nonwoven, extrusion coating is carried out at high speed. The required porosity is achieved by using calcium carbonate as an additive, followed by biaxial stretching of the composite. This nonwoven composite has properties that predestine it for use as a protective coverstock for hygiene products [7]. The production of polypropylene staple-fibre nonwovens with subsequent thermal calender bonding made it necessary to refine web formation, especially carding and web laying, taking into account the processing of fine fibres and reduced web weights (see also Section 4.1.2).

## References to Chapter 10

[1]   White C (1999) Baby diapers and Training Pants, Market Overview Nonwovens Industry, 1: 26–39
[2]   The Absorber 5, Information der Chemischen Fabrik Stockhausen GmbH Krefeld (1994)
[3]   Herrmann U (1993) Nutzung der Adsorptions- und Absorptionseigenschaften von Faserstoffen in Polmaschenstoffen zum Inkontinenzschutz, Forschungsbericht AIF Nr. 218D, FIFT Chemnitz GmbH
[4]   European nonwoven statistics EDANA Brüssel (2001)
[5]   Ruzek E (1997) Baby Diapers – Continuous Challenge to the Nonwoven Industry, XXV Intern Nonwovens Colloquium Brno/CS
[6]   Kunze B (1997) Vliesstoffherstellung Feinfaservliese oder SMS?, XXV Intern Nonwovens Colloquium Brno/CZ
[7]   Kunze B (1999) A New approach to breathable structures, VAPORWEB – lightweight breathable nonwoven/film product in line, Part B: The Process. INDEX'99, Session Hygiene 2 Geneva/CH

# 11

# The use of nonwovens in medicine – safety aspects

J. Hoborn

The possibilities to treat or cure human disease have increased tremendously and continuously during the last fifty-year period. Diseases or insufficiencies, having been impossible to intervene with are today open for successful treatment. Examples are open heart surgery, total hip or knee replacements, kidney and other organ transplants. Also the possibilities to diagnose diseases have increased resulting in enhanced effectivity and precision in the treatment. The developments both within the field of imaging and the field of in vivo and in vitro diagnostics have made this possible. The results of the above have been that new categories of patients are now under treatment and that the quality of life for the patients has increased.

The above described changes in the health care panorama have been made possible because medical research has been very intense during the post war period and that a wide range of new medical devices have been developed, very often in co-operation between medical professionals and industry. Up to the forties medical devices were made mainly of classical materials such as cotton, glass, natural rubber and metal. These types of materials typically resulted in restrictions in device design and product function. Ways out of this situation were offered as a result of developments in polymer technology, both concerning types of material and manufacturing techniques. Using these new materials, however, caused another problem. The classical device materials could all be sterilized with saturated steam at 121 °C or 134 °C. Very few of the new materials could withstand these temperatures without deterioration and ruined product function. The possible use of biological warfare during the Second World War had identified the need to decontaminate material and equipment that had been contaminated with biological agents. A number of known fumigants were investigated for their antimicrobial properties but also for their gentleness to various materials. As a result of this research ethylene oxide was chosen because it was found to be highly antimicrobial and very gentle to materials subjected to it. These properties in combination with the fact that the treatment could be performed at room temperature or there about made ethylene oxide ideal as a sterilant for the new generations of medical devices containing heat labile materials. Ionizing radiation was successfully investigated as a sterilant, later than ethylene oxide. Then, treatment of medical devices with $\gamma$-rays started in the fifties. Later on also $\beta$-rays, accelerated electrons, have been added to the arsenal of ionizing sterilization methods. Common for $\beta$- and $\gamma$-ray sterilization is that even if the treatment temperature is low, room temperature, the irradiation very often has strong negative effects on the materials used in the devises. Reduction in strength and a pungent odour have often been noted.

It can be noted that ethylene oxide sterilization is performed both in hospitals and primarily in industry whereas sterilization with ionizing radiation only is used more and more commonly in industry.

Medical research, new materials and new efficient sterilization methods together have given industry totally new possibilities to design the new medical devices which were necessary to realize the changes which were described in the beginning of this paper. One of the technological fields which has been opened up by this development is the intensive generation of nonwovens. I would like to conclude this part of the chapter stating that we see an ongoing change in the spectrum of diseases now possible to cure. This change has been based on the results of intensive medical research but to a very high extent on the development of new medical devices developed by industry, often in conjunction with medical professionals.

The new generations and types of medical devices are obviously designed to fit into new medical treatment schemes. The success of device related medical treatment as a consequence depends both on the reliable performance of the device and the professional use of it by the medical practitioner. The treatment result now becomes a shared responsibility for industry and the medical profession.

The European Community realized, during the 80 s, that the "reproducible" performance of medical devices ought to be regulated by a harmonized legislation throughout the European Union and the EFTA countries. Central reasons for this ambition then was both that the health care spectrum in Europe is reasonably homogenous and that all European countries depend to some 80% on import of medical devices. Therefore, despite the company or country of origin for a device, the safety of the user and the patient require that a medical device reliably performs every time as the manufacturer intends and claims it to perform and that the user expects it to perform.

In cases where the Member States of the European Community have agreed to find and apply a common way to deal with a specific issue that way has to be described in a contractual way to all Member States and to all citizens and official bodies of the Member States. An often used way to do this is by means of a European Directive agreed in and issued by the Council of Ministers. During later years the Directives have often been of the New Approach type where only requirements on safety and working environment are stated in what is called Essential Requirements, ER's. ER's describe in a generic way the aspects on safety, which Member States have agreed to be relevant for a specific type of products. The technical contents for each aspect are normally not specified. An adopted European Directive has to be transposed into national legislation in all Member States in order to be effective.

A parallel organization for European co-operation is the European Standardization Organization, CEN. Members of CEN are all the standardization bodies of all

EU and EFTA countries plus a few of the former east European countries. All CEN member states have the right to participate in a European Standards project if they so wish. A further right given to CEN member states is to comment and vote on the new or revised European Standards being the result of the drafting work. Assuming that a new text has been approved in the voting process, then the new standard has to be published in an unchanged form in all countries and all national standards on the same subject have to be withdrawn. European Norms (ENs) therefore become an important tool for technical harmonization as the Directives are for the political harmonization.

The European Commission has been given the possibility to request CEN to draft necessary European Safety Standards for New Approach Directives and to harmonize them through publication in the Official Journal where also new European Directives are published. A further means to increase the strength of Harmonized European Standards is to include text in a Directive stating that compliance with such a relevant Standard shall be regarded as a way to show compliance with the Essential Requirements of that Directive. The Harmonized European Standards thereby form the language by which manufacturers, users and competent authorities may communicate with each other that a product satisfies the requirements on performance and safety, described in a European Directive. An applied CE marking is the way for the manufacturer to signal compliance with safety requirements to the user.

The element of communication is essential in the European safety and performance legislation. The manufacturer shall communicate to the presumptive user what performance and intended purpose a particular product has. The user then has to match these with the needs he has. If the claims of the manufacturer match user needs then that product can be considered for use. The responsibilities following the use of the product for other purposes than those stated by the manufacturer fall entirely under the responsibility of the user.

In the light of the recognized needs in the European countries for ensuring medical devices consistently exhibit the claimed properties and performance, the European Council in 1993 agreed on a Directive on Medical Devices to be mandatory after a transitional period of five years, 1998. The Directive is of the New Approach type and therefore requires that products, which comply with the essential requirements and therefore may be placed on the market, shall carry the CE marking. The Medical Device Directive (93/42/EEC) basically describes three areas: (1) an administrative system for the manufacturer, (2) allocation of responsibilities and, for this context the most important part, (3) the essential requirements (ERs) to be fulfilled by all medical devices to be placed on the European market from June 1998.

A Medical Device is defined in the Directive as:

"Medical devices are all single or used in combination instruments apparatuses, equipment, materials and substances or any other objects including software that serve the purpose to be used in human medicine as follows:

- to diagnose, prevent, control, treat or cure diseases
- to diagnose, control, treat, cure or compensate injuries or handicaps
- to examine, replace or to change the anatomic situation or a physiological process
- to control perception

With regard to there defined effects in or at the human body, the sad medical devices cannot be substituted by any pharmacologically or immunologically effective means or by metabolism, the way of how they effect, however, can be supported by such means. Regenerated medical devices are equal to novel medical devices."

This definition obviously encompasses a very wide range of devices. It harbours everything from computer the tomographs to the tung spatula. In between in complexity of risk spectrum such typical nonwoven products as drapes and gowns for the operating theatre and many wound management products, e.g. dressings, can be found.

The *administrative system* shall be complied with by all parties placing medical devices on the European market. To "place a product on the market" for the first time is considered to be done by somebody who is a separate legal entity than the user. An industry, in the case of operating theatre barrier textiles, a leasing laundry or a hospital with its own reprocessing services are regarded as manufacturers in the eyes of the regulations in most European countries. An operating department reprocessing devices for its own use is not regarded as a manufacturer in most countries.

The administrative system describes the way a manufacturer shall proceed to satisfy himself that a product can be CE marked and put on the European market. The *allocation of responsibilities* stated by the Directive is quite clear The manufacturer shall follow the administrative procedures outlined above. In these he shall ensure/ declare that all safety documentation and risk analysis performed by him or in co-operation with an independent third party, a notified body, has concluded in a decision that the product complies with the ER's of the Directive and therefore may first be CE-marked and then be placed on the European market. *Responsibilities* are shared by the four parties of the supply chain: the manufacturer, the notified body, the competent authority and the user of the product.

As said above, the manufacturer shall state and be able to present documentation for the performance of the product. Products may be classified in one of four classes depending on complexity of the risk spectrum associated with their use. This is to ascertain an optimal use of surveillance capacity and costs. In one end of the scale we find a situation of self certification. This applies e.g. to manufacturers of operating theatre barrier materials. In the other end we will find implants, where type testing and approval by a notified body are required. The more complex the risk spectrum is for a certain product, the more involvement and consultation with a notified body the manufacturer needs. A notified body is an independent consultant or test house that has been approved by the appropriate national official body for certain type/-es of medical devices. The guardian in each Member State that the system operates as intended is the competent authority. This body maintains the links to the manufacturers within its territory, to other competent authorities and to the European Commission. As described before Essential Requirements are the properties that medical devices must exhibit in order to be placed on the European market and then freely sold in the whole European area. Again, the signal to the user that the manufacturer and the product comply with the requirements of the Medical Device Directive is that the product or its package carries the CE-marking. It must be underlined that the Essential requirements focus on three areas: safety for the user, i.e. the hospital personnel, the safety of the patient and the product performance as claimed by the manufacturer. The various aspects on safety and performance can be seen in following:

– design and solutions
– performance
– chemical, physical and biological properties
– infection and microbial contamination
– construction and environmental properties
– measuring accuracy
– protection against radiation
– energy sources
– information supplied by the manufacturer
– reliability of clinical data

The requirements may be divided into three areas: general, design and construction and information to the user.

## 11.1
## General requirements

All forms of health care are associated with risks for the user, the patient and in some cases a third party. Therefore risk analysis and risk management are fundamental principles of the Directive. The principle is that avoidable risks must be avoided and the remaining risk must be acceptable when compared with the positive outcome of the treatment. Documented risk analysis is the procedure that a manufacturer must go through to identify possible risks for the above groups of persons. Typical questions to be asked are: Could any aspect of product design lead to any risks when the product is used generally or on any particular patient group? Can any of the stages of manufacture lead to risks to clinical condition or safety?

Risk analysis also comprises risk assessment. Here the question is: What would be the seriousness if an identified risk turned into an injury?

Documented risk management is the process used to find ways to change product design, ingoing materials or components or process that would result in the risk being avoided. If the risk cannot be avoided, then it is the obligation of the manufacturer to issue a warning to the user to point out that a certain risk exists and to give advice on how the probability for injury could be minimized. An example could be a latex glove. We know that allergy against natural latex exists. If the risk is indicated in labelling, e.g. "Contains natural latex", the user might choose a vinyl glove instead. We also know that the glove powders are important carriers of the allergen. Therefore washing off the glove powder before a procedure would reduce the risk.

The manufacturer must be aware of the life time of his entire product, and state it. This is logical as the claimed performance of a product shall fit into the total treatment scheme whenever the product is used during the stated life time of the product. All materials are subjected to ageing. Sometimes a manufacturing process such as sterilization speeds up the ageing. Heat treatment might do the same. Product performance such as strength, product softness or packaging integrity might be seriously affected by ageing. Therefore changes of properties over time must be monitored and be the base when the life time of a product is determined. A word of warning must be issued here concerning accelerated ageing test. Ageing is a very complex process in many cases. It means that the manufacturer must convince himself that the parameters he uses to accelerate the ageing are those which apply to the ageing of his product under the conditions which apply throughout the storage and handling and under normal use conditions and stress.

## 11.2
## Design and construction

Local or generalized toxicity and noncompatibility with the tissues and the media with which the various parts of the product comes into contact may critically and negatively affect the performance of a medical device. Agents causing these effects might be knowingly added to the materials of the product, be formed during the manufacture or be external contaminants having come in during transport and handling along the supply chain. This poses a particular problem to the nonwovens industry. The fundamental advantage with the nonwoven material designs is that the necessary properties may be added to the material one by one via fibre properties build up and finishing of the completed construction. This, however, brings advantages as well as disadvantages. Two examples, the biological effect of and the chemical composition of fibre spinning aids seem often to be the secret of the fibre manufacturer, even if the producer of the final converted product ultimately has the responsibility for final product safety. The binding of the fibres might positively increase hydrophilicity if a detergent is added. Such an additive, however, often shows negative effects in biological test systems.

Medical devices often come into contact with either biological liquid systems or with liquids administered to a patient. These liquid systems have properties that are very different from those of water. Various forms of biological activity from e.g. enzymes can be seen as well as different surface tension and ion strength. Leakage of substances as deterioration of the product must therefore not take place when the product comes into contact with the surroundings relevant to the intended use of the product.

A special phenomenon that often applies to nonwovens and that must be taken into consideration is the risk that products may catch fire. This is for example known with use of operating theatre barrier products like drapes and gowns and must be considered in product design or as a warning.

The use of medical devices often takes place in situations when the normal defence barriers of the body, such as skin or mucosa, have been breached. As a consequence there is risk of ingress of bacteria and therefore risk of infection. Those micro-organisms might either come from the device itself or from other sources. To avoid transfer of deviceborne organisms, medical devices are delivered in sterile state and the sterility is preserved by a system of packages. Commonly, innermost a sterility barrier is an integrated part of the device. A number of such devices are then packed in a shelf container which is suitable for internal distribution and handling in the hospital. Lastly a number of shelf containers are packed in a transport box that has the task to protect the devices during distribution and storage. A device might lose its sterility if the package system integrity is not maintained until the use of the product. To demonstrate that package integrity is maintained during transport and storage and handling is therefore as essential as it is to demonstrate that the product is rendered sterile in the sterilization process.

The other infection risk is associated with the design of the device itself. An example of this is patient drapes used only to shield off non disinfected areas of the patient's skin in order to prevent skin micro-organisms from contaminating the operating wound. Single-use drapes are made form nonwovens. To effectively isolate the wound nonwoven drapes are either laminated with a plastic film or treated with water/blood repellent. The edges of the drapes near the wound are equipped with self adhesive edges to prevent contaminated air from under the drapes penetrate through the drape fenestration and to reach the operative field. Now, if design or manufacture of surgical drapes are not such that the above described impermeability can be reliably ensured, then the patient on the operating table is subjected to a serious risk of developing an infection. It is the obligation of the manufacturer to consider and ensure these properties.

The Essential Requirements also list a number of physical properties to be considered. These include such as compatibility with other devices, electrical properties, radiation, temperature, mechanical risks and energy supply. Most of these do not apply to nonwoven products and are therefore not elucidated in this chapter.

## 11.3
### Information to the user

It has been pointed out several times that medical devices now are integrated parts of medical treatment. It is therefore highly important that the manufacturer makes it clear to the user what the intended use of the product is. This is done by two means: instructions for use and labelling. Instruction for use is not needed for products like drapes, gowns and simpler dressings if they can be safely used without it. The example with patient drapes above, however, shows that instructions are very often needed to ensure that the product is used as the manufacturer intends.

The labelling shall give information that is complementary to the instructions for use. The scope of the obligatory labelling:

– name of the manufacturer
– contents of package
– STERILE, when appropriate
– LOT + batch code
– expiry date, year and month
– single-use, when appropriate
– custom-made device, when appropriate
– exclusively for clinical investigation, when appropriate
– special storage and/or handling conditions
– special operating instructions
– warnings and/or precautions
– year of manufacture, for active devices
– method of sterilization

For a number of the requirements there are approved symbols.

## 11.4
## Conclusion

Medical treatment today has opened up totally new horizons in treatment results and in quality of life for patients. Medical devices play a fundamental role to achieve this change. Medical devices made from nonwovens, such as theatre drapes, gowns, masks, head wear or dressings contribute a great deal to better treatment.

The harmonized European safety legislation for Medical Devices as specified in the Medical Device Directive 93/42/EEC also applies to these nonwoven products. Essential requirements to be fulfilled by all CE marked products are given in the Directive.

Medical device manufacturers often rely on supply of raw materials and components from external suppliers. To ensure an unbroken chain of safety measures resulting in a medical device that reliably performs as claimed, it is of the utmost importance that also these suppliers understand their role in the totality of work and that they co-operate with the device manufacturers in an open and trustful way.

# 12
# Nonwovens for cleaning and household products
J. WIRSCHING

Nonwovens for household products are used for a wide range of applications: though mainly for cleaning purposes, i.e. in the field of wiping and floor cleaning, they are also employed as scouring, grinding and polishing media, as filter materials in coffee filters, tea and vacuum-cleaner bags, and for vapour extractors as well. Linings for garden furniture, table cloths, sunshades and fine-wash bags made of nonwovens can also be subsumed under household articles, but these will not be dealt with here.

The spectrum of the nonwoven materials under discussion here is concomitantly varied, both in terms of their mass per unit area (ranging between 25 g/m$^2$ for disposable wiping cloths and 2,500 g/m$^2$ for scouring nonwovens used in mechanical floor cleaning), of their visual appearance and of the manufacturing processes employed to make them. All nonwoven production processes are used and will in each case be assigned below to the product categories concerned, in the ranking of their significance. Furthermore, the intention is to highlight the fact that nonwoven-based cleaning products are used not only by private consumers but also in the field of commercial routine cleaning. Here, long-term product success can frequently be achieved only when taking out-of-the-ordinary requirements into account at the product development stage, such as the cleaning products' autoclavability for use in hospitals, a high level of absorbency in the catering and restaurant sector, resistance to chemicals and solvents for cleaning sanitary facilities and facades, or a high dry-cleaning performance for office cleaning.

## 12.1
## Market situation

For Western Europe, the EDANA (European Disposables And Nonwovens Association) shows a nonwovens consumption of 89,600 tons (corresponding to 1,659.1 million m$^2$) in the cleaning-cloth sector for 1998, thus accounting for 10.7% of total nonwoven consumption in Western Europe, a substantial increase compared to the 6.4% in 1986.

In purely mathematical terms, this produces a low average mass per unit area of approx. 50 g/m$^2$, entailed by the very overproportional production of disposable

**Table 12-1** Percentage distribution of manufacturing technologies for the cleaning-cloth-sector (annual nonwovens production in Western Europe, 1996. Source: EDANA)

|  | *Mass-referenced* | *Area-referenced* |
| --- | --- | --- |
| Wet-laid nonwovens | 16.7% | 30.5% |
| Dry-laid nonwovens, total | 64.5% | 54.6% |
| – thermally bonded | 9.8% | 6.3% |
| – chemically bonded | 30.7% | 24.9% |
| – mechanically needled | 0.7% | 0.3% |
| – hydroentangled | 23.3% | 23.1% |
| Air-laid short-fibre | 18.8% | 14.9% |
| Total | 100% | 100% |

nonwovens in terms of yardage. Re-usable nonwovens, however, whose masses per unit area are usually about 75 g/m$^2$ or above, are traditionally predominant in terms of product value and tonnage. Table 12-1 illustrates the degree of equality for the various manufacturing technologies used for producing cleaning cloths.

What's most conspicuous, apart from the large proportion of area accounted for by the often fully impregnated wet-laid nonwovens, is the enormous increase in hydroentangled or spunlaced products. Even as late as 1981, spunlaced nonwovens accounted for only 6% of the total. Moreover, you can see that the traditional drylaid nonwoven processes (except spunlace), which usually result in the production of durables, continue to constitute the dominant manufacturing technology for cleaning products.

## 12.2
### Wet and damp-cleaning products

This category subsumes not only the classical wiping and floor cloths but also a multiplicity of products given a special finish for particular application purposes (e.g. nonwoven stripes of wet mops, constituent parts of pads for floor flat-wiping systems, scouring pads, etc.).

Composite materials based on nonwovens are increasingly moving into special application categories whose cleaning requirements cannot be satisfactorily met by classical nonwovens made of single or multi-fibre mixtures. The combination of advantages offered by products using different manufacturing technologies to create an improved cleaning article results in a continuous stream of new and interesting product innovations. In this context, several bonding technologies like lamination, sewing or ultrasonic fixing are available to the nonwovens industry. To mention just one example: laminates made of microfibre nonwovens and synthetic leather which combine high cleaning performance on glossy surfaces with the option of smear-free final drying (Fig. 12-1).

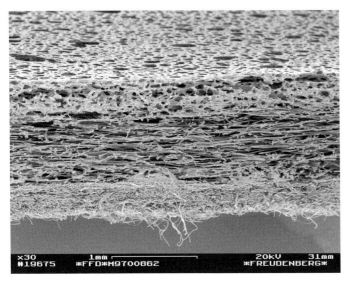

**Fig. 12-1** Cross-sectional scanning electron microscope micrograph of a laminated nonwovens product composed of synthetic leather and a microfibre nonwoven (layer sequence from top to bottom: 1. macroporous synthetic rubber surface, 2. nonwoven absorbent core capacity, 3. microfibre nonwoven)

Traditionally, clothing textiles which had gone past their suitability were used for wet cleaning. The triumphant progress of cleaning products made of nonwovens began at approximately the same time as commercialization of nonwovens in general back in the 50 s and was essentially based on the following properties:

– high water absorption, water storage and dirt retention capacity, thanks to the open structure of the nonwoven when compared to other textiles
– low mass per unit area coupled with a high cleaning performance
– low gliding and static friction forces during cleaning
– good hygienic characteristics, thanks to fast drying speed
– good self-cleaning properties
– inexpensive to manufacture
– easy to wring out
– enhanced absorbency due to capillary forces and the large free inner volume
– product haptics settable within broad limits, thanks to selection of suitable process parameters (choice of fibre and bonding agent, mass per unit area, bonding technique)
– good washability

12.2.1
## Floor cloths and materials for floor cleaning systems

Nonwovens for floor cleaning are usually staple fibre nonwovens with masses per unit area of above 150 g/m². This lower limit is entailed by the heavy mechanical stress to which this product category is exposed in a wet condition, both due to the process of wringing them out and by using them for cleaning uneven, perhaps even sharp-edged surfaces. Their fibre content is essentially made up of cellulosic and polyester, less frequently of polyolefine, polyamide or polyacrylic nitrile fibres.

The cellulosic fibres concerned (cotton, viscose, Lyocell) are responsible for high water retention and storage capacities, the pleasant soft feel when wet, plus a constant water release rate, while the thermoplastic fibres ensure a high wet tensile strength. To achieve a structure that is mechanically stable while at the same time as light and open as possible, the starting material used is carded, mostly cross-laid and slightly mechanically preneedled webs.

An isotropic fibre layer in the web can be achieved by stretching the cross-laid web of fibres in a longitudinal direction. Mechanically needled nonwovens are additionally adhesive-bonded by applying polymer dispersions, mostly as impregnating foam, or by thermal bonding of thermoplastic bonding fibres in convection dryers. The bonding agent may be applied to the mechanically preneedled material either by doctor-blading or by padding. The setting of a bonding-agent gradient through the cross-section enables pilling-resistant qualities to be produced while simultaneously guaranteeing high absorbency. For the same reason, the aim is to achieve spot-gluing of the individual fibres at the respective fibre intersections. The nature and degree of cross-linkage of the polymer dispersion used exerts a marked influence on the product's properties, especially on mechanical strength, resilience, touch, resistance to solvents, washing and cleaning agents, and on aging stability. Vulcanization or auto-cross-linking butadiene copolymers, or polyacrylate bonding agents are frequently used, while polyurethanes or polyvinyl acetates are less preferred options.

Final printing, either on one or both sides, with colour-pigment-loaded bonding-agent printing pastes not only creates an attractive colour design for the product but also constitutes an additional adhesive bonding of the nonwoven's surface-vicinity zones.

For many application categories, the finishing and upgrading steps employed for cleaning cloths must result in washability, and in most cases even in washability at 95 °C. Furthermore, another targeted property is maximized resistance to chemicals, e.g. to heavily acidic sanitary cleaners, alkaline-oxidative bleach (chlorine bleach), aldehydic disinfectants or organic solvents. This puts a marked limit to the number of expedient finishing and upgrading processes.

## 12.2.2
### Durable wiping cloths

They are the "workhorses" among the wet cleaning products and are used in almost every application category encountered in the household.

Essentially, the remarks made in the previous section also apply to this product category, with the exception of a mass per unit area reduced to approx. 100–150 g/m$^2$. This produces values for maximum tensile strength of about 80 N/50 mm, which have proved to be sufficient to achieve a product lifetime of several weeks in daily use.

For these product categories, it is significantly more important to obtain a soft, pleasant, textile-like feel than for the floor cleaning products; equally important is a high absorbency, so as to be able to absorb liquid soiling "in one sweep" if at all possible, and to function as an intermediate water store during washing-up, for example. All of this, in conjunction with the stipulation for low fibre loss and hygienic behaviour due to fast drying will as a rule result in a large proportion of cellulosic fibres being used in the mixture.

Homogenous colouring of the products is usually achieved by using spun-dyed synthetic fibres, or by dying beforehand in the fibre flock.

When highly viscous printing pastes, or those with increased cross-linkage, are applied in an embossing process, the resultant cross-linked polymer ribs can quite definitely be said to possess a slight abrasiveness, which can be used to remove dried-on or firmly adhering residues.

## 12.2.3
### Disposable and semi-disposable wiping cloths

This category subsumes an extraordinarily large number of products from most varied origins. The salient criterion for the selection of suitable fibre blends is the option for using inexpensive raw materials like polyolefine fibres or cellulose. Shared characteristic is, besides the frequently low durability, only the small mass per unit area of less than 80 g/m$^2$. For this reason, typical representatives of this category are listed below, though without any claim to completeness:

a) *Longitudinally laid staple fibre nonwovens, bonded by pressure or spunlacing, with or without perforation*

These cloths often contain a high proportion of polyester or polypropylene fibres and therefore possess only a limited water storage capacity. They are essentially competing with the higher-quality durables without reaching the latters' broad range of properties, and are therefore only suitable for simple cleaning jobs. Their useful lifetime seldom exceeds one week only in exceptional cases, since their tear strength (especially in the transverse direction) is low, due to insufficient bonding.

b) *Pre-moistened wet wipes and tissue*

This product category was commonly referred to cosmetical or hygiene articles rather than to hard surface cleaning cloths. As ready-to-use convenience products they became very popular just recently. Materials are often soft and absorbent and contain in many cases cellulosic pulp or short-cut fibres and are based on longitudinally laid, random-laid, air-laid or hydrodynamic wet-laid nonwovens technologies. They are usually hydroentangled or chemically bonded and can contain a monofilamentary fabric or a spunbonded nonwoven base for purposes of stability. Since they are usually disposables pure and simple, they are produced in large quantities and frequently packed in dispensers. As pre-moistened wet wipes, products are impregnated with detergents or cleaning solutions. These products are often positioned against paper, which is inexpensive but significantly harder in terms of touch, for example in the fields of cosmetics, hygiene or baby wipes.

c) *Electrostatic dedusting cloths*

A completely new product category has evolved in the late 1990's: single usage cloths which are technically similar to the above mentioned wet wipes, but predominantly based on synthetic polymer fibres such as polyester or polypropylene. They are intended to be used simply by hand or as single-use floor cloth by means of a floor cleaning tool with pad holder. Mass per unit area is normally in the range of 50–100 g/m². Prerequisite for the dust-catching effect is that fibres are made of highly insulating polymer material. If charges are put on these fibres via the wiping motion, they will get trapped there (triboelectric effect). The electrical field of the so-called "electret" material can thus guide charged dust particles to the fibre surfaces where they are firmly hold by opposite charges.

Interestingly, the spunlacing production process enhances the insulation properties of the synthetic fibres as it washes off polar spin finishes.

d) *Impregnated cloths*

These are special products, and can be used for a multifaceted range of special applications: for industrial and machine cleaning, as towelettes, as dust-collecting or antistatic cloths, as demisting cloths, lens cleaners, polishing, hygiene or cosmetic cloths.

The respective nonwoven base materials are produced using the familiar technologies and then finished with solvents, paraffin oils, tensides, grinding pastes, fragrances, silicones, or other chemicals. This kind of finish is not wash permanent, so that after having been used a few times, the cloth's properties will have been exhausted.

## 12.2.4
### Synthetic leather (Chamois type)

As a composite nonwoven, these cloth products emulating chamois leather consist of a basic nonwoven as the absorbent core and outer layers of synthetic rubber or polyurethane, with resultant masses per unit area of more than 150 g/m².

The macroporous nature of the rubber or thermoplastic elastomer outer layer
permits not only smear-free doctor-blading of water drops from target surfaces but
also the latter's transport into the absorbent core of the cloth. This property of
smear-free wet cleaning is a stipulation particularly in the field of window, glass
and mirror cleaning, and in the commercial sector quite generally for office and
hard surface cleaning.

When compared to chamois leather, these products are usually more absorbent,
do not become slippery, and can be washed at 95 °C, which renders them suitable
for use in hospital hygiene, for example. One disadvantage compared to natural
leather is their lower resistance to chemicals and aging, which is due to oxidation,
which in turn makes rubber more brittle.

## 12.3
### Dry and damp-cleaning products

For these products, a high cleaning performance is required in special application
categories. High water absorption capacity, by contrast, is usually not among the
stipulations.

## 12.3.1
### Microfibre nonwovens

By definition, microfibres possess a fibre fineness of <1 dtex. A particularly high
cleaning performance for greasy soiling like finger prints or ultra-fine dust particles but feature only microfibres in the range of 0.1 to 0.2 dtex. These can be created by fibrillation of thermoplastic bicomponent split fibres.

This extraordinary cleaning efficiency in a dry or damp condition in some cases
enables tensides to be entirely dispensed with when removing soiling with a
grease content; it is essentially based on the large inner hydrophobic surface of
the cloth, which becomes fully effective only by the capillary force action due to
the high number of small cavities in the micrometer range. In turn, these capillary forces are responsible for leaving behind ultra-fine water droplets in the sub-
micrometer range on the surfaces cleaned, which then evaporate without forming
any visible lime stains.

Modern nonwoven products made of these bicomponent fibres are produced by
carding and subsequent hydroentanglement processes. The high water jet pressures used in the latter process step cause efficacious fibre interlacing and simultaneous fibrillation of the mutually incompatible split fibre components (e.g. poly-

**Fig. 12-2** Cross-sectional scanning electron microscope micrograph of an Evolon™ endless microfilament nonwoven after splitting of the bicomponent fibres by the spunlacing process

ester and polyamide-6). This can be followed by further chemical or mechanical finishing steps. The Freudenberg Nonwovens Group has recently introduced the Evolon™ technology which basically combines fibre spinning, web-formation, fibrillation and web-bonding in one continuous process. The resulting endless microfilament fabrics (Fig. 12-2) combine multi-directional high tensile strength and very good tear-resistance with a textile-like drapability.

**Table 12-2** Selected quality features of spunlaced bicomponent microfibre nonwovens: a) staple microfibre nonwoven, b) Evolon™ endless microfilament nonwoven (mass per unit area: 85 g/m²; composition: 70% polyester, 30% polyamide)

| Properties[1] | Staple microfibre nonwoven | Evolon™ endless microfilament nonwoven |
|---|---|---|
| Thickness | 0.6 mm | 0.5 |
| Maximum tensile strength, MD | 200 N/50 mm | 285 N/50 mm |
| Maximum tensile strength, CD | 90 N/50 mm | 230 N/50 mm |
| Elongation at break, MD | 40% | 40% |
| Elongation at break, CD | 60% | 50% |
| Modulus 10%, MD | 30 N/50 mm | 130 N/50 mm |
| Modulus 10%, CD | 9 N/50 mm | 60 N/50 mm |
| Water release, moist | 0.9 g/m² | |

1) MD, CD: machine direction, cross direction.

Microfibre nonwovens, consisting solely of thermoplastic fibres, are extraordinarily hard-wearing and aging-resistant. Not least due to their high mechanical stability (Table 12-2), they have no problems in outlasting several hundred machine washes. Apart from household and office cleaning, such products are also used in industrial cleanroom cleaning.

## 12.3.2
**Polyvinyl alcohol nonwovens**

Haptics and visual appearance of these mostly brown nonwoven products based on polyvinyl alcohol are largely similar to the natural model, chamois leather. In addition to the very high resistance to a multiplicity of even aggressive cleaning media, the most notable characteristic of PVA nonwovens is their good sliding properties of the damp cloth on most of the surfaces to be cleaned. For human-engineering reasons, therefore, they are the preferred choice of commercial cleaning firms for office cleaning.

The manufacturing process is based either on impregnation of a basic web made of polyvinyl alcohol fibres with aldehydic cross-linking systems or on thermal bonding by means of crystal formation.

In the first case, the bonding liquid contains, among other things, liquid or soluble aldehydes which cross-link the individual fibres through the PVA hydroxyl groups into a firm matrix by means of acetalization.

The salient characteristics of PVA nonwovens are firstly the hard, cardboard-like touch of the material in dry condition and secondly the extraordinarily high shrinkage at washing temperatures above 70 °C which renders the product unusable.

## 12.3.3
**Impregnated cloths**

Some of the impregnated cloths already dealt with in Section 12.2.3 cannot be regarded as disposable cloths, in view of their high mass per unit area, sometimes >100 g/m². The cloths in question are mostly antistatic, dust-collecting or demisting nonwoven cloths. Since the application categories are more or less the same, however, these products will not be dealt with in further detail here.

## 12.4
**Scouring media** [1]

Whereas absorbent, fine cellulose fibres and soft, elastic bonding agents are used for wiping and drying cloths, raw materials with exactly the opposite properties are employed for scouring materials: coarse synthetic fibres made of polyamides, polyester, and less frequently polypropylene of 15 to >200 dtex. Very hard and tough phenolic formaldehyde resins (and sometimes also polyacrylates, urea or melamine resins) are used as bonding agents. To this are added substances with an abrasive effect: minerals ground to a fine or not so fine grain size in various hardnesses, e.g. chalk, quartz, corundum and the extremely hard silicon carbide. They are incorporated into the synthetic-resin bonding agent and in conjunction with the entire nonwoven's elasticity permit a broad variation in scouring intensity, ranging from mildly polishing (scratch-free) to coarse-roughening.

A colour code has found wide international acceptance for identifying their aggression and their purpose of use.

### 12.4.1
### Pan scourers, scouring sponges and pads

For sensitive surfaces like Teflon-coated pans, car paintwork, fittings or plastic surfaces, scouring sponges with a mild abrasive effect are mostly white or blue, for aluminum or steel pans mostly green or black with a more aggressive abrasive grain. As a consequence of their open structure (caused by coarse-titered fibres), scouring sponges are also able to absorb scoured-off particles for easy removal when rinsed. Hand pads consist solely of the scouring nonwoven, whereas scouring pads laminated onto sponges are meant to be used for treating large surfaces. The sponge serves as a reservoir for water and cleaning agent while simultaneously protecting the hand from the scratchy scouring nonwoven.

### 12.4.2
### Floor pads

Another field of application is commercial floor cleaning of office and sales areas, hospitals, airports, etc. Here, pad-shaped scouring nonwovens used under rotating cleaning machines serve to remove coarse dirt (cleaning after conclusion of building work, basic cleaning), daily soiling (heel marks, walking traces, dust) and for mat and high-gloss polishing. The structure of the nonwoven used for these pads is correspondingly varied (ranging from open to dense), as is their scouring effect (from coarse to mild).

Here, the nonwoven colours denote the following meanings:
– white for high-gloss polishing (extremely mild and dense structure)
– yellow or light-brown for gloss-polishing (mild and dense structure)
– red for daily care, given a normal degree of soiling (cleaners), with or without spray wax (less mild and an already more open structure)
– green for wet scouring using water and cleaning agent(s), without removing a coating (coarser and open structure)
– black for the removal of coarse dirt in cleaning procedures after conclusion of building work and during routine cleaning (very aggressive and open structure)

Since the cleaning machines on offer feature driving plates of various sizes, an equally varied number of pad diameters is necessary (127–650 mm). Moreover, the pads are available in different thicknesses, ranging from 6–25 mm, with masses per unit area of about 250–2,500 g/m². Relatively thick pads (from 20 mm upwards) are usually referred to as "super pads" or "long-life pads", with their advantages including not only the over proportionally longer useful lifetime but also a higher thickness elasticity, making them better suited to compensate for any unevenness in the floor covering. This means more even cleaning and polishing effects are achieved over the surface as a whole.

Furthermore, special pads with a particularly open structure are also available for what are called "high-speed machines" (400–1,000 rpm), which enable higher performance per unit area and/or more intensive polishing and cleaning effects to be achieved.

With regard to the arrangement of the fibres in the scouring nonwoven pads, a distinction is made between two manufacturing processes. In the aerodynamic weblaying process, fibres are arranged not in parallel to the nonwoven plane but at an angle to it (roof-tile structure). As the floor pad rotates, the friction forces act on one half of the pad at an acute angle, and on the other half at an obtuse angle to the fibre axis. With cross-laid nonwovens, the fibres are arranged in parallel to the nonwoven plane, and thus in parallel to the floor surface as well. The consequences of this are not only a uniform polishing and scouring effect but also wear and tear that is evenly distributed over the entire area, especially over the border zones, which means the pad will in most cases have a longer lifetime.

**References to Chapter 12**

[1]   Päßler M (1982) in Vliesstoffe (eds.: Lünenschloß J, Albrecht W), Georg Thieme Verlag, Stuttgart 1982, 298 f

# 13
# Nonwovens for home textiles
W. STEIN

More than $2.0\times10^3$ million m$^2$ nonwovens are used in home textiles worldwide. Measured in terms of roll goods, this represents a fraction of approximately 6.5%. According to information from EDANA, approximately 5% of the nonwovens produced in Western Europe are used in interior design, including upholstery material. Nonwovens for home textiles relate to the durable products group. In particular, textiles including felts from wool, horsehair, jute and cotton were previously processed in the upholstery industry.

Today nonwoven fabrics, which have been produced by different methods, have acquired a significant share of the market in the home textiles sector. In the USA the consumption of nonwovens in room fittings, including upholstery materials, comes to approximately 11.5% of the durable products. This gives the following ranking for the products which are produced by different technological processes:

– high-bulk nonwovens
– needled nonwovens
– filament nonwovens
– random-web nonwovens

The use of nonwovens in interior design is concentrated chiefly on upholstery materials, floorcoverings, wallcoverings and furnishings. The task of the interior designer is to use the nonwovens, which are supplied as roll goods, in

– the production, repair and restoration of upholstered furniture
– the laying of textile and flexible floorcoverings
– the production of wallcoverings
– the production and fitting of window furnishings

## 13.1
### Nonwovens for upholstery material

In this case nonwovens are used as upholstery supports and as upholstery covers. They are particularly used in the upholstered furniture industry and in the production of mattresses. Nonwovens are used in the case of approximately 80–90% of foambacked upholstery and mattresses. Whereas plain-weave cotton fabric were previously used for upholstering and mattress covers, today filament nonwovens based on polyester or polypropylene are being increasingly used in this sector.

The processor makes the following demands on these nonwovens. They must have equally high strength in all directions, be processable in any direction and not fray at the cut edges. They must have a high area and dimensional stability.

Nonwovens for covering are known in the specialist language as polyester wadding or polyester nonwovens. Depending on the type of bonding, a distinction is drawn between:

– bonded polyester wadding or polyester nonwovens for covering upholstery materials in cases parts of the upholstered furniture are not subjected to much stress
– thermobonded nonwovens as covers for foambacked upholsters
– nonwovens laminated or quilted with cheesecloth or locknit on one or both sides as covers for foambacked upholstery with high dimensional stability

Until recently the main parts of upholstered furniture and mattresses were produced from foam, as it has good elasticity, compressibility and recovery properties. More recently nonwovens have also been produced as 'fibrous webs' from bicomponent fibres which form good bonding points when heat and pressure are applied. These are stable under dynamic loads and ensure that the thick, bulky nonwovens have high compressibility and a good recovery capacity. Such nonwovens are produced from high-bulk fibrefill with a bicomponent fraction of up to 25% of the same polymer. After web forming and hot-air bonding, the nonwovens are compressed to their required thickness and cooled down between steel belts in a cooling zone. These nonwovens are equal in quality to the foams at the same thickness. They have good air permeability and ensure a high level of comfort in the case of upholstered furniture.

Nonwovens, e.g. filament nonwovens, have to be very strong and have a high load-bearing capacity. No special requirements are set with respect to the elastic recovery. Fig. 13-1 shows a section through nonwovens for upholstery with a spring core. The spring core is simultaneously used for insulating against any noise which may occur as a result of spring movement in the core and to prevent the core springs from penetrating through the foam.

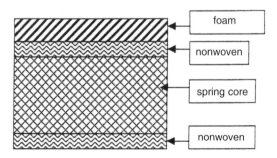

**Fig. 13-1** Nonwoven fabrics as support material and covering for spring cores

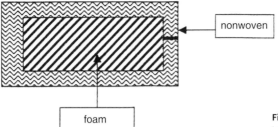

**Fig. 13-2** Wrapping a foam core in a nonwoven fabric

Nonwovens for covering foams cores are either laid around (Fig. 13-2) or only on the foam part.

These nonwovens must be slip-resistant. In some cases they are bonded to or sewn to the padding. Wrapping of the foam part ensures that the padding be drawn more easily into the outer covering. Further, this nonwoven provides protection against the fibres falling out in the case of plush fabrics. This can occur if the creases in the upholstery fabrics produced by sitting coincide with the same creases in the foam.

## 13.2
## Nonwovens for floorcoverings

Floorcoverings play a significant role in the home textiles sector. Needled nonwoven fabrics are the ones which are usually processed. They constitute approximately 20–25% of all the floorcoverings produced in Germany.

As a result of further developments in needling technology, especially in patterning when the two-stage method is used, these textile floorcoverings have held their own in the home textiles sector, both with regard to both indoor and outdoor use. Rib or velours effects can be achieved by structuring the needled fabric surface.

In some cases, particularly in the household sector, two-colour patterns are available with high-low effects. Needlefelt floorcoverings are mainly produced from polypropylene or polyamide fibres. In the production of textile floorcoverings with a smooth surface, using the two-stage process, the covering layer can be produced from primary fibres and the bottom layer from waste fibres. High dimensional stability is achieved by using reinforcing fabrics or back coating. Needled nonwoven fabrics are preferably supplied as roll goods in widths of 2,000 mm. Widths of up to 5,000 mm are possible. Fabrics which have been converted into carpet tiles mostly have the standard dimensions 500×500 mm or 330×330 mm. In addition to needled nonwovens being used in the household sector, they can also be employed in the public sector if the floorcoverings have a high wear resistance. These end-use sectors may be trade fairs and exhibitions, schools, hospitals, offices, tennis arenas and the departure lounges at airports, e.g. the departure lounge at Duesseldorf Airport with an area of 10,000 m$^2$. The main demand is for high wear resistance. An antistatic and/or flame-resistant finish is required for certain parts of the public sector. The comfort index must also be adapted to the given part of the public sector. Coloured floorcoverings must have a high light resistance. Needlefelt carpets normally require to be stuck down to the floor over the whole of their surface. Therefore the consumer expects that the carpet can be laid free of bubbles and creases during the first coarse laying. It is also expected to have a certain 'own weight' the consumer also expects this property if laying is effected with adhesive tape which is sticky on both sides and with the so-called 'bonding grid'. Neither of these laying methods is very durable and only permit very limited adhesion between the floor and the fitted carpet. In total-area adhesion the floorcovering must not undergo any dimensional changes during cutting or adhesion.

Easy-care properties and stain resistance are important for the consumer. In the case of the majority of needlefelt floorcoverings these properties are usually arrived at using the appropriate finishing treatment. In order to be able to give optimum advice the consumer when selecting a floorcovering the processor can sometimes refer to the RAL certificate which is granted by the German Institute for Quality Assurance and Labelling to product manufacturers on a voluntary basis. This also gives information about the suitability and comfort of needlefelt carpets and about additional properties, e.g. whether they are antistatic, whether they pass the castor chair test, etc.

More recent floorcoverings are those produced by the Starfloor method, which is a combination of yarn and nonwoven technology. These are wall-to-wall goods. This product, which was introduced in 1998 by DLW (Deutsche Linoleumwerke), has a three-dimensional structure which a needled nonwoven does not have and thus opens up new horizons for the interior designer.

In addition to the needled nonwovens already mentioned, the floorcovering sector also makes use of adhesive-bonded webs. In the majority of cases these adhesive-bonded webs are provided on one side with a web of strong, stable synthetic fibres and with a covered adhesive layer on the other. These adhesive-bonded webs are mostly used as 'skid-resistant carpets' when one textile covering is laid over another. The adhesive-coated side is bonded to the base of the top carpet. The web side faces the bottom carpet. The fibres of the adhesive-bonded web mesh with the knops or the pile of the lower carpet and form a bond which is not very permanent.

However, the adhesive-bonded webs can also be bonded to a smooth, non-textile surface by the adhesive-coated side. The textile goods are rolled on to the top web side and rubbed or pressed on. This method of laying fitted textile carpets is used when nothing must be stuck onto the carpet or if the textile carpet so laid has to be removed rapidly and in a non-worn condition.

As the laying of textile floorcoverings on adhesive-bonded webs is not a long-term solution and bonding means have to exit which can be removed again, these adhesive-bonded webs are being superseded on the market by other, more long-lasting products.

A further end-use sector for nonwovens in the home textiles sector is as support material for tufted carpets (Section 13.5).

## 13.3
## Nonwoven fabrics for wallcoverings

Nonwoven fabrics are only used to a limited degree in wallcoverings. Thus, for example, use is made of textile wallpaper, where medium- to heavyweight paper is laminated with, amongst others, fibre webs and then coated with polyethylene to form wallcoverings. Further, nonwoven wall hangings, so-called 'furnishing felts', are used as wallcoverings. These wall hangings are needled nonwoven fabrics from synthetic fibres. Walls can be suitably decorated by using the appropriate surface patterning. During processing these nonwoven fabrics, which have an areal density in the range 150–300 g/m², can be stuck down over their entire surface or stretch-laid.

In addition nonwoven fabrics are used as backings for textile wallcoverings. As far as production and properties are concerned, these nonwovens resemble needlefelts or stitch-bonded fabrics. The nonwovens which are used as backings for textile wallcoverings are used as thermal and acoustic insulation. Additionally their purpose is to fill the space between the wall and the tensioned textile (Fig. 13-3).

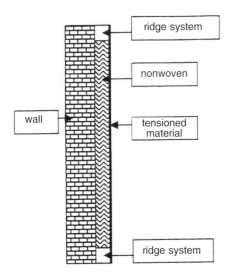

**Fig. 13-3** Nonwoven as backing for wall hangings

The demands which are made on these nonwoven fabrics depend mainly on the end-use. The thickness of the nonwovens may not appreciably exceed or be less than the distance between the wall and the textile. It is thus dependent on the type of insulation required. The back of the nonwoven fabric should be smooth and relatively strong, as it must be accurately bonded to the wall.

## 13.4
### Nonwoven fabrics as furnishing fabrics

The manufacturer and consumer of furnishing and curtain fabrics expects, amongst other things, that these will be dimensionally stable, light- and water-resistant, slip-resistant and easy to care for. Nowadays furnishing fabrics are still predominantly woven or knitted. Nonwovens are used only as means for stiffening transverse hangings (lambrequin/saddlecloth). In this case they are mostly nonwovens which are coated on one or both sides with hot-melt adhesives and ironed on to the furnishing fabric as a reinforcing lining. However, the nonwoven linings described here are also frequently replaced by plain-weave fabric strips which are therefore known as stiffenings.

In the USA tanglelaced webs have appeared on the market as net and heavy curtains. The spun-laced products have a good tear strength, a soft handle and a good appearance. Further processing can be effected as in the case of other 100% polyester products. They have easy-care properties, are stable and do not fray. The crease recovery is good and shrinkage during washing is 1.5%. Besides being used as heavy curtains, these materials can be given a water-resistant finish on one side and used as roller blinds or shower curtains. As a rule the spunlaced products are bonded with approximately 3–5% acrylate binder, which can have a negative effect on the drape. The washability is not good because of the relatively poor binder adhesion. Products from polyamide which are bonded with binder fibres – mostly bicomponent fibres – have somewhat superior properties, e.g. the so-called 'melded fabrics'.

Up to now nonwoven fabrics have not been used to give protection against the sun (awnings, sun-blinds, vertical strips, etc.).

## 13.5
### Tufting substrate
J. M. SLOVACEK

Tufted carpets consist of 3 elements: pile, tufting substrate (or primary backing), and secondary backing. Of these three, the primary backing is sandwiched in between pile and secondary backing: invisible to the enduser, it is nevertheless an important factor in determining carpet properties and influences some aesthetic aspects. The substrate's role goes beyond enduse properties, however, and is also quite significant in the various processing stages: tufting, dyeing, coating, and – where applicable – molding.

Substrate properties depend on the choice of raw material and manufacturing specifics. The principal raw materials in current use are polypropylene (PP) and polyester (PET). Of these two, PP has the lower strength. Its thermal resistance is also lower, and it is harder to stabilize against the effects of climate change, more flammable, and capable of only moderate adhesion to coating materials.

Table 13-1 Carpet manufacturing requirements on tufting substrates

|  | Woven fabrics | | Nonwovens | | |
|---|---|---|---|---|---|
|  | PP woven | PP woven + fleece | PP | PET/coPET | PET/PA bico |
| Tuftability | ++ | ++ | ++ | + | + |
| Necking | ++ | ++ | – | - | - |
| Needle wear | + | + | + | + | - |
| Tear resistance | ++ | ++ | + | + | + |
| Dyeability | - | +[1] | - | - | + |
| Stitch holding | ++ | ++ | + | + | + |
| Bowing | – | – | + | ++ | ++ |
| Skewing | – | – | + | ++ | ++ |
| Pile evenness | - | - | ++ | ++ | ++ |
| Adhesion to coatings | - | +[1] | - | + | ++ |
| Moldability | - | - | - | ++ | ++ |

Ratings:
– very poor     + good
- poor     ++ very good

[1] In contrast to the PP woven fabric, the PP woven + fleece is dyeable and features good adhesion to coatings.

From the manufacturing perspective, there are two fundamentally different tufting substrates: woven substrates and nonwovens (such as spunbonds). Woven substrates are composed of a warp and a weft at right angles to one another, which is why the fabric is not isotropic. Since the warp and weft do not bond, dimensional stability is not very high, but the resulting mobility of the threads has a positive effect on pile anchorage and tear resistance.

Table 13-2 Enduse requirements on tufting substrates

|  | Woven fabrics | | Nonwovens | | |
|---|---|---|---|---|---|
|  | PP woven | PP woven + fleece | PP | PET/coPET | PET/PA bico |
| Dimensional stability | – | - | + | ++ | ++ |
| Bowing/skewing | – | - | + | ++ | ++ |
| Tear resistance | ++ | ++ | + | + | + |
| Molded stability | – | – | - | ++ | ++ |
| Nonfraying behaviour | – | - | + | + | ++ |
| Flame retardancy | – | - | - | ++ | ++ |
| Pile anchorage | + | + | + | + | ++ |

Ratings:
– very poor     + good
- poor     ++ very good

**Table 13-3**  Recommended tufting substrates for various enduses

|  | Broadloom | | Tiles | | Car carpeting mats | Dust control mats | Bathroom mats |
|---|---|---|---|---|---|---|---|
|  | C | D | C | D |  |  |  |
| PP woven | × | × |  |  |  |  | × |
| PP woven + fleece | × | × |  |  |  |  | × |
| PP nonwoven | × | × |  |  |  |  | × |
| PET/coPET nonwoven | ×[1,2,4] | ×[1,2,4] | ×[3] | ×[3] | ×[5,3] | × | × |
| PET/PA bico | ×[1,2,4] | ×[1,2,4] | ×[3] | ×[3] | ×[5,3] | × | × |

C=contract, D=domestic.
1) For high flame retardancy
2) For high pattern definition
3) For high thermal and dimensional stability
4) For low bowing/skewing
5) For high moldability

Spunbonds consist of multidirectionally laid filaments which, for tufting purposes, are made to bond to one another by thermal, mechanical, or chemical means, or a combination thereof. Given the filaments' multidirectional distribution, spunbonds are practically isotropic. Dimensional stability and nonfraying behaviour are a factor of the number of points at which the filaments bond. They are always better in spunbonds than in woven fabrics. On the down side, pile anchorage and tear resistance are lower than in a woven fabric.

The primary carpet backings in most frequent use are: PP woven fabric, PP woven fabric combined with a fleece, PP spunbonds, PET/coPET spunbonds, and PET/PA bicomponent spunbonds. Tables 13-1 and 13-2 rate performance of these substrates from the carpet manufacturing and enduse angles.

The requirements on tufting substrates are a function of the envisaged use for the carpeting. There are five different applications: broadloom carpeting, tiles, car carpeting, dust control mats, and bathroom mats. For broadloom and tiles, a further distinction can be made between contract and domestic uses. In spite of these clearcut categories, it is possible only to give some general suggestions in regard to substrate suitability for particular uses. This is due to the fact that there are so many variables to account for. Nevertheless Table 13-3 suggests broad end use areas for the various tufting substrates.

# 14
# Nonwovens for apparel

## 14.1
## Nonwoven interlinings
H.-Cl. Assent

### 14.1.1
### Introduction

The generic term "nonwoven interlinings" defines materials based on nonwovens that are incorporated into articles of clothing during production to fulfil a range of functions.

The processing methods used can be divided into sewn and bonded interlinings (fusible interlinings).

Sewn interlinings are incorporated between the shell and the lining material during the sewing process. Bonded interlinings are fused to the shell, lining or another inlay material by a bonding process (heat sealing process). The ratio of sewn to bonded interlinings is currently approximately 20:80.

### 14.1.2
### History of nonwoven interlinings

The use of nonwovens as interlinings goes back to the years 1947/48. While the first sewable nonwoven interlinings were available in sheet form in 1947, in 1948 production began of yard goods, the form commonly used today. These were fibrous nonwovens bonded by means of an aqueous binder.

Nonwoven interlinings are therefore one of the oldest successful applications of nonwovens. Even by 1960, they were dominant on the nonwovens market in the Federal Republic of Germany, with a share of over 60%.

In the mid to late 1950 s, the winning streak of fusible nonwoven interlinings began, which today, as we have seen, have a share of around 80% of the total market. The first fusible products were nonwovens bonded with binders until at the start of the 1960 s, the first binder-free nonwoven interlinings were developed. They were bonded with thermoplastic fibres by full-width calendering and had a stiff, rather brittle hand. Spunlaid nonwovens, which appeared in the mid-1960 s, gained importance in the interlinings sector as "adhesive nonwovens". They were made from spun melded filaments and served as a processing aid for bonding textile fabrics. At the start of the 1970 s, the first binder-free nonwoven interlinings came onto the market that were bonded by the thermobonding principle. Unlike the previous flat calendered products, they had a soft, plump hand. In 1973, the development of the spot calender bonded nonwoven enabled interlinings to break through to other end-use sectors. With this technology, it was possible to expand the possible variations in the construction of nonwovens, giving nonwoven interlinings a previously impossible soft, plump, textile hand. So-called spunlaced nonwovens, developed around the same time, had similar aims. These are conventionally laid card webs that are bonded without binders by means of water jets. The first wet-laid nonwovens for use in the interlinings sector also go back to 1973. Here, the fibres are deposited from an aqueous suspension onto a screen fabric in a similar way to paper manufacture and then bonded like a dry-laid nonwoven using binders. In 1988 came the breakthrough to warp-knitted interlinings produced by knitting a pillar stitch construction into a nonwoven. Here, a heat-bonded nonwoven is normally fed into a warp knitting machine and stabilized in the longitudinal direction.

Practically the whole range of possible and required nonwoven interlinings is now produced using the two main technologies for this purpose, full-width bonding using binders and binder-free spot bonding.

### 14.1.3
### Functions of nonwoven interlinings

Every nonwoven interlining has a range of functions to fulfil, related to its end-use, that should satisfy both the processor (garment manufacturer) and the purchaser of the garment (consumer). Due to the complexity of the factors affecting the production and use of garments, a universal nonwoven interlining is inconceivable. This means that it is necessary for the manufacturer to establish the required properties of the made-up article by close communication with the shell manufacturers and garment manufacturers and by monitoring the consumer market, and from this to determine the specification of the nonwoven interlining.

By using the resulting specification, a suitable nonwoven interlining can be developed and constructed for almost any end-use. In general, the functions of a nonwoven interlining can be divided into three main groups:

#### 14.1.3.1 Interlining fabric for shaping and support
Shaping and support are the traditional tasks of an interlining fabric. It forms the internal frame of garments (for example jackets and coats) and helps to absorb and bear the static and dynamic stresses to which the garment is subjected in use. The shape given to the clothing for anatomical or fashion reasons should be maintained permanently by the interlining without changing the textile properties

**Fig. 14-1** Shaping and support with nonwoven interlinings

of the shell. A nonwoven interlining of this type is primarily used over a large area ("front fusing interlining").

### 14.1.3.2 Nonwoven interlining for stabilizing and/or stiffening

The task of a nonwoven interlining used for stabilizing is to reinforce or stiffen certain parts of a garment in the desired way. Moreover, these areas, often the most visible on a garment (for example collars and cuffs on shirts and blouses), should look good and should not lose their appearance after the care cycle. In terms of their application, these nonwoven interlinings are primarily for use over small areas and aid rationalization as punched and narrow fabrics.

### 14.1.3.3 Nonwoven interlinings for providing bulk

So-called quilting nonwovens can fulfil two different tasks in garments. The first, as a backing for quilting or embroidery to create a decorative look is determined by fashion. These are normally used over a small area.

The second task of providing heat insulation conforms to the rules of clothing physiology. With an entrapment of air of over 90%, these nonwovens make ideal heat insulators and stand out from other textile fabrics in this respect. In this case, the filling material is used over a large or the whole area. The boundaries between the two tasks can be fluid.

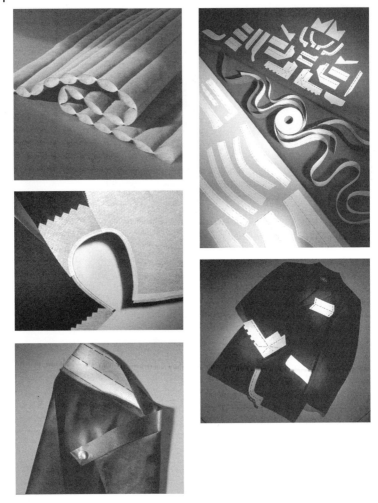

**Fig. 14-2** Nonwovens to aid rationalization (punched and narrow fabrics)

**Fig. 14-3** Nonwoven interlinings for providing bulk and heat insulation

| Functions / Properties / Functional elements | Ensure coverage — Weight per unit area, Thickness | Ensure resistance to mechanical stresses — Tensile strength, Abrasion resistance | Ensure suitability for apparel — Hand, Surface smoothness | Ensure processability — Cutting, Setting, Sewing, Ironing, Appropriate dimensional change | Ensure serviceability — Resistance to care processes, Colourfastness |
|---|---|---|---|---|---|
| Fibres* | ● | ● | ● | ● | ● |
| Web formation | ● | ● | ● | | |
| Nonwoven bonding | | ● | ● | | ● |
| Dyeing finishing | | | ● | | ● |
| Adhesive | | | | ● | ● |
| Application technique and formulation of the adhesive | | | ● | | ● |

*This includes all the material properties

Fig. 14-4  Relationship matrix for nonwoven interlinings

## 14.1.4
**Properties of nonwoven interlinings**

The property characteristics of a nonwoven interlining ensure that it fulfils the required function. Here, it should be noted that

– for one function several properties can sometimes be decisive
– individual properties have effects on various functions
– not all possible properties are relevant for every nonwoven interlining

## 14.1.5
**Functional elements of nonwoven interlinings**

By using the basic elements of a nonwoven interlining, that is the raw materials (fibres, binders, finishes, hot melt adhesives), by combining them together and using the different manufacturing possibilities, such as web formation, finishing, application and formulation of the hot melt adhesive, it is possible to design the individual properties and therefore to fulfil any function. These basic elements are therefore functional elements for the nonwoven interlining. The relationship between the functional elements and the properties can be represented in a matrix.

The aim of such a matrix is:

– to supply data on which functional elements are relevant for a property
– to provide the basis for specifications for stages carried out by external manufacturers
– to provide the stimulus for new or improved technical equipment

14.1.6
**Outlook**

In conclusion, here are some comments on the economic significance and future prospects for nonwoven interlinings.

In 1997, approximately 190 million m² of interlining materials were sold in Germany. Of these, approximately 76% or 145 million m² were nonwoven interlinings. Ten years before, the proportion of nonwovens was just 66%. In light of this success, it is easy to predict that the economic development potential of nonwoven interlinings has not yet been exhausted by far. This applies particularly to countries where the proportion of nonwovens is currently significantly lower than that of Germany.

However, if this positive development is to continue, it is essential that the manufacturers of nonwoven interlinings maintain their ability to comply with the requirements resulting from rapidly changing fashion trends in clothing and to follow garment manufacturers into production markets that are constantly changing geographically.

14.2
**Nonwovens for protective clothing**
J. HAASE

Protective and safety textiles of the most diverse kinds are classed as technical textiles with a high-tech character. They have a growing market importance. Protective clothing occupies first place among technical textiles in Europe [1]. It is used to protect people and/or property. For protection at work, protective textiles are used mainly in personal protective equipment (PPE) in the following areas:

– protective clothing (body protection)
– protective gloves
– protective headgear
– protective footwear
– protection against falling
– protection against drowning

Textiles for protecting objects serve purposes in property protection such as fireproofing and flameproofing, protection against vandalism (protection against cutting), moisture protection and protection for equipment/structural elements, clean room textiles as well as protection against electrostatic and electromagnetic fields.

Protective clothing is required in many workplaces in industry, in public companies and in the armed forces. Demand is also growing in the areas of leisure activities, sports, medicine and clean room technology. Statistical market analyses for 1996 in Western Europe (excluding rainwear and sportswear) according to [3] show a production volume of more than 200 million $m^2$ (see Table 14-1). According to this, the proportion of nonwovens is more than 50% and they are dominant in the area of particle and chemical protection.

The size of the European market for protective clothing is estimated to be around 3 billion Euro per annum [2]. Stricter legal requirements for work and health protection, increasing product liability requirements and the development of new technological processes call for constant further development and innovation in protective clothing. An annual growth in volume of 5–6% is predicted for Western Europe up to 2005. The use of nonwovens will increase disproportionately rapidly compared to woven and knitted fabrics. In the developing countries, an annual growth of approximately 10% is assumed [3]. Protective clothing is worn for protection against [4]

– mechanical influences
– being caught by moving parts
– thermal, climatic influences, for example cold, heat, moisture, wind
– other damaging environmental influence, for example dusts, gases, hot fumes
– electrical energy
– heat: flames, sparks, radiant heat, molten masses
– chemical substances: acids, alkalis, solvents, fats, oils, solid chemicals and
  so forth
– micro-organisms
– danger from vehicles in the form of high-visibility clothing
– contamination by radioactive emissions

**Requirements for protective clothing**
Protective clothing should protect the torso, the arms and the legs from dangerous influences at work. The different types of protective clothing can protect against one or more influences. The performance specifications, conditions of use, testing and certification of protective clothing come under European Directive 89/686/EEC on the approximation of the laws of the Member States relating to personal protective equipment [5]. This directive is translated into national law in Germany by the 8th regulation of the Equipment Safety Law [6]. Manufacturers and retailers should only offer protective clothing for sale if it conforms to the aforementioned directive and the product is labeled externally with the CE mark. For types of protective clothing from category II, a type test is necessary at a notified testing and certification centre as a declaration of conformity. For types of clothing from category III, which protect against life-threatening or serious health risks, it is also necessary to ensure the quality of the production series.

The directive formulates the fundamental requirements for protective equipment with regard to health protection and safety function as well as design and the information obligations of the manufacturer. The requirements refer to an ergonomically functional design of the clothing with the highest possible level of protection. In addition, the wearer should be subjected to the lowest possible thermal-physiological stresses. The protective clothing should not itself be a source of additional danger or obstructions under the intended end-use conditions. Test and product standards specify the general requirements for product types. There are currently more than 200 harmonized standards in the area of PPE.

The most important general standards for PPE protective clothing, which have to be observed in practically every case, are:

– DIN EN 340: 1993-09 Protective Clothing – General Requirements
– DIN EN 420: 1994-06 General Requirements for Gloves

Nonwovens are a component of protective clothing systems. They vary in the fibrous materials used and in their textile construction. In particular it is worth mentioning their use

– in disposable clothing for chemical/aerosol/dust protection
– as liner components in moisture and cold protection composites
– as bulky insulation linings with a protective function against the effects of heat from a radiant heat source, flames or hot objects

### Chemical/aerosol/dust protective clothing

As Table 14-1 shows, clothing for protection against dangerous chemicals, dusts, aerosols, gases and solid materials is clearly in first place in terms of quantity.
Protective clothing against chemicals is generally assigned to category III.

The European Standard [20] differentiates between: chemical protective clothing for

– "parts of the body" (see EN 467 [7] and EN 1513 [8]). This refers to for example protective aprons, sleeves, leg protectors
– "whole body" protective clothing, for example overalls, two-piece protective suits with and without a hood, face protectors and protective shoes

They are subdivided into six types, according to the protective aim:

– Type 1: impervious to gases (EN 943-1 [9])
– Type 2: not impervious to gases
– Type 3: impervious to liquids (EN 466 [10], EN 1511 [11])
– Type 4: impervious to sprays (EN 465 [12], EN 1512 [13])
– Type 5: impervious to particles (prEN ISO 13982-1 [14])
– Type 6: limited imperviousness to sprays (prEN ISO 13034 [15])

According to this subdivision, dust protective clothing made from nonwovens corresponds to type 5 chemical protective clothing. End-uses for these products are found in the following sectors:

**Table 14-1** Estimated material consumption for 1996 in million m² for protective clothing in Western Europe [3]

| Product function | End-use | Public utilities | Armed forces | Medicine | Industry, construction, agriculture | Total |
|---|---|---|---|---|---|---|
| Fireproof/high-temperature resistant | Woven/knitted fabric | 5 | 2 | – | 15 | 22 |
| | Nonwoven | – | – | – | – | – |
| Dust and particle protection | Woven/knitted fabric | – | – | 12 | 22 | 34 |
| | Nonwoven | – | – | 62 | 10 | 72 |
| Gas and chemical protection | Woven/knitted fabric | 1 | 1 | – | 4 | 6 |
| | Nonwoven | 3 | – | – | 47 | 50 |
| Radioactive, biological and chemical protection | Woven/knitted fabric | – | 2 | | – | 2 |
| | Nonwoven | – | 2 | | – | 2 |
| Extreme cold protection | Woven/knitted fabric | | 1 | | 2 | 3 |
| | Nonwoven | | – | | – | – |
| Fluorescent, reflective clothing | Woven/knitted fabric | 11 | 1 | | 3 | 15 |
| | Nonwoven | – | – | | – | – |
| Overall | Woven/knitted fabric | 17 | 7 | 12 | 46 | 82 |
| | Nonwoven | 3 | 2 | 62 | 57 | 124 |
| Total | | 20 | 9 | 74 | 103 | 206 |

- chemical industry
- construction and renovation
- mining
- wood-working
- spraying and painting work
- food industry, mills, for example
- insulation work, for example installing, removing piping
- removal of asbestos and clearing old pollutants
- pest control, including agriculture and forestry

This list does not claim to be complete, but gives an impression of the diversity of applications, the associated requirements and conditions of use. Table 14-2 summarizes the complex property requirements for the different types of chemical protective clothing.

Table 14-2 Requirement profile for chemical protective clothing [16]

| Characteristic/test property | Type of protective clothing | | |
|---|---|---|---|
| | 3/4 | 5 | 6 |
| Abrasion resistance | × | × | × |
| Block resistance | × | × | − |
| Flexural strength | × | × | − |
| Piercing resistance | × | × | − |
| Tear propagation resistance | × | × | × |
| Burst or tear strength | × | × | × |
| Peeling resistance of coatings | × | × | − |
| Imperviousness to particles | − | × | × |
| Imperviousness to chemicals | × | − | × |
| Flameproofing | × | × | × |

Chemical protective clothing with a "limited period of use" is normally worn until it has to be removed and disposed of for reasons of hygiene or due to contamination with dangerous substances. This definition covers single-use clothing and clothing for limited re-use according to the manufacturer's instructions. "Multiple-use" chemical protective clothing must be able to be cleaned and decontaminated and is designed for multiple re-use [4].

The most diverse nonwovens are used for single-use protective clothing. Spunlaid nonwovens made from polypropylene or polyester are favoured. However, nonwoven composites, which can consist of up to three layers and vary in terms of the fibrous materials used and in their structure, are also used. The individual layers take on specific protective functions [16]. Thus, especially for dust protective clothing, the mechanical properties of a spunlaid nonwoven have been combined with the barrier effect of a meltblown nonwoven while retaining the textile properties of the individual products.

The most important aspects of chemical protection are the properties of barrier protection against dangerous substances. A differentiation should be made between two kinds of barrier effect, penetration and permeation. Penetration is a physical "flow" process, whereby gases, liquids or solid particles pass through "pores" or "holes" in the material. EN 368 [17] specifies a method, known as a "gutter test", which measures the penetration of liquids through a material and also determines the liquid-repellent effect of the fabric. In the case of solid particles, for example dust, the particle size and exposure quantity have to be taken into account, as does the pressure during exposure. Permeation can best be described as a three-layer process. It comprises the absorption into the material on the contact surface, the diffusion through the material and the desorption out of the material onto the opposite surface. Permeation rates are dependent on the type, concentration and temperature of the chemicals. Permeation is measured according to EN 369 [18] or EN 374-3 [19] with a permeation test cell. The results are expressed in minutes as a "standardized breakthrough time". This breakthrough time is the average time between "first contact on the outside" and the time after which the chemicals are found on the inner surface of the material. It is referenced to a velocity of permeation of $1 \, \mu g/\text{min} \cdot \text{cm}^2$. The permeation times of individual chemicals and materials naturally vary within a wide range. When choosing chemical protective clothing, the user must know precisely the conditions of use and ensure that the manufacturer of nonwovens for chemical protection provides the relevant properties. Databases are also available for this [for example 21, 22].

Additional property requirements for chemical protection can be met using specialist finishes. Thus, for example, TYVEK® 1431N, TYVEK® C and TYVEK® F nonwovens from DuPont used for TYVEK-PRO.TECH® disposable protective clothing, have antistatic, moisture absorbing finishes. Where explosive powders, fumes and gases are present, sparks caused by static electricity on unfinished nonwovens made from synthetic fibres can be enough to set off an explosion. High electrostatic charges can be generated during the routine transfer of liquids and powders and during filtration, drying, grinding and pulverization tasks. A shock of weak intensity caused by static electricity (for example a charged person touches an earthed object with his finger) can lead to a moment of surprise. For workers who handle chemicals or work on machines, this can be dangerous and lead to mistakes. Antistatic clothing that guarantees a conductive discharge capacity, according to EN 1149-1 with a surface resistivity of less than $5 \cdot 10^{10}$ Ohms, eliminates dangerous charges. It is necessary for the entire system to be earthed. The person must wear conductive footwear, the protective suit must have contact with the skin and the floor must be earthed [23–25].

### Moisture and cold protective clothing

Liner components in moisture and cold protection composites provide protection against moisture, wind and cold environments in weatherproof clothing according to ENV 343 [26]. The clothing must be designed in such a way that it supports the heat regulation process of the human body. This includes the highest possible water vapour permeability with simultaneous waterproofing and suitable thermal insulation. The requirements for cold protective clothing according to ENV 342 [27] include a temperature range of –5 °C to –50 °C, for example for workplaces in cold-storage houses. To provide appropriate thermal insulation, light thermal linings made from nonwovens with a high resistivity to heat are normally used. The resistivity to heat $R_{ct}$ of the thermal lining for weatherproof clothing from class 2 must be greater than $0.15 \, \text{m}^2\text{K/W}$. Thermal nonwovens, made from 3M Thinsulate thermal insulation webs [28], for example, are extremely light, have a low heat conductivity and therefore a high thermal insulation.

Novel heat-insulating or, under high-temperature conditions, cooling materials are now finding entry into clothing materials. They are based on so-called "phase change materials" (PCM materials) [29–31]. Phase change technology is based on a NASA research programme from the 1980 s. The aim was to develop new materials for space travel to protect the astronauts and instruments better from the extreme temperature fluctuations in space. The name "phase change materials" is a collective term for materials that possess the ability to change their

**Table 14-3** Temperatures and stored heat of PCM materials [31]

| PCM material | Crystallization point (°C) | Melting point (°C) | Stored heat (cal/g) |
|---|---|---|---|
| Licosane | 30.6 | 36.1 | 59 |
| Octadecane | 25.4 | 28.2 | 58 |
| Heptadecane | 21.5 | 22.5 | 51 |
| Hexadecane | 16.2 | 18.5 | 57 |

state of aggregation within an adjustable temperature range and to store or release considerable amounts of energy during the phase transition according to the direction of the temperature gradient. These are mostly certain paraffins (see Table 14-3), which are enclosed inside a protective shell in microcapsules a few micrometres in diameter using specialist technology.

The microcapsules can be incorporated into fibres, nonwovens, coatings or in foams. Garments treated in this way remain resistant to washing, cleaning and weatherproof. The materials are also used in winter and summer clothing, shoes, boots, protective clothing, seat covers, in textile construction, for technical insulation and so forth. Outlast phase compensation insulation [30] was developed as a completely new type of effective cold protection insulation. Tests have shown that Outlast materials are up to 390% more heat insulating than foam insulation of the same thickness. The phase change, accompanied by the release of the stored

**Table 14-4** Comparison of resistivity to heat of materials with and without PCM materials (taken from [31])

| Test sample | Fibre batting A | Fibre batting B | Fibre batting C | Coated nonwoven |
|---|---|---|---|---|
| Thickness (mm) | 16.1 | 7.9 | 3.9 | 0.6 |
| Fabric weight (g/m$^2$) | 360 | 180 | 90 | 170 |
| Test value | Standard resistivity to heat $R_{ct}$ | Additional dynamic resistivity to heat $R_{dyn}$ | Total resistivity to heat $R_{total}$ | |
| | m$^2$K/W | m$^2$K/W | m$^2$K/W | |
| Batting A | 0.431 | – | 0.431 | |
| Batting B | 0.201 | – | 0.201 | |
| Batting C | 0.063 | – | 0.063 | |
| Batting B+50 g/m$^2$ PCM | 0.212 | 0.148 | 0.360 | |
| Batting B+85 g/m$^2$ PCM | 0.220 | 0.241 | 0.461 | |
| Coated nonwoven +85 g/m$^2$ PCM | 0.007 | 0.235 | 0.242 | |

heat, considerably increases the normal resistivity to heat $R_{ct}$ of the material (without PCM) via a "dynamic" resistivity to heat $R_{dyn}$. Table 14-4 contains numerical examples.

### Heat protective clothing

The latest developments are nonwovens made from temperature and flame-resistant fibrous materials with a high protective function against the effects of heat and flames. They protect people and property against thermal risks and are being used increasingly in firefighters' clothing, welders' protective clothing, combined fire and chemical protective clothing, on public transport, for example planes, trains, in seating and beds as well as in the public and private sector. An increasing number of inherently flame-resistant fibrous materials are available for producing such nonwovens. An important criterium is the LOI value (Limited Oxygen Index), see Table 14-5 on the characterization of flammability.

By blending appropriate fibres, the different requirements of the respective end-use sector can be met. Chemical methods are not suitable for reinforcing nonwovens as they reduce the LOI value. Such webs are usually reinforced mechanically (see Sections 6.1 to 6.3).

The company Freudenberg in Weinheim has developed the Vilene Fire Blocker product range [32], based on water-jet reinforcement, consisting of an optimal blend of melamine fibres, meta-aramid (Nomex) and para-aramid fibres (Twaron). They are heat- and flame-resistant, non-melting or dripping, dimensionally stable, air permeable, soft, with a good drape as well as abrasion-resistant. They are used for lightweight single or multi-layer insulation linings and as substrates for water-proof barriers (SYMPATEX® and GORETEX®). Table 14-6 shows characteristic test properties for a heat insulation web in a typical composite system.

**Table 14-5** LOI value of fibrous materials [32]

| Fibre | LOI value |
| --- | --- |
| Meta-aramid | 28–32 |
| Para-aramid | 29–32 |
| Preox fibre | 56–58 |
| Melamine fibre | 32 |
| Phenolic fibre | 31–33 |
| Modacrylic fibre | 28–30 |
| Viscose FR | 28 |
| PBI | 40 |
| Polyamide imide | 32 |
| Polyimide | 38 |

**Table 14-6** Vilene® Fire Blocker system in structural components: shell Nomex Delta TA 195 g/m², waterproof barrier Vilene Sympatex 110 g/m²; insulation lining Vilene quilted composite 250 g/m² [32]

| Test property | Unit | Results | Requirement for firefighters' apparel EN 469 [36] | Test standard |
|---|---|---|---|---|
| *Combustion behaviour* | | outside inside | | |
| Continued burning to top or side edge | | no | no | EN 532 [33] |
| Hole formation | | no | no | |
| Burning melted drops | s | 0 | <2 | |
| Average duration of flame | s | 0 | <2 | |
| Average duration of afterglow | | | | |
| *Heat transmission on exposure to flame* | | | | |
| Heat transfer index | | | | |
| HTI 24 | s | 19 | >13 | EN 367 [34] |
| HTI 24-HTI 12 | s | 6 | >4 | |
| *Heat transmission on exposure to a radiant heat source* | | | | |
| $t_2$ | s | 28 | >22 | |
| $t_2-t_1$ | s | 8 | >6 | EN 366 [35] |
| Heat transmission factor TF | % | 35 | <60 | |

## 14.3
## Nonwoven support materials for footwear
M. STOLL, M. BRODTKA

Coated textile support materials – summed up under the term "artificial leathers" – compete with "genuine leather" in most diverse applications, e.g. as shoe upper and upholstery cover materials.

The structures of leather and traditional artificial leather differ from each other fundamentally. While leather consists of a collagen fibre tissue, the density of which increases continuously towards the grain side (Fig. 14-5), artificial leathers are layered materials composed of textile supports and – mostly several – polymeric layers (Fig. 14-6).

Natural leather prevails in application, in spite of multiple and diverse efforts made in the last decades in order to use plane materials produced from high-grade synthetic products, e.g. for footwear manufacture (shoe upper materials and linings). Its dominance does not only result from promotion of "genuine leather", but also from the beneficial hygienic properties in wear, which are expressed by high values of water vapour permeability and water vapour absorption as well as by the expansion behaviour, as shown in Table 14-7.

**Fig. 14-5** Cross-section of cattlehide leather

Because of the insufficient quantities of leather expected to be available it is necessary to offer the market substitutes, the properties of which correspond to those of leather. They should be produced as web materials, thus permitting more efficient processing by automation of the manufacturing processes.

A statistical study of EDANA reported that in 1998 1.3% of the total area of nonwovens manufactured in Western Europe were produced for footwear and leather goods. According to a FAO (UNO Food and Agriculture Organization) report for 1996, plane substitute materials ("synthetics"), which have a safe position among the materials for shoe uppers and linings, garments, and fancy goods, were produced in an amount of approx. 2.5 mrd $m^2$ per year. Approx. 100 million $m^2$ of this quantity are high-grade leather substitutes ("poromerics") supported by fibrous and microfibrous nonwovens.

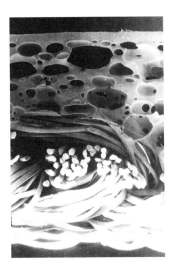

**Fig. 14-6** Artificial leather with foamed interlayer

**Table 14-7** Physical characteristics of shoe upper materials

|  |  | Leather | Synthetic leather |
| --- | --- | --- | --- |
| Water vapour permeability | $(mg/24\,h \times 10\,cm^2)$ | 350–500 | 250–350 |
| Water vapour absorption | (%) | >25 | 3–3.5 |
| Elongation at 100 N/cm², dry | (%) | 13/11 | 10/4 |
| Plane deformation | (%) | 20–35 | 50–60 |
| Permanent plane deformation after 10-fold deformation in a tensometer at 3 bar | (%) | 9 | 4 |
| Area increase after raise of air humidity from 65 to 100% | (%) | 7–8 | 3–4 |

Already after the Second World War first synthetic products became known, the quality of which was insufficient. Woven and knit fabrics coated with polyvinyl chloride or polyurethane prevailed among them, together with most different versions of conventional nonwovens. As known, woven and knit fabrics exhibit different strength values in weft and warp directions. This must be taken into account when these materials are processed. Therefore they are mainly used for large-area leather goods and upholstery. For the footwear industry they are less suitable, also because of the fraying risk. Nonwovens have mainly been successful as support materials, because of their cutting-edge stability. Nonwoven materials reinforced by water jets are used for lining materials, while needled nonwovens are applied for shoe uppers.

In 1958 a method of forming microporous coatings from polyurethane solutions by precoagulation and subsequent coagulation was published. According to it, a film of polyurethane solution in an organic solvent was applied onto a supporting web, subjected to precoagulation in a conditioning channel (air humidity >90%), and coagulated in a precipitation bath consisting of a blend of solvent/non-solvent (dimethylformamide/water) (see Fig. 14-7). Thereafter the residual solvent was washed. The application of this technological principle resulted in the formation of a layer with clearly higher water vapour permeability. The basic process mentioned above was improved already at the beginning of the 60 ies [37]. In a first evaluation, a product of the DuPont company, USA, which appeared on the market under the brand of Corfam, was considered to be an alternative to natural leather. However, already first comparative studies revealed the disadvantages of the products in comparison with leather [38]. The shortcomings concerned the extension properties in processing and wear as well as the hygienic properties in wear. Permanent further development of these product classes resulted in quality improvement and in a rationalization of the manufacturing technologies (e.g. elimination of the expensive precoagulation step) [39, 40].

At the end of the 70 ies and at the beginning of the 80 ies the range of poromeric materials was extended to impregnated woven fabrics (so-called wet-coagulated materials) for light shoe upper materials and garments [41].

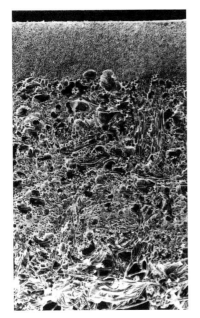
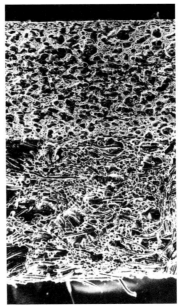

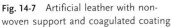

**Fig. 14-7** Artificial leather with non-woven support and coagulated coating

**Fig. 14-8** Porous artificial leather with non-woven support and microporous foamed layer

In Forschungsinstitut für Leder- und Kunstledertechnologie a method of using a mechanically foamed polyurethane solution in dimethylformamide for the coagulation process was developed [42]. By the formation of a so-called twin structure, as shown in Fig. 14-8, it became possible to form a stable porous layer with high water vapour permeability.

In spite of extensive studies aiming at further improvement of products and their manufacturing technologies, it proved impossible to achieve the required leap in quality (new material generation) on the base of classical methods (use of needled nonwovens produced on traditional equipment). This led to a stagnation or reduction of annual outputs, respectively.

Processing of nonwovens with use of microfibril fibres resulted in a clear improvement of material quality [43]. Therefore these products are named leatherlike materials [44]. Brands, such as Alcantara, Clarino, Sofrina, and Lorica, became well-known. Such finest fibres having a diameter <1 mm are produced by forming matrix/fibril fibres. This matrix/fibril structure permits to process such fibres also on traditional web-forming equipment. To ensure the excellent material properties (strength, force/extension behaviour, softness), free mobility of the finest-fibre fibrils is required. It is provided by extraction of the matrix (see Fig. 14-9).

However, high prices of the leatherlike materials are a consequence of the great deal of time and energy required for this process step. Therefore the range of their use has been limited so far. The characteristics proposed by Prüf- und Forschungsinstitut für die Schuhherstellung e.V. (PFI) apply to shoe upper and lin-

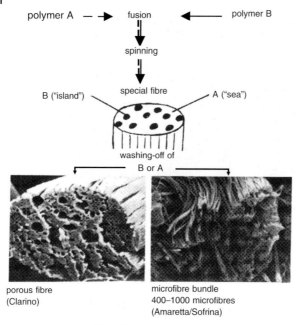

polymer A — → fusion ← polymer B

spinning

B ("island") special fibre A ("sea")

washing-off of
B or A

porous fibre
(Clarino)

microfibre bundle
400–1000 microfibres
(Amaretta/Sofrina)

**Fig. 14-9** Scheme of fibre formation

ing materials. For example, the minimum tensile force amounts to 600 N for unlined and 400 N for lined materials (longitudinal and transverse directions). The
required value of the elongation achieved by maximum tensile force ranges from
50 to 120%, and the thickness range of shoe upper materials amounts from 1.1 to
1.4 mm. 1.8 mm are required for unlined materials, whereas the maximum thickness of shoe linings should be 0.8 mm [45].

Due to the high demands made on the strength characteristics, in particular,
the support structure often comprises a large portion of synthetic fibres (predominantly polyester). Such supports are not up to the demand for composite
materials based on natural raw materials, which provide a high comfort in wear,
or match this demand only in an insufficient degree.

Natural fibres are used for manufacturing of coated supports in a limited scale.
Nonwovens can be produced from viscose fibres by needling as well as by water-
jet compression. However, their wet strength is especially low, and so they are not
suitable for the application of certain coating methods (e.g. the coagulation process).

Various fibre manufacturers have introduced modifications of viscose fibres.
The so-called Lyocell process provides fibres with higher wet strength, which can
be used for processing of nonwovens to be coated by the coagulation method,
therefore.

In recent time, multiple nonwoven composite materials have been described in
literature, which permit the manufacture of high-grade novel products. The basic
idea for the use of nonwovens in composite structures consists in a symbiosis of
properties and in using the benefits of the nonwoven by incorporation of func-

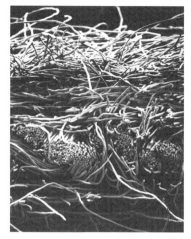

**Fig. 14-10** Composite consisting of cloth and needled nonwoven

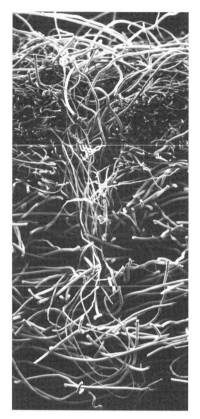

**Fig. 14-11** Composite consisting of fine entangled and coarse needled nonwovens

tional (e.g. superabsorbing) fibres or by combination with other plane materials, e.g. cloths, knitwear, layed textiles, sheeting, etc., to meet the demands made on the textile support, which are often very heterogeneous [46]. The fibre fineness of a fabric essentially influences such properties of the final product, as appearance, handle, and drape [47], as well as the comfort in wear [48].

To ensure adequate strength characteristics, needled composite materials with incorporated plane materials, such as cloths or knitwear, mostly have a thickness >2.0 mm and a weight per unit area >300 g/m² (see Fig. 14-10).

Practical experiments have confirmed that a pre-needled nonwoven can be needled on a woven support of appropriate fibre density in such a manner that the fibre plugs are bound in the cloth. In a final needling process, in which the stitch depth is low, the fibre tufts protruding from the cloth are fixed in the non-woven/cloth composite structure [49].

Action of high-energy water jets under high pressure forms entangled nonwovens with a weight per unit area ranging from 25 to 150 g/m². The use of such entangled nonwovens permits to form combined supports with a thickness <2.00 mm and a weight per unit area < 300 g/m², which are applicable as supports of coated materials.

Joining of needled and entangled nonwovens (see Fig. 14-11) by needling gives nonwoven composites with good and nearly isotropic strength/elongation properties, permitting PUR coating by the coagulation process. Fibres of different fineness used in the single nonwoven layers and a combination of entangling and needling processes make it possible to produce a nonwoven composite providing a density gradient over the cross-section [50].

In this case the stitch depth chosen is decisive for the smoothness of the composite surface and, thus, for an even appearance after coating and finishing. Optimum stitch depth, stitch density, and the share of Lyocell fibres (CLY-F) in the nonwoven composite ensure the required adhesion between support material and coating. An incorporation of woven fabrics made from Lyocell or mixture cloth consisting of Lyocell and polyester fibres results in particularly even surfaces after coating and in thickness values < 1.3 mm.

### References to Chapter 14

[1] Manufactures of Technical Textiles in Europe (1997) Techn Textilien 40, 11: 248

[2] Dokument Draft Business Plan (V2.1) of CENT/TC 162 (1998) – Protective clothing including hand and arm protection and lifejackets, Dec. 98

[3] Davies B (1997/1998) Westeuropäischer Markt für Schutzkleidung: Zukunftsaussichten nach Segmenten, Vortr 36, Intern Chemiefasertag Dornbirn 09/97, Techn Textilien 41, 2: 43–46

[4] Noetel K-H (1999) Handb Persönliche Schutzausrüstungen, ecomed-Verlag, Periodicum 42. Erg.

[5] RICHTLINIE DES RATES zur Angleichung der Rechtsvorschriften d Mitgliedsstaaten f persönl Schutzausrüstungen (Dok. 89/686/EWG) v 21.12.89 (1989) Amtsbl d EG No L 399/18 v 30.12.89, see also RICHTLINIE DES RATES 89/656/EWG v 30.11.89, ABl No L 393 v 30.12.89, S 18 zur Bereitstellung und Benutzung von PSA am Arbeitsplatz

[6] 8. VO zum Gerätesicherheitsgesetz (VO über das Inverkehrbringen von persönl Schutzausrüstungen (1992) – 8. GSGV), BGBl Issue 92, Part 1, No. 26, 17.06.92

[7] DIN EN 467:1999-01 (1999) Schutzkleidung – Schutz gegen flüssige Chemikalien – Leistungsanforderungen an Kleidungsstücke, die für Teile des Körpers einen Schutz gegen Chemikalien gewähren

[8] E DIN EN 1 513:1994-10 (1994) Schutzkleidung – Schutz gegen flüssige Chemikalien – Leistungsanforderungen an Kleidungsstücke zum begrenzten Einsatz, die für Teile des Körpers einen Schutz gegen Chemikalien bieten (Teilkörperschutz zum begrenzten Einsatz)

[9] prEN 943-1:1997-10 (1997) Schutzkleidung gegen flüssige und gasförmige Chemikalien, einschließlich Flüssigkeitsaerosole und feste Partikel – Part 1: Leistungsanforderungen für belüftete und unbelüftete "gasdichte" (Typ 1) und "nicht gasdichte" (Typ 2) Chemikalienschutzkleidung

[10] DIN EN 466-1:1999-01 (1999) Schutzkleidung – Schutz gegen flüssige Chemikalien – Part 1: Leistungsanforderungen an Chemikalienschutzkleidung mit flüssigkeitsdichten Verbindungen zwischen den verschiedenen Teilen der Kleidung (Ausrüstung Typ 3); DIN EN 466-2:1966-11: Schutzkleidung – Schutz gegen flüssige Chemikalien – Part 2: Leistungsanforderungen an Chemikalienschutzkleidung mit flüssigkeitsdichten Verbindungen zwischen den verschiedenen Teilen der Chemikalienschutzanzüge für Notfallteams (Ausrüstung Typ 3ET)

[11] E DIN EN 1 511:1994-10 (1994) Schutzkleidung – Schutz gegen flüssige Chemikalien – Leistungsanforderungen an Chemikalienschutzkleidung zum begrenzten Einsatz mit flüs-

sigkeitsdichten Verbindungen zwischen den verschiedenen Teilen der Kleidung (Ausrüstung Typ 3 zum begrenzten Einsatz)

[12] DIN EN 465:1999-01 (1999) Schutzkleidung – Schutz gegen flüssige Chemikalien – Leistungsanforderungen an Chemikalienschutzkleidung mit spraydichten Verbindungen zwischen den verschiedenen Teilen der Kleidung (Ausrüstung Typ 4)

[13] E DIN EN 1 512:1994-10 (1994) Schutzkleidung – Schutz gegen flüssige Chemikalien – Leistungsanforderungen an Chemikalienschutzkleidung zum begrenzten Einsatz mit spraydichten Verbindungen zwischen den verschiedenen Teilen der Kleidung (Ausrüstung Typ 4 zum begrenzten Einsatz)

[14] prEN 13982-1:1999-08 (1999) Protective clothing for use against solid particulate chemicals – Performance requirements for chemical protective clothing providing protection to the full body against solid particulate chemicals (type 5 clothing)

[15] E DIN EN 13 034:1998-01 (1998) Schutzkleidung gegen flüssige Chemikalien – Leistungsanforderungen an Chemikalienschutzanzüge mit eingeschränkter Schutzleistung gegen flüssige Chemikalien (Ausrüstung Typ 6)

[16] Bernstein U (1996) Microfasertextilien für Schutzbekleidung mit Barriereeigenschaften, ITB Vliesstoffe-Techn Textilien 3: 9–12

[17] DIN EN 368:1993-01 (1993) Schutzkleidung – Schutz gegen flüssige Chemikalien – Prüfverfahren; Widerstand von Materialien gegen die Durchdringung von Flüssigkeiten

[18] EN 369:1993-03 (1993) Schutzkleidung – Schutz gegen flüssige Chemikalien – Prüfverfahren: Widerstand von Materialien gegen Permeation von Flüssigkeiten

[19] DIN EN 374-3:1994-04 (1994) Schutzhandschuhe gegen Chemikalien und Mikroorganismen, Part 3: Bestimmung des Widerstandes gegen Permeation von Chemikalien

[20] Eichinger H (1998) Anforderungsprofile und Typisierung von Kleidung mit Schutzwirkung gegen feste, flüssige und/oder gasförmige Chemikalien und Gefahrstoffe, Vortr 4. Dresdner Textiltagung (DuPont International S A, Geneva)

[21] Ist der Schutzanzug, den Sie tragen, auch der richtige? (1999) Firmenprospekt DuPont Nonwovens, Tyvek-Pro-tech® Informationsservice 43, Informationsschrift Ausgabe L-11862-2 04/99, Datenbanksystem TYDAT

[22] Fiedler H-B (1999) VOICE 3.1 – Gefahrstoffdatenbank für Chemikalienschutzkleidung, Dräger Sicherheitstechnik GmbH Lübeck

[23] Statische Elektrizität, Richtl No 4, Ausg Okt 1989 (1989) Berufsgenossenschaft der chem Industrie, Richtl f d Vermeidung von Zündgefahren infolge elektrostatischer Aufladungen

[24] Safety of machinery (1999) Guidance and recommendations for the avoidance of hazards due to static electricity, CENELEC Report No RO 44-001, Feb 1999

[25] (1997) Barrierekleidung und elektrostatische Aufladung, Firmenprospekt DuPont Nonwovens, Tyvek-Pro-tech® Informationsservice No 36, Informationsschrift Ausg L-11149-2 09/97

[26] DIN V ENV 343:1998-04 (1998) Schutzkleidung – Schutz gegen schlechtes Wetter

[27] DIN V ENV 342:1998-04 (1998) Schutzkleidung – Kleidungssysteme zum Schutz gegen Kälte

[28] Thinsulate Thermal Insulation Types B, C, CS, CDS; Lite Loft, Ultra Insulation (1998) Insulation Products Project Europe c/o 3M Deutschland GmbH Neuss (Firmenprospekt)

[29] Rupp J (1999) Aktive Textilien regulieren die Körpertemperatur, ITB Intern Text Bull 1: 58–59

[30] Pause B (1995) Development of the first cold protective clothing with microcapsulated PCM, Lecture No 241, 7. Intern Techn Textil Symposium Frankfurt/Main

[31] Cox R (1998) Synopsis of the new thermal regulating fiber Outlast, Chem Fibers Intern 48: 475–479

[32] Schäfer W (1999) Hitze- und flammbeständige Vliesstoffe aus high-tech-Fasern für Schutzkleidung und technische Anwendungen, Vortr The Freudenberg Nonwovens Group Techtextil-Symposium 04/99 Frankfurt/Main

[33] DIN EN 532:1995-01 (1995) Schutzkleidung – Schutz gegen Hitze und Flammen – Prüfverfahren für die begrenzte Flammenausbreitung

[34] DIN EN 367:1992-11 (1992) Schutzkleidung – Schutz gegen Wärme und Flammen – Prüfverfahren, Bestimmung des Wärmedurchgangs bei Flammeneinwirkung

[35] DIN EN 366:1993-05 (1993) Schutzkleidung – Schutz gegen Hitze und Feuer – Prüfverfahren, Beurteilung von Materialien und Materialkombinationen, die einer Hitze-Strahlungsquelle ausgesetzt sind

[36] DIN EN 469:1996-01 (1996) Schutzkleidung für die Feuerwehr, Anforderungen und Prüfverfahren für Schutzkleidung für die Brandbekämpfung

[37] Wittke W (1983) Die Herstellung von Polyurethan-Kunstleder (PUR-Kunstleder) nach dem Koagulationsverfahren, Coating St Gallen 16, 4: 9–93

[38] Reich G (1991) Leder und synthetische Austauschmaterialien heute – eine vergleichende Betrachtung, Leder- und Häutemarkt 43, 2: 1–7 and 43, 5: 6–9

[39] Markle R, Tackenberg W (1984) Alternatives to Leather: Laif Poromerics, Journ of Coated Fabrics 13, 4: 228–238

[40] Freitag H (1983) Über die Herstellung von synthetischem Leder – Vlieskunstleder, Coating St Gallen 16, 8: 210–214

[41] Lomax R (1984/85) Recent Developments in Coated Apparel, Journ of Coated Fabrics 14, 2: 91–99

[42] Stoll M (1994) Verfahren zur Herstellung poröser Polymerschichten nach dem Koagulationsverfahren, Coating 1: 9–11

[43] What is a Microfibre (1988) Textile Horizons 4: 49–50

[44] Nagoshi K (1987) Leatherlike Materials, European Polymer Journal 8: 677–697

[45] Mädler A (1998) Schlußbericht PUR-Beschichtung von Lyocellträgern, FILK gGmbH Freiberg

[46] Schäfer W (1992) Vliesverbundprodukte für Anwendungen in der Elektrotechnik, Melliand Textilber 73, 112: 866–869

[47] Ninow H, Heidenreich I (1993) Viscose-Feinfilamentgarne – quo vadis? Melliand Textilber 74, 2: 107–110

[48] Umbach K-H (1993) Feuchtetransport und Tragekomfort in Mikrofaser-Textilien, Melliand Textilber 74, 2: 174–178

[49] Brodtka M, Mädler A (1997) Nadelvliesstoff-Verbundstrukturen für die Kunstlederindustrie unter Einsatz von Lyocell-Fasern, Vortr 12, Hofer Vliesstoffseminar

[50] Brodtka M (1998) Entwicklung von Vliesstoffen unter Einsatz von schadstoffarm hergestellten Cellulosefasern (Typ Lyocell) für technische Anwendungsbereiche, insbesondere Produkte mit lederähnlichen Eigenschaften, Forschungsbericht 292/96 des STFI e.V. Chemnitz

# 15
# Nonwovens for technical applications
J. HAASE

## 15.1
## Insulation

In engineering, the term "insulation" refers to measures and equipment for restricting losses or the unwanted admission of energy or media such as fire/radiant heat, thermal energy, sound, electricity, moisture etc. Likewise, the term also refers to the insulation and barrier materials used for this.

### 15.1.1
### Fire, heat, sound

**Insulation against fire/heat**

Frequent and spectacular accidents in air and sea travel as well as fire disasters, for example in discos, have recently contributed towards increased efforts to find ways of protecting people, property and equipment from fires. Schäfer refers to the fact [1] that the increased attention being paid to environmental issues, heightened safety consciousness and related new regulations and standards open up new fields of application in the aforementioned sectors for nonwovens made from inherently heat and flame-resistant fibres. The end-use spectrum is wide-ranging. It ranges from preventive fireproofing in means of transport and in the property sector, flameproofed cables and wiring, to the use of inorganic heat insulation materials on parts of cars exposed to heat.

On airliners, all the components of the cabin fittings are subject to strict fire regulations. If cushions made from flammable polyurethane foams are used in passenger seats, they must be covered with appropriate fire blockers. Woven fabrics, knitted fabrics, needle-punched nonwovens or composite materials made from meta- and para-aramid blends, melamine fibres and PBI with weights per unit area of up to 380 g/m$^2$ are normally used. Weight savings in aircraft construction are of great interest in terms of reducing fuel consumption and emissions. This has led to the development of lighter water-jet bonded nonwovens made from preox and phenolic resin fibres reinforced with aramid yarns, and from para-aramid/melamine resin fibres reinforced with lightweight knitted fabrics made from FR viscose. Despite a comparatively low weight per unit area of 200 g/m$^2$, they fulfil all the requirements for flame resistance, reduced fume emissions, minimum release of heat and their capacity to withstand mechanical stresses [1]. In the USA, developments are under way to increase the burn-through resistance of the aircraft hull in a post crash fire on the ground. Burning kerosene with a flame temperature of over 1,000 °C currently leads to burn-through times of only 1.5–2 minutes, which is too short for adequate measures to save passengers. Trials are being carried out on various new systems for thermal and acoustic cabin insulation, including nonwovens made from high-performance fibres that are also finished with fire retardants based on mica [1]. Water-jet bonded nonwovens and nonwoven composites made from melamine fibres, meta-aramid, para-aramid and polyamide imide fibres in weight per unit area ranges of 50–150 g/m$^2$ fulfil all the requirements for fireproofing seating and mattresses (BS EN-1021/1, BS EN-10121/2, BS 58 5852, BS 7176, US Fullscale Test according to Cal 133) [1].

The most commonly used heat insulation materials are needle-punched nonwovens. The temperature application range depends on the type of inherently flame-resistant fibres used, whereby the following ranges are normally quoted [2, 4]: aramid fibres 250–350 °C; E-glass fibres 480–550 °C, PBO fibres (poly-p-phenylene-2.6-benzosoxazole) [3] 650 °C, basalt mineral fibres 815–1,000 °C; ceramic fibres up to 1,150 °C; silicate fibres 1100–1200 °C and alumina fibres up to 1,400 °C.

Fire-retardant finishing of nonwovens can be carried out using flameproofing agents. Flameproofing agents containing halogens increase in their fire-retardant effect in the order fluorine < chlorine < bromine < iodine. However, in practical terms, chlorine and bromine compounds are most important. In case of fire, these agents release halogen radicals, which interfere in the mechanism of the combustion process in the vapour phase. This occurs due to the formation of hydrogen halides, which neutralize the high-energy hydrogen and hydroxyl radicals, the actual vehicles of the free radical chain reaction as the halogen radicals degrade. In this way they "cool" the flame, thereby reducing the flame propagation rate. Flameproofing agents containing phosphorus, on the other hand, work primarily in the condensed phase. In the case of fire, these inorganic or organic products release polyphosphoric acid-like products as a result of disintegration and/or oxidation. These promote carbonization by dehydrating the pyrolysing substrate, for example, and reduce the output of flammable pyrolysis products. Insulation materials made from cotton or jute are flameproofed with borax, ammonium sulphate or boron salt.

Another possibility for flameproofing is offered by inorganic, endothermic products such as aluminium or magnesium hydroxide. An example is APYRAL filler with an energy consumption of 1,211 kJ/kg Al(OH)$_3$ and phase transition in the region of 200–400 °C [5]. These products decompose into the relevant oxides at elevated temperatures, absorbing energy and simultaneously release water. The heat consumption of the decomposition reaction causes a "cooling" of the substrate and reduces the amount of energy available for pyrolysis. Furthermore, the steam released brings about a dilution of the flammable gases. Moreover, the oxidic decomposition products together with the carbonization cleavage products can form a fire-isolating protective layer.

A special and new form of fireproofing are so-called intumescent systems (lat.: intumescere = swell) based on phosphoric compounds, which will become even more important in future. By means of the combined action of three components, a carbon donor, an acid donor and a propellant, from a defined temperature during the combustion process, these systems produce "swelling" carbon-rich voluminous protective layers on the burning substrate, which protect the material underneath from further heat effects. During the swelling process, an increase in volume/thickness of 100 times can occur in the ideal case. These systems are already widely used as fireproof coatings for steel, plastics and wood. Recently, intensive applied studies have begun on their use in textile products, including nonwovens [6]. Horrocks et al. [7, 8] studied different nonwoven structures (see Fig. 15-1) made from FR viscose into which intumescent powder substances were introduced.

Powders based on ammonium polyphosphates or melamine phosphates, for example products MPC 1000 or MPC 2000 from the company Albright & Wilson, were used. Two-layer nonwoven composites of 200 g/m² with the addition of up to 125% MPC were produced, as well as FR cotton or FR viscose woven fabrics. In burning tests in which they were exposed to flames with a primary energy flow of 50 kW/m² for 10 minutes, corresponding to a "flash over" fire, they resulted in a significantly improved residual strength and thermal barrier effect comparable with standard commercial heat-insulating needle-punched nonwovens.

A nonwoven that increases in volume is available for fireproofing on door seals, pipe elements, glass surfaces and panelling [9]. On contact with fire, the nonwoven expands in its cross-section to 3 to 20 times the initial thickness. If this expansion is limited mechanically, an excellent sealing and insulation layer develops due to the resulting pressure.

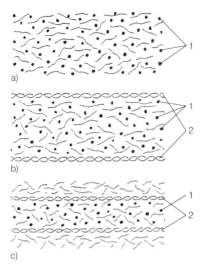

**Fig. 15-1** Examples of nonwovens and composites, treated with intumescent powders (taken from [6, 7])
a) Nonwoven, b) nonwoven composite with scrim covering, c) nonwoven composite with embedded scrims.
1 Intumescent powder, 2 scrim

**Thermal insulation**

The thermal insulation properties of nonwovens play a significant role above all in the construction of houses and industrial and public buildings. In these areas, nonwovens have developed from a less important position at first to standard construction elements.

Nonwovens are now manufactured that are rated in building material classes up to Class A2 (non-flammable) according to DIN 4102 [10] and can therefore be enlisted for fireproofing tasks.

Thermal insulation materials for the construction industry are a big development market. To achieve the measures required by law (low energy standard, $CO_2$ reduction, environmental protection and so forth) the insulation market is predicted to grow in Germany from a current level of 30 million $m^3$ per annum (see Fig. 15-2) to 60–100 million $m^3$ per annum by around 2050 [11].

Most commonly used are chemically and thermally bonded nonwoven insulating mats made from mineral fibres such as, for example, glass, mineral, and panels made from EPS hard foam, PUR hard foam and polystyrene XPS. The proportion of alternative insulating materials such as mineral granulate insulation or nonwovens made from natural materials (cork, cotton – for example Isocotton – wood, sheep's wool, flax – for example Isoflax – used paper) is around 5%. Table 15-1 gives a comparison of thermal insulation materials.

The basic evaluation criteria for thermal insulation materials are presented in Table 15-2.

Processing technology has been developed for the stages of fibre opening – binder feeding – web formation – thermal bonding of the web to produce combination insulation materials made from wool and wood corresponding to building material class B2. They are distinguished by their good elasticity, improved installation behaviour and low weight [12].

Thermal insulation materials are multicomponent systems. Basically, they consist of a solid component and insulating gas/air. With an appropriate structure for the

$10^6$ m$^3$

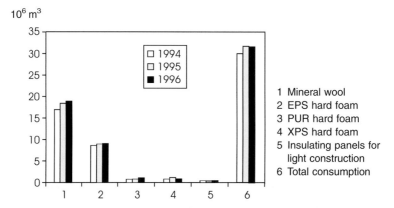

**Fig. 15-2** Statistical data on the production of thermal insulation materials in Germany (taken from [14], according to data from the Trade Association for the Mineral Fibre Industry e.V.)

**Table 15-1** Selection of thermal insulation materials according to [19]

| Insulation materials | Raw materials | | | Primary energy content | Harmful substances released | | Building material class/ Thermal conductivity group | Disposal |
|---|---|---|---|---|---|---|---|---|
| | Base | Binder | Additives | | Production | Use | | |
| Mineral fibre panels | Glas, mineral | Phenolic or urea formaldehyde resin | Mineral oils, if required silicon emulsion | Average | Release of fibres possible | Release of fibres possible | A1, A2/035 to 050 | Rubble |
| Polyurethane hard foam | Diisocyanate and polyol propellent gas, preferably $CO_2$ | None | Catalysts, stabilizers, FPA: phosphoric acid ester | High | Toxic reaction products released | Diffusion of cell gases possible, in a fire toxic gases | B1, B2/020 to 035 | Recycling unknown, landfill problematic |
| Polystyrene (EPS) | Styrene with pentane as propellant | None | FPA: hexabromocyclododecane | High | Emitted styrene | Release of styrene possible, in a fire toxic gases | B1/025 to 040 | Recycling rare, problematic for landfill due to additives |
| Expanded perlite | Expanded volcanic mineral | None | None | Average | None | None | A1/055 to 060 | Recyclable |
| Cork insulation panels | Cork bark | Own natural resins | None | Low | None | None | B2/045 to 055 | Return to manufacturer |
| Soft wood fibre panels | Conifer wood fibres | Wood's own binders | Aluminium sulphate | Very high | None | None | B2/040 to 070 | As wood |
| Isocotton | Cotton | None | FPA: borax | Low | No | No | B2/040 | Recycling |
| Isoflax | Flax fibres | Starch or sodium silicate possible | | Low | No | No | B2/040 | Recycling |
| Homatherm | Newspaper, jute | Lignin, natural resin | FPA: ammonium sulphate or boron salt | Low | No | No | B2/040 | Return guarantee |

**Table 15-2** Evaluation criteria for insulation materials according to [19]

| General characteristics | Strength properties | Thermal data | Moisture behaviour | Behaviour under the influence of heat | Resistance to |
|---|---|---|---|---|---|
| Raw material basis, production technology, supply format | Compressive strength, Bending strength, Rupture strength, Tensile strength | Bulk density, thermal conductivity | Reference moisture content, steam diffusion behaviour | Dimensional stability on exposure to heat, Combustion behaviour | Mould fungi, damage by animals |

gas enclosures, thermal conductivities of <0.1 W/m K can be achieved. Systematic structural analyses [13] led to the development of a super-insulating film material, SIFM composite, consisting of polyamide fibres, aluminium-coated polyester film as the reflective internal membrane, adhesive and flock. The insulation material, which has an extremely low bulk density of 12 kg/m³, has a thermal conductivity of 0.028 W/m K at normal room temperature [14]. Similarly constructed is a thermal insulation nonwoven for flexible pipe insulation made from polyester web with aluminium-coated polyester film and wool or wood components, which has a bulk density of 30–60 kg/m³, $\lambda = 0.035$ W/m K and a fire of rating B1 [15].

As Table 15-3 shows, the thermal conductivity of thermal insulation materials is not a constant, but increases with the temperature.

For ceiling and wall coverings, nonwoven mats made from mineral fibres (glass wool, rock wool) are predominantly used. An overview of all matters concerning mineral wool is provided in [18]. Insulation materials are commonly referred to, especially in the construction industry, as mineral wool. Inorganic fibrous materials include those made of slag, rock, glass and ceramic. These fibres are produced from their melts by means of an air-laying or centrifugal process. The production of panels or mats has similarities to the production of webs by the aerodynamic method. The fibres are blown onto a conveyor belt and binders are simultaneously sprayed on. After subsequent drying, the chemically bonded webs are made into panels or rolled up. The structure depends on the fibre properties of fibre thickness, length, crimp and on the fibre layering. The basis for the thermal and acoustic properties is the density, the open pore structure with a low passage resistance.

By establishing the fibre properties, such as geometry, chemical composition, solubility behaviour, biodegradability and so forth, it is possible to evaluate and classify the toxicology of mineral fibres. This evaluation is carried out in Germany on the basis of the Technical Rules for Dangerous Substances (TRGS 905 [20]) and the Dangerous Substances Directive (GefstoffV, Appendix V No. 7.1 [21]).

**Table 15-3** Thermal conductivity of thermal insulation materials

| Product | Raw material | Bulk density (kg/m³) | Thermal conductivity (W/m K) | Source |
|---------|--------------|----------------------|------------------------------|--------|
| "Glanamat" needle-punched nonwoven | Glass filaments made of E-glass, mechanically bonded without chemical binders | 70–170 Web thicknesses 6–25 mm | 0.032 (10 °C) 0.041 (100 °C) 0.056 (200 °C) 0.076 (300 °C) | [16] Tests according to DIN 52 612 [47] |
| ThermProtect silica needle-punched nonwoven mats | Silica glass fibres, 9 µm, application temperature up to 1,000 °C | 130 | 0.04 (20 °C) 0.34 (1,000 °C) | [17] Test according to DIN 51 046 [48] |
| Nonwoven mats | Manmade mineral fibres, glass wool | 15–60 Average fibre diameter 4–5.5 µm | 0.03–0.04 (20 °C) 0.045–0.07 (100 °C) 0.06–0.1 (200 °C) | [18] |
| Nonwoven mats | Manmade mineral fibres, rock wool | 30–200 Average fibre diameter 6.5 µm | 0.03–0.04 (20 °C) 0.04–0.06 (100 °C) 0.05–0.1 (200 °C) 0.08–0.2 (400 °C) | [18] |
| Isocotton | Cotton | | Thermal conductivity group 040 | [19] |
| Thermolana nonwoven thermal insulation | Sheep's wool | | 0.037 (20 °C) | [49] |

Thus, a mineral fibre is cleared of suspicion of causing cancer if it has a carcinogenic index ≥40 (CI concept of the TRGS 905) or if it has a half-life of <40 days in the relevant animal test (GefstoffV). To identify mineral fibre products that do not pose a threat with regards to TRGS 905 and GefstoffV, the "Quality Association for Mineral Wool e.V. (GGM)" was set up on the initiative of the Trade Association for the Mineral Fibre Industry e.V. The GGM is proprietor of an official quality mark for "Products made from Mineral Wool", which is exclusively awarded to products that fulfil the exoneration criteria of GefstoffV. The GGM represents mineral wool manufacturers whose exonerated products make up a share of approximately 90% of the German mineral wool market [22].

Sarking felts made from mineral wool, laminated on one side with a glass web, provide extra thermal insulation for sloping roofs that have insulation between or over the rafters [23]. Robust, water-repellent glass wool or rock wool insulation panels laminated on one side with a black protective glass web are used for external thermal insulation on exterior walls in back-ventilated non-transparent front facings [24]. Wind and waterproof, but at the same time breathable roof protection materials based on nonwovens have proved themselves in practice. The first "water vapour-permeable" roof coverings appeared about 20 years ago based on polyurethane or acrylate-coated nonwovens with steam diffusion resistance values $S_d$ of more than 30 cm equivalent air layer thickness. Today, specialist spunlaid nonwovens have values of $S_d < 2$ cm [25, 26].

In the flooring sector, appropriate nonwovens are used under floating stone floors and under parquet floors and carpeting. In the case of carpets, these underlays prolong the service life according to their resilience. The good thermal insulation of nonwoven underlays makes the carpet feel comfortably warm under foot.

**Sound insulation**

Soundproofing in buildings and for road traffic is another very important end-use sector for appropriate nonwovens. The general level of noise has risen many times over as more technology has been introduced into the environment. Millions of people have to tolerate loud noises in the workplace. These become unpleasant from 55 dB and dangerous to health from 120 dB [27]. Primary soundproofing measures tackle the sound source directly. However, many soundproofing problems cannot be solved adequately in this way. Secondary soundproofing is therefore also necessary, which interrupts the transmission paths of the sound energy emitted.

Sound absorption is the conversion of sound energy into heat. Porous, open-cell fabrics, such as mineral wool, are especially well suited to this. The energy conversion essentially occurs due to frictional processes in the absorber material [28].

Sound attenuation is preventing sound from being transmitted between two separate rooms. In most cases, sound-reflecting barriers carry out the insulating function. In principle, however, a reduction in sound transmission can also be achieved by sound-damping measures [28].

Footfall sound, but also noise, can also be transmitted via ceilings and walls and be radiated in neighbouring rooms as sound transmitted by air. If the cause cannot be removed by optimizing the vibrations in the construction of the source of the problem, it must be elastically decoupled from the surroundings by mounting or connection via a "spring-weight system". For footfall soundproofing, insulation materials with a low dynamic stiffness are suitable, which are tested according to DIN 52 214 [29].

European harmonization of technical regulations increasingly also includes soundproofing, whereby different types of proofing can be required for soundproofing products, including proofing from outside noise and proofing from transmission by air and footfall sound transmission in building interiors [30].

A common way of establishing sound levels is to measure in an impedance tube, also called a Kundt's tube, according to DIN 52 215 [31]. The method works with vertical sound incidence and can measure very precisely the impedance, the reflection factor and the absorption coefficients. Other measuring techniques are, for example, the Alpha Cabin (sound absorption with diffuse sound field radiation), the Apamat device (measurement of the sound attenuation of materials and insulation superstructures) or the Artificial Head (acoustic recordings of sound events with reference to perception factors such as sound intensity, orientation angle and length of time sustained) [32]. Secondary soundproofing in building interiors is important for soundproofed cabin walls and screens, for suspended ceiling constructions, so-called acoustic ceilings and freely suspended room elements for noise attenuation and damping. Comfortable noise levels in vehicles, both cars and lorries, are a high priority [32, 33]. The use of nonwovens in vehicle acoustics can range from bonnet and end plate absorption, dashboard attenuation and floor coverings to side panelling and headlinings (see also Section 15.5).

Fundamental theoretical studies on the sound absorption behaviour of nonwovens are carried out by Shopshani and Yakubov [34]. Based on the theory of Zwikker and Kosten [35] on the propagation of sound in porous media, numerical methods for calculating the degree of sound absorption by nonwovens are presented. Good agreement with impedance pipe measurements are obtained in the frequency range from 100 to 4,000 kHz for needle-punched nonwovens and chemically bonded nonwovens with varying fibre orientation, a thickness of up to 10 cm and high porosity made from acrylic, cotton and polyester fibres, especially in the higher frequency range and with a thickness >1.5 cm.

Very fine fibre structures in nonwovens provide a high level of absorption of sound transmitted by air with a low weight and therefore a highly efficient absorber system, which can be used as mats or shaped parts. By combining them with heavy layers it is possible to produce efficient spring-weight systems that are 30 to 50% lighter compared to conventional materials. In [36], it was shown that the sound absorption of tufted carpets can be significantly increased with needle-punched nonwoven backings. The main influencing factor is the nonwoven thickness and less relevant is the type of fibre. The nonwoven "Paraphon" [37] is condensed with two areas of different densities in the nonwoven cross-section using a special technique. In the frequency range of many technical and stereoacoustic noises, with an appropriate design, the nonwoven has an average sound absorption coefficient of $a(0) > 0.8$ between 250 and 2,000 Hz and, with its low weight, it is suitable for light construction. The damping properties based on absorption in the high-frequency range are combined with those of resonators in the low-frequency range in a single nonwoven. In this way, broad-band damping properties could be achieved which would otherwise only be possible with bulky porous absorbers, such as mineral fibre mats.

Specially developed nonwovens can carry out several sound-absorbing functions. When light acoustic nonwovens are used under flat roofs, it is found that they demonstrate good physical behaviour. The risk of condensation is diminished and thermal expansions and tensions caused by changes in temperature are reduced.

15.1.2
**Electrical insulation**

Electrical insulation serves to physically separate conductive live elements in an electrical plant or appliance from each other or from the earth. Insulation prevents failing currents and energy losses, while at the same time providing protection against electric shocks. The insulation materials must have a very low conductivity and be resistant to penetration. In addition, mechanical and thermal functions have to be fulfilled, such as minimum ageing and protection from moisture and dirt.

Schäfer gives a comprehensive overview in [39] of how the most diverse types of nonwovens, produced by different processing methods, are used in the electrical sector from the smallest relays to telecommunications cables, standard motors and traction machines, to large generators with 25 kV nominal voltage. In electrical engineering and electric motor construction, nonwovens are used primarily for the production of dielectrics, that is insulation materials. This can occur directly by impregnating with insulating resins or varnishes to obtain high dielectric strengths. Alternatively, further processing can be carried out by the manufacturers of sheet insulation materials. Nonwovens for the electrical sector must meet the following fundamental requirements:

– good absorbency and impregnation capacity
– normally high mechanical strength values
– low moisture absorption
– high degree of uniformity
– good compatibility with insulating resins and varnishes
– absolute purity
– low dielectric losses
– and as a criterion of increasing importance, high long-term thermal
  stability in relation to their electrical properties

Table 15-4 contains the dielectric coefficients for fibres and other commonly used materials [40].

As the trend in electric motors is leading to ever lower weights per horsepower and smaller dimensions, electrical heat is produced that the insulation material used must withstand. Nonwoven insulating materials are therefore primarily produced on the basis of polyester or aromatic polyamide fibrous materials. These fibrous materials essentially fulfil the required dielectric and thermal property specifications. The nonwovens made from them can be rated as high-temperature resistant sheet insulation materials according to VDE 0530, Appendix II [41] and, depending on the composition, are approved for insulation material classes B, F and H for maximum temperatures of 130 °C, 155 °C and 180 °C respectively [39]. The requirement for purity in the electrolytic sense can only be fulfilled when only a small amount or no binder or chemical agents are used to bond the nonwoven. Thermal bonding methods and needle-punching are used. In this way, products can be obtained that result in a conductivity of the aqueous extract (ac-

**Table 15-4** Dielectric coefficients of fibres and other commonly used dielectrics taken from [40]

| Material | $\varepsilon$ | kHz | Additional information |
|---|---|---|---|
| Air (0°C) | 1.00059 | | |
| Water (18°C) | 81.1 | | |
| Cotton | 18 [1] | 1 | 3.2 (0% RH) 7.1 (45% RH) |
| Cotton | 5 | 1,500 | |
| Viscose | 8.4 [1] | 1 | 3.6 (0% RH) 5.4 (45% RH) |
| Acetate | 3.5 [1] | 1 | 2.6 (0% RH) 3.0 (45% RH) |
| Wool | 5.5 [1] | 1 | 2.7 (0% RH) 3.5 (45% RH) |
| Polyamide 6.66 | 3.7 [1] | 1 | 2.5 (0% RH) 2.9 (45% RH) |
| Polyamide 6.66 | 3.8–4.3 | 0.05 | |
| Polyamide 6.66 | 3.4–3.8 | 1,000 | |
| Polyimide | 3.6 | 0.06 | |
| Polyacrylic (Orlon) | 4.2 [1] | 1 | 2.8 (0% RH) 3.3 (45% RH) |
| Polyester (Dacron) | 2.3 [1] | 1 | 2.3 (0% RH) 2.3 (45% RH) |
| Polyvinylidene chloride (Saran) | 2.9 [1] | 1 | 2.9 (0% RH) 2.9 (45% RH) |
| Polypropylene | 2.38 [1] | | across the lines of force 2.0 |
| Fluorofibres | 2.00 | | across the lines of force 1.73 |
| Polyacrylic (Nitron) | 4.53 | | across the lines of force 3.17 |
| Polystyrene | 2–3 | 1,500 | |
| Polyvinyl chloride | 3.3–3.8 | | |
| Glass filament | 4.4 [1] | 1 | 3.7 (0% RH) 3.7 (45% RH) |
| E-glass filament | 6.1 | 1,000 | |
| A-glass filament | 6.5 | 1,000 | |
| D-glass filament | 3.9 | 1,000 | for high-frequency applications |
| Glass | 2–16 | 1,500 | |
| Rubber, raw | 3.0 | 0.8 | |
| Porcelain | 6 | 1,500 | |
| Phenolic moulding resin + asbestos | 10 | 0.8 | |
| Phenolic moulding resin + textile fibres | 8 | 0.8 | |
| Phenolic moulding resin without filling | 5 | 0.8 | |
| Specialist ceramic materials | ≤5,000 | | |

1) At 65% RH (RH = relative atmospheric humidity).

cording to DIN 7743 [38]) of 10 µS/cm [39]. No adhesion agents are required as both fibrous materials are distinguished by good chemical affinity to the normal impregnation agents such as unsaturated polyester resins, epoxy resins, ester imide resins or polyimide varnishes. When nonwovens based on E-glass fibres are used, which are especially suitable for the production of laminates for certain applications, it is not possible to dispense with coupling sizes. Compared to other fabrics, nonwovens are distinguished in the field of electrical insulation by an excellent impregnation capacity, which, after the parts to be insulated have been impregnated and hardened, results in a high layer strength and therefore also a high dielectric strength. Examples of applications for nonwovens include the manufacture of

- small and medium-sized three-phase motors (insulation material classes B and F) for the phase separation of stators, as coil end binding or as impregnatable linings for rotors
- direct current electrical machines for coil insulation
- large motors and generators for spacing in the coil end region, as carriers and coverings for mica papers and split mica tapes in high voltage windings
- transformers for insulating primary and secondary windings
- large magnets for layer insulation of the magnetizing coil
- cables for keeping a distance between conductive strands and as spacers in sockets
- high-voltage cables for equipotential bonding between shielding and insulation

Nonwovens in non-impregnated form are not regarded as actual insulating materials as the dielectric strength is relatively low without impregnation. For the production of insulating materials, they are very important, for example in the manufacture of

- flexible multilayer insulation materials
- prepregs
- mica tapes

Information on nonwoven composites for use in electrical engineering can be found in [42] by Schäfer. Flexible multilayer insulators, so-called DMDs, are three-layer composites produced by the double-sided adhesion of two nonwoven layers onto a polyester film. This type of composite combines the properties of impregnatable nonwovens with those of dielectric polyester film for use in electric motors and transformers. In the cable industry (production of energy, telephone or lighting wiring cables), nonwovens fulfil fixing, separation and cushioning tasks. Swelling nonwovens, so-called "water blocking tapes", three-layer composites made from a nonwoven polyester carrier with a swellable layer inside and tissue covering are extremely important as water blocking tapes under the layer sheathing in communication cables or in insulated medium and high-voltage cables. If the cable sheathing is damaged, allowing water to enter, they seal off cavities inside the cable by immediately swelling up, thereby preventing water from progressing along the longitudinal axis of the cable. The damage remains limited to a few metres of cable. In the battery industry, nonwoven composites are used as separators for specialist applications. Wet-laid nonwovens made from polyvinyl alcohol fibres have proved themselves in particular [42]. Finally, nonwoven composites are used as solder-resistant, flexible base materials in the production of printed circuit boards.

Nonwovens are not only processed as insulation materials in the electrical sector, but can also be treated with conductive finishes. Such electrically conductive nonwovens are used, for example as corona shielding in cables, in high-voltage insulation materials to discharge static electricity or as surface heat conductors.

With electronics entering into increasingly wide areas of our daily life, so the need increases for protective measures against electromagnetic interference (EMC protection, EMC-electromagnetic compatibility). This not only concerns installations that are relevant in terms of security, for example those of the authorities or military establishments, but also hospitals, banks and computing centres. For this, it is possible to shield a room according to the Faraday's cage principle using highly conductive materials. Very successful in this area are also nonwovens treated with conductive finishes, for example by copper, silver or aluminium coating. Also used are surface-modified fibres in nonwovens and woven fabrics, such as copper or nickel-coated polyacrylic or polyamide fibres, fibres with external carbon coating, and fibres modified with copper sulphide [43]. The room shielding system Shieldex is based on a nylon 66 nonwoven that has been permanently coated with copper in a special process. Shielding efficiency values from 40 to over 80 dB are achieved in the frequency range from 4 MHz to 1 GHz (tested by E-field measurement according to MIL-STD285). Here, the properties of the nylon remain intact. The material is flexible and breathable. The surface resistance is 0.04 Ohm/m$^2$. At only 60 g/m$^2$ and with a wall thickness of 0.1 mm, the material is very light. It is applied in a similar way to normal wallpaper [44, 45]. For EMC shielding of round and flat cables an adhesive-finished, flexible polyamide web with a series of evaporated films of copper and tin has been developed. The acrylic adhesive used is designed to have good electrical conductivity, thereby guaranteeing total conductive casing of cables with no contact resistance. In the frequency range from 100 MHz to 1 GHz, a shielding efficiency of over 60 dB is achieved [46].

## 15.2
## Filtration
E. Schmalz, M. Sauer-Kunze, L. Bergmann

The field of filtration is tremendously diversified. There are more than 1,000 different applications characterized by different profiles and conditions and consequently requiring different filter materials. Also in addition to industrial fabrics paper, soft foam, sand, sintered materials and ceramics are being used just to name a few. One of the most important segments of filter materials are nonwovens. Due to their variability and their economical manufacture they can be easily adapted to nearly all kinds of filtration jobs. This explains that already in 1994 such products had a share of the world filtration market of 89% (836 million m$^2$) Table 15-5) [50].

The annual sales in 1995 for all nonwovens amounted to US $ 1.4 billion worldwide. This represents growth of approximately US $ 100 million for the previous year and more than US $ 600 million als compared to 1988. By the year 2000 with an annual increase of 6% the market should represent US $ 2 billion. Tables 15-5 and 15-6 illustrate how nonwoven markets differ and can be divided into individual markets on different continents.

For the number of square meters of nonwovens in filtration according the EDANA statistics, Table 15-7 shows that liquid filtration is, based on square meters, about 3 times as large as dry filtration.

**Table 15-5**  Filtration market 1994/1995 – Annual roll goods sales US $ [50, 51]

| Fabric type | Share (%) | Sales 1994 (million $) | Sales 1995 (million $) |
|---|---|---|---|
| Needle punched nonwovens | 30 | 400 | 415 |
| Other drylaid nonwovens | 14 | 180 | 180 |
| Wetlaid nonwovens | 21 | 270 | 275 |
| Meltblown nonwovens | 18 | 230 | 250 |
| Spunbonded nonwovens | 6 | 80 | 85 |
| Other fabrics (composites, hybrid-like flash-spinnings, membranes, specials) | 11 | 140 | 200 |
| Total | 100 | 1,300 | 1,400 |

**Table 15-6**  Nonwoven market for filtration 1995 [51]

| Continents | Share 1988 (%) | Share 1995 (%) | Sales 1995 (mill. $) |
|---|---|---|---|
| Europe | 41.5 | 39.5 | 531 |
| USA/Canada | 38.5 | 37.5 | 505 |
| Japan & Far East | 20.0 | 18.5 | 247 |
| Central and South America, South Africa and others | | 4.5 | 61 |
| Total | 100.0 | 100.0 | 1,344 |

**Table 15-7**  Deliveries of nonwovens for filtration in Western Europe or outside [52]

| Year | Production of nonwovens total (million m²) | Liquid filtration (million m²) | Dry filtration (million m²) |
|---|---|---|---|
| 1991 | 12,481.5 | 728.1 | 237.4 |
| 1996 | 17,570.6 | 901.9 | 250.1 |
| 1998 | 21,304.8 | 1,091.1 | 173.0 |
| 2000 | 25,271.8 | 1,155.5 | 351.1 |

To manufacture filter media economically it is important to consider a number of properties which are present in the carrier or scrim fabric and also related to the type of particles being separated. Factors related to the carrier or scrim material are among others:

– temperature
– humidity
– degree of turbulence
– mass flow
– chemical composition

Factors related to the particles are among others:

– their size
– particle composition
– type of material
– particle concentration

Based on these demands a suitable filter material is selected. Usually this often includes a compromise due to the many different requirements which may include high pressure drop or sometimes relatively short running time. To tailor the filter medium to these requirements a wide range of designs of textile elements are available. One of the first aspects is to choose the right fibre material. The kind of fibre has to do with the thermal, physical, chemical and biological conditions. Fibre fineness as well as cross section are important properties which influence the performance of the material. The type of fibre structure and also effective filter surface may influence the efficiency. To manufacture such filter materials a wide range of fibres showing different properties are available. Selecting filter materials for the different applications depends also on the type of equipment being used, let alone economical aspects. In the early years natural fibres like wood, cotton, cellulose and asbestos were used in filtration. They are virtually all substituted by manmade and glass fibres.

In the family of synthetic fibres we see more and more finer fibres as well as bi-component fibres. Fibres with different cross sections to enlarge the fibre surface enhancing the performance are increasingly being used. Very important is the worldwide trend to use finer and even micro fibres with a variety of cross sections. Fibres which show a profiled cross section possess a wider specific surface which makes the separation of particle smaller than 5 μ more effective [53]. Factors influencing these characteristics are the cross section, shape and micro fibrillation of such materials.

Both finest and micro fibres are, because of their filter surface, preferably arranged and used to enlarge the effective filter surface. With depth filter media they should be found at the flow-out surface in order to refine the pile labyrinth separating finest dust particles. Static charge is sometimes found a problem in filtration which could create danger when parks ignite a dust cloud which may lead to an explosion. To neutralize and sometimes 'ground' such materials metallic fibres and carbon fibres or fibres with metallic coating are mixed with such synthetic fibres and being used [54].

In the case of a hazardous gas or to attack odours or taste carrying substances activated carbon fibres are increasingly being used in so-called combination filter media. Due to their filtering properties mineral fibres are sometimes of great importance for such filter media. In particular they are applied in hot gas applications. For special uses high performance fibres such as polyamide, aramide, polyphenylen-sulfide and melamine resin are available. After selecting the fibre material the principle of web formation and the mode of bonding is determined which will influence the consistency, thickness, permeability, tenacity and strength of the nonwoven. One has to take into consideration the economic aspects as well in manufacturing such products. As the design of the filter medium largely depends on its practical application, it makes sense to discuss the structure of the filter material in the context of filtration systems available. Filtration is a process by which particles are separated from dirty incoming air or liquids often described as solid/gaseous, gaseous/gaseous, liquid/gaseous, solid/liquid or liquid/liquid filtration. Depending on the carrier medium, one distinguishes between dry and liquid filtration.

### 15.2.1
### Dry filtration

#### 15.2.1.1 General

Under dry filtration we understand a separation process in which a contaminated carrier medium such as air or process gas is separated from hazardous particles, the filter medium being the separator between the gas and any particle. Of course we differentiate between separation of solid, liquid and gaseous particle (Fig. 15-3).

Independent of the mode of separation we distinguish between surface and depth filtration. Surface filters show a separation effect on the surface of the filter medium. The filter effect as such is often achieved through the filter cake which builds up during the process. Such surface filtration may not a economical when used with dust concentration below $5 \text{ mg/m}^3$. Sometimes in filtration dust concentration may be as low as $1 \text{ mg/m}^3$. Depth filters mostly function as storage filters which

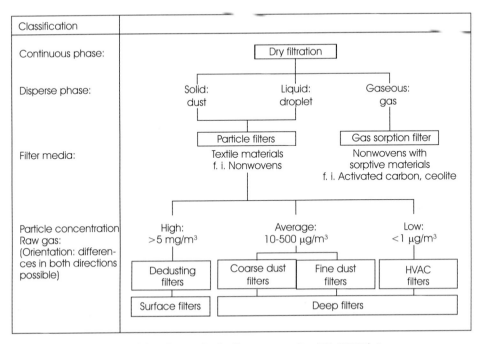

**Fig. 15-3** Criteria of classification for dry filtration according VDI 3677/Bl. 2

cannot be regenerated and have to be disposed of after use. On the other hand filtration takes place in three dimensional due to the fibre labyrinth if the material is highly porous. In most cases their use is not economical with dust concentrations higher than 1 mg/m$^3$ up to 5 mg/m$^3$ maximum. Within this space described (1–5 mg/m$^3$) the selection of the best suitable systems depends on filter specific parameters. Table 15-8 names some of the essential filter technological values of different

**Table 15-8** Differentiation between deep and surface filtration with regard to typical characteristics, filter shape and properties as soon as fibre types and nonwoven constructions

|  | *Deep filtration* | *Surface filtration* |
| --- | --- | --- |
| Fields of application | HVAC and process air technology | Process air and dedusting technology |
| Characteristics | Separation of particles within the filter media – filter has to be replaced when clogged | Separation after a starting phase on the media surface or, respectively, in the dustcake which is created – this is why cleaning is necessary |
| Typical parameters Dust concentration Particle size Temperature | <1 mg/m$^3$ mainly <1–5 µm –20 °C to +60 °C, partly +350 °C | >5 mg/m$^3$ mainly >5–10 µm +20 to >+800 °C |
| Fibre types | Glass fibres, standard synthetic fibres (PET, PP), micro-fibres, Meltblown, natural fibres, cellulose fibres | Standard synthetic fibre types, high-tech fibres, micro-fibres, mineral fibres, ceramic fibres, metallic fibres |
| Nonwoven types | Voluminous nonwovens, paper-like wetlaid nonwovens, composite nonwovens | cloths, needle punched nonwovens, spunbonded nonwovens, thermal bonded nonwovens, composite nonwovens |
| Porosity | >90 to >99% | 45 to 80% |
| Fibre volume | low | high |
| Mass per unit area | 50 to 800 g/m$^2$ | 80 to >700 g/m$^2$ |
| Filter design | Filter mats, pocket filters, panel filters, rigid filters, filter cartridges | Hose filters, filter cartridges, filter candles, compact filters, pocket filters, rigid filters |
| Typical characteristics | not cleanable | cleanable |
| Duct air velocity | 0.1–3 m/s | 5–50 m/s |
| Media velocity | 100–10,000 m$^3$/m$^2$h | 60–300 m$^3$/m$^2$h |
| Pressure difference | 20–500 Pa | 500–3,000 Pa |

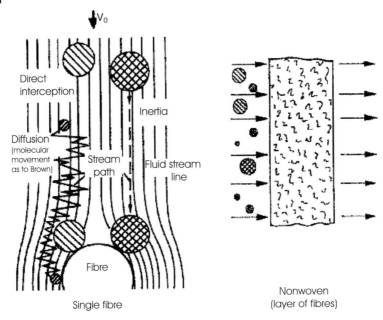

**Fig. 15-4** Sketch of physical separation mechanism according to [56]

types of filter media and different filter designs which are most widely used [55]. The use of nonwovens in filter technology as well as the range of technological problems to be solved is so wide that we would only like to concentrate on a few examples of applications, designs and types of nonwovens.

In dry filtration as in industrial textiles in general the functionality of nonwovens depends on how the single fibres are designed and arranged. Desirable is an isotropic position of the fibres or filaments in the web. The essential physical separation mechanisms are inertia separation, interception, diffusion and electrostatic forces (Fig. 15-4) [56]. They are strongly influenced by velocity, the fibre and particle diameters, the density and thickness of the filter medium.

Fig. 15-5 shows the range of particle sizes of certain substances collected.

The flow-through velocity being sufficiently small, the separation performance is improved with antipole surface charge if available on fibres and/or particles. This is achieved by electrostatic forces with particles finer than $5 \mu$ ($1 \mu$). The sieve effect is only of interest for course particles.

### 15.2.1.2 **Functional requirements properties**
The different filter designs are of great importance for the selection of suitable nonwovens. Process related properties and textile technological requirements often lead to so-called combination nonwovens. Fig. 15-6 shows the properties which may influence depending on the technology of manufacturing being used. There may be

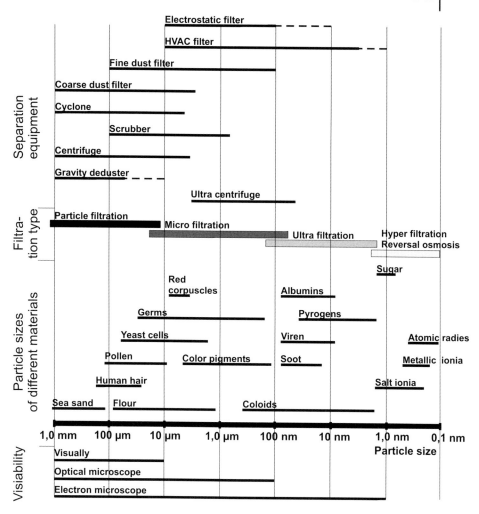

**Fig. 15-5**   Particle sizes of different substances

process or design specific requirements which have to be met in the use of non-wovens in filtration. For more information see the relevant chapters of this book.

Due to the wide range of applications and their attributable design of non-wovens, the following are differences between surface and depth filtration taking into consideration the different types of separation and their process and the design specific requirements.

### 15.2.1.3 Surface filters

Surface filters are mainly used in the context of industrial exhaust gas with a high mass concentration (larger than 5–1 mg/m$^3$) minimum (dust separation equip-

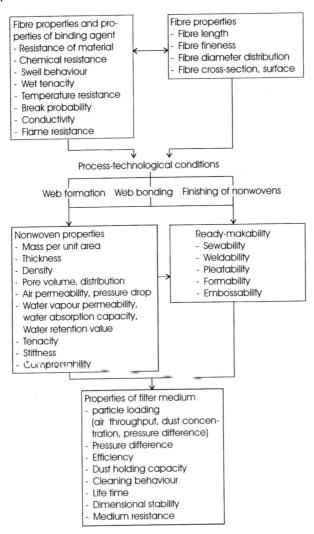

**Fig. 15-6** Influence of raw material and conditions of manufacturing process on properties of filter material

ment). Equipment currently being used includes centrifuges, electrostatic precipitators, wet scrubbers and fabric filters. The most efficient variation is basically the design of a fabric filter using filter bags and also dust cartridge filters.

**Filter bags**

Worldwide needle punched nonwovens predominantly are being used for filter bags [57]. They can be used in a wide variety of designs, they are very flexible and can be combined using all kinds of different textiles including nonwovens

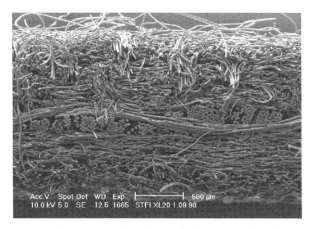

**Fig. 15-7** Cross section of needle punched nonwoven for surface filtration

(Fig. 15-7). To improve the integrity of the material physical requirements during cleaning most needled felts, notably in Europe, use a scrim. In the United States approximately 50–75% of the filter materials from needled felts are used without scrim [50]. The incorporation of a scrim in needle punched nonwovens increases its stability. Some people feel that a scrim also contributes to the better efficiency of certain dusts resulting in better process technology. It is important to realize, however, that the dust cake which is periodically removed should be as close to the surface as possible. Such dust cake structure is crucial for the separation performance.

In order to achieve the highest possible efficiency it is preferable to use fine fibres, preferably microfibres, finer than 1 dtex or, respectively, apply a special surface to the needle punched product. Certain characteristics are determined by the applications in which filter bags are being used. Table 15-9 lists a range of values concerning such needle punched nonwovens.

**Table 15-9** Textile-technological characteristics of needle punched nonwovens for filtration [50]

| Characteristics | | Unit | Value |
| --- | --- | --- | --- |
| Mass per unit area | | g/m$^2$ | 200–2,200 |
| Thickness | | mm | 0.7–3.5 |
| Density | | g/cm$^3$ | 0.11–0.72 |
| Porosity | | % | 60–93 |
| Permeability | | l/m$^2$s | 40–500 |
| Tenacity | longitudinal | N | 650–2,600 |
| | cross | | 500–2,600 |
| Elongation | longitudinal | % | 11–64 |
| | cross | | 12–79 |

In addition nonwovens are used for filters which are chemically bonded but also thermally bonded normally in weight ranges between 200–600 g/m². The relatively thin layer of such nonwovens is being used collecting dust particles in its entire structure. They come in different densities and weight ranges according to the application. The gradient density and compactness depend on fibres blends used. One advantage can be achieved with a fuzz-free smooth surface. Certain materials need chemical agents to determine a certain stiffness which is sometimes important for proper shape stability, notably pleats. Such high shape stability can be achieved by Latex binders. Also antistatic chemicals can be used and blended in. Bonding by means of bi-component fibres is becoming more and more popular. Their share of these materials is constantly increasing. Bonding webs containing bi-component fibres are normally accomplished by thermal bonding. Polyester fibres are often used. Again, the kind of fibre is determined by the end-use of its application.

Flow-in velocity is mostly less than 3 cm/s. This may amount to a filter capacity load of 180 m³/m²h. The result is that these nonwovens accomplish a very high separation performance particularly in the early process, building eventually a filter cake during use. To compare such materials the VDI 3926 [58] is being used (see Section 18.3.5). A very efficient surface filtration process allows a high cleaning performance, also with sometimes adhesive dusts which are mostly achieved by means of pulse jet cleaning, occasionally by shaking. In many cases such high cleaning performance is reached through surface finishing. Adequate surface finishing may highly influence the life time, the type of filter cake, cleaning behaviour and finally filtration performance. Finishing may be based on thermal, chemical and physical processes as well as a combination of the above (Fig. 15-8) [59].

A heat procedure stabilizes any nonwoven and leads to a more stable product. In general terms thermal treatment in connection with pressure changes fibre morphology, thus raising the resistance of the filter media against aggressive dusts. The surface of such textile fabrics may be even and smooth, structured, uneven of fibrous. Modifying the initial surface, the cleaning behaviour can be im-

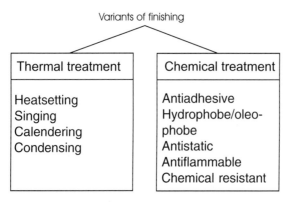

**Fig. 15-8** Finishing processes for filter nonwovens [59]

proved and influenced. Such modification may be carried out by means of calendering, singeing, or chemical processes. In order to achieve improved emission values one can choose chemical finishing, but also membrane laminating is currently being performed (for example Gore-Tex®).

### Cartridge filters

In order to use filter cartridge the filter medium may be pleated or folded sometimes into a star. This allows for a larger filter area in a given dimension resulting in a more compact filter element [60].

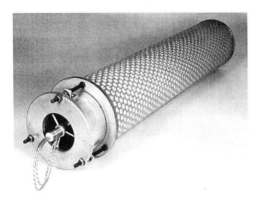

**Fig. 15-9** Combined filter cartridge, general view

Consequently the filter media need to be pleatable. Important quality criteria are high quality pleats, high pleat stability and sufficient mechanical and thermal resistance, preferably thin, stiff nonwovens of mass per unit areas arranging from 80–300 g/m² are often used as filter materials. Currently the use of paper-like wet laid nonwovens which sometimes are thermally bonded and in some cases from polyester or polypropylene and can also be made from impregnated cellulose papers. They provide high stability at very low thickness (from 0.1 to 2.0 mm). They should have high stability which is important for the processing into any pleated filter element. Sometimes carrier cloth or wire is needed for large depth of pleats. Those properties also make them well suited for carrier material. Sometimes also meltblown material is used, notably in liquid disposal cartridges. Some current developments are based on combinations of stable layers of carrier material and other layers of nonwovens, thus being meltblown or other filament nonwovens sometimes supported by metal structures, sometimes hydroentangled. Special after-treatments may result in high quality and make them usable as filter material. In liquid filtration to collect particles such as oil emulsion special finished materials are being used.

### 15.2.1.4 **Depth filters**

Depth filters are used in the context of low mass concentration from larger than 1–5 mg/m³ which means mainly in the field of general HVAC air filtration. In the field of such application filter media use is completely different from dust filtration. Above all the major reason is substantially lower dust concentration and also particle size present. The air-flow velocity, however, is much higher, filter media show a much more open structure and surface and elements are often in V-shape and also pleated. This results in a considerably higher flow velocity within the filter material. This filter type is not being cleaned due to the dust fineness and low mass of dust (see Table 15-8). In addition to dust separation and HVAC applications there is the field of process air technology. The latter may require special solutions often addressing surface filtration and depth filtration depending on dust concentration and particle size. In process air applications nearly all different filter media are being used. A progressive structure of the filter media leads to longer life cycle often requiring several layers of fibres in varying fineness across the cross section (Fig. 15-10). Particle separation is provided by means of a systematic structuring of layers of the nonwoven of varying fibre fineness. The linear decrease in fibre fineness across the thickness of the composite balance out

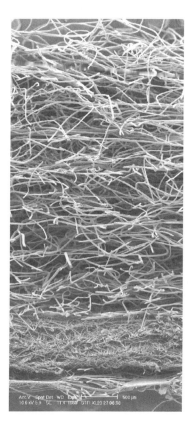

**Fig. 15-10** Nonwoven for deep filtration (filter class F6) with progressive structure of cross-section

particle deposition. The separation conditions stay nearly the same all across the depth of the media.

Thus, blocking effects may largely be avoided. An important aspect is the rise in pressure drop, and to keep the pressure drop as low as possible at the same time offering high filter efficiency to separate the aerosol. Such filter materials are often represented by filter mats, filter cells, pocket type filters for coarse and also fine dust, and compact filter elements as well as HEPA and ULPA filters. Figs. 15-11 to 15-14 show designs by GEA Delbag Luftfilter GmbH.

In order to compare and classify filters into coarse and fine filtration classes of G1 to G4 and F5 to F9 are based DIN EN 779 [61]. To classify HEPA and ULPA filters classes H10 to H14 and U15 to U17 DIN EN 1822 is applicable [62]. It is expected in the year 2000 revised editions of these standards may be published

**Fig. 15-11**  Filter panel

**Fig. 15-12**  Filter element

**Fig. 15-13**  Pocket filter

**Fig. 15-14**  S-Filter

from the filter classes G1 to F5. Usual face velocities are 1–3 m/s, highly voluminous nonwovens from synthetic fibres and sometimes glass fibres show mass per
unit areas between 100 and 800 g/m$^2$ and thicknesses between 5–40 mm or sometimes up to 100 mm. In most cases thermal bonding is achieved by using either
bi-component fibres or binding agents. Filter mats are being used, also filter cells
particularly for coarse dust filtration and so-called pocket-type filters which offer
enlarged filter surface.

Filter media for fine dust filtration (filter classes F5–F9) are manufactured from
fibres of 0.5–20 dtex. As fibre materials polypropylene, polyester and glass fibre
are preferably used. In order to achieve the performance required, flow-in velocities as compared to coarse filtration are reduced to 0.05–0.3 m/s. This may be accomplished by using pocket-type filters and sometimes pleated filter elements. To
make nonwovens used for these filters are between 70–180 g/m$^2$. Thicknesses
vary between 2 and 10 mm. Wet laid nonwovens may range from 60–120 g/m$^2$,
thicknesses varying from 0.5–1 mm. State-of-the-art materials include composite
nonwovens which are manufactured by a variety of web bonding and web forming processes. Current nonwoven composites could be for example:

– one of several layers of fibre nonwoven plus a layer of spunbonded
– several layers of meltblown of varying fibre fineness with a carrier material
  mostly heavier than the filter layer from filament-spunbonded nonwoven
– thermally bonded nonwovens plus meltblown layer [63–65]
– mechanically manufactured stitch-bonded nonwoven Kunit with compacted loop
  surface [66, 67]

In order to produce filter media for fineness filtration (HEPA and ULPA) filter
materials of classes between H10 – U17 are being used with microfibre glass
materials. In this field wet laid nonwovens are most frequently found. These paper-like materials show mass per unit areas between 60 and 120 g/m$^2$ with thicknesses of 0.4–1 mm. When prepared these materials are often pleated and, therefore, offered with enlarged surfaces, i.e., as a plain filter element, a cartridge filter
or V-shaped. These velocities range from 0.01 to 0.1 m/s and are considerably below those being used for coarse and fine filtration.

15.2.2
**Liquid filtration**

In liquid filtration we understand the separation of solid substances being removed
from suspensions whereby the filter media divides the particles from the fluids. The
systems to be dispersed may consist of solid substances in fluids, which is called
suspension. The fluid dispersed is named a filtrate (see Fig. 15-15). For a separating
agent, a filter medium is used similar to dry filtration. Except for nonwovens, filtering layers may be dense, wet-laid filter cloths, membranes, ceramic materials, and
sometimes bulk goods. Sand, for instance, and porous sintered materials are also
being used as filter media materials. Fibre materials in liquid filtration are cellulose, also cotton, viscose, polyamide, polyester and polypropylene.

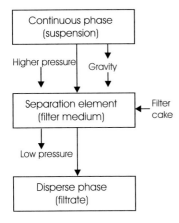

**Fig. 15-15** Liquid filtration

The difference in pressure, which drives the filtrate fluid, the filter medium and filter cake deposited may be created by means of gravity, low pressure on the filtration side and by means of high pressure on the suspension side. Depending on the process, in liquid filtration we differentiate between surface filtration and cake filtration (see Fig. 15-15). The separation of particles from fluids with a low mass concentration is called 'purity filtration'. One typical example is sterile filtration. High mass concentration in the suspension requires cake filtration. Furthermore, depending on the procedure of how the solid and liquid masses are separated, we distinguish between three types of filtration (see Fig. 15-16).

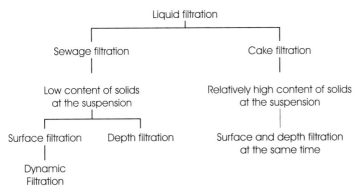

**Fig. 15-16** Classification of liquid filtration

**Surface filtration**

In surface filtration the separation process is mechanical, the solid substances being separated from the suspension on the surface of the filter medium. One variation is sieve filtration in which solid substance is held back at the opening of the filter medium, no filter cake being built up. This type of filtration allows for particles being held back which are normally larger than the openings of the filter medium. The particles separated are removed from the filter medium sooner or later by cake build up. Sieve filtration is used to separate coarse solid substances from fluids. The filter media being used are sometimes punched sheets or sometimes sieves are used.

### Depth filtration

One special variation of surface filtration is dynamic filtration or cross-flow filtration. If the solid substance to be separated is held back in the interior of the filter medium, we speak of depth filtration. In this process the solid substance enters into the hollow spaces present in the filter medium and his held back both mechanically and by forces on the surface. Depth filtration may affect particles into the sub-micron range. Depth filtration is used with medium filtration velocities and medium pressures such as in water treatment. Regeneration is achieved by means of re-rinsing. Depth filters often consist of beds of bulk material, of varying grain sizes.

### Cake filtration

With cake filtration, a filter cake builds up on the filter medium, which contributes to the filtration. The filter cake, in the case of filtration becomes more and more compact, thus with decreasing velocity. The separation performance goes hand in hand with the filtration pressure. Typical applications include syrup or sand. Selecting the type of filter design and the filter medium is influenced by many factors. By the process such as throughput, the amount of solid particles in the suspension, the degree of cleanness of the filtrate required particle size. The chemical composition of the suspension, temperature, density and viscosity of the suspension, as well as the stability required of the filter medium is part of this.

Essential parameters to assess the quality of filter media for us in liquid filtration are the physical, chemical and structural properties such as porosity, pore diameter, water permeability and separation performance.

#### 15.2.2.1 Liquid filters based on nonwovens

Selecting a suitable nonwoven for use in liquid filtration, the filter-technical properties are, in addition to the chemical and thermal resistance of the fibre materials essential.

In addition to that, the type of filter designs, the wear and tear behaviour of the nonwoven when in operation and economy have to be taken into consideration. The technical requirements have to be met by the nonwoven materials. They must provide high separation performance at a pressure differential as low as possible in relation to the flow characteristics. Separation performance and efficiency largely depend on the fibre fineness as well as the density of the nonwoven.

Starting parameters, fibre material and mass per unit area remaining constant, the separation efficiency will rise with the use of finer fibres to the nonwoven. Pores are becoming smaller. Higher nonwoven density, too, results in higher separation efficiency. The specific performance characteristics depend on different applications.

Depth filter media are used both in liquid and dry filtration in order to separate waste in low concentration and a large throughput through the filter medium. This for instance is true in milk filtration and the filtration of drinking water. High concentrations of solid substances in the suspension require of nonwovens, which allow for surface filtration. These applications can be found in industrial fields. One example given is the filtration of emulsions or coolants in metal machining, a process in which belt filters are often used together with nonwovens.

Other applications of such nonwovens in different types of designs and filter equipment are for instance wash and degreasing baths, industrial effluence, phosphating baths, cycling water from lacquer-separating equipment, lubricants, and hydraulic oils. Often needle-punched fabrics are used in liquid filtration. They are found both in depth and surface filtration. They are being used with and without scrim. In this process homogeneous needle punched nonwovens or needle punched nonwovens with carrier cloth are used. Filtration properties are determined by the nonwoven. The requirements to be met with regard to stability are secured by cloths which are centrally or asymmetrically incorporated. In order to improve the separation performance of the nonwoven composites finishing processes including singeing, calendering or chemically modifying surfaces are being used.

### 15.2.2.2 Variables in design of liquid filters
We distinguish between belt – bag – candle – and drum filters.

### Vacuum belt filters
Depending on the way work (Fig. 15-17), we make a distinction between belt filters with and without vacuum. Vacuum belt filters RT (reciprocating tray) shows a sliding vacuum tray. Filter belts are continuously loaded and drained. The dry filter cake is then thrown off. Prior to using the belt again, it will be washed. Application can be seen in the chemical and pharmaceutical industries, in production

**Fig. 15-17**   Belt filter equipment of Pannevis Wiesbaden

of catalytic converters (powder type). Vacuum belt filters are used with inlet concentrations in the suspensions ranging from 50–500 g/l. Filtration efficiency is between 30 and 1,500 kg/m²h, remaining moisture content may be 10–80%.

### Gravity belt filters

These simple, often small types of filters and the filter media is on a roll and rolls off in one direction. The suspension is loaded onto the belt and solid particles remain on the belt. Gravity allows for the fluid to escape through the filter medium. The filter medium is eventually rolled off and waste is disposed. A typical case is coolant filtration.

### Liquid bag filters (micron rated)

Liquid bag filters are simple types of filters in which the suspension flows into the center of the bag; the solid particles are being held back and the filtrate flows to the outside. Bag filters are usually placed in a supporting basket (Fig. 15-18).

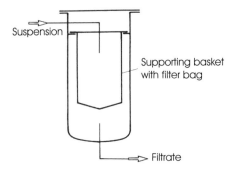

Suspension

Supporting basket
with filter bag

Filtrate

**Fig. 15-18** Bag filter according to [73]

As filter media, needle punched nonwovens are used, sometimes also composites, above 100 microns. Woven fabrics of natural or synthetic fibre showing thickness of 0.5 mm are also found. Typical materials are polyester, polypropylene, polyamide, and viscose fibres in standard fineness and also sometimes fine fibres.

Using needle punched nonwovens, which distribute pores with different pore sizes over a wide area, filtration, contrary to woven fabrics takes place in the interior of the filter media. Regeneration by means of washing or rinsing is, therefore, only possible to a very limited extent. The liquid bag filters are offered by a number of different companies. There are filters with a single bag as well as multiple bags for larger flows of fluid, and can be as many as 20 filter bags in one vessel.

The relatively low purchase price, simple maintenance, easy access to these filter bags are the main features of these filters. Liquid bag filters are used with suspension ppm range and with a throughput $\leq 80$ m³/m²h. They are often used as prefilters for other depth filters and sometimes as police filters in numerous different processes.

## Candle filters

The filter vessel is similar to the previously mentioned liquid bag filters. Filter elements, however, are completely different. Candle or disposable cartridge filters are available in a wide variety. They can contain activated carbon, string wound, resin-treated, membrane candles. They can also be made from metallic filter media and they can often come in a pleated filter cartridge. Sometimes nonwovens are being used as filter media. Spunbonded and meltblown materials are well established.

In addition, novel hydro-entangled composite nonwovens are also available (Fig. 15-19). "Spray-spun" candles, which have become more popular in recent years, are also manufactured using meltblown technology.

Pleated cartridge/candle filters are very well suited for surface and sieve filtration. Their large filtration surface allows low initial pressure differential at the same time throughput. As bag filters, candle filters are available for a large range of applications. The filtration throughput as compared to liquid bag filters, however, is considerably lower with candle cartridge filters. Experience determines which filter type to be used. Often applications overlap. Their use in the industry is expected to grow.

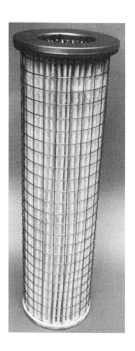

**Fig. 15-19** Pleated filter candle with a nonwoven-metal composite

## Vacuum drum filters

Of all continuous filters, drum filters (Fig. 15-20) are most widely used. Main fields of application are seen, for instance, in basic industry, including the chemical industry, hydrometallurgy and related food producing industries, often where high concentrations of solids have to be treated.

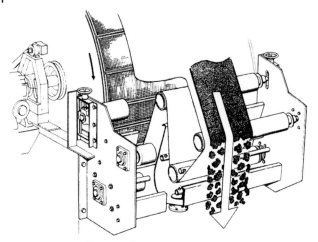

**Fig. 15-20** Drum filter according [73]

The filter medium material for instance, a needle felt, covers the drum, which is located in a bowl. This piece of equipment allows stirring and is filled with a suspension while in operation. The filtration surface is exposed to a vacuum, which builds up and builds the filter cake. Rotation speed is determined by the process parameters. The filter cake, which is built up by suspension, is washed by the vacuum and drained. Filtrate and wash emulsion are separately disposed of. The drum filters for cake filtration are manufactured, depending on their use. Different take off methods are being used including knife, rope, and role take off.

Filtrate cleaning is accomplished by separation filtration equipment.

Vacuum drum filters are available with filter surfaces between 1 and 120 m$^2$ and the drum diameter may vary and be as large as 4.5 m.

## 15.3
## Building and construction industry

### 15.3.1
### Geononwovens
K. Lieberenz

Nonwovens are being used increasingly in geotechnology; for example, they are used in earth engineering and for laying foundations in the construction of roads and buildings and on land reclamation sites, and as structural, permanent or temporary elements. They therefore belong to the sector of geotextiles which, together with geogrids and sealing substrates, form the group of products known as geosynthetics (Fig. 15-21).

A common feature of these materials is that they are all used in geotechnology, i.e. they come into contact with the ground or rock. However, soil has certain

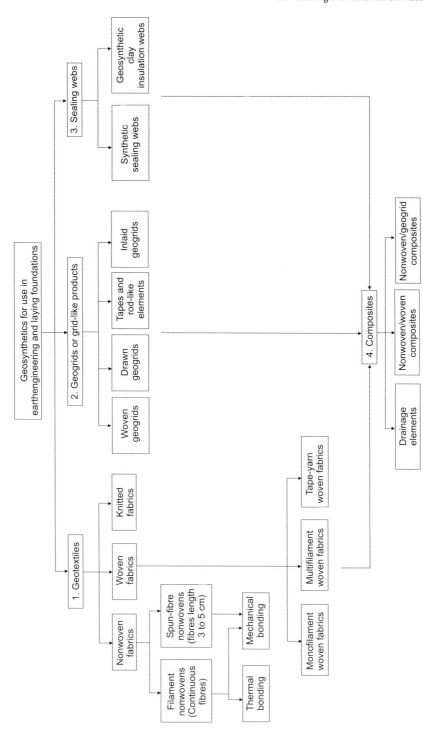

**Fig. 15-21** Classification of geosynthetic materials

drawbacks when used as a basis for building (e.g. laying foundations for buildings) or as a building material (embankment materials for dams, etc.). This is because it can only absorb tensile and shear forces to a limited extent; it is also easily deformable under the effect of water flow, and shifts under a hydraulic and/or dynamic load because it has insufficient structural stability. This is why our ancestors used to use reeds, straw, wood, etc., to stabilize the soil and to assist in reinforcement, filtration, drainage, and foundation stabilization.

Using geosynthetics is just one way of continuing these old traditional building techniques by using new, modern materials. In the last few decades, manufacturing techniques for producing the nonwovens have developed rapidly, so that textile constructions having a wide range of different characteristics are now available. Nonwovens for use as geotextiles are mainly made from polypropylene or polyester in the form of fibre or filament nonwovens. In Western Europe and North America, 80% of the fabrics are made from polypropylene. The main type of bonding technique used is needling technology, and subsequent heat setting improves the strength characteristics and dimensional stability. The world consumption of geotextiles for the period covering 1990–2005 is shown in Table 15-10 [91]. The percentage of nonwovens compared to woven fabrics, warp-knitted fabrics and grid constructions has increased both in absolute as well as percentage terms.

**Table 15-10** Consumption of geotextiles

|  | *1990* | *1995* | *2000* | *2005* |
| --- | --- | --- | --- | --- |
| Geotextiles in kt | 178 | 251 | 400 | 574 |
| Proportion of nonwovens in % | 63 | 65 | 72 | 75 |

When used in contact with the ground, the following characteristics of the nonwovens are particularly important: their planar, thin, stable and uniform structure, deformation capacity, ability to restrain granular material whilst being water-permeable at the same time, their low mass per unit area, and their high strength and elongation characteristics.

### 15.3.1.1 **Functions and requirements**

When used under the ground as a foundation and building material, geosynthetics carry out a number of functions. Because they are in contact with the ground, they have to meet specific requirements. The main functions are shown in Fig. 15-22.

Nonwovens are used mainly for filtration, separation, drainage, protection and reinforcement.

When used as a *filter*, the nonwoven is laid in the contact zone between two types of soil having different grain sizes, or between the soil and a drainage ele-

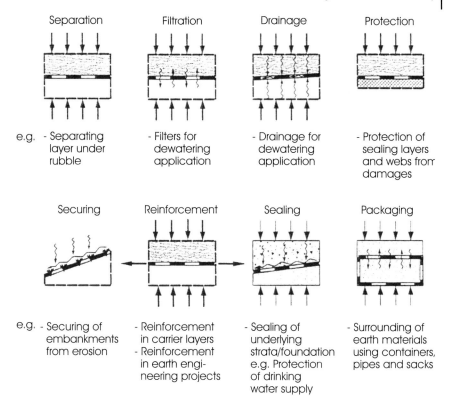

**Fig. 15-22**  Functions of geosynthetic materials

ment, such as a drainage pipe or some other drainage element. It should prevent detrimental shifting of the granular material under the influence of a water flow, and guarantee a high water permeability over a long period of time.

A suitable nonwoven must therefore
– have a high water permeability over its entire surface
– have an effective aperture width, which holds back any eroding granules but allows suffoding granules to pass through
– have a mass per unit area and strength which can resist mechanical loading from the soil and during laying, without being damaged

When used as a *separating layer*, the nonwoven is laid between two types of soil having a different grain size to separate the layers. It should stop the soils from mixing and penetrating into each other under a dynamic load and water flow, and thus ensure that the mechanical characteristics of the coarse-grained soil are maintained (Fig. 15-23).

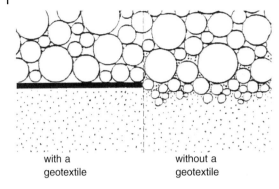

with a
geotextile

without a
geotextile

**Fig. 15-23** Effect when used as a separating element

A suitable nonwoven must therefore
– be stable yet deformable
– have a mass per unit area and a strength which can resist mechanical stresses during installation and use, without being damaged
– act as a filter at the same time

When used for *reinforcement,* the geosynthetic is laid between two or more layers of soil.

This improves the load-bearing capacity and stability of the soil, since the geosynthetic material absorbs any tensile loads generated and transfers them into the soil by friction.

A suitable geosynthetic must therefore
– have a high tensile strength at low elongation levels
– activate a high coefficient of friction when in contact with the soil
– exhibit the relevant creep strength for the entire service life of the product
– be resistant to stresses (see section on use as a separating layer)

When used as a *protection and drainage layer,* the nonwoven or nonwoven composite is laid in contact between a building component/sealing web and the soil. The building component/sealing web is protected from mechanical stresses during the period of building and use, and any flowing water is diverted throughout the plane of the textile.

A suitable nonwoven or nonwoven composite must therefore
– have an adequate thickness
– be dimensionally stable and able to distribute stresses
– have a high water permeability over its planar surface
– be resistant to stresses (see filters)

In general, the material has to carry out a range of functions, such as separation, filtration and reinforcement, so that it is possible to combine all of these requirements. The textile has to have specific characteristics, depending on the particular end-use, and these can be classified on the basis of their importance. The applications and textile characteristics are classified in order to importance in [92] (Table 2).

## 15.3.1.2 Selected examples of the uses of nonwovens

### Filters for use in drainage systems for roads

The nonwoven is laid into the pipeline trench to act as a trench filter, and sur-

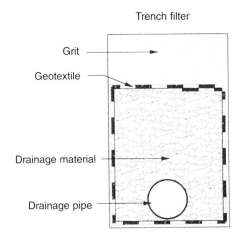

**Fig. 15-24** Nonwoven used as a trench filter

rounds the drainage material as well as the drainage pipe (Fig. 15-24).

It keeps the exposed soil that has to be drained separate from the drainage material and, as a filter, has to

- allow the flowing water to pass through unrestricted (water permeability)
- protect the exposed soil from erosion, i.e. shifting and flushing-out of the stabilizing granules as the water flows through (erosion protection)
- allow the fine components to pass through during suffosion (shifting and flushing-out of fine granules in the soil as the water flows through) in order to prevent colmation (clogging) of the apertures in the geotextile (colmation protection)

The aperture width therefore has to be measured in accordance with the filter specifications. The characteristic aperture width of a nonwoven is determined in accordance with E DIN EN ISO 12 956. It characterizes the granule size of a testing surface at which 90% of the granules are retained by the nonwoven, whilst 10% pass through during a wet sieving test. The following requirements are made of the filter efficiency and its resistance to mechanical stresses in accordance with [93]:

- The permanent filtration efficiency (i.e. its long-term behaviour) is guaranteed by
  - a water permeability of $k_v > 1{-}10^{-3}$ m/s at a load of 20 kPa
  - an aperture width of $O_{90,w}$ obtained by measuring the filtration efficiency
  - a filter length or thickness of $d \geq 10 \cdot O_{90,w}$ (mm) at a load of 20 kPa

- The resistance values during installation are guaranteed by
  - a mass per unit area of            $\geq 150 \text{ g/m}^2$
  - a stamping penetration force of    $\geq 1{,}500 \text{ N}$
  - a thickness of                  $\geq 10 \cdot O_{90,w} \text{ (mm)}$

and therefore by a geotextile toughness class of GTC 3 in accordance with [92]

A filter having adequate stability is produced by using mechanically bonded non-wovens having aperture widths of $O_{90,w} = 0.08$–$0.16$ mm. Since a nonwoven used as a trench filter also protects the soil from erosion, a drainage material having a high pore volume can be used in pipeline trenches, which also improves the drainage efficiency.

### Use as a separating layer in the building of roads

The nonwoven is laid between the exposed soil, which has a limited load-bearing capacity, and the carrier material, such as rubble. It keeps the two layers separate, prevents them from becoming mixed together, and reduces deformation and the formation of channels and grooves (Fig. 15-23).

A decisive factor in selecting what type of nonwoven to use is the capacity of the nonwoven to resist stresses during installation from the rubble, and whilst the building work is in progress, i.e. its toughness. These conditions are difficult to quantify.

For this reason, nonwovens can be classified within geotextile toughness classes on the basis of empirically obtained figures for the stamping penetration force and mass per unit area (Table 15-11) in accordance with [92].

**Table 15-11** Geotextile toughness classes for nonwovens [92]

| Geotextile toughness class (GTC) | Stamping penetration force (x–s) (kN) | Mass per unit area (x–s) (g/m²) |
|---|---|---|
| 1 | ≥0.5 | ≥80 |
| 2 | ≥1.0 | ≥100 |
| 3 | ≥1.5 | ≥150 |
| 4 | ≥2.5 | ≥250 |
| 5 | ≥3.5 | ≥300 |

Explanations: (x–s) – Mean value minus standard deviation.
When using the breaking strength as a basis for classification, the lower value obtained for the lengthwise or crosswise direction in each case is used for the calculation.

The geotextile toughness class required (Table 15-12) can be determined, and the relevant nonwoven can be selected on the basis of the stamping penetration force and mass per unit area for a specific amount of loading by the rubble (applications AS 1 to 5) and specific loading during installation and whilst the building work is in progress (loading conditions AB 1 to 4).

**Table 15-12** Classification in accordance with geotextile toughness classes [92]

| Application | Loading conditions | | | |
|---|---|---|---|---|
| | **AB 1** | **AB 2** | **AB 3** | **AB 4** |
| AS 1 | GRK 1 | | | |
| AS 2 | GRK 2 | GRK 2 | GRK 3 | GRK 4 |
| AS 3 | GRK 3 | GRK 3 | GRK 4 | GRK 5 |
| AS 4 | GRK 4 | GRK 4 | GRK 5 | 1) |
| AS 5 | GRK 5 | GRK 5 | 1) | 1) |

1) For these applications, trials should either be carried out on the construction site, or the thickness of the layer of rubble should be increased.

**Use as a reinforcement element with additional separation and filtration functions when used in railway track foundation systems**

Composite materials are laid between the exposed ground, which has a low load-bearing capacity, and the carrier layer (mixture of chippings) (Fig. 15-25).

When used as a reinforcement element, they
– prevent deformation, distribute loads and therefore decrease stress
– act as a bridge between adjacent zones having a low load-bearing capacity, and thus guarantee a higher and more uniform load-bearing capacity

When used as a filtration and separation element, they prevent long-term mixing and shifting of the granules, which could cause problems, and thus maintain the strength properties of the carrier layer.

When used as a filtration and drainage element, they permit the release of the ground water, and divert it into the crossfall, and thus maintain the load-bearing

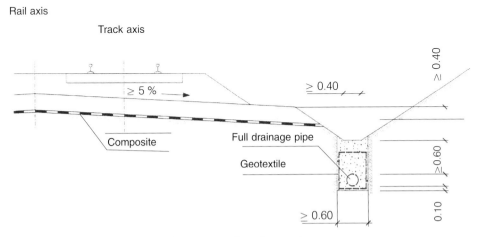

**Fig. 15-25** Composite used in a foundation system for railway tracks

capacity of the exposed ground. The textile is also stabilized by deposition of the fine granules and this promotes a mutually beneficial relationship between the soil and the geosynthetic material [94, 95].

Nonwovens on their own cannot usually fulfil these functions, and their properties have to be modified by incorporating yarn layers having a high tensile strength in the lengthwise and crosswise directions, or else the nonwovens have to be used in composites in combination with geogrids. A suitable composite material must also exhibit planar load-bearing characteristics and be able to absorb tensile stresses instantly, even at low elongation levels [96].

Ref. [93] contains a profile of the requirements for composite materials.

When used for reinforcement, they must have
– a breaking strength in both main directions of          $\geq 40$ kN/m
– a tensile strength in both directions at 3% elongation of      $\geq 10$ kN/m
– isotropy of the tensile forces in both main directions of      $1:1–1:1.25$
– a mesh width of                          $\leq 10$ mm

A further parameter, needed to describe the frictional behaviour between the soil and the carrier material, still has to be developed.

The material must exhibit the following values, if it is to be used as a separation, filtration and drainage element:
– type of material                          nonwoven
– mass per unit area                        $\geq 250$ g/m$^2$
– vertical and horizontal water permeability at a testing
  pressure of 20 kPa $k_v/k_H$                  $\geq 5–10^{-4}$ m/s
– effective aperture width $O_{90,w}$              0.06–0.20 mm

Both components must be joined together so that they are as flat as possible, and they must remain joined together for the entire service life of the product.

### 15.3.1.3 Outlook

The use of technical textiles in the building and construction industry has solved a number of technical building problems, and enabled new working practices to be introduced. Only a few examples can be given here of the wide range of possible applications for technical textiles in the building and construction industry. Only in a few cases is sufficient information available on the interaction between the soil and the building material and the technical textile to preclude carrying out measurements to enable specific requirements to be drawn up for the technical textile. In most cases, the civil engineer only has empirically obtained findings to fall back on to explain the requirements of the technical textile and the development specifications to the textile engineer. There exists a huge amount of scope here for cooperation between the two sectors.

As far as the long-term behaviour is concerned, questions arise as regards ageing and the interrelationship between the soil and the nonwoven.

The types of nonwovens currently used, which are made from synthetic fibres, particularly polyester and polypropylene, have a high resistance to ageing, as long as they have been able to withstand any damage during the installation stage, which could impair their functions, and as long as they are adequately protected from the effects of light.

The sensitivity of certain polymers to the effects of strong alkalis, as well as their different creep behaviours, have to be taken into account when deciding which one to use

For certain applications, such as erosion protection or the laying of foundations for loose, rocky embankments, degradable fibrous materials must be used, such as natural fibres or blends of recycled fibres.

Some interesting results have been obtained as far as the interaction between the ground and the nonwoven over a long period of time is concerned. Granules become embedded in the nonwoven fabric as a result of dynamic loading and hydraulic effects. This deposition of granules changes the hydraulic and textile/physical characteristics of the nonwoven. The structure of the nonwoven is stabilized, and the stress/strain behaviour changes. Nonwovens containing granules absorb tensile stresses, even at low levels of deformation (Fig. 15-26). It would be worth investigating this, as well as other effects resulting from this interrelationship, in order to use the results for solving further geotechnical problems.

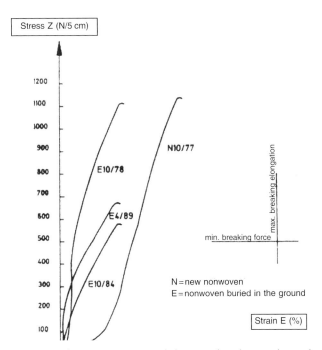

**Fig. 15-26** Change in stress/strain behaviour when the samples are buried in the ground (the figures refer to months/years)

Granules and particles can also be incorporated with the express intention of achieving certain characteristics, e.g. sealing, protection, damming, damping, fixing, bonding, etc. For example, granules and particles can be incorporated during manufacture of the nonwoven/composite materials to create:

– a protective effect by incorporating chippings as spacer elements
– a sealing effect by incorporating granules, fine bonding agents, or chemicals
– a plant-growth-promoting effect by incorporating fertilizers
– a binding effect for anchoring harmful substances by incorporating chemicals

For example, a sealing effect could also be achieved by constructing the nonwoven/composite so that clogging occurs at the feed side of the filter as a result of suffoding particles becoming deposited, thus decreasing the water permeability (colmation filter) [97].

## 15.3.2
### Bituminous roofing membranes
J. J. FRIJLINK

#### 15.3.2.1 Introduction
Nonwovens used as carriers for bituminous roofing membranes constitute a major market, especially for those materials that maintain a sufficient level of mechanical properties at the bitumen impregnation temperature, roughly between 180 and 200 °C.

It is easily understood that – from the eaíly days     natural fibres, cellulose based, were the preferred material; limited of performance but cheap and still important in those regions where low prices are a must.

Later on, glass fibre and glass filament webs became an option, featuring improvements like high modulus, flame retardancy and being rot free. Glass web still has an important position, although properties like puncture resistance, flexibility and elongation are relatively weak.

It is to some extent surprising that polyester (mainly polyethyleneterephthalate fiber and filament), being a thermoplastic polymer, has obtained such a strong position in the high end of the market in the second half of the 20th century. We will illustrate this development but turn first to the requirements that need to be met on a roof.

In Western Europe, improved thermal isolation of roofs and high labour costs imposed new demands on roofing membranes from 1970 onward: the number of layers on the roof had to be reduced, at the same time watertightness guarantee was extended to 10 years or more. Mechanically fixed membranes became an issue and all aspects of shrinkage and shrinkage force of the membrane on the roof had to be reconsidered.

Usually the roof construction will thermally isolate the bituminous membrane from the interior of the building below. Therefore its temperature swing will be determined by atmospheric conditions, in northern Europe it may range from +80 °C to –30 °C easily. If the membrane were free to expand and shrink this would not cause any problem, however, the membrane is fixed to the upper surface of the roof. Thermal expansion is imposed by the supporting construction and generally this means that the elongation of the membrane is not "free" but has to follow the roof construction via forces exerted by the upper surface of the roof (via nails, glue or tar). This situation should not lead to wrinkling or cracking, nor should the membrane itself exhibit relaxation phenomena on the roof. The worst case is tearing of overlapping joints, caused by shrinkage forces developing inside the carrier.

Historically a high E-modulus of the carrier is preferred in order to reduce necking-in during the bituminization process. Today it is realized that a high modulus introduces large forces in the membrane on the roof and this is undesirable as explained above. Also, in order to prevent tearing, a high elongation at break is preferred. So, on the roof a low modulus, high elongation carrier is preferred today (see Section 15.3.2.4 for solutions to these contradictory demands).

In summary, the main function of the carrier on the roof is "keeping the bitumen together". The carrier should not develop large forces when shrinking or expanding. Ideally, the membrane should easily follow all periodical deformations of the construction, showing a high elasticity and tearing resistance. It is here where polyester nonwovens appeared to do an excellent job.

Highest local deformations occur at dilatation joints of the construction, where glass carriers often break and polyester easily survives. The same holds for traffic on the roof and falling objects.

These accentuated demands with respect to elongation require improved bitumen properties as well. Today mainly polymer modified bitumens (based on APP, atactic polypropylene or SBS, styrene butadiene styrene) are used in high quality products. They show less ageing and contribute significantly to fatigue and heat resistance as well as flexibility of the membrane. Oxidized bitumen is still used for commodities. Sometimes a sandwich concept is applied, using SBS as sheath and oxidized bitumen, being less expensive, as core material.

In summary, the nonwoven substrate acts both as a processing aid during bituminization and as a significant part of the final product. A compromise has to be found between the demands in both cases.

### 15.3.2.2 Market overview

The annual production volume of roofing membranes is difficult to estimate for at least two reasons:

- Traditional roofing membranes, using paper, felt, jute etc. as a carrier, are produced everywhere in the world, mostly in small factories, and the volumes are not listed as part of any inventory. Nevertheless it is an enormous potential market for high tech products.

- Roll goods, factory produced, are not the only type of bituminous roof covering. In the USA, for instance, BUR, build-up roofing (laying reinforcement in liquid bitumen on the roof) is an important technology, competitive to the roll goods technology.

**Table 15-13** World market 1998 of bituminous roofing membranes (roll goods only)

| Type of carrier | World market (million m²) | EC production (million m²) |
|---|---|---|
| Traditional (paper, felt, jute) | 1,000–2,000 | 200 |
| Glass fleece | 500 | 200 |
| Polyethyleneterephthalate (polyester) | 700 | 400 |

The data for polyester include so-called composite webs, featuring some combination of glass fibre or filament with thermoplastic nonwovens.

**Types of polyester nonwovens**

Polyester nonwovens can be produced via filament or via fibre technology. Roughly 60% is spunbond filament today, but the competition of modern drylaid (carded) webs is significant.

The mainstream comprises spunlaid filament, needle punched and chemically bonded. The bonding agents applied are mainly aqueous dispersions of acrylic copolymers. Some manufacturers add thermosetting monomers to the binder in order to improve stability during bituminization. Much emphasis is on reduction of formaldehyde emission. Curing of the binder system takes place during the drying step, following the binder application. The same bonding technology is used for carded webs.

Filament properties are tuned to the demands of bituminization: mechanical specifications like a high initial modulus at high temperatures require spinning speeds of about 5,000 m/min. Titer is coarser than with disposables and ranges mostly from 3 to 10 dtex in order to facilitate bitumen impregnation. Specific mass is usually between 120 and 300 g/m², with a clear trend towards lower weights.

The American BUR technology, mentioned above, requires lower specifications of the nonwoven mat because it does not have to be processed on the bitumen line. Needle punched nonwovens without chemical binder are frequently applied.

A number of special products, including high performance thermally bonded webs, play a strong part where lighter webs are applied, often with glass scrims or glass warp threads, or functional combinations of glass fleeces and polyester (see Section 15.3.2.5 "Trends and developments").

### 15.3.2.3 The bituminization of polyester nonwovens

**Technology**

The amount of bitumen applied per square meter is usually more than a tenfold of the nonwovens specific mass.

The application can be achieved by dipping, followed by calibration or by coating directly the required amount of molten bitumen on the nonwoven. Still submerged processing is the leading technology. Coating techniques, with hardly any mechanical load on the carrier during bituminization are expected to gain importance. It is clear that pulling a polyester nonwoven through a bitumen bath at temperatures around 180 °C may introduce necking-in and elongation during processing at the risk of dimensional instability on the roof. The higher the processing speed, the larger the problem. Therefore, during the past decades, as dimensional stability became more and more important, drive systems have been improved and pulling forces applied to the nonwoven have been reduced. The problem is easily illustrated by the fact that a force of 500 N/m may elongate a certain type of nonwoven by about 0.1% at 20 °C but by nearly 2% at 180 °C.

These problems have played a role in the trend towards composite carriers, featuring glass threads or scrims (see Section 15.3.2.5 "Trends and developments").

**Processing stability**
The forces exerted during processing will generally have two effects:

– a macroscopic deformation of the web structure, strongly depending on the web construction (and partly maintained when the bitumen solidifies) and
– a stretching of part of the fibre material, leading to increased orientation of the amorphous phase at the molecular level, which is "fixed and frozen" when the polyester glass transition temperature is passed during cooling down.

The residence time at elevated temperature during bituminization may introduce changes in physical molecular structure as well; this contribution is dependent on the thermal history of the polyester (spinning, thermal bonding, or stabilization and drying during chemical bonding). All these steps affect the micro structure of the fibre and consequently the properties of the final product.

**Testing processing stability**
Common laboratory tests for determining the stability of the nonwoven during bituminization comprise:

– preparing a sufficiently large web sample, indicating machine direction (MD) on the sample and attaching markers for accurate measurement of dimensional changes in both machine and cross machine direction (CMD)
– mounting the material in an oven, pre-heated to the required bituminization temperature and applying a load in the range of 500 to 800 N/m in MD of the nonwoven
– exposing the loaded sample to the required temperature for typically 10 minutes
– measuring dimensional changes in MD and CMD at elevated temperature without removing the load
– after cooling down, the load is removed and dimensions may be measured again if required (confusion with dimensional stability on the roof is to be avoided, however, see next section)

15.3.2.4 **The end use requirements of membranes with polyester inlay**

In the introduction it was explained that the membrane on the roof should easily follow all movements of the construction and in fact exhibit a low modulus of elasticity. From the previous section it has become apparent however that a high modulus is preferred for most bitumen impregnation processes. This is the contradictory situation with traditional polyester carriers.

**Dimensional stability**

Test procedures on the ready membrane are slightly different in most countries having a national standardization. However, the same principles apply everywhere and always comprise measurement of dimensional changes in MD and CMD after exposure of unconstrained samples to a temperature of 80 or 90 °C (i.e. well above the glass transition temperature of the thermoplastic polyester) for a number of hours (e.g. 24 hours).

The free shrinkage (or stability) specification found is an important characteristic for the membrane and may range in MD from –0.5% for commodities to –0.1% for top qualities. In CMD these numbers usually are zero or slightly positive (i.e. the nonwoven may slightly expand in CMD). Target is of course a zero shrinkage.

In recent years emphasis has increasingly been put on shrinkage force rather than shrinkage because the membrane on the roof is pretty much fixed dimensionally and consequently free shrinkage will not take place. Since its dimensions are more or less fixed the question is whether the carrier will develop shrinkage forces of such magnitude that overlapping seams may start tearing. In order to measure such shrinkage forces, tests have been developed that simulate the actual conditions. In such a test, a specimen, typically a few meters long and one meter wide, is mounted in a clamping frame in an oven. Through the frame (using a load cell arrangement) the shrinkage force is measured that develops while the temperature is either kept constant at, for instance, 80 °C, or cycled between 20 °C and 90 °C. The temperature cycles simulate the actual conditions where sunshine may be followed by a rainshower. It is expected that results of these R&D efforts will lead to further improvements of membranes in the next couple of years.

**Isotropy**

Often it is preferable to have the same mechanical properties values in MD and CMD. Due to the manufacturing process, yarn orientation and mass regularity of the nonwoven carrier will often be different in MD and CMD, resulting in a MD/CMD ratio of mechanical properties that is not equal to one. Values may range from 0.9 to 1.9. Usually the MD/CMD ratio of carriers is between 1.0 and 1.6 resulting in similar ratio's in the finished article.

**Tear and nail testing**

Two types of tear tests can be distinguished. In the one type of test, called the single rip or tongue tear test, the tearing loads are perpendicular to the plane of the nonwoven. In the second type of test, often referred to as trapezoidal specimen tear test, the tearing forces are applied in the plane of the nonwoven.

When shrinkage forces develop in a roofing membrane that is nailed to the roof structure, high loads will be induced at the nails. This condition can be simulated with the nail test in which a nonwoven test specimen is pulled while being attached to a test set-up by means of a pin having a diameter of typically 2.5 mm.

**Puncture resistance**

While in place on a roof, the membrane could be exposed to highly concentrated loads, for instance caused by a ladder or dropped hammer. These loading conditions can be simulated by tests in which a weight of prescribed shape and dimensions is dropped on to the test specimen from a prescribed height, with the test specimen laying on a substrate of prescribed hardness and stiffness. In some instances an additional requirement may stipulate that the specimen be cooled down to, for instance, $-10\,°C$ before the weight is dropped.

In addition to this dynamic test a static test may be specified in which a roofing membrane test specimen is statically loaded perpendicular to its plane by a prescribed weight of prescribed dimensions. Test duration may be 24 hours at $20\,°C$.

**Fire retardency**

Over the years, a lot of attention has been given to flammability of bituminous membranes.

Sparks of fire from, for instance, a chimney may ignite the material if no precautions are taken. Emphasis is in general on

– reduction of flammability of the bitumen
– carrier constructions that prevent burn-through of the membrane
– carrier constructions that prevent flow of molten bitumen to the fire

Tests existing in various countries differ strongly today, but standardization is strove for. In one such a test a basket with beech wood shavings is put on a full size model-roof. There are, however, various laboratory test configurations as well. The results of these different tests often show a lot of scatter and conclusions are sometimes contradictory.

Among the products that specify increased fire retardency two types will be mentioned here.

• Carriers with a thin metal foil on one side, needle punched to such an extent that the objectives mentioned above are maintained and delamination is prevented. These products are chemically bonded as a final step.

- Carriers having a glass fleece on one side or between polyester web layers, sometimes needle punched, sometimes thermally bonded via binder fibres added to the glass fleece.

### 15.3.2.5 **Trends and developments**

A number of reasons can be given for the development of glass reinforced polyester carriers from 1990 onward. Always separation of functions is the name of the game. Stability on the roof requires minimal deformation of the polyester during the bituminization process, so a high modulus component, sufficient to pass the impregnation line at less than 0.5% elongation, was added. This development was quite beneficial, because an increase of bituminization speed improves the economy of the production line, making the higher costs of glass reinforced carriers acceptable. Often the polyester weight could be reduced as well.

Successful improvements are the following:

- Incorporating a glass scrim of approx. $10\,g/m^2$ at e.g. $10\times10\,mm$ mesh between two polyester layers. Isotropy of properties is largely maintained in this way, while necking in during processing was reduced from a few percentage points to less than 0.5%. The polyester weight in some types went down from over 200 to values below $150\,g/m^2$.

- Unidirectional reinforcement. The addition of glass warp threads, approx. $10\,g/m^2$, usually 7 or 10 mm apart, is less expensive than scrim. However, the production of warp thread reinforced nonwovens (keeping yarns equally spaced and at equal tension) is more difficult than fabricating scrim reinforced nonwoven. Often ondulations are observed during bituminization.

- Laminates of polyester nonwovens with glass fleeces. A variety of products has been proposed for high end applications like "single layer" roofs. The same products also show improved flame retardancy, because the glass fleece inhibits burn-through of the carrier. Excellent dimensional stability specifications of –0.1% (MD) and +0.05% (CMD) have been obtained by various manufacturers.

An interesting phenomenon appears in the stress-elongation curve of nonwovens combining glass and polyester:

At low elongation, the force increases steeply because of the presence of glass yarn. At about 3% elongation these yarns will break and the force drops nearly to the level of the unreinforced polyester web. At high elongation the effect of the glass threads becomes negligible.

The amount of glass influences the magnitude of this "glass peak", it may be even higher than the final breaking strength of the polyester layers.

At the impregnation temperature of about 180 °C, the picture is basically the same, except that the forces are much lower as a consequence of the modulus of polyester being low at this relatively high temperature. The amount of glass has to be chosen in such a way that no breakage of glass threads occurs during the bituminization process.

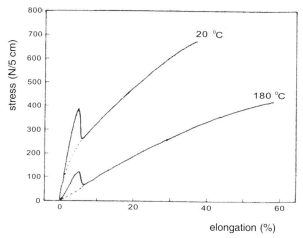

**Fig. 15-27** Stress-elongation curves of 175 g/m² polyester nonwovens at 20 °C and 180 °C with and without glass scrim.
Full line: glass reinforced. Dashed part of curve: without glass

## 15.4
## Agriculture
H. FUCHS

### 15.4.1
### Introduction

Nonwovens for use in agriculture – often called agricultural nonwovens – are found in agriculture, forestry and gardening. Their use aims at

– early harvests
– a better quality of fruit and vegetables
– higher yields
– protection from erosion
– greening
– re-cultivation

The nonwovens are applied at varying distances from the ground. This is done manually by unrolling them or by means of special equipment.

With agricultural nonwovens, the material often serves as a storage and transportation medium for functional fillers such as water, water-storing material, fertilizers or plant seed. The main characteristics of agricultural nonwovens are evenness, sufficient stability and elasticity, bio-degradability or recyclability. They also allow to control thermal effects.

15.4.2
**Requirements to be met by agricultural nonwovens**

Depending on the strain they are exposed to, the requirements to be met by agricultural nonwovens vary.

- High potential to retain water so seeds can germinate and plants can grow. This is achieved by means of fibre materials which allow to take in much water and by filling in super-absorbers. While nonwovens meant for the covering of plants show a mass per unit area of 15 to 60 $g/m^2$, values between 100 and 500 $g/m^2$ are reached with materials for use on embankments and slopes.

- Protection from wind and the creation of a micro-climate between the ground and the nonwoven, which results in temperature and humidity being balanced out. At the same time, temperatures in the root area rise. This is what causes earlier harvests.

- Effective protection of late crops from the first night frosts

- Good potential to slow down the growth of weeds

- Bringing plant seed and fertilizers (depository fertilizers) into the nonwoven, probably in combination with super-absorbers and pesticides. Such agricultural nonwovens are often structured as multi-layer composites.

- Good potential to reduce the impact of UV radiation on plants by light-absorbing or light-reflecting nonwovens (light permeability: 80 to 90% to allow photosynthesis to take place).

- Optimal vertical water distribution by means of nonwovens that show high transmissivity and porosity. Porosity is influenced by fibre blends ranging from 1.0 to 15 dtex and by the nonwoven density.

- Sufficient stiffness, flexibility, shape stability, dimensional stability and resistance to wetness.

- Fungicidal finish (up to 2% of the total mass), which avoids soil contamination.

- Protection from seed losses caused by wind and birds as well as protection from insects

- With agricultural nonwovens that root in together with the plants, fibre materials are required which rot after 2–3 years.

- With lengths of nonwoven material wider than 5 m, weldability is necessary, which is achieved by means of fibres of polyolefins, polyamid, copolymerisates or by bonding the webs using binding agents suitable to be hot-sealed.

15.4.3
**Technological processes**

Agricultural nonwovens may be produced to the full range of technological processes known in the manufacture of nonwovens:

- needle-punched nonwovens
- stitch-bonded nonwovens (Maliwatt, Malivlies)
- thermally bonded nonwovens
- hydroentangled nonwovens
- spunbonded nonwovens
- wet nonwovens

15.4.4
**Application**

- A wet nonwoven used to grow plants (root-taking aid) with a mass per unit area of 25 g/m$^2$ and a thickness of 0.23 mm will show the characteristics as contained in Table 15-14.

- Spunbonded nonwovens of polypropylene or polyester with masses ranging from 10 to 30 g/m$^2$ are used to guide germs and to protect the plants from frost. These nonwovens are manufactured on machines which are 5 m wide or more. With field crop tunnels, UV-stabilized spunbonded nonwovens ranging from 40 to 70 g/m$^2$ are used. They allow both water and air to permeate [105].

- Needle-punched nonwovens from rotting reclaimed fibres or from flax/viscose blends with mass per unit areas ranging from 250 to 600 g/m$^2$ are used to cover up the ground and thus to avoid the growth of weeds. Such nonwovens are also used to wrap up the root bales of both fruit and forest trees or to re-cultivate open-cast mines and industrial ruins [106].

**Table 15-14** Wet nonwovens for growing plants [104]

| Fibre material | cellulose, PAN | |
|---|---|---|
| Binding agent | self-netting acrylate | |
| Fungicidal finish | benzimidazole derivate | |
| Dry strength (N/5 cm) | longitudinally | 45 |
| | crosswise | 35 |
| Wet strength (N/5 cm) | longitudinally | 20 |
| | crosswise | 15 |
| Dry elongation (%) | longitudinally | 3 |
| | crosswise | 5 |
| Wet elongation (%) | longitudinally | 3 |
| | crosswise | 5 |
| Tear growth resistance (N) | longitudinally | 1.2 |
| | crosswise | 1.4 |
| Air permeability (l/s/m$^2$) | 1,300–1,600 | |

- Thermally bonded nonwovens with masses ranging from 8 to 25 $g/m^2$ lead to a rise in micro-temperature of 2 to 4 °C. They are used to cover up the ground. At the same time, they serve as protective filters against parasites.

- Little sacks of nonwoven material are also used to protect high-quality fruits (peaches, pears, grapes) from insects.

- Polyester spunbonded nonwovens coated with super-absorber gel may be used to protect young plants. For super-absorber agents, polyelectrolytes and polyacrylonitrile derivates are applied. The mass per unit areas range from 35 to 60 $g/m^2$. They are used to cover up young seeds and to keep root systems wet. Spinning starch derivates into the polyester spunbonded nonwovens results in bio-degradability.

- Farmers in China growing ginseng use spunbonded nonwoven material from PP to avoid direct sun [107]. To achieve a sufficient stability in mountainous regions exposed to strong winds, a tensile strength of 120 to 140 N/5 cm is necessary in both the longitudinal and the crosswise directions. For this purpose, a mass per unit area of 70 to 80 $g/m^2$ is required. Optimal light permeability can be reached by means of blue spunbonded nonwoven material. To stop water running through the nonwoven material, it is laminated with polyethylene film. So far, thatch covering has been used, which, however, does not last long and is expensive.

- In England, tests were made with cabbage, using nonwoven covering to achieve early harvests. They resulted in the ripening period being shortened by 10 days. With celery, the nonwoven covering allowed harvesting 20 days early [108].

- Thermally bonded polypropylene and polyester spunbonded nonwovens show good water- and air-permeability (e.g. Lutrasil® and Lutradur®). They are UV-proof and can thus be used several times. Such spunbonded nonwovens whose colour is photo-selective are also used to bleach lettuce. The mass per unit areas of PP spunbonded nonwovens used to protect from wind, hail, damage caused by game and pests range from 15 to 20 $g/m^2$. At the same time, the covering allows irrigation. Below the covering, protection from frost is available down to –5 °C. At night, the spunbonded nonwoven material retains ground warmth. By day, the good potential for air permeability avoids too high temperatures, thus protecting from damage caused by heat.

- Needle-punched and stitch-bonded nonwovens from rotting primary and/or reclaimed fibres are used with greening mats protecting from erosion as well as with air-conditioning. During the manufacturing process, they are filled with plant seed (mostly grass), super absorber and a depository fertilizer. The mass per unit areas of such nonwovens range from 200 to 500 $g/m^2$ (e.g. erosion protection mat from Vlifotex). The different functions may be assigned to different layers of nonwoven. In that case, multi-layer agricultural nonwovens are achieved [109]. Bonding can also be reached by means of hot calendering [110].

**Table 15-15** Fields of application of agricultural nonwovens as depending on mass per unit area

| Mass per unit area | Field of application |
|---|---|
| 10 to 30 g/m$^2$ | Protection from frost |
| | Protection from light |
| | Protection of seed from damage, caused by wild animals |
| | Protection from pests |
| | Protection from hard rain or hail |
| | Early harvest through micro-climate |
| 30 to 150 g/m$^2$ | Storing wetness so as to feet roots |
| | Protection from light |
| | Root sheathing in transit |
| 150 to 600 g/m$^2$ | Ground covering to avoid the growth of weeds |
| | Plant carrier mat |
| | Greening mat |
| | Mats to protect from erosion |
| | Re-vitalization |

- Chemical-bonded nonwovens from cotton, viscose fibres or blends of these fibre materials with mass per unit areas ranging from 10 to 45 g/m$^2$ are used as cover materials. For a binding agent, polyacrylate is found. Web formation is executed on a carding machine. Doubling at laying equipment ranges from 2 to 7 [111].

Different applications require nonwovens of different mass per unit areas (Table 15-15).

## 15.4.5
## Market trends

Agricultural nonwovens are becoming more and more important. They are now well-known to agriculture and horticulture. There is worldwide growth potential with the following fields of application:

- restoration in open-cast mines, quarries and gravel-extraction
- creation of water reservoirs
- roll-back of steppization and deserts

Agricultural nonwovens compete with products of film of polypropylene and polyethylene. They will be used in case film properties do not meet the requirements. Such properties are water permeability, water storage capacity, breathability, insulation to light, flexibility. Films may, in addition, cause mechanical damage to the plants.

In the period between 2000 and 2005, about 10,000 tons of agricultural nonwovens will be manufactured and used in Western Europe. In 1995, the figure amounted to only 5,000 tons. Worldwide application was 19,000 tons in 1995 as compared to 26,000 tons in 2000. Very probably, some 40,000 tons of agricultural nonwovens will be used in 2005 [91].

## 15.5
## The motor vehicle industry
G. SCHMIDT

### 15.5.1
### The market

Reclaimed fibres have been used for some time now in car seat cushioning and as insulating materials in the floor area. Even as late at the 1970s, very few non-woven fabrics were used in the motor vehicle or transport sectors.

Since the 1980s, however, the use of technical textiles in general and nonwoven fabrics in particular has grown steadily. The car industry has led this trend, but the railways sector and the aerospace and shipbuilding industries have also been involved in this development. During the 1980s, the use of nonwovens in a number of applications grew by more than ten percent annually. Current estimates of further growth are said to be good to moderate, indicating an 8.5% increase in the area of fabric consumed [112], and a 0.6% increase in weight consumption [91]. There is also a tendency to concentrate on lighter-weight nonwovens. It should also be noted that there are marked regional differences and/or wide variations in terms of end-uses. The amount of material used in each vehicle will also vary. The size of the vehicle, i.e. whether it is a convertible, tourer or bus, and its class, i.e. whether it is a standard or a luxury model, will also dictate how much textile and nonwoven material is used. At the beginning of the 1980s, most car interiors were still mainly covered with synthetic leather and film, with only a small amount of textile material being used. Nowadays, however, most of them are covered with textiles and some real leather as well. Synthetic leather upholstery is not used in Europe at all now, and very little of it is used in the USA. An average of 40 m$^2$ textile substrates are currently used in every new car [113]:

- 20 m$^2$ nonwoven fabrics
- at least 5 m$^2$ seat upholstery fabrics
- just under 4 m$^2$ carpeting and
- 11 m$^2$ other textile materials for airbags, seatbelts, side protection elements, covered components, possibly the headliner, etc.

The amount of nonwoven fabric used will also depend on whether the floorcovering is a needled or a tufted fabric, and whether the roof is lined with Malivlies or a knitted fabric. EDANA estimates that 20 m$^2$ nonwoven fabrics are used in every motor vehicle [114]. According to 'Nonwovens Markets' [112], worldwide consumption of nonwoven fabrics in the car industry amounted to approximately 420 million m$^2$ in 1996. With 38.5 million private cars being produced, this only amounts to about 11 m$^2$ per vehicle. Just under 40 million cars were produced in 2000. According to statistics compiled by the Organisation Internationale des Constructeurs d'Automobiles (OICA), 54 million vehicles were produced worldwide in 1997 and 38.8 million of these were private cars. Heavy goods vehicles, buses and coaches accounted for the remainder [115]. Between 16.7 [91] and 24 kg [113] fibres are used in each vehicle. Fibre consumption in the whole of the motor vehicle industry was estimated at 2,220,000 t for the year 2000 [91]. Out of this total amount, the proportion of nonwovens produced for use in the motor vehicle industry was produced by just a few nonwoven fabric manufacturers. A number of companies are involved in various countries worldwide in processing and supplying nonwovens to the car industry [112, 116–118].

Koslowski has prepared a list of the 40 largest firms active in the nonwovens sector [119]. It should be noted that the globalization of the world economy also extends to the globalization of the nonwovens industry, whereby a recent example in the motor vehicle industry is DaimlerChrysler. Over the last few years, vehicle suppliers in particular have built up a worldwide network of subsidiaries and have become involved in a number of cooperative projects, i.e. joint ventures.

## 15.5.2
### The car industry

Nonwoven fabrics have many applications in cars, and there are several reasons for the increased use of nonwoven fabrics:

- Nonwoven fabrics offer a better price/performance ratio than textiles made out of yarns for many applications. A patent search would show that a great deal of work was carried out in the 1980s into developing nonwoven fabrics for the car industry [120].

- By using specific types of fibres and optimized processing techniques, it is possible to produce nonwoven fabrics with a lower mass per unit area than substrates made out of yarns, but this is only relevant if all the performance requirements can be met. This may contribute to reducing fuel consumption, a factor that is of major importance in the car industry.

- Nonwovens containing thermoplastic fibres exhibit favourable thermal deformation characteristics, yet still retain their three-dimensional deformation characteristics when cold.

- Nonwovens have a particular advantage as far as recycling is concerned, as long as the same material is used for the web component and the composite. Nonwovens used in motor vehicles can usually also be removed very easily.

Fig. 15-28 shows the types of nonwovens used in cars and the areas where they are used; it does not include woven, warp-knitted or weft-knitted fabrics. The latter are used in tyre fabrics, airbags and seatbelts, and composite nonwovens made from glass fibres, for example, for use as load-bearing elements. Work is currently being carried out into replacing other substrates made from yarns with nonwoven fabrics.

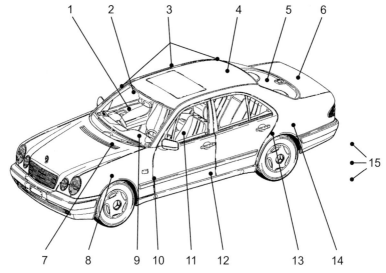

1 Door lining
– Edge trim
– Door mirror
– Arm-rest
– Lower part (door pocket)

2 Sun visors

3 ABC-pillar covering (covering of seat belt)

4 Headliner (moulded roof)
Roof insulation
Sun roof (cover)
Hood
Hood padding

5 Parcel shelf
Speaker covering

6 Boot lining
– Floor mat
– Sides (wheel casings)
– Rear cover
– Back seat wall
– Spare wheel case

7 Filters
– Air filter
– Cabin filter
– Fuel filter
– Oil filter
– during car manufacture (lacquering) etc.

8 Engine housing
Bumper felts
Bonnet lining
Rear side (dashboard)
Battery separators
Other insulation points

9 Instrument panel ( insulation)
Instrument panel (lower part)

10 Dashboard mat

11 Seats:
– Lining for backs of seats
– Laminated padding for seat covers and bottom of seats
– Upholstered wadding
– Upholstery cover, reverse sides
– Head-rest cushioning
– Seat sub-padding
– Foam reinforcement
– Padding for centre arm-rest

12 Floor mats with tunnel cladding
Sub-upholstery (insulating material, stuffing)

13 Interior rear wall lining
Floor of the car body, un-

der the back seats (exterior wheel-case)

14 Estate cars and convertibles:
– Side wall covering (lining for the wheel-case)
– Boot floor
– Lining for the hood-case
– Cover for the hood-case

15 General:
– Covering and transport tarpaulins
– Seat covers
– Tool pockets
– Pockets for holding vehicle documentation
– Cleaning cloths
– Child seat
– Sleeping compartment in lorries
– Protective clothing worn during car manufacture/ maintenance
– Backing substrates for imitation leather and microfibre nonwovens for seats and/or all types of covering components

**Fig. 15-28** Uses of nonwovens in cars (diagram of the car by kind permission of DaimlerChrysler AG)

### 15.5.2.1 **Required characteristics**

The demands made of nonwoven fabrics for use in cars will depend on the loads and stresses to which they are subjected during use, and the effects of long-term usage.

The VDA specification published in 1999 on behalf of the Car Industry Federation includes an overview of the standard textile testing methods and properties used in the German motor industry. It covers specific properties' requirements, which should also be taken into account when using nonwoven fabrics in the car industry, such as temperature and light resistance. During the summer months, the effects of normal outdoor temperatures plus radiant heat on the car windows and metal mean that temperatures of e.g. 120 °C are reached on the instrument panel and parcel shelf, up to 105 °C on the seats and even 80 °C in the boot (Table 15-16). It is therefore extremely important to evaluate the changes that occur in nonwoven fabrics under the influence of heat by carrying out continuous and cyclical tests.

Increased exposure to light accelerates ageing and colour fading. A rule of thumb suggests that every ten-degree increase in temperature doubles the rate of ageing, i.e. both the service life and colour of the component are reduced. The determination of fastness to and ageing in light at high temperatures is covered in DIN 75 202. As Fig. 15-29 indicates, the introduction of this standard has led to a significant improvement in product quality.

Cyclical exposure to light at high temperatures enables useful conclusions to be drawn regarding colour change. According to Schmidtmann [121], colour damage

**Table 15-16** Qualitative temperature and light requirements for automotive lining components and moulded, upholstered components

| Requirement | Effect | | Application Examples |
|---|---|---|---|
| | Temperature | Light | |
| Low | 80 °C | Without | Boot lining |
| | | | Interior floor |
| | | | Cover for back of drivers seat |
| | | | Door pockets, central zones of doors |
| Normal | 90 °C | Partial | Centre arm-rests |
| | | | Tunnel lining |
| | | | Edge protection |
| | | | Boot lining for estate cars |
| Normal with short term increase | 90 °C short term 105 °C | Partial | Seat cushioning/seat backs |
| | | | Moulded roof/sun roof; A, B, C pillar cladding |
| | | | Door lining, edge trim |
| High | 120 °C | Full | Upper part of instrument panel – Parcel shelf – base of head-rest – leather steering wheel |

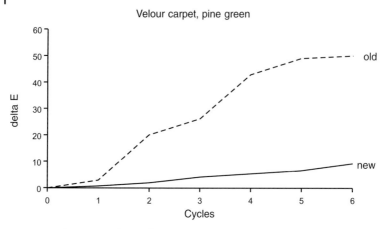

**Fig. 15-29** Changes in colour difference, Delta E, after testing of lightfastness in accordance with DIN 75 202 as a function of the number of cycles
- - - - delta E, old; ——— delta E, new

**Table 15-17** Changes in colour difference, Delta E, for the lightfastness of velour carpet samples in accordance with DIN 75 202, measured and calculated for the 5th cyclus

| Sample | Change in colour difference | Cycles | | | | | |
|---|---|---|---|---|---|---|---|
| | | *0* | *1* | *2* | *3* | *4* | *5* |
| | Delta E, measured | 0 | 1.4 | 2.6 | 3.7 | 4.9 | 6.2 |
| Pine green | Delta E, calculated after 2 cycles | 0 | 1.4 | 2.6 | | | 6.5 |
| | Delta E, calculated after 3 cycles | 0 | 1.4 | 2.6 | 3.7 | | 6.2 |
| | Delta E, measured | 0 | 0.8 | 1.3 | 1.8 | 2.5 | 2.9 |
| Beige | Delta E, calculated after 2 cycles | 0 | 0.8 | 1.3 | | | 3.3 |
| | Delta E, calculated after 3 cycles | 0 | 0.8 | 1.3 | 1.8 | | 3.0 |

follows a linear path, so that the result can be estimated after two to three exposure cycles.

The basic model for drawing up the requirements and carrying out the tests, which simulate the conditions occurring in practice, was also applied to large components. The study of the ageing behaviour was extended to various climatic conditions in accordance with DIN 75 220, 'Ageing of components in motor vehicles in instruments which simulate sunlight'. This allows the test sample to be subjected to high and low temperatures as well as humidity under constant or cyclical conditions. Various standardized procedures enable every climatic and lighting condition and their interrelationships to be simulated under time-lapse conditions; these include humid, tropical heat, hot, dry heat and the effects of cold temperatures at high altitudes.

Placing the sample behind glass can simulate the climatic conditions likely to occur in practice. In order to determine the capacity of the material to withstand continuous exposure, tensile and abrasion tests are carried out after the climatic tests, in order to be able to estimate the service life of the products. This is important since, under normal circumstances, nonwovens must be able to last as long as the life of the car, i.e. they must be able to resist cold, heat, light, perspiration, moisture, vibration and dust, as well as tensile, compression and abrasion loads for a period of 12–15 years [122].

The emissions and their effect on fogging and odours are also another important factor. The car industry is particularly concerned that the products it uses during manufacture and processing are low-emission and environmentally friendly. An even more important factor is the emission potential of products bought in by the car manufacturer [123]. These substances can cause emissions inside the car, either as fumes in their own right or as reaction products, and will depend greatly on the temperature and the partial pressure drops. The gases given off generate odours and are undesirable, since they also put the health and safety of the passengers at risk. Odour testing is carried out in accordance with VDA specification no. 270. Many substances are only emitted from the interior furnishings once a relatively high temperature is reached. They condense on the inside of the windscreen at relatively low temperatures to form tiny droplets or a film, which is known as fogging. This can make driving hazardous at night as reflected light shines on the windscreen and impairs driving visibility; consequently, it must be avoided or at least reduced. Fogging testing is carried out in accordance with DIN 75 201. Improving the fogging behaviour of nonwovens and reducing the generation of odours is expensive and time-consuming, and has to be tackled in the long term by various ways and means [123–125]. The method which uses head-space gas chromatography to measure the emission of organic substances from plastics, fibres and other non-metallic materials has proved to be particularly successful [126]. The amount of formaldehyde released also has to be determined when using formaldehyde-containing products.

Another important factor is the ease with which the nonwovens can be processed on an industrial scale and their suitability for their intended end-use. This requirement is based on a combination of technical feasibility and sound economic principles. Depending on the final end-use, it involves optimizing the characteristics of elongation, deformation capacity, tear propagation resistance, sewability, weldability, adhesion properties, stitch tear resistance, seam strength, formation and removal of lint, generation of dust during processing, and the formation of scratch marks during assembly. The products must also be offered in a range of colours, depending on their end-use. The same colour must be maintained from the beginning to the end of every batch, even over a period of several years. Tolerances of only 0.4 to 0.8 dE are permitted, depending on the colour and the area where the product is to be used.

The initial region of the stress/strain curve is an important factor in assessing the processability of the material. The length changes that occur at low force levels enable predictions to be made regarding the draping behaviour of the fabric in the upholstery or during moulding in a press, for example. Based on experience, the percentage elongation values are set at 25, 50 and 100 N [127]. Fig. 15-30 shows the initial paths of the stress/strain curves for a number of typical automotive fabrics. It should also be noted that, when processing nonwovens at high temperatures, the change in length is greater than under normal temperatures. Another factor is the different elongation behaviour in the lengthwise and crosswise directions at a load of 50 N. A ratio of 1:1 is ideal and, even up to a ratio of 2:1, there are not usually any problems. However, difficulties may arise during processing if the difference is any greater than this.

Of particular importance is that *every* type of material used in car interiors must be flame-resistant. Testing is carried out in accordance with DIN 75 200, and the threshold level is set at 100 mm/min maximum, in accordance with US Standard

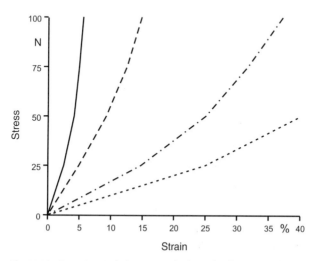

**Fig. 15-30** Stress/strain behaviour in the lower loading region for automotive textiles
―― nonwovens (lower level), ―·―·―· for pronounced deformation,
― ― ― nonwovens (upper level) · · · for simple deformation

FMVSS 302. Buses are subject to additional conditions laid down in the French specifications, UTAC-NR ST 18-502/1 and 18-502/2. The high levels of flame-resistance agents used run the risk of increasing the emission levels. Only in a limited number of cases is it possible to calculate this level in advance and meet it theoretically.

Another important aspect is the level of cooperation between the nonwovens manufacturing industry and the car industry. In addition to the increased level of globalization that has taken place in the last few years, the range of components produced by the car manufacturers themselves has declined, and this has been an intentional development. Suppliers are now producing many of the components that the car manufacturers used to produce themselves. The sector dealing with components for car interiors was the first to be affected and was particularly badly hit. Transferring the production of particular parts or entire components has meant that the overall responsibility and development work have also been transferred. So-called system suppliers emerged who, with the agreement of the car manufacturer, began to cooperate with those suppliers who had once dealt directly with the car manufacturer, and who took over the responsibility for producing the components. The number of suppliers to the car manufacturers declined, but not to the car itself. A system supplier who is also involved in product development must have at his disposal all the relevant testing equipment, e.g. acoustic chamber, and the company must also have the expertise to use them. These system suppliers form an integral part of the supply chain involved in producing a particular model of car, and may even be responsible for carrying out some of the assembly work. However, as suppliers, they have to operate along the principles of 'just in time'. The relevant component must be made available at just the right moment to suit the production cycle of the car. Since it is no longer possible to carry out any checks on the products entering the production cycle, the system supplier must guarantee the quality of the product by checking products as they leave. For this to be achieved, everyone involved in the production chain must be able to apply and make full use of all the regulations and specifications laid down in ISO 90000–90004 [128].

Recycling legislation obliges the car industry to develop strategies for disposing of materials and to ensure that the components can be taken apart easily so that the materials can be reclaimed and recycled. This means that the various alternatives must be taken into account when designing and producing the components. This also applies to nonwovens. It is becoming increasingly important for all the disposal strategies to be harmonized with each other. Ensuring that the covering components are all made from the same material [129] has encouraged this development, and this strategy has won widespread acceptance in recent years.

### 15.5.2.2 Acoustic and thermal insulation – inside the car

One of the first applications of nonwovens in cars was for acoustic and thermal insulation inside the car body, but they are now used in more varied and complex areas. Many of the fabrics are finished so that they are self-adhesive, either all over or in parts, and they can be applied by means of ultrasonics or hot emboss-ing to a thermoplastic synthetic carrier, or else simply laid on top of it. However, some processors also use robots to apply thermoplastic adhesives to parts of the insulating components, so that they are only stuck down at certain adhesion points. In most cases, the components are not usually visible, but are positioned between the covering components and the car body.

In addition to being used inside the car, nonwoven fabrics can also be used in the boot. The main insulated components inside the car include the dashboard, the instrument panel, the roof, the parcel shelf, the tailgate, the floor area, the spare wheel, the rear wall, the doors, the ABC pillars, the side walls, the sides of the boot, the air-conditioning conduit, the floor space (underneath the rear seats), the footwell (under the carpet) and the tunnel.

Nonwovens are still competing with PUR foam for use as insulation, e.g. visco-elastic PUR foam in the footwell. Most of the fabric is made from fibrous material incorporating reclaimed fibres from industrial punching waste, and more expensive new fibres may also be included. Fibrous nonwoven material may also be blended with a proportion of PUR foam flock, which greatly improves its resili-ence in the area around the footwell, for example. Nonwovens made from recov-ered cotton fibres, preferably cellulose fibres, which are bonded with powdered phenoplasts, have been particularly important over the last 35 years.

Webs having a density of 50–150 kg/m$^3$ [130–133] and with acoustic absorption and strength values suitable for use in practice can be produced by adhesively bonding the web at a binder content of 30±10%. At the same time, the density can be increased to 600–1,000 kg/m$^3$ by compression, enabling carrier compo-nents (tunnel, door, seat backs) and pillar coverings or spare wheel covers to be produced. This improves the bending resistance values for an increased thermal resistance and reduced acoustic absorption levels. Nonwoven fabrics having densi-ties in the range of 150–1,000 kg/m$^3$ can be used to produce components having the dual function of carrier/insulator, such as the parcel shelf, the roof, the mats for use in the boot, or the wheel covering. Strong-smelling amines (trimethyl amine) may be produced. Ammonia is always produced and formaldehyde is also an inherent component of the system. The emission values are extremely high immediately after production, but they decrease rapidly after a few days [124]. Thermally bonded nonwoven insulating materials have been developed using bin-der fibres, preferably made from polypropylene, in order to avoid any possible emissions which may not be detectable until this moment. Nonwovens having densities of 40–110 kg/m$^3$ can be produced. The nonwoven fabrics weigh less than nonwovens made from recovered cotton fibres bonded with phenoplasts, which is an enormous advantage as far as the weight is concerned. Another rea-son for the success of these nonwoven insulating materials is that they can be treated with hot air much more intensively at temperatures of 170 °C during pro-duction, which enables more of the fugitive components to be removed. This keeps fogging values to below 0.5 mg and emission values to less than 5 µgC/g, which is exceptionally good. These types of nonwovens can be produced from re-cycled cotton and reclaimed fibres comprising PET, PAN or WO and PP or CO/

PET binder fibres. They are usually covered with a thin nonwoven on both sides to facilitate the application of a self-adhesive material, or to make it easier to weld them to a backing substrate, or purely for optical reasons. Nissan Motor and Ka-nebo have developed a special thermally bonded nonwoven on the basis of US patent no. 5286926, which comprises 80% modified PET fibres and 20% CO/PET binder fibres. The acoustic absorption has been improved by modifying the shape of the fibre cross-sections so that they may be triangular, rectangular, star-shaped or even more complex in shape [134]. US patent no. 5068001 also discloses a multilayer laminate which can be thermally moulded; it is porous, lightweight and has good acoustic properties [135]. Kurary, Du Pont, 3M, Hobbs and Cascade have also developed modern acoustic materials by combining meltblown microfibres with bicomponent fibres [112]. They can be recycled, do not generate dust and fly, and do not cause irritation. They are also clean and resilient and, of course, are suitable for making acoustic materials. They can be used to improve the sound absorption and insulating properties of most types of hollow spaces.

### 15.5.2.3 Acoustic and thermal insulation – the engine housing

Apart from the car interior, the biggest components are the front wall of the engine housing and the bonnet. This includes materials for insulating the catalytic converter from the bottom floor, the wheel-case coverings, insulation for the fuel, hydraulic and air-conditioning pipes, heat exchangers, etc. The latter are often used in combination with glass fibres, mineral wool, aramid, or metallized ceramic fibres, and may also be combined with metal films, either in the form of tubes or mats [136]. The large components in the engine housing are either covered with nonwovens made from recycled fibres bonded with phenoplasts, or PUR foam in combination with a glass fibre web. Bicomponent spunbonded webs (e.g. Colback) are being used more and more instead of glass fibres, because they can be moulded more easily [137, 138]. A nonwoven fabric made from oxidized acrylic fibres (Paramoll NIF) has been developed for the bonnet [139]. It is non-flammable, does not melt, plasticise or glow, is chemical-resistant, and exhibits no electrical conductivity and only a slight thermal conductivity. It can be treated with a water- or oil-repellent finish, or spot-coated with a melamine resin.

### 15.5.2.4 Covering materials – interior of the car

The interior covering materials can either be applied by laminating or welding, or by some other method, to a carrier component, which may also be made from a nonwoven material, or else it may be manufactured as a self-supporting element. They can either be produced as a continuous web or else combined with a moulding process. Needled nonwovens are used for preference. Development work on meltblown nonwovens has already begun. The main fibres used are PET or PP in a range of finenesses, with the trend being to use finer and finer fibres. The most important areas are the entire boot space, together with the wheel coverings, the floorcovering, together with the tunnel covering, the linings for the doors, side walls, seat backs, sides of the seat backs, the ABC pillars, the parcel shelf (and perhaps also the speaker housing), and the roof with the sun-roof, sun-roof cover and blind. For example, a nonwoven fabric with a two-colour effect, which has been velour-finished to create a flannel-like look, made from e.g. spin-dyed PP, can be laminated onto a carrier for producing the parcel shelf [140]. The parcel shelf can also be made in the form of a self-supporting, pressed, moulded component from a needled fabric containing a proportion of bicomponent fibres [141–144]. Other alternatives may include fabrics made from PET with CO/PET fibres or from PET with a reinforcement layer comprising a spunbonded nonwoven, such as Lutradur [145] or a thermally bonded nonwoven [146, 147]. These examples illustrate the wide range of practical possibilities available for each individual component, whereby the development of stretch meltblown nonwovens has extended these possibilities even further. The traditional end-uses for bonded and subsequently moulded needled nonwovens are all the moulded components used in the boot, including the wheel-case and side linings. Initial development work has been carried out on polyethylene sintered layers, and polyolefin layers have also been extruded between two nonwovens. The fabrics have to meet all the important characteristics already mentioned, and they should also be easy to mould and quick to process; they should also exhibit good dimensional stability and stiffness properties.

A large proportion of the floorcovering materials are still currently made from tufted fabrics, and good quality types can be produced having pile weights of $300 \text{ g/m}^2$. However, nonwovens having a weight of $100 \text{ g/m}^2$ are also used as the backing substrates for the tufted fabrics They can be used with Lutradur (100% PET) or Colback (PET/PA) mouldable spunbonds, which can also be used in lower weights than secondary backing materials. Most of the nonwovens used as floorcoverings are needled fabrics. Standard car models are usually furnished with fabrics having a smooth surface, but the 'velour look' is the most desirable. The structure of Tara Look nonwoven fabric, which is made from PET/PP, is such that only a few melt-bonding fibres can be seen on the surface, with the majority of them being concentrated in the base layer. The result is a mouldable carpet with an excellent appearance and good abrasion resistance values, which guarantees that the fibres do not work loose from the fabric [112]. Such mouldable nonwovens, which meet recycling requirements, are made from 100% polyester with Fosshape in the USA and Japan for car floorcoverings and for furnishing exhibition stands [148]. Development work on improving the quality of the nonwoven floorcoverings is still continuing.

As far as covering the roof is concerned, the distinction must be made between purely decorative furnishing materials made from warp- or weft-knitted fabrics or a nonwoven, and the stiff sub-assembly of the roof construction, which acts as a stabilizing element. Until now, the latter has usually been made from a semi-hard PUR foam core, which was either prefabricated or else produced from an impregnated, post-cured flexible PUR foam. The two cover faces of this or of a similar foam core were then covered with glass fibre webs weighing between 110 and $160 \text{ g/m}^2$. A thin nonwoven covering made from PP or PP/CV was laminated onto the reverse side. A warp- or weft-knitted fabric, made from PET/PA or PET, combined with flexible PUR foam, and laminated on the front (using an adhesive) was used as the furnishing fabric. Höflich describes a construction which comprises up to 100% PET fibres, which is a real advantage for recycling [149]. A hot moulding and a cold moulding type are described. With hot moulding, the required shape is achieved by compressing a needled nonwoven under the influence of heat and pressure, with simultaneous activation of the binder fibres. The advantage of this is that a high inherent stiffness is achieved for a low weight. A fine, polyester needled nonwoven, bonded with a 'minibond' (a nonwoven made from polyester), is recommended for use as the furnishing material. This construction is shown in Fig. 15-31.

A substantial amount of fine needled and stitch-bonded fabric is also used for lining the roof of the car; these types of fabrics were used mainly in small cars initially. The most suitable fibre is polyester because of its low mass per unit area, good acoustic and thermal insulation properties, the low forces required for the moulding process, and the fact that it does not crease during moulding. JPS Auto Products and Kimberley Clark Corp. in the USA have developed a new nonwoven fabric for the roof, which is particularly extensible and easy to process. It is used in those cases where the knitted fabrics previously used cannot be moulded deeply enough [112]. Pile material from JPS is laminated with Kimberley Clark's meltblown or hydraulically entangled 'Demique' (the thicker version) as the stretch nonwoven carrier. The adhesive bond has no adverse effect on the extensibility. The amount of nonwoven fabric use in the roof area will depend on how much it is able to improve the design and appearance.

Lining for the doors can be divided into the side trim, the central area and the foot area, which all have their own particularly design characteristics. Door linings are often particularly complicated, and different processes are used to produce the various different elements. The upper edge and side trim must match the parcel shelf and the padded upholstery material. The central zone usually carries the seat upholstery fabric as an upholstered panel, finished off with a welded seam. The weldability of the materials in this area is a particular problem, and this also applies to the nonwovens. Carpeting is used mainly for the floor area, and a

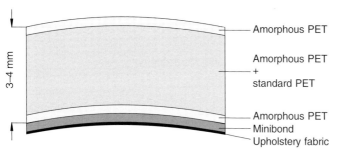

**Fig. 15-31** Processing of nonwovens by the hot moulding process (in accordance with [149], by kind permission of Sandler GmbH & Co. KG, Schwarzenbach).

needled fabric may be used in the footwell. The door linings of simple cars can also be moulded and laminated with a single nonwoven over the entire area. Various techniques are available for producing the door linings.

Two of these special techniques are rear injection and rear moulding, which can be used to produce covered components for the floor, the door panels or the entire door, and especially pillar coverings, boot linings for tourers, side wall linings, the backs of the seats, flaps, lids, and many other small components. The main driving force behind developing rear injection and subsequently rear moulding technology was to combine, in a single processing step, the previously separate processes of injection moulding of the carrier, followed by subsequent laminating with the covering material. The processing time can be reduced by up to 15 to 45% [150–153], reducing costs by about a third on average. The adhesive no longer has to be applied by spraying, which is more environmentally friendly. However, the rear injection process makes special demands of the mould and processing technology [151, 154, 155]. Whereas rear injection is regarded as being the most suitable for components having a size of about 0.4 m$^2$, with relatively thin wall thickness of up to 2.2 mm being possible, rear moulding technology is more suitable for processing larger areas or very flat lining components, and the moulds used are also cheaper [154, 156]. The melt is deposited into the open tool in the form of strands. Introducing the melt via a hot channel distribution system has proved to be the most suitable method when the geometry of the components is relatively complicated. Both procedures involve using a separating film or, even better, a separating nonwoven as a dividing layer between the actual surface material (warp-knitted fabric, weft-knitted fabric, woven fabric, synthetic film, leather, nonwoven, etc.) and the hot mass of synthetic material. The most suitable material is a needled nonwoven weighing roughly 100–130 g/m$^2$, made from polyester fibres of e.g. 1.7 dtex, which has been specially compressed and treated to make the surface smooth. It is laminated to the outer material to form a firm bond. During processing, this nonwoven presents a pliable, buffering labyrinth of tiny hollow spaces to the hot melt which seeps into it on cooling. This enables the displaced air to escape through the porous surface of the material and the tool itself. The system can also be used for incorporating an additional, stretch, PUR foam intermediate layer to create an outer material/flexible PUR foam film/separating nonwoven assembly. Soft lining components, produced by the rear injection or rear moulding technique, are the end product.

### 15.5.2.5 **Nonwovens for seat cushioning and laminated nonwovens**

The vast majority of all seat cushioning is made from PUR foam nowadays, which may be hot or cold crosslinked. They may be produced as single- or two-zone foams, and a small amount is produced from 'rubberized hair mats', consisting of coir fibres and animal hair, which are chemically bonded with a latex. Moulded padding and seat cushioning made from nonwoven polyester fabrics using particularly elastic and durable bicomponent fibres [157, 158] are a recent development. Experiments [159] carried out to compare moulded padding made from rubberized hair, polyurethane foam and nonwoven polyester for car seats have shown that, to achieve similar characteristics, i.e. between a PUR foam padding and a padding made from a nonwoven polyester fabric, the latter would be too heavy and expensive. Sitting on it at high temperatures is also a problem. However, nonwoven padding was found to be suitable for the seat backs, although the cost is still too high. However the development work was carried out with special regard to the advantages offered for recycling and to avoiding the use of flame resistance agents, e.g. with Densatil from Kem-Wove or Du Pont's Fiber Clusters [160, 161]. The latter are produced from a spirally crimped, bicomponent fibre blend. When processed into seat cushions for the car, they weigh a third less, but offer similar support characteristics to PUR foam; their density can also be varied. Sackner (USA) is also planning to produce seat cushioning up to 10 cm thick using its fibre/foam material [112]. A great deal of development work is being carried out on seat cushioning, but nonwoven cushioning has not been particularly successful so far. However, its properties and overall economic aspects will decide whether a breakthrough is likely in the near future.

However, considerably more progress has been made with laminated nonwovens. Although most composite upholstery fabrics are made from the conventional arrangement of outer fabric/PUR foam/lower fabric [162], which has been used successfully for some time now, Malivlies fabrics have also been used for some years now for this purpose [129]. Their lower thickness of between 2 and 16 mm makes them easier to handle. On average, they have a thickness of about 4.5 mm [163]. The most difficult characteristic to achieve is that of temporary deformation and this can only be brought about to a very limited extent. This is virtually impossible to achieve with nonwovens whose fibres are positioned in a mainly horizontal alignment. Nevertheless, investigations are being carried out in this area; they involve carrying out slight pre-needling plus heat-bonding using a high proportion of reclaimed fibres, on the one hand, and spirally crimped primary fibres, on the other hand, e.g. Minibond from Sandler [149] or a fibre/foam material from Sackner [112]. The development of nonwoven fabrics with a vertical fibre arrangement would seem appropriate in this case. Kunit stitch-bonded fabric appeared in 1991 [129] and, more recently, Cosmopolitan (UK) has combined it with an integrated, lightweight, double-faced fabric as a replacement for PUR foam [112]. The so-called Kalitherm technique is another development in this area [164]. This is a new process, whereby a mechanically preformed nonwoven consisting of stitches and protruding loops is simultaneously laminated with the outer fabric and thus calibrated; the assembly is then heat-bonded and treated to make it smooth. Such laminated nonwovens are referred to as Caliweb, and so far have achieved the best values compared to foam-backed laminates (Table 15-18) [127].

The advantages of this type of assembly are that an additional double-faced fabric is not required, it can be recycled easily because the materials are pure, it reduces emissions, and it is possible to use recycled fibre material. The thickness of the Caliweb nonwoven may be between 2 and 8 mm (Fig. 15-32).

Small amounts of so-called upholstery wadding, which is usually about 20 mm thick, are still used in top-of-the-range vehicles; it is made up of a needled nonwo-

**Table 15-18** Comparison of a PUR foam/warp-knitted reverse side with Caliweb A and B

|  |  |  | PUR foam combined with a warp-knitted fabric | Caliweb A | Caliweb B |
|---|---|---|---|---|---|
| Density |  | g/m³ | 60 | 70 | 80 |
| Compression strength |  | kPa | 4.6 | 3.6 | 6,0 |
| Compressive behaviour | $a_{30}5H$ | % | 28 | 39 | 42 |
|  | $a_{30}E3$ | % | 96 | 93 | 92 |
| Recovery capacity | lengthwise | % | 96 | 84 | 82 |
|  | crosswise | % | 93 | 90 | 88 |
| Permanent deformation | Dry | % | 8 | 18.5 | 16,5 |
|  | Wet | % | 19 | 20.5 | 16,5 |
| Flammability testing | lengthwise | mm/min | A* | A* | A* |
|  | crosswise | mm/min | A* | A* | A* |

A* = is extinguished when the ignition flame is removed.

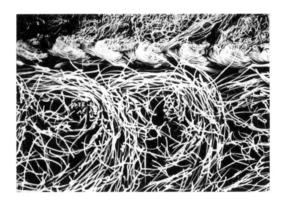

**Fig. 15-32** Cross-section through a nonwoven composite: weft-knitted fabric with Caliweb® (photograph courtesy of ACORDIS, Microlaboratory) Caliweb®: calibrated nonwoven, e.g. Kunit

ven/PUR foam/needled nonwoven, and may also include a woven jute insert. Further development work is continuing with producing nonwovens by the Struto process (see Section 4.1.4); they have a vertical fibre arrangement and a thickness range of 8 to 50 mm.

### 15.5.2.6 Automotive filter fabrics
There are between 8 to 10 filter points in every car. The role of the filter is to keep back, capture, collect and restrain impurities from a range of media (air, fuel, oil, exhaust gases) in its pores. The design of the filter will vary, depending on the country and/or state of development achieved. The following types of filter

can be found in the car [165], i.e. filters for the airbags, recirculated air, ABS wheel brakes, carburettor air, car interior, crankcase, soot, engine oil, fuel tank, fuel line, steering gear, gearbox and the wiper unit.

Soot filters are made in a similar way to exhaust filters from high-temperature-resistant ceramic materials, and airbag filters and ABS wheel brake filters are made from metal fibres, i.e. a woven metal wire mesh, and are therefore not so interesting as far as nonwovens are concerned.

Most of the filter materials are, or at least have been until now, made from synthetic-resin-impregnated modified paper, e.g. all of the oil filters and some of the air filters. Activated carbon, arranged in an appropriate assembly, is used to trap fuel fumes (tank, filling points). Woven screen fabrics made from monofilaments are used for the gear filters (steering gear filters) and wiper unit filters, and some nonwovens may be used in the form of needled webs for the gear filters. The needled webs are specially impregnated; they are usually made from polyester fibres and are fixed into a metal frame. They are able to filter out more impurities than the old type of gear filters made from paper or woven fabric. The driving force behind their development has been the increased use of and the high demands made of automatic gear systems.

Fuel filters are located in the tank as well as in the fuel line, just in front of the engine. In conventional cars fitted with carburettors, these filters are usually made from monofilament woven fabrics consisting of nylon, polyester, polypropylene or polyvinylidene chloride, as well as from porous ceramic material, sintered bronze and impregnated paper. For a filter to keep the fuel, and consequently the combustion chamber in the engine, free from dust, water, rust and other impurities, it must be able to meet the following requirements: long service life/low maintenance, good wettability – constant flow, no moisture absorption – no swelling, separation of up to 98% of particles down to a size of $<15\,\mu\text{m}$, resistance to methanol – no swelling, high degree of softness and elasticity, good insulation properties, and high pickup capacity.

The first and last requirements correlate particularly well with each other, and have become especially important in recent times. With modern, fuel injection cars, the fuel pump in the tank is always operating to maximum capacity. The reverse of the filter is hardly ever washed, and the filter medium becomes heavily loaded. This problem was solved by developing 'in tank filters' with depth filtration (polymeric depth media = PDM). Examples of this include sandwich arrangements consisting of polyester meltblown webs between thin polyester spunbonded layers (e.g. PPDM31 from Cuss Corp.) or the 'Strata Pore' density gradient nonwoven made from SMS nonwoven fabric, in which the meltblown nonwoven core is based on three layers of different densities [166].

The filtration efficiency is characterized by the parameters of absolute and nominal filtration effect, filtration ratio, and impurity retention capacity. Modern techniques and sophisticated testing technology are currently being used in the continued development of fuel filters.

Air filters are made mainly from wet-laid paper media impregnated with synthetic resin. Nonwoven filters are able to offer a number of important advantages in terms of their efficiency (97 to 99%), space requirements (0.73 to 0.44 m$^2$) and capacity (108 to 319 g/m$^2$) [165]. In Japan, air filters based on needled nonwovens only need to be changed roughly every 40,000 km [167]. The impurity retention capacity is particularly important, which is why filters made from composite nonwovens based on both needled and thermally bonded nonwovens (e.g. Qualiflo from Reemay) have been developed [168]. For the filter to be suitable for both surface and depth filtration, it must be tailor-made to suit the end-use conditions exactly, and these are laid down in the various test standards. The relationship between the pore size distribution of the filter material and the particle size distribution in the intake air is the most important factor [169].

The German company Freudenberg did not begin the development and production of filters for the car interior until the second half of the 1980s. The driving force behind this development was the need for ventilation, heating and air-conditioning in the cabin, on the one hand, and the increasing level of air pollution, traffic, and incidence of allergies, on the other hand. In 1992, 3.8 million filter units were built into cars, but by 1996, the figure was 28.6 million [170]. It has been established that, in the year 2000, cabin filters were fitted in 90% of cars in Europe and the USA, and in 50% of cars in Japan [171]. Good filters should last at least one year, preferably two, or at least 30,000 km. Initially, they were used solely as particle filters for filtering out dust, soot particles (from diesel engines), pollen, fungal spores, bacteria and particles down to a size of 3 μm. However, the combination filter appeared at the beginning of the 1990s, which consisted of a very fine particle filter plus an activated carbon filter, enabling small particles from industrial waste gases, tobacco smoke, oil fumes, and smog down to a size of 0.01 μm to be filtered out. The activated carbon layer also absorbs odours and some noxious gases (diesel exhaust, sewage/manure smells, nitrous oxide, sulphur compounds, ozone, hydrocarbons, ammonia, formaldehyde, etc.) in possible contact times of between 10 and 50 ms. However, to be effective on all counts, filters have to exhibit the relevant physical and chemical adsorption characteristics and catalytic degradation [172, 173].

In addition to a high degree of separation (e.g. 90% for particles of <0.3 μm), the requirements' profile for cabin filters include a high particle retention capacity, a long service life, and resistance to temperatures, chemicals, fire and water. They must also be odourless, have a low flow resistance, be usable at high relative humidities, have the capacity to filter out odours and toxic substances, and be mechanically stable and recyclable. Investigations to compare their characteristics can be carried out in accordance with DIN 71 460. The geometry of the filter has a marked influence on its efficiency [174–176]. A time-lapse test method, known as 'accelerated simulation of filter ageing', which simulates the conditions occurring in practice, can be compared with the practical test using similar products [177].

Modern cabin filters are made up of a system of layers. The outer layer comprises a rigid, highly porous nonwoven, which acts as a pre-filter and guarantees that the product is mechanically stable. The main filter is a three-dimensional, layered microfibre web which has both mechanical and electrostatic properties, and which is adjacent to or embedded in the absorbing activated carbon layer. The permanent polarization of the constant electrostatic charge creates an electrical field which can also filter out and permanently bind the many smaller particles, e.g. down to 0.1 µm, which the microfibres would not be able to do by mechanical means alone. Suitable polymeric materials, e.g. polycarbonate or polyester, are used as the electret, extruded nonwovens [165, 178–180].

### 15.5.2.7 Nonwovens as backing substrates, microfibre webs and other nonwovens

Needled nonwovens are the most suitable for coating. Roofing materials for convertibles are made from acrylic and acrylic/polyester woven fabrics, but can also be made from polyester needled nonwovens coated with PVC. They have to exhibit long-term resistance to temperature fluctuations, UV radiation, sub-tropical humidity, mildew, fire, ageing and shrinkage, and they must not lose their strength or elasticity [181]. Artificial leather based on nonwovens, which is used for covering the seats, must also meet these criteria. PVC is used as the coating agent and, increasingly, polyurethane. Artificial leather based on nonwoven fabric is also being used for tool bags, pockets for holding a vehicle's documentation, and ski bags. Microfibre nonwovens, such as Amaretta or Alkantara, are being used increasingly as high-quality furnishings instead of woven fabrics, warp-knitted or weft-knitted fabrics for seat upholstery fabrics, on covered components, for the door lining, and/or for the headliner. They are made from matrix fibres based on PET, which are processed into a needled nonwoven, and compressed by shrinking. The micro-structure is produced by removing one of the fibre components. The final, characteristic, suede-like microfibre web is produced by impregnating it with a polyurethane, and precipitating it within the web structure, in combination with splitting and special finishing (including tumbling, dyeing and raising). These types of nonwovens are breathable and soft, and are usually laminated with a thin woven fabric and treated with a flame-resistant finish on the reverse side to increase their strength and to enable them to meet the flame resistance requirements specified for car seats.

In addition to using thin, synthetic-fibre woven fabrics, nonwoven fabrics can also be used as covering tarpaulins. 'Evolution', type 4, developed by Kimberly Clark, is a particularly well-known example, and is made up of four layers. The two outer layers are thermally bonded spunbonds comprising drawn polypropylene fibres, having different masses per unit area. The inner layers comprise two polypropylene meltblown nonwovens having low masses per unit area. The outer nonwovens are responsible for the strength and abrasion resistance, whilst the meltblown nonwovens are responsible for the volume, the absorption characteristics and the barrier properties. The layers are joined together by means of ultrasonic welding. The tarpaulin is breathable and washable, and can be printed and sewn. It is also resistant to temperature, UV light, water and bacteria. Sentrex is a similar type of laminate, and consists of a three-layer material made from a meltblown nonwoven sandwiched between two outer spunbonded webs [182].

A less breathable, fully recyclable, waterproof material consists of a polypropylene Maliwatt web with a polyolefin melt coating.

Seat covers for the driver's seat of cars that are to be exported (if they are not covered with film) are also made from nonwovens (spunbonds or Maliwatt). Normal covers are usually made from weft-knitted fabrics rather than nonwoven fabrics.

It should also be noted that almost all the seals used in cars are made from impregnated and compression moulded nonwovens.

### 15.5.3
### The aircraft, shipbuilding and railways industries

Fibre-reinforced composites made from reinforcing fibres and a matrix material are used throughout the entire vehicle construction industry for components that are to be subjected to high levels of stress. The main fibres used are glass, aramid and carbon. Polyester, vinylester and epoxy resins have proved to be particularly successful as the matrix materials [183]. One particular type is a fibre-reinforced, thermoplastic composite, reinforced with natural fibres. They are temperature-resistant, lightweight, stable, breakage-resistant and non-splitting; they also have acoustic-insulating properties and can be used as lining components in car interiors. Hemp, kenaf, flax, jute or sisal is usually processed along with polypropylene in the nonwovens [184].

Weight reduction is particularly important in the aircraft building industry. An Airbus usually contains between 800 and 1,600 m$^2$ textiles for use in the seat covers, fire blockers, curtains, carpeting and wallcoverings. Nonwovens are currently only used for the wall and floorcoverings. They have to meet strict specifications in terms of their burning behaviour, mass, abrasion resistance, shrinkage and stretch behaviour, light- and colourfastness ratings, felting, seam distortion, corrosion effects, electrostatic charge and bonding properties [185]. The main requirement is that they should meet the relevant non-flammability standards [186–188].

Floor- and wallcoverings made from nonwovens are also used in the shipbuilding sector.

The main uses of textile materials in railway carriages are similar to those in aircraft and ships, i.e. they are used as coverings for padded seats, fire blockers in the seat cushioning, curtains, floorcoverings and wallcoverings. Nonwovens can only be used for the floor- and wallcoverings. All textiles used in passenger trains must have a service life of at least eight years. Wallcoverings must be easy to clean and dry quickly, and they must also be flame-resistant. They become dirty through 'passenger carelessness', metal abrasion, sand and soot (diesel). The Federal Railways Authority (Eisenbahnbundesamt) lays down specifications for all materials for use on the railways [189]. In this case, the regulations concerning flame resistance are particularly important. In the UK and France, these standards and regulations go even further, and cover flammability when in contact with a small flame, a smouldering source, and large-scale fires. The following standards are applicable: BS 6853; CP/DDE 101; ASTM-E 648; BS 476, Part 1; BS 5852; UIC-Regulation 564-20 R; NF F 16-101; NF C 20-455; NF G 07-128; P 92-

507; NF X 10-702; NF X 70-100 and oxygen consumption calorimetry, together with flammability and heat evolution velocity [190].

### 15.5.4
### Outlook

In the future, nonwovens will be used more and more for technical applications and rather less as fashion furnishings, which means that the great leaps forward made in the last 20 years have come to an end. However, the prospects for non-wovens remain good, thanks to the new developments that are emerging all the time, and the discovery of innovative solutions to particular problems.

The aspect of recyclability is of paramount importance when tackling any prob-lem. Multilayered composite constructions, in which the layers are combined to achieve certain characteristics, can often be constructed easily and cheaply using a combination of nonwovens. Using the same polymers plays an important part in this, despite conflicting requirements. This is where the opportunities lie for sub-sequent, economical reprocessing, as long as the components can be taken apart easily. The way is open to producing a headliner made solely from polyethylene terephthalate, a carpet from polyamide, or a lining component from polyolefin. Even if this can be achieved, recycling may be restricted by the level of dirt gener-ated during use and polymer decomposition [191]. In some cases, downcycling and combustion are the best solutions as far as ecology and economics are con-cerned. New solutions to problems also need to be examined within the context of improved comfort, not only in terms of air filtration, but also with regard to re-ducing odours, for example, by using textiles having desorbing characteristics for the seats [192], or for creating an electrostatic equalizing discharge across the pas-sengers. Further reductions in weight to save energy and the constantly increas-ing number or requirements demand sophisticated constructions, which in turn paves the way for multilayered nonwovens.

There is still some scope left in the field of developing fibrous cushioning, and work on using nonwoven fabrics for covering seat cushioning is still at the develop-ment stage. Perhaps water jet technology or some other fibre entangling process will present new opportunities for extending the use of nonwovens in the car industry.

### 15.6
### Papermaking fabrics
W. Best

Papermaking machines are some of the largest machines used to continuously produce a ready-to-use product, in this case paper, from a raw material. The aim is to operate at production speeds of 2,000 m/min on today's high-speed ma-chines. The textile papermaking fabrics form the heart of the machine. Polyester woven fabrics, on which the sheet or web of paper is formed, are mainly used at the wet end of the machine. Nonwoven composite fabrics are used in the press section for mechanical dewatering of the web of paper, and these so-called press felts are made up of a combination of woven fabrics onto which fibrous webs are needled. Although woven fabrics are mainly used here, inlaid constructions can also be used, and these have also proved very successful. This mechanical dewatering process is extremely important for the papermaking process, since it is more cost-effective than to subsequently dry the paper web in the drying section, where screen fabrics made from polyester (PET) and, in certain special cases, from polyphenylene sulphide (PPS) are used. Press felts may be between 10 m and 60 m long, and may be up to 10.5 m wide. During their life on the papermaking machine, they travel a distance of roughly 70,000 km, and are pressed several million times between two nip rollers at specific pressures which, in some cases, may be even higher than 10 Mpa.

Press felts carry out a range of important tasks. Their construction (Fig. 15-33 a and b) must be such that they take-up the fragile paper web from the paper-forming screen and transport it gently through the papermaking machine's press system. They are normally produced as continuous webs, apart from press felts which are fitted with a closable seam on the papermaking machine. The textile carrier for the press felt is made from woven fabrics, made up of one, two or three layers. Combinations of one, two and, in some cases, three carriers are being used more and more for press felts. During use, the woven carrier must absorb the tensile forces generated and make an almost non-compressible, internal, free volume of water available in the press zone; if possible, this must remain unchanged over the entire period of use. Combinations of carrier substrates are particularly good at meeting this requirement. The lower woven fabric is a very stable construction and makes the required volume of water available. The upper, finer woven fabric acts as a cover for equalizing the transfer of pressure, and for preventing the paper web from being marked.

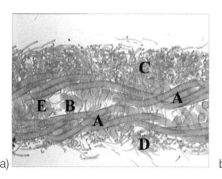 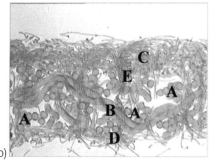

a)           b)

**Fig. 15-33** Structure of a double-layer press felt
a) Press felt: microtome cut in the lengthwise direction, b) press felt: microtome cut in the crosswise direction
A: lengthwise yarns, monofilament twists, B: crosswise yarns, monofilament twists, C: web layer on the paper side, D: web layer on the roller side, E: fibres from the web that have been needled into the carrier fabric

The press felts play an important role in dewatering the paper web, and must ensure that the quality of the upper and lower surfaces of the paper is suitable for printing. In order to ensure that they can meet these requirements without any problems, a suitable fibrous web layer is applied by means of a needling process to the carrier or carrier assembly. The fibrous layers can be applied to both sides of the carrier, and the mass of the layers can be varied, depending on the particular application. Needling technology is also used to join combinations of carrier substrates to each other. The web, which has viscoelastic properties, forms a labyrinth of very fine pore channels, through which the water can flow into the carrier, and transfers the mechanical pressure forces as uniformly as possible to the paper web. The elastic fraction of this behaviour must be maintained for as long as possible under the applied pressure load.

Three different processes are used to produce these web layers. The oldest process is the on-line needling technique. The fibres used to make the web are first of all carded and the web, which is produced using a cross-lapper, is fed continuously to the needling machine in the required width. The carrier or carrier assembly is joined to the fibrous layer in the needling zone, and the layers are needled together. The web is applied to the circulating carrier on the needling machine until the fibrous layer reaches the required mass. The web assembly is compressed by further needling passes, without any more web being supplied, and this improves bonding with the carrier. The pre-needling technique refers to a process in which the cross-laid web is bonded loosely to itself on a pre-needling machine before being rolled up. This pre-needled web roll is then laid onto the carrier on a finishing needling machine. The advantage of this technique is that the quality of the web can be checked before it is needled to the carrier. In order to ensure that the fibrous layer has the required mass, the relevant number of web layers are laid on top of each other by a rolling technique. The third alternative is the Beltex process (see Section 6.1). The entire web assembly is produced as a continuous web tube having the same circumferential length as the press felt, and bonded firmly to itself by needling. It is then joined to the carrier. The structure of the web assembly varies, depending on whether the fibres are needled directly into the carrier or whether the fibrous assembly is compressed firmly first, before being applied to the carrier. Of course, it is also usual to apply layers having different fibre fineness on top of each other. In general, the fibres are finer, the nearer the fibrous layer is to the paper web.

Selecting the right type of fibre to use is extremely important. The fibres must be able to withstand the high compression forces without splitting. They must also have a high abrasion resistance in order to guarantee a long service life. Ideal polymers are semi-crystalline polyamide with average and high molecular weights, especially polyamide 6 or polyamide 6.6, as well as polyamide 6.10, polyamide 6.12 and polyamide 12. It is possible to choose between various degrees of set, i.e. non-set, medium-set and super-set, which is also an indication of the remaining shrinkage capacity of the fibres as a function of temperature; this is an important factor for final setting of the press felt. The fineness of the fibrous layer is selected on the basis technical considerations. The type and quality requirements of the paper, and the geometry of the press system all influence the choice of fibre fineness. Fibres are available in the range of 2.8 to 67 dtex. Polyamide 6 fibres may have diameters of 17.7 to 86.5 μm. The fibre lengths may range from 38 to 120 mm, depending on the fibre fineness. Blends of fibres having different finenesses and lengths are used to obtain uniformly carded webs. Fibre producers have been supplying fibres for some time now with 'Vario-Schnitt' (variable staple lengths). These fibres are actually produced in different lengths. This enables web assemblies to be produced having a high degree of uniformity from fibres which all have the same fineness value.

What criteria are used to select the fibrous layers? One criterion relates to the type of paper and water volume in the press. Extremely fine fibres are needed in the fibrous layer to produce papers for hygiene applications. As the volume of water in the press increases, the fibrous layer should have a lower flow resistance. This can be achieved by using the coarsest possible fibres. The transport of the paper web also has a considerable influence on the fineness of the layer. The fineness of the fibrous layer, the uniformity of the felt surface and the compactness of the fibrous layer can all be used to control peeling of the paper web, sticking of the paper web to the press felt, and the formation of bubbles between the felt and the paper web. The finer and more compact the web layer is, and the more uniform the surface of the press felt, the firmer is the paper web fixed to the felt.

Using recycled paper as a raw material to produce paper has led to contamination becoming an even bigger problem. However, by carefully selecting the fibrous web, problem-free running of the press felt can be achieved even in this case. The properties of the felt, e.g. the soiling behaviour or hydrophilic properties, or the starting-up behaviour, can also be improved by applying chemical finishes.

As textile machine components, heavy-duty press felts are involved in a highly specialized operation. Simply by the way they operate, they can have a decisive effect on the quality of the paper produced and the efficiency of the paper machine.

### References to Chapter 15

[1] Schäfer W (1999) Hitze- und flammbeständige Vliesstoffe aus high-tech-Fasern für Schutzkleidung und technische Anwendungen, Vortr The Freudenberg Nonwovens Group Techtextil-Symp Frankfurt/Main
[2] Chou H, Forsythe J, Kotz P (1999) Needled Glass Fiber Mats for Automative Applications, Vortr Techtextilsymp, Block 4
[3] New PBO fiber for technical applications (1998) Man made fiber year book: 34
[4] Prospektangaben Fa ME Schupp Industriekeramik, Würselen, Germany
[5] Prospektangaben Fa Nabaltec GmbH, Germany
[6] Edited by Le Bras M, Camino G, Bourbigot S, Delobel R (1998) Fire Retardancy of Polymers – The use of Intumescence, The Royal Society of Chemistry, Cambridge Vortr 6. Europ Konf f d Flammschutz von Polymermaterialien, Lille, Frankreich Sept 1997
[7] Horrocks R, Kandola BK (1999) Complex Char Formation in Flame Retardened Fiber/ Intumescent Combinations: Physical and Chemical nature of Char, Text Res J 69, 5: 374–381

[8] Horrocks R et al (1999) Novel Flame and Heat Resistant Nonwoven Textiles, Vortr INDEX'99

[9] Volumenvergrößernder Vliesstoff für den Feuerschutz (1997) ITB Vliesstoffe – Techn Textilien 3: 20–28

[10] DIN 4102-1: 1998-05-00; Brandverhalten von Baustoffen und Bauteilen, Part 1: Baustoffe, Begriffe, Anforderungen und Prüfungen

[11] Hennig W (1997) Neue Dämmstoffe – (k)eine Alternative? bau-zeitung 51, 5: 6164

[12] Wolle und Holz im Verbund – ein Kombinationsdämmstoff, Produktinformationsschrift (1999) Inst f Textil- u Bekleidungstechnik TU Dresden

[13] Freudenberg CH, Offermann P (1998) Systematik zur Analyse und Synthese von Dämmstoffstrukturen, Techn Textilien 41, 11: 218–219

[14] Bershev E, Bershev N, Dörfel A, Freudenberg CH, Hoffmann, Lobova L, Lorenz I, Offermann P, Schierz CL (1999) Super light structures for thermal insulation materials, Vortr INDEX'99

[15] Rohrschalen aus Polyesterfasern (1999) Produktinformationsschrift, Inst f Textil- und Bekleidungstechnik TU Dresden

[16] Emissionsarme Glasnadelmatten (1998) AVR – Allgem Vliesstoffrep 1: 3233

[17] Prospektangaben Fa Culimeta Textil-Glas-Technologie, Bersenbrück, Germany

[18] Ullmann's Encyclopedia of Industrial Chemistry (1988) VCH-Verlag Weinheim, Vol. A 11, Reprint 1988: 20–27

[19] Konventionelle und alternative Dämmstoffe im Vergleich (1997) bau-zeitung 51, 11: 52–54

[20] TRGS 905 (1997/98) Verzeichnis krebserzeugender, erbgutverändernder oder fortpflanzungsgefährdender Stoffe, Ausg Juni 1997, BArbBl No. 6/1997: 40, geändert und ergänzt durch BArbBl 11/1997: 42 and No 5/1998: 72

[21] GefstoffV-Verordnung zum Schutz vor gefährlichen Stoffen v 26.10.1993 (1993/1999) zuletzt geändert durch Art 2 Gesetz v 29. Jan 1999 (BGBl I, S. 50)

[22] Persönliche Mitteilung aus dem Unternehmen G+H ISOVER Ludwigshafen 09/99 (1999)

[23] Produktblatt Isophen-Plus, Informationsschrift „Das Warmdach-Dämmsystem" Fa G+H ISOVER Ludwigshafen

[24] Produktblatt, Informationsschrift „Hinterlüftete Fassaden", Fa G+H ISOVER Ludwigshafen

[25] Gilbert J (1999) New under-roof structure with moisture-controlling nonwovens sheets, Vortr 10. Intern Techtextilsymp

[26] Gilbert J (1999) Vapor open insulated systems for pitched roofs in Europe, Vortr INDEX'99

[27] Schallschutz im Reihen- und Geschoßwohnungsbau (1999) bau-zeitung 7/8: 4041

[28] Schallschutz-Kompendium zur schalltechnischen Planung bei Wohnhäusern und Verwaltungsgebäuden, Informationsschrift Fa G+H ISOVER Ludwigshafen

[29] DIN 52214: 1976-09-00- Bauakustische Prüfungen; Bestimmung der dynamischen Steifigkeit von Dämmschichten für schwimmende Estriche

[30] Neumann R (1996) DIN 4109 als Erfüllung bauaufsichtlicher Anforderungen an den Schallschutz, bau-zeitung 50, 3: 5961

[31] DIN 52215: 1963-12-00- Bauakustische Prüfungen; Bestimmung des Schallabsorptionsgrades und der Impedanz im Rohr

[32] Schuster D, Becker U (1999) Neue Konzepte für Pkw-Innenverkleidungen zur verbesserten Geräuschreduzierung, Vortr 38 Intern Chemiefasertag Dornbirn

[33] Funck G (1998) Einfluß technischer Textilien auf den Geräuschkomfort in Fahrzeugen, Vliesstoff Nonwoven Intern No 140/1998

[34] Shopshani Y, Yakubov Y (1999) A Model for Calculating the Noise Absorption Capacity of Nonwoven Fiber Webs, Text Res J 69, 7: 519–526

[35] Zwikker C, Kosten C (1949) Sound Absorbing Materials, Verlag Elsevier Publ Comp NY

[36] Shopshani Y (1990) Effect of Nonwoven Backings on the Noise Absorption Capacitiy of Tufted B9, Text Res J 8: 452–456

[37] Riediger W, Jochim G, Sinambari GhR (1997) Paraphon, ein neuer Vliesstoff als „Breit-band-Resonanzabsorber" für den sekundären Schallschutz, Vortr Techtextil-Symp Block 6 (Fa. Lohmann GmbH); ITB Vliesstoffe, Techn Textilien 3: 28

[38] DIN ISO 7743-1991-10-00, Elastomere oder thermoplastische Elastomere, Bestimmung des Druck-Verformungsverhaltens

[39] Schäfer W (1982) Vliesstoffe für die Elektroindustrie, Chap 7.2 in: Lünenschloß J, Albrecht W (1982) Vliesstoffe, Georg Thieme Verlag Stuttgart/New York

[40] Bobeth W (ed) (1993) Textile Faserstoffe – Beschaffenheit und Eigenschaften, Springer-Verlag Berlin/Heidelberg/New York

[41] DIN VDE 0530-1, 07.91, Umlaufende elektrische Maschinen, Bemessungsdaten

[42] Schäfer W (1992) Vliesstoff-Verbundprodukte für Anwendungen in der Elektrotechnik. Vortr No 3.24 (Fa Freudenberg Vliesstoffe KG Weinheim) Techtextil-Symp Frankfurt

[43] Koprowska J, Vogel Chr (1998) Neue leitfähige Fasern zum Schutz gegen elektrostatische und elektromagnetische Gefahren, Vortr 4. Dresdner Textiltagung

[44] Mangstl H (1999) Schutz vor elektromagnetischen Störungen in Gebäuden – das architek-tonische Raumschirmungssystem Shieldex, EMC-Journal 3: 38–39

[45] Burteleit K (1994) Neue EMV-Abschirmung von Räumen durch metallisierte Textilien, Melliand Textilber 5: 368–373

[46] EMV-Schirmung von Kabelbäumen mit Spezial-Polyamidvlies (1999) EMC-Journal 3: 23

[47] DIN 52 612-1: 1979-09-00; Wärmeschutztechnische Prüfungen; Bestimmung der Wärme-leitfähigkeit mit dem Plattengerät

[48] DIN 51 046-1: 1976-08-00; Prüfung keramischer Roh- und Werkstoffe; Bestimmung der Wärmeleitfähigkeit bei Temperaturen bis 1600 Grad nach dem Heißdrahtverfahren, er-setzt durch EN 993-1 (1998)

[49] Neuer Wärmedämmstoff Thermolana (1996) bau-zeitung 50, 3: 65

[50] Applications: Filtration (1996) Nonwovens Markets, Internat Factbook and Directory: 135–140

[51] Burgmann I, (1995/96) Overview of filtration media worldwide, Technical Textiles Interna-tional: 14

[52] EDANA Nonwoven Statistics (1996)

[53] Filternde Abscheider – Oberflächenfilter, VDI-Richtlinie VDI 3677 (1997)

[54] Antistatische Spezialfasern zur Trockenfiltration (1992) Techn Textilien 35, 3: T9

[55] Sauer-Kunze M (1996) Anwendungstechnische Beispiele für den Einsatz unterschiedlich-ster Filtermaterialien in den Bereichen Prozeßluft, Raumluft- u Klimatechnik, 3. Symp Textile Filter Dresden

[56] Sauer-Kunze M (1993) Schwebstoffilter für die Reinraumtechnik, Handb Technik für Reine Räume, Vulkan Verlag Essen

[57] Bergmann L (1998) Latest developments and major trends of filtration towards the year 2000, Vortr 4. Symp Textile Filter Dresden

[58] Prüfung von Filtermedien (1994) Verein Deutscher Ingenieure, VDI-Richtlinie VDI 3926

[59] Siersch E (1996) Nadelfilze mit Spezialausrüstung für niedrige Reststaubgehalte in Ver-brennungsanlagen und Gießereien, Vortr 3. Symp Textile Filter Dresden

[60] Löffler F, Staubabscheiden, Georg Thieme Verlag Stuttgart/New York

[61] Ritscher G (1998) Kombinierte Abscheidung, Zwischenber SMWA-Verbundförderprojekt No 1822/377

[62] DIN EN 779 (Sep 1994) Partikel-Luftfilter für die allgemeine Raumlufttechnik, Beuth Ver-lag Berlin

[63] DIN EN 1822-1 (Juli 1998) Schwebstoffilter (HEPA und ULPA) Beuth Verlag Berlin

[64] Struto Strutis, Voluminöse Textilien – neuer Konstruktion – neuer Eigenschaften, Pros-pekt I.N.T. s.r.o. Liberec

[65] Krčma R (1997) Fortschritte bei der Entwicklung hochvoluminöser senkrecht verlegter Textilien, Techn Textilien 40, 3: 32–34

[66] Krčma R (1990) Heißluftgebundene Textilien mit neuer Struktur, Chemiefasern Text Ind 40, 3: 272–273

[67] Schmalz E (1998) Polvlies-Nähwirkverbunde als Tiefenfiltermedium, Vortr 4. Symp Textile Filter Dresden

[68] Schmalz E (1998) Polvlies-Nähwirkverbunde als Tiefenfiltermedium, kettenwirk-praxis 3: 148–151

[69] Aktuelle Markttrends für textile Filter (1994) Techn Textilien 37: 5

[70] Böttcher P (1992) Vliesstoffe und Freudenberg; Tradition, Internationalität, Verfahrensvielfalt, Innovationskraft, INB Nonwovens 3: 22–25

[71] Schollmeyr E, Bahners Th (1993) Strukturierung von Filteroberflächen durch UV-Laserstrahlung, Vortr INDEX '93

[72] Schwan R (1989) Einfluß der Morphologie und Elektrostatik von Synthesefasern auf die filtertechnischen Eigenschaften moderner Filtermedien, Melliand Textilber, 8: 577–581

[73] Toray Industries Inc (1990) Textile Elektret-Materialien eröffnen neue Perspektiven in der Filtration, Techtex Forum 10: 11–12

[74] Gasper H (1990) Handb der industriellen Fest/Flüssig-Filtration, Hüthig Buch Verlag GmbH Heidelberg

[75] Löffler F, Dietrich H, Flatt W (1991) Staubabscheidung mit Schlauchfiltern und Taschenfiltern, Vieweg Verlag Braunschweig

[76] Grießer H, Wimmer A (1992) Fasern und Garne für den Einsatz in der Heißgasfiltration, Melliand Textilber 9: 689–692

[77] Spurny KR (1993) Aerosolfilter aus aktivierten Kohlenstoffasern, F & S Filtrieren und Separieren, 7 1: 34

[78] Fasertabellen (1989) Filter Media Consulting Corp

[79] Kraft G (1992) Heizungs- und Raumlufttechnik Günter Kraft Berlin, München, Verlag Technik, Vol 2: Raumlufttechnik, 1st edition

[80] Webster C (1998) Spunbonded and Nonwoven Composite Materials for Filtration, Vortr 1st Internat East European Filtration Conf

[81] Hirschberg HG (1999) Handb Verfahrenstechnik und Anlagenbau, Chemie, Technik, Wirtschaftlichkeit, Springer-Verlag Berlin/Heidelberg/New York: 895–910

[82] Media for Liquid Filtration (1995) TUT Textiles a Usages Techniques 18: 17–18

[83] Stibal W, Kemp U, Ensinger J (1996) Filtration for modern polymer melt spin process, Internat Polymer Melt Filtration Conference Proc Stuttgart: 5–12

[84] Wierzbowska T (1997) Application of Polypropylene Fibres to Needled filtration Materials, Fibres and Textiles in Eastern Europe 5

[85] Schäfer T (1997) Härtetest für Bioöl. Schnell abbaubare Hydrauliköle im Filtrationstest, Instandhaltung 1: 42–45

[86] Alexander J (1996) Demands on Filtration Systems for Different Polymer Processing Applications, Internat Polymer Melt Filtration Conference, Proc Stuttgart, 47–51

[87] Houghton J, Anand SC, Purdy AT (1997) The Characterization of Fabrics Used for Wet Filtration, Techtextil-Symposium 97, Block 6: Neue Produkte 6.2 Filtertextilien für morgen, Frankfurt: 1–14

[88] Kullik W (1997) Besorgniserregender Beta-Salat, Neue Maßstäbe bei Hydraulikfiltern, Fluid-Technik 10: 32–34

[89] Wnuk R, Alexander J (1997) Filtrationsversuch mit Polymer und Teststaub, Barmag AG: Entwicklungen, Trends, Technologien 23: 23D1–23D2

[90] Superhigh-Purity Water Manufacturing System Requiring no Maintenance Work (1997), New Technology Japan 25, 1: 27–28

[91] David Rigby Associates (1997) The World Technical Textile Industry and its Market, Prospekts to 2005, Techtextil Messe Frankfurt GmbH 4

[92] Merkblatt für die Anwendung von Geotextilien und Geogittern im Erdbau des Straßenbaues (1994) FGSV

[93] Technische Lieferbedingungen „Geokunststoffe" (1997) DB AG-TL 918 039, 4

[94] Göbel, Lieberenz, Richter (1996) Der Eisenbahnunterbau, DB-Fachbuch, Vol 8/20, Eisenbahn-Fachverlag 1996

[95] Göbel, Lieberenz, Nietzsch (1995) Untersuchungen zu Verhalten von Tragsystemen mit Geokunststoffen an der Ausbaustrecke Leipzig–Dresden, Geotechnik, Sonderh: 136–140

[96] Lieberenz K (1997) Verbundstoffe für Eisenbahnstrecken, Kettenwirkpraxis 4: 49–51

[97] Blechschmidt D (1998) Mechanisch verfestigte Verbundstrukturen für das textile Bauen, vorzugsweise unter Verwendung von Filamentvliesstoffen, AiF-Forschungsvorhaben No 10544 B/V

[98] Ruzek I (1998) Polyester spunlaid nonwovens, EDANA Internat Nonwovens Symp

[99] Zerfass KC (1992) Vliesverbundstoffe als Einlage für Bitumenbahnen, Techtextil-Symp, Vorlesung 326

[100] Gassel T van (1993) High stability Nonwovens, Techtextil-Symp, Vorlesung 326

[101] Jahn U (1996) Bituminous roofing felts, what future for carriers?, INDEX'96, construction session

[102] Gassel T van (1997) Composite carriers for modified bitumen membranes, 4. Internat Symp on Roofing Technology Gaithersburg USA

[103] Baravian J (1993) New developments in reinforced polyester spunlaid nonwovens, INDEX'93, session 6A

[104] Wagner R (1993) Vliesstoffe für die Pflanzenanzucht, Techtexil-Symp Frankfurt

[105] Lennox-Kerr P (1992) Nonwovens for covering crops, Elsevier Science Publisher Ltd, UK: 1–2

[106] Fuchs H, Arnold R, Bartl A-M, Hufnagl E, Schmalz E (1995) Grobe Gewirke, Stränge und Vliesstoffe als Pflanzenträger, Techtextil-Symp Frankfurt

[107] Pan, Z-D, Cai F, Zhu H (1991) 3. International Nonwovens Conference Polymer to Web Nonwovens, Bordeaux Nov 91

[108] Antill D (1990) Use of nonwovens on field crops in the United Kingdom, INDEX-Kongr Geneva

[109] Döttcher P (1999) Mutterbodenlose Begrünung, Forschungsber Sächsisches Textilforschungsinstitut e.V. Chemnitz

[110] Bershev IN, Prosvirnizyn AW, Mogilny AN (1995) Die Entwicklung neuer Agrotextil-Strukturen auf Vliesstoff-Basis, Techtextil-Symp Frankfurt

[111] Cheng KPS, How YL, Cheung YP (1996) Preliminary investigation of the Effects of types of nonwovens on the growth of grass, Journ of China Textile University, Vol 13: 31–41

[112] Nonwovens Market, International Factbook and Directory (1996) Verlag Miller Freeman Inc San Francisco

[113] Fuchs H, Böttcher P (1994) Einsatz von Textilabfällen im Auto – Möglichkeiten und Grenzen, text praxis internat 49, 4: 236, 238, 240

[114] Nonwovens gaining ground in automotive end-uses, Markets (1993) Nonwovens Rep Internat 265: 7–8

[115] Going up a gear: getting to grips with interior aesthetics (1998) Nonwovens Rep Internat May 1998: 22

[116] Conroux J-J (1994) Des emergences espagnoles, TUT Textiles a Usages Techniques 13: 14–16

[117] Rewald FG (1994) Nonwovens gaining ground in the Brazilian automotive industry. Nonwovens are advancing in more and more automotive end use applications, Nonwovens Ind 25, 6: 26–29

[118] Needlepunched nonwovens for use in automotive interiors (1994) Nonwovens Ind 25, 1: 32–34

[119] Koslowski H-J (1998) Vliesstoffe in Zahlen, 13. Hofer Vliesstoffseminar

[120] Mansfield RG (1992) Automobiles: Growing applications for nonwovens, Konferenz-Einzelbericht IDEA'92, INDA's Internat Nonwovens Conf Expos Washington USA, INDA, Association of Nonwoven Fabrics Ind: 15–30

[121] Schmidtmann JU (1994) Untersuchungen über Korrelation der Farbechtheit von Textilien im Automobilsektor bei Frei- und Laborbewitterungen, Inst f Textil- u Faserchemie der Uni Stuttgart (Diplomarbeit bei Mercedes-Benz, Sindelfingen)

[122] Ehrler P, Mavely J, Schreiber H (1995) Spezielle Aspekte der Gebrauchseigenschaften aktueller textiler Autoinnenausstattung, Konf Einzelber: 34. Internat Chemiefasertag Dornbirn mit 18. Intercarpet, österr Chemiefaser-Inst Dornbirn, A: 1–14

[123] Eisele D (1992) Emissionspotential am Beispiel eines mit PVC beschichteten Polvliesbelages, Melliand Textilber 73, 1: 84–87

[124] Marutzky R, Meyer B (1992) Emissionen aus duroplastisch gebundenen Formteilen für den Kfz-Innenraum, Konf Einzelber: Anwendungen von duroplastischen Kunststoffen im Automobilbau, SKZ-Fachtag, Süddeutsch Kunststoff-Zentrum Würzburg: 215–229

[125] Ehrler P, Schreiber H, Haller S (1993) Emission textiler Automobil-Innenausstattung: Ursachen und Beurteilung des Kurzzeit- und Langzeit-Foggingverhaltens, Konf Einzelber 32. Int Chemiefasertag mit 17. Intercarpet 93, Dornbirn, A: 1–20

[126] Schmidt H, Lüßmann-Geiger H (1992) Emissionsarme Kunststoffe – eine neue Forderung im Automobilbau, Kunststoffe 82, 8: 685

[127] Schmidt GF (1995) Anforderungsprofile von Autotextilien und Erkenntnisse zu einigen Prüfkriterien, 34. Internat Chemiefasertag Dornbirn, A

[128] Bongartz M (1993) Von der Qualitätskontrolle zur Qualitätssicherung bei der Herstellung textiler Flächen für technische Anwendungen, Kettenwirk-Prax 27, 1: 35–40

[129] Schmidt GF (1992) Kaschierung von Autopolsterstoffen mit Faservliesen, Melliand Textilber 56, 6: 479–486

[130] Eisele D (1996) Recyclate aus Reißbaumwolle plus Phenoplast, Firmenschrift Borgers

[131] Eisele D (1990) text praxis 45: 1057–1063

[132] Eisele D (1992) Melliand Textilber 56, 11: 873–878

[133] Gardiziella A (1992) ATZ 94: 17–19

[134] Improved sound absorbency for automobiles (1994) High Performance Text Nov: 8–9

[135] Composite for use in car interiors (1992) High Performance Text Oct: 3–4

[136] Lilani HN (1994) Flexible thermal barrier materials for underhood automotive insulation systems, Konf Einzelber: Hi-Tech Textiles Exhibition & Conf, Textile World and INDA Assoc of the Nonwoven Fabrics Ind, Greenville USA, Vol 1: 87–101

[137] Dijkema J (1992) Protections acoustiques pour l'automobile, TUT Textile a Usages Techniques 6: 47–49

[138] Gassel T van (1993) New headliner and bonnet liner concepts, Konf Einzelber: Index 93, Internat Congr for the nonwovens and disposables ind, Automotive, Geneva CH, EDA-NA Vol 3A: II1–II8

[139] Feuerfestes Dämmaterial für Automobile (1995) Internat Text Bull Vliesst – Techn Textilien 41, 3: 14

[140] Caldwell KG (1989) Non-woven flannel fabric, United States Patent 4 828 914, May 9, 1989

[141] Aimone JH (1994) Challenges and opportunities for designing and manufacturing molded and needled automotive products, Konf Einzelber: Hi-Tech Text Exhibition & Conf, Text World and INDA Assoc of the Nonwoven Fabr Ind, Greenville USA, Vol 1: 103–107

[142] Recyclingfähige Preßformteile aus Grilene Nadelvlies-Matten (1992) Problemlose Wiederverarbeitung, Techtext Telegr 26: 3

[143] Aimone JH (1993) Challenges and opportunities for designing and manufacturing molded and needled automotive products, Konf Einzelber: Text in Automotives Conf, Text World Atlanta USA: 1–4

[144] Parsons R (1992) Thermoforming and cutting molded needlepunch fabrics, Konf Einzelber: Needlepunch Fabr – Durable Needlepunch Conf INDA Association of the Nonwoven Fabr Ind, Charlotte USA: 135–148

[145] Wooten HL (1994) Spunbond support scrims for needlepunched fabrics, Konf Einzelber: Int Durable Needlepunch Conf, Book of Papers, Charlotte USA: 73–78

[146] Firmenschrift: Vliesstoffe für die Automobilindustrie, Vliesstoffwerk Christian Heinrich Sandler GmbH & Co KG

[147] Recyclingfähige Vliesmatte für verformbare Auto-Innenteile (1991) Techtext-Telegr 21: 3

[148] Nishio H, Aono M (1994) Application of a composite fiber for floor covering, Konf Einzelber: ISF 94, Proc of the Int Symp on Fiber Sci and Technol, Yokohama, J: 551

[149] Höflich W (1998) Polyester- und Naturfasern im Automobilbereich, Techn Textilien 41: 154

[150] Mischke J, Bagusche G (1991) Hinterspritzen von Textilien, Teppichen und Folien, Kunststoffe 81, 3: 199

[151] Schäfer M (1993) Technik des Hinterspritzens von Textilien und Dekormaterialien für das Automobil, Firmenschr dura automobil comfort der DURA Tufting GmbH Großenlüder

[152] Barisani KR von (1995) Sortenreine Kunststoffverkleidungsteile mit textilen Oberflächen aus 100% Polypropylen, Konf Einzelber: 34. Chemiefasertag mit 18. Intercarpet, Österr Chemiefaserinst Dornbirn, A: 1–18

[153] Hettinga S (1992) New composite molding method, America's Text Internat 21, 9: FW2, FW4, FW8–FW9

[154] Barisani KRV (1993) Textilien für die Automobilinnenausstattung – Neue Techniken für die Innenraumgestaltung von PKWs: Speziell Hinterspritzen und Hinterpressen von Verkleidungsteilen, Konf Einzelber: 32. Internat Chemiefasertag mit 17. Intercarpet 93, Dornbirn, A: 1–10

[155] Kaufmann G (1996) Hinterspritz- und Hinterpreßtechnik von der Beratung bis zum Serienteil, Konf Einzelber: Polypropylen, erfolgreich im Kraftfahrzeug-Bereich, Süddeutsch Kunstst-Zentrum, Würzburg: 1–10

[156] Dorner P, Hahnekamp R (1994) Neue Technologien im Bereich der Innenverkleidungen, Konf Einzelber: Textilien im Automobil, Trends erkennen, Entwicklungszeiten verkürzen, Produktsysteme entwickeln, Kongr d VDI-Gs Text u Bekl, Düsseldorf: 33–47

[157] Kmitta S (1992) Anwendung von Formpolstern aus Polyesterfaser-Vlies im Sitzbereich, Konf Einzelber: Text im Automobil – Ideen entwickeln, Konzepte anwenden, VDI-Kongr, VDI-Ges Text u Bekl, Düsseldorf: 88–104

[158] Yoshida M, Takahashi N, Sasaki Y (1993) Cushion fibre structure, Konf Einzelber: INDEX'93, Intern Congr for the nonwovens and disposables ind, New Ideas, Geneva CH, Vol 6A: IV-1–IV-13

[159] Kmitta S (1995) Polyester-Faservlies – ein alternativer Polsterwerkstoff für PKW-Sitze?, Konf Einzelber: 34. Chemiefasertag mit 18. Intercarpet, Österr Chemiefaserinst, Dornbirn, A: 1–31

[160] Ekman L (1993) Fire safety in cars, some aspects of fire causes, materials and requirements, Konf Einzelber: Flammability 93 – The Inside Story, Conf Proc, BTTG, London, GB: 1–10

[161] Mansfield RG (1994) Automotive molded products from nonwovens, Konf Einzelber: INDA-TEC 94, Book of Papers, Int Conf and Showcase, Nonwovens Technol for Disposable and Durable Applications, Baltimore, USA: 107–111

[162] Lampe T, Bachor M (1993) Anforderungen an Automobiltextilien – am Beispiel von Sitzbezugsstoffen, Konf Einzelber: 32. Internat Chemiefasertag mit 17. Intercarpet 93, Dornbirn, A: 1–28

[163] Schmidt GF (1993) Abstandsgewirke ohne und mit Dekorseite für Anwendungen im Automobil, Melliand Textilber 1 and 2: 37–39, 129–134

[164] Schmidt GF, Schmitz M, Böttcher P, Offenlegungsschrift DE 195 34 252 A 1, Offenl 20.03.97

[165] Bergmann L (1993) Automotive filtration. 8 different filters in each automobile. A growing application, Konf Einzelber: Index 93 Automotive, Geneva/CH, EDANA, Vol 3A: V1–V8

[166] Combest JF (1996) Recent developments in automotive fuel filtration, Konf Einzelber: Filtration 96 Baltimore USA: 16.0–16.7

[167] Bergmann L (1994) Automotive Air filter how does Japan compare with Europe? Konf Einzelber: 6. Internat Techtextil Symp, Technische Textilien – neue Märkte u Zukunfts-Chancen, Vol 1.0: 1–8

[168] Mansouri B (1995) Nonwoven composite structures in filtration, Konf Einzelber: Nonwovens in Filtration, Internat Conf, Filter Media Consulting, Stuttgart: 94–98

[169] Schaefer JW, Olson LM (1996) Air filtration media for transportation applications, Konf Einzelber: Filtration 96 Baltimore USA: 11.0–11.15

[170] Bittermann H, Rosenberg G (1995) The use of nonwovens in cabin air filtration and the related requirements, Konf Einzelber: Nonwovens in Filtration, Internat Conf, Filter Media Consulting Stuttgart: 43–49

[171] Whitehouse A (1996) Cabin air filtration: The pollution filter, Konf Einzelber Filtration 96 Baltimore USA: 13.0–13.9

[172] Neveling V (1995) Automobil-Innenraumfiltration: Leistungsanforderungen, konstruktive Gegebenheiten, Möglichkeiten, Konf Einzelber: 34. Internat Chemiefasertag mit 18. Intercarpet Dornbirn, A: 1–8

[173] Cashin A (1995) Automotive cabin air filters: No longer a future issue, Konf Einzelber: Filtration 95 Chicago USA: 189–197

[174] Mohr U (1996) De l'air pur pour les voitures, TUT Textiles a Usages Techniques 22: 47–49

[175] Moser N (1995) Automotive cabin air filters, Konf Einzelber: Nonwovens in Filtration, Internat Conf, Filter Media Consultig, Stuttgart: 67–72

[176] Automotive cabin filters (1996) Technical Textiles Internat 4, No 10: 16—18

[177] Bittermann H (1996) Cabin air combination filter aging in field and test conditions, Konf Einzelber: Filtration 96 Baltimore USA: 14.0–14.8

[178] Reinhardt H (1996) Experience – Driven cabin air filter development, Konf Einzelber: Filtration 96 Baltimore USA: 12.0–12.20

[179] Groh W (1994) A new approach to high-performance dust filters, Konf Einzelber: 6. Int Techtextil Symp, Neue Textilien – neue Technologien – neue Produkte, Vol 2.2: 1–5

[180] Bergmann L (1993) Automotive cabin air filters show potential for nonwovens filters, Konf Einzelber: Textile World Atlanta USA: 1–2

[181] Alling P (1993) The evolution of automotive convertible topping and other exterior decorative trim fabrics, Konf Einzelber: Textiles in Automotives Conf, Textile World Atlanta: 1–9

[182] Gardner C (1995) Kimberley-Clark improves a winner, America's Textiles Internat, Vol 24, 2: 44–46

[183] Schmidtschneider R (1994) Hochbelastbare Faserverbundwerkstoffe im Automobilbau, Gummi Fasern Kunststoffe 47, 5: 306–308

[184] Hahnekamp R (1996) Polypropylen und Naturfasereinbindungen kombiniert mit Spritzgussmaterial: Neue wirtschaftliche Verarbeitungstechnologien, Konf Einzelber: Polypropylen, erfolgreich im Kraftfahrzeug-Bereich, Süddeutsch Kunstst Zentrum Würzburg: 1–9

[185] Berg HD (1993) Anwendung von Textilien in der Innenausstattung von Verkehrsflugzeugen am Beispiel des Airbus, Konf Einzelber: Textilien im Automobil, Kongreß der VDI-Gesell Textil u Bekleid Düsseldorf: 60–74

[186] Cortizo D (1993) Standards for textiles used in commercial air crafts, Konf Einzelber: Flammability 93 – The Inside Story, Conf Proc BTTG London GB: 1–5, Paper No 7

[187] Saville N, Squires M (1994) Fire protection – from the ridiculous to the sublime, Konf Einzelber: Fibres 94 Coventry GB: 1–7

[188] Beare S (1995) Resistant to fire and to washing, TUT Textiles a Usages Techniques 17: 31–32

[189] Kruegel W (1994) Einsatzbereiche von textilen Werkstoffen für den Innenausbau von Reisezugwagen, Konf Einzelber: Textilien im Automobil, Trends erkennen, Entwicklungszeiten verkürzen, Produktsysteme entwickeln, Kongr d VDI Ges Textil u Bekl Düsseldorf: 106–115

[190] Jones HR (1993) Textiles in the railway passenger environment, Konf Einzelber: Flammability 93 – The Inside Story, Conf Proc, BTTG London GB: 1–10

[191] Kiefer A, Bohnhoff A, Ehrler P, Klingenberger H, Schreiber H (1995) Untersuchungen zur Wiederverwendbarkeit von Altautotextilien in Neufahrzeugen, Konf Einzelber: 34. Chemiefasertag mit Intercarpet, Österr Chemiefaser Inst Dornbirn, A: 1–17

[192] Yamada Y, Araki O, Sugiura M, Horii M, Sekihara T, Matsuyama A (1992) Advanced seat fabrics with deodorant function, ISAE Review 13, 4: 82–87

# 16
# Re-utilization of nonwovens
B. Gulich

The re-utilization of nonwovens can be seen from different aspects. As with other textile waste, we can distinguish between waste resulting from the production of nonwovens and from products being used.

Nonwovens are used for many different purposes in almost all areas of the textile industry. Accordingly, there is much diversity with regard to raw materials, processes and designs. Exploitability and the processes to accomplish exploitation, known as recyclability – largely depend on these factors. The growing importance of the environmental compatibility of the raw materials used and eco-friendly processes as well as the ever louder call for better recyclability have led to the development of appropriate machines and novel production technologies. It remains a question of high rank that all enterprises concerned with the development, production and application of nonwovens see to waste being avoided or recycled, thus taking full responsibility for their products.

Legal regulations becoming ever stricter and economical production being of highest interest, the disposal of production waste in landfills will play a minor role in the future.

It is well worth mentioning nonwovens offer outstanding opportunities to use raw materials already recycled, which is best proven by the wide range of nonwovens manufactured from reclaimed fibres. It is estimated that approximately 95% of the reclaimed fibres produced in Germany are processed to manufacture nonwovens [1].

The formula to remember in the future is waste is equal to valuable material and is worth being exploited several times in ever new products, preferably to generate heat.

## 16.1
### Waste from the production of nonwovens

The utilization of waste is a problem already solved in many areas of the production of nonwovens. If possible, edges remaining in the production process as well as products of inferior quality and products returned to the manufacturer are reused in the same production process or another suitable way. Above all, mechanically compacted nonwovens offer proper characteristics to do so. One more possibility is to process production waste in enterprises specializing in recycling, which may entail the manufacture of defined blends of recycled materials meant for the original or other applications.

In case nonwovens are produced in processes different from the well-known mechanical ways of compacting, the traditional mechanical opening-up of fibres by means of breaking them down will quickly reach the limits of its applicability. The intensified development of sandwich materials and composites from both textile and plastic materials also results in waste whose utilization cannot be mastered by means of the traditional processes. With such products, process-related solutions are necessary. It is more sensible to develop clean-sorted composites from one and the same polymer, whose recycling is possible the same way as plastic materials are recycled.

### 16.1.1
**Measures to reduce waste**

Generally, it is in the manufacturer's interest to keep production waste as little as possible. Easy-to-take measures as seen from the technical/technological point of view are

- optimization of available production plants to better exploit the material in the production process
- optimization of the products with regard to recyclability (choosing the right materials and technologies)
- optimization of the production technology, e.g. choosing the optimum point of time to cut edges or process control when changing quality or assortment.

### 16.2
**Waste of used nonwoven materials**

### 16.2.1
**Disposables**

Nonwovens being distinguished by long-life and short-life ones, nonwoven waste is subject to the same classification.

With regard to current processes, it does not make much sense to recycle short-life disposables used for medical or hygienic purposes. Manufacturers are trying to make them biodegradable (so they can be composted). One problem is the classical synthetic polymers contained in these products are not degradable, which is why degradable plastics for application in fibres and films are being intensively developed. Disposables account for no more than 2% of the total volume of landfills. Incineration might, in most cases, be the most sensible way to dispose of them.

## 16.2.2
## Long-life products

Long-life nonwovens are expected to meet highly diversified requirements, their original characteristics and/or functionality being partially or totally lost while they are in use (e.g. floor covering, filters, packing material).

In addition, nonwovens will become waste if the product in which they are used to fulfil certain functions loses its value so it is disposed of (e.g. motor vehicles, garments, upholstery furniture, mattresses). Such nonwovens (as well as other textile materials) being integrated in the product in question, they often cannot be recovered very easily or, at least, not in a well-sorted way. In the future, legal regulations will expect the seller to take back such products (e.g. German regulations on the disposal of used cars). The call for designs that are easy to recycle will rise. This will also affect the textile industry.

The recycling of used nonwovens having been of minor importance so far, there will be much need for action in this field in the future.

## 16.3
## Utilization of nonwoven waste

### 16.3.1
### Mechanical processes to recover fibres

With mechanically compacted nonwovens of blends of chemical and natural fibres or with pure natural-fibre nonwovens, the mechanical opening-up of the textile structure by means of breaking them down is frequently found (see Section 1.3 "Reclaimed fibres"). The manufacture of reclaimed fibres is wide-spread and economical. Although the fibres, to a certain degree, are physically damaged in this process, the functional components of the fibrous material are maintained.

One advantage of the process is it can be applied with both production waste and nonwovens after their use. However, contamination, bad soiling as well as the presence of non-textile materials may disturb recovery.

### 16.3.2
### Re-granulation

Nonwoven waste being available from selected synthetic fibrous materials (pure or polymer-sorted), it can be processed to fibres in one spinning process. The basic prerequisite is sorted material so currently, this process is limited to production waste [2–4].

All nonwoven waste from thermoplastic fibres such as polyethylene, polypropylene, polyamide, polyester etc. can be processed on agglomeration plants so as to make free-flowing granulates. The granulate can also be used to produce fibres (generally, for lower-value application). Important characteristics for the workability of the granulates are sufficient melt viscosity, bulk density and flowability.

Adding re-granulate from nonwoven waste to achieve the new product re-presents an ideal case. Mostly, this is impossible for different reasons (fibrous material moistened by means of finishing substances, nonwoven materials bonded by means of binder agents, nonwoven materials blended with plastic materials so the blended melt needs to be granulated). Here, it is important the best-value application variant is found (e.g. injection moulding, extruding, production of a minor-quality nonwoven from 100% re-granulate) [2, 5].

The processing of nonwoven waste from blends of synthetic (thermoplastic) and natural fibrous materials represents a special case, which is accomplished by means of compacting. Depending on the melting point of the thermoplastic component, up to 60% of the non-melting fibres can be embedded in the agglomerate in a matrix way. Such agglomerates can be made from floor covering waste or from waste of moulded parts, to give a few examples. Among other applications, they can be used as heavy-insulation layers (sprinkled onto or sintered onto the backs of moulded parts or floor covering) or as a powdery binder agent to substitute phenolic resin when producing thermally bonded nonwovens and mats [6].

### 16.3.3
#### Production of textile chips and their application

Nonwoven waste may be made into textile chips. One may cut, mill or shred it [7]. Most preferably, textile chips can be made of edges of material in the place where they occur. Above all, edges of thermally bonded nonwovens, of nonwovens used to produce moulded parts or of coated nonwovens are well suitable for the purpose. Used nonwovens or textile composites for application in the automotive industry have also been examined for their recyclability [8].

For example, the textile chips being available in approximately homogeneous size distribution, a pouring process can be used to make mats of them. The mats can then be made into boards (pressure >100 bar, temperature >180 °C), which may be used as semi-finished material for the production of moulded parts [6].

Textile chips can also be added as auxiliary material to produce textile concrete. This requires to modify the surface of the textile fibre in such a way that it can combine with the concrete to form mineral cement. This is achieved by means of a particular physical process. Textile concrete can be used in a number cases where it is advantageous [9, 10].

### 16.3.4
#### Processing nonwoven waste on KEMAFIL machines

Nonwoven waste in the form of material edges, section bobbins or refuse material can be used in the KEMAFIL process as a valuable textile material for the production of a huge range of cord products [11, 12]. Prerequisites for the processing of the edges are availability on drums, coils or hawsers and large lengths.

In the KEMAFIL process, such rope-like waste is embedded as core material in the centre of a coat of loop threads which is created by means of special tools. Utilizing the nonwoven characteristics (e.g. absorption properties, wrinkle recovery, tenacity), ropes are achieved of diversified functionality and up to 130 mm in diameter for uses in agriculture, industry and the building industry.

In combination with other core materials, the nonwoven edges are used to make irrigation and drainage ropes, sensor lines, welts, verbound protection ropes and wicks.

On a modified KEMAFIL flocculation line, fibres and chips (also granulates) can continuously be processed into ropes [13]. Thus, fields of application for a wide variety of waste structures are received. Flocculation and chipping is preferable with edges of insufficient running lengths as well as with textile waste (tailor clippings) of different materials.

The subsequent treatment of the cord- and rope-like structures as weft threads to produce rough mat structures is found at the Saxon Textile Research Institute. The mats are used as heat-insulating mats, plant-growing carriers and extremely rough embankment grids.

## 16.3.5
### Re-use of nonwoven waste

Re-use is the use of a product no more suitable for the original purpose without any or just small material modification for a new application. The basic idea is to exploit certain functions still available in the worn product for a purpose different from the original one.

One well-proven example is the re-use of textile covers of paper-making machines. Every year, approximately 500,000 m$^2$ of these covers are disposed of in Germany. These are fabrics of reinforced nonwoven materials and woven sieves. They do not rot, are permeable to water and show high strength (mostly more than 60 kN/m). Due to these characteristics, they may be economically used as geotextiles. Preferably, these materials are applied to improve foundations in road construction and civil engineering and to provide mineral sealing when building landfills. With the materials examined, loads of harmful substances resulting from paper-making were so small they could be neglected.

One essential prerequisite to effectively re-use materials is a well-designed system of collection, sorting and processing, via which the transformation is achieved from the waste to a new product. At present, investigation is going on to re-use textile floor covering [14, 15].

## References to Chapter 16

[1] Böttcher P, Gulich B, Schilde W (1995) Reißfasern in Technischen Textilien – Grenzen und Möglichkeiten, Techtextil-Symposium Frankfurt/Main, Vortr No 237
[2] Watzl A (1992) Vom Textilabfall zum Nonwovenprodukt – Nutzen durch Recycling, Melliand Textilber 73, 5: 397–401; 73, 6: 487–495; 73, 7: 561–563
[3] Bacher A (1995) Wirtschaftliches Recycling von thermoplastischen Primärabfällen, ITB Vliesstoffe, Techn Textilien 41, 3: 48–50

[4] Wendelin G (1996) Recycling von Vliesstoffabfällen, INDEX'96, Environment Session, Geneva/CH: 1–8

[5] Watzl A (1994) Rohstoffkreislauf in der Textilindustrie durch Recyclen, Taschenbuch für die Textilind: 320–340, Fachverlag Schiele & Schön, ISBN 3-7949-0566-0

[6] Eisele D (1993) Recycling von Textilien – Vliesstoffen, INDEX'93 Geneva/CH, Sektion 4C Recycling

[7] Böttcher P, Gulich B (1997) Grundlagen des Schneidens und Reißens von Textilabfällen, Freiberger Forschungshefte A 840: 45–59, Technische Universität Bergakademie Freiberg

[8] Kiefer A, Bohnhoff A, Ehrler P, Klingenberger H, Schreiber H (1995) Untersuchungen zur Wiederverwendbarkeit von Altautotextilien in Neufahrzeugen, 34. Internat Chemiefasertag, Dornbirn/A

[9] Anders F, Liebscher U (1995) Verbundstoff aus mineralisierten Reststoffen, Techtextil-Symposium Frankfurt/Main, Vortr No 313

[10] Fuchs H, Buchfeld M, Schollmeyer E, Knittel D (1996) Verwendbarkeit von Textilabfällen als Betonzuschlagstoff, Techn Textilien 39, 2: 88

[11] Arnold R, Bartl A-M, Hufnagl E (1993/94) Herstellung von Kordel- und Banderzeugnissen nach der KEMAFIL-Technologie, Band- und Flechtindustrie 30: 4–10, 76–81, 31: 48–52

[12] Arnold R, Bartl A-M, Hufnagl E (1994) KEMAFIL: universelle Ummantelungstechnologie, Techn Textilien/Technical Textiles 37: T 85

[13] Arnold R, Bartl A-M, Hufnagl E (1996) Recyclingverwertung nach der KEMAFIL-Technologie, Techn Textilien/Technical Textiles 39: 24–28

[14] Böttcher P, Hoy G (1997) Müssen gebrauchte Papiermaschinenbespannungen verbrannt werden? Recycling Magazin 15: 14–15

[15] Böttcher P, Hoy G (1997) Untersuchungen zur ökologisch und wirtschaftlich effektiven Entsorgung von Naßfilzen und Trockensieben, Techn Textilien 8: 178

# Part VI
# Regulations to be observed and processes to test raw materials and nonwoven products

# 17
# General principles
M. MÄGEL, B. BIEBER

## 17.1
## Sampling and statistics

DIN 53 803-1 covers the statistical principles for the sampling of textile materials. This standard forms the basis for evaluating the relevance of the evidence provided by the test results and, in addition to containing terms relating to sampling, also includes a list of the statistical terms and the associated algorithms. Sampling must be carried out in accordance with this standard when carrying out any type of investigation involving the use of random samples to evaluate the piece as a whole.

The choice of the sampling process will depend on the aims of the test and the amount of processing carried out on the material being examined. DIN EN 12 751 contains the specifications for carrying out the sampling process.

In general, when carrying out experiments on textile materials, samples are usually taken from a whole piece of material. The parameters relating to a single characteristic of the sample must enable conclusions to be made relating to the whole piece. The samples must therefore be taken in such a way that the difference between the parameters determined for the sample and those of the whole piece is purely a matter of chance. In other words, the sample taken must be representative of the whole piece.

The chance variations must also be evaluated quantitatively, and it should also be possible to specify regions of confidence for the parameters that have to be determined for the whole piece for a prescribed level of confidence.

The following standards can be used in the sampling and evaluation of measurement results:

- Sampling                                                    DIN 53 803-1: 1991-03
  Statistical principles for simple sectioning

- Sampling                                                    DIN EN 12751: 1999-10
  Practical execution of the procedure

- Statistical evaluation                                      DIN 53 804-1: 1981-09
  Measurable (continuous) parameters

- Statistical evaluation                                    DIN 53 804-3: 1982-01
  Ordinal characteristics
  (used only when making evaluations and allocating points)


## 17.2
## Testing climate

The characteristics of textile materials are greatly affected by moisture and temperature. In order to guarantee the reproducibility and comparability of the textile/physical parameters, the experiments must be carried out on samples having a specified initial condition under normal climatic conditions. DIN EN 20 139 contains specifications covering this.


## 17.3
## Standards and specifications

A number of national and international standards organizations are involved in drawing up standards relating to usage and testing on the basis of the latest state of technology. Examples of such organizations include:

| | |
|---|---|
| DIN Deutsches Institut für Normung e.V. | (DIN Standards) |
| BSI British Standards Institution | (BS Standards) |
| AFNOR Association Française de Normalisation | (NF Standards) |
| CEN Europäisches Institut für Normung | (EN Standards) |
| ISO International Organization for Standardization | (ISO Standards) |
| ASTM American Society for Testing and Materials | (ASTM Standards) |

EDANA also provides its members with recommendations for carrying out tests. The addresses of the various standards institutes are included in Appendix 1. EDANA's recommendations are included in Appendix 2. These can be obtained from EDANA.

In the interests of globalization on the European market, European standards are becoming more and more important. In accordance with contractual regulations, the EN standards can be incorporated within the national standards systems (for example, the standard would then be designated DIN EN...). This is not compulsory for ISO standards. According to the Vienna Convention signed between CEN and ISO, the various standards can be assimilated within the various different systems (DIN EN ISO...).

Basically speaking, the use of standards is quite voluntary. The exception to this is if the standards are incorporated in legal documents. Since the standards are drawn up by 'interested parties', i.e. technologists, they correspond to the latest state-of-the-art and, in principle, it is advisable to use them.

The standards applicable to nonwovens are listed in Appendix 3. Chapter 18 contains only the main tests used for testing nonwovens and, in some cases, a comparison is also made with processes and methods for testing other textile substrates.

## Appendix 1

Addresses of the standards institutes:

**DIN**
Deutsches Institut für Normung e.V.                    D-10772 Berlin

**BSI**
British Standards Institution                    389, Chiswick High Road
                                                 London, W4 4AL, UK

**AFNOR**
Association Française de Normalisation           Tour Europe, Cedex 7
                                                 F-92049 Paris la Défense

**CEN**
Europäisches Institut für Normung                Rue de Stassart 36
                                                 B-1050 Bruxelles

**ISO**
International Organization for Standardization    1 rue de Varembé 56
                                                  CH-1211 Genève 20

**ASTM**
American Society for Testing and Materials       1916 Race Street
                                                 Philadelphia, PA 19103-1187,
                                                 USA

**EDANA**
European Disposables and Nonwovens               157 avenue Eugène Plasky,
Association                                      Bte 4, B-1030 Bruxelles

## Appendix 2

Test methods recommended by EDANA: ERT, April 1999

INDEX: Nonwovens

| | | | |
|---|---|---|---|
| 0.0-89 | definition | 152.0-99 | run-off |
| 1.3-99 | vocabulary | 160.0-89 | wet barrier-hydrostatic head |
| 10.3-99 | absorption | 170.0-89 | wet barrier-mason jar |
| 20.2-89 | tensile strength | 180.0-89 | bacterial filtration efficiency |
| 30.5-99 | thickness | 190.0-89 | dry bacterial penetration |
| 40.3-90 | mass per unit area | 200.0-89 | wet bacterial penetration |
| 50.5-99 | bending length | 210.1-99 | free formaldehyde I |
| 60.2-99 | conditioning | 211.1-99 | free formaldehyde II |
| 70.4-99 | tear resistance | | (under stressed conditions) |
| 80.3-99 | burst | 212.0-96 | free formaldehyde III |
| 90.4-99 | drape | | (determination by HPLC) |
| 100.1-78 | brightness | 213.0-99 | free formaldehyde IV |
| 110.1.78 | opacity | | (in processing) |
| 120.1-80 | repellency | 220.0-96 | linting – dry state |
| 130.2-89 | sampling | 230.0-99 | demand absorbency |
| 140.2-99 | air permeability | 300.0-84 [1] | useful method surface |
| 150.4-99 | liquid strike-through time | | linting |
| 151.2-99 | coverstock-wetback | | |

Related products

**Superabsorbent materials – Polyacrylate superabsorbent powders**

| | | | |
|---|---|---|---|
| 400.1-99 | pH | 442.1-99 | absorbency against pressure |
| 410.1-99 | residual monomers | 450.1-99 | flow rate |
| 420.1-99 | particle size distribution | 460.1-99 | density |
| 430.1-99 | moisture content | 470.1.99 | extractables |
| 440.1-99 | free swell capacity | 481.1-99 | respirable particles |
| 441.1-99 | centrifuge retention capacity | 490.1-99 | dust |

1) No. 160-84 of the first issue
English terms are used in all the test procedures published by EDANA

**Appendix 3**

Selected standards/specifications for nonwovens raw materials and nonwovens, and selected processes for making comparisons with other products, situation as at 01/1999

| Standard | Title |
|---|---|
| DIN EN 71-2: 1994-01 | Safety of toys – Part 2: flammability |
| ISO 105-B02: 1994-09 | Textiles – Tests for colour fastness – Part B02: Colour fastness to artificial light: Xenon arc fading lamp test |
| DIN EN ISO 105-B03: 1997-05 | Textiles – Tests for colour fastness – Part B03: Colour fastness to weathering: Outdoor exposure |
| DIN EN ISO 105-B04: 1997-05 | Textiles – Tests for colour fastness – Part B04: Colour fastness to artificial weathering: Xenon arc fading lamp |
| DIN EN ISO 105-B05: 1995-12 | Textiles – Tests for colour fastness – Part B05: detection and assessment of photochromism |
| DIN EN ISO 105-C06: 1997-05 | Textiles – Tests for colour fastness – Part C06: Colour fastness to domestic and commercial laundering |
| DIN EN ISO 105-D01: 1995-04 | Textiles – Tests for colour fastness – Part D01: colour fastness to dry cleaning |
| DIN EN ISO 10-D02: 1995-12 | Textiles –Tests for colour fastness – Part D02: colour fastness to rubbing: organic solvents |
| DIN EN ISO 105-E01: 1996-08 | Textiles – Tests for colour fastness – Part E01: Colour fastness to water |
| DIN EN ISO 105-E02: 1996-08 | Textiles – Tests for colour fastness – Part E02: Colour fastness to sea water |
| DIN EN ISO 105-E03: 1996-10 | Textiles – Tests for colour fastness – Part E03: Colour fastness to chlorinated water (swimming-pool water) |
| DIN EN ISO 105-E04: 1996-08 | Textiles – Tests for colour fastness – Part E04: Colour fastness to perspiration |

| Standard | Title |
|---|---|
| DIN EN ISO 105-E05: 1997-05 | Textiles – Tests for colour fastness – Part E05: colour fastness to spotting: acid |
| DIN EN ISO 105-E06: 1997-05 | Textiles – Tests for colour fastness – Part E06: colour fastness to spotting: alkali |
| DIN EN ISO 105-E07: 1997-05 | Textiles – Tests for colour fastness – Part E07: colour fastness to spotting: water |
| DIN EN ISO 105-E08: 1996-10 | Textiles – Tests for colour fastness – Part E08: Colour fastness to hot water |
| DIN EN ISO 105-N02: 1995-05 | Textiles – Tests for colour fastness – Part N02: colour fastness to bleaching – Peroxide |
| DIN EN ISO 105-N03: 1995-05 | Textiles – Tests for colour fastness – Part N03: colour fastness to bleaching – Sodium chlorite (mild) |
| DIN EN ISO 105-N04: 1995-05 | Textiles – Tests for colour fastness – Part N04: colour fastness to bleaching – Sodium chlorite (severe) |
| DIN EN ISO 105-P01: 1995-04 | Textiles – Tests for colour fastness – Part P01: colour fastness to dry heat (excluding pressing) |
| DIN EN ISO 105-X05: 1997-05 | Textiles – Tests for colour fastness – Part X05: Colour fastness to organic solvents |
| DIN EN ISO 105-X06: 1997-05 | Textiles – Tests for colour fastness – Part X06: Colour fastness to soda boiling |
| DIN EN ISO 105-X11: 1996-10 | Textiles – Tests for colour fastness – Part X11: Colour fastness to hot pressing |
| DIN EN ISO 105-X12: 1995-06 | Textiles – Tests for colour fastness – Part X12: Colour fastness to rubbing |
| DIN EN ISO 186: 1996-02 | Paper and board – Sampling to determine average quality |
| EN ISO 186: 1996-01 | Paper and board – Sampling to determine average quality |
| ISO 186: 1994-11 | Paper and board – Sampling to determine average quality |

| Standard | Title |
|---|---|
| DIN V ENV 343: 1998-04 | Protective clothing – Protection against foul weather |
| DIN EN 348: 1992-11 | Protective clothing – Test method – Determination of behaviour of materials on impact of small splashes of molten metal |
| DIN EN 366: 1993-5 | Protective clothing – Protection against heat and fire – Method of test: evaluation of materials and material assemblies when exposed to a source of radiant heat |
| DIN EN 367: 1992-11 | Protective clothing – protection against heat and flames – Test method: determining of the heat transmission on exposure to flame |
| DIN EN 368: 1993-01 | Protective clothing for use against liquid chemicals – Test method: resistance of materials to penetration by liquids |
| DIN EN 369: 1993-04 | Protective clothing – Protection against liquid chemicals – Test method: resistance of materials to permeation by liquids |
| DIN EN 373: 1993-04 | Protective clothing – Assessment of resistance of materials to molten metal splash |
| DIN EN 388: 1994-08 | Protective gloves against mechanical risks |
| DIN EN 469: 1996-01 | Protective clothing for firefighters – Requirements and test methods for protective clothing for firefighting |
| DIN EN 530: 1995-01 | Abrasion resistance of protective clothing material – Test methods |
| DIN EN 531: 1995-04 | Protective clothing for workers exposed to heat |
| DIN EN 532: 1995-01 | Protective clothing – Protection against heat and flame – Method of test for limited flame spread |
| DIN EN 533: 1997-02 | Protective clothing – Protection against heat and flame – Limited flame spread materials and material assemblies |

| Standard | Title |
| --- | --- |
| DIN EN 597-1: 1995-01 | Furniture – Assessment of the ignitability of mattresses and upholstered bed bases – Part 1: Ignition source: Smouldering cigarette |
| DIN EN 597-2: 1995-01 | Furniture – Assessment of the ignitability of mattresses and upholstered bed bases – Part 2: Ignition source: Match flame equivalent |
| DIN EN 702: 1995-01 | Protective clothing – Protection against heat and flame – Test method: Determination of the contact heat transmission through protective clothing or its materials |
| DIN EN 863: 1995-11 | Protective clothing – Mechanical properties – Test method: Puncture resistance |
| DIN EN 918: 1996-02 | Geotextiles and geotextile-related products – Dynamic perforation test (cone drop test) |
| EN 918: 1995-12 | Geotextiles and geotextile-related products – Dynamic perforation test (cone drop test) |
| DIN EN 963: 1995-05 | Geotextiles and geotextile-related products – Sampling and preparation of test specimens |
| EN 963: 1995-03 | Geotextiles and geotextile-related products – Sampling and preparation of test specimens |
| DIN EN 964-1: 1995-05 | Geotextiles and geotextile-related products – Determination of thickness at specified pressures – Part 1: Single layers |
| EN 964-1: 1995-03 | Geotextiles and geotextile-related products – Determination of thickness at specified pressures – Part 1: Single layers |
| DIN EN 965: 1995-05 | Geotextiles and geotextile-related products – Determination of mass per unit area |
| EN 965: 1995-03 | Geotextiles and geotextile-related products – Determination of mass per unit area |
| DIN EN 1021-1: 1994-01 | Furniture – Assessment of the ignitability of upholstered furniture – Part 1: Ignition source: smouldering cigarette |

| Standard | Title |
|---|---|
| DIN EN 1021-2: 1994-01 | Furniture – Assessment of the ignitability of upholstered furniture – Part 2: Ignition source: match flame equivalent |
| DIN EN 1103: 1996-01 | Textiles – Burning behaviour – fabrics for apparel – Detailed procedure to determine the burning behaviour of fabrics for apparel |
| DIN EN 1146: 1997-05 | Respiratory protective devices – Self-contained open-circuit compressed air breathing apparatus incorporating a hood (compressed air escape apparatus with hood) – Requirements, testing, marking |
| DIN EN 1149-1: 1996-01 | Protective clothing – Electrostatic properties – Part 1: Surface resistivity (Test methods and requirements) |
| DIN EN 1149-2: 1997-11 | Protective clothing – Electrostatic properties – Part 2: Test method for measurement of the electrical resistance through a material (vertical resistance) |
| DIN EN 1413: 1998-05 | Textiles – Determination of pH of aqueous extract |
| DIN EN 1773: 1997-03 | Textiles – Fabrics – Determination of width and length |
| DIN V ENV 1897: 1996-03 | Geotextiles and geotextiles-related products – Determination of the compressive creep properties |
| ENV 1897: 1996-01 | Geotextiles and geotextiles-related products – Determination of the compressive creep properties |
| DIN EN ISO 1973: 1995-12 | Textiles – Determination of linear density – Gravimetric method and vibroscope method |
| DIN EN ISO 3175: 1995-10 | Textiles – Evaluation of stability to machine dry-cleaning |
| DIN EN ISO 3759: 1995-04 | Textiles – Preparation, marking and measuring of fabric specimens and garments in tests for determination of dimensional change |

| Standard | Title |
| --- | --- |
| DIN 4102-1: 1998-05 | Fire behaviour of building materials and building components – Part 1: Building materials – Concepts, requirements and tests |
| DIN EN ISO 5079: 1996-02 | Textiles – Fibres – Determination of breaking force and elongation at break of individual fibres |
| ISO 5081: 1997-03 | Textiles – Woven fabrics – Determination of breaking strength and elongation (Strip method) |
| DIN EN ISO 6941: 1995-04 | Textile fabrics – Burning behaviour – Measurement of flame spread properties of vertically oriented specimens |
| DIN EN ISO 9073-2: 1997-02 | Textiles – Test methods for nonwovens – Part 2: Determination of thickness |
| ISO 9073-3: 1989-07 | Textiles – Test methods for nonwovens – Part 3: Determination of tensile strength and elongation |
| DIN EN ISO 9073-4: 1997-09 | Textiles – Test methods for nonwovens – Part 4: Determination of tear resistance |
| E DIN EN ISO 9073-7: 1998-02 | Textiles – Test methods for nonwovens – Part 7: Determination of bending length |
| DIN EN ISO 9073-9: 1998-10 | Textiles – Test methods for nonwovens – Part 9: Determination of drape coefficient |
| DIN EN ISO 9237: 1995-12 | Textiles – Determination of the permeability of fabrics to air |
| ISO 9862: 1990-08 | Geotextiles – Sampling and preparation of test specimens |
| DIN EN ISO 9863-2: 1996-10 | Geotextiles and geotextile-related products – Determination of thickness at specified pressures – Part 2: Procedure for determination of thickness of single layers of multilayer products |
| EN ISO 9863-2: 1996-08 | Geotextiles and geotextile-related products – Determination of thickness at specified pressures – Part 2: Procedure for determination of thickness of single layers of multilayer products |

| Standard | Title |
|---|---|
| ISO 9863-2: 1996-08 | Geotextiles and geotextile-related products – Determination of thickness at specified pressures – Part 2: Procedure for determination of thickness of single layers of multilayer products |
| ISO 9864: 1990-09 | Geotextiles – Determination of mass per unit area |
| ISO 10318: 1990-11 | Geotextiles – Vocabulary |
| DIN EN ISO 10319: 1996-06 | Geotextiles – Wide-width tensile test |
| EN ISO 10319: 1996-05 | Geotextiles – Wide-width tensile test |
| ISO 10319: 1993-04 | Geotextiles – Wide-width tensile test |
| DIN EN ISO 10320: 1999-04 | Geotextiles and geotextile-related products – Identification on site |
| EN ISO 10320: 1999-02 | Geotextiles and geotextile-related products – Identification on site |
| ISO 10320: 1999-02 | Geotextiles and geotextile-related products – Identification on site |
| DIN EN ISO 10321: 1996-06 | Geotextiles – Tensile test for joints/seams by wide-width method |
| EN ISO 10321: 1996-05 | Geotextiles – Tensile test for joints/seams by wide-width method |
| ISO 10321: 1992-12 | Geotextiles – Tensile test for joints/seams by wide-width method |
| DIN V ENV ISO 10722-1: 1998-05 | Geotextiles and geotextile-related products – Procedure for simulation damage during installation – Part 1: Installation in granular materials |
| ENV ISO 10722-1: 1998-03 | Geotextiles and geotextile-related products – Procedure for simulation damage during installation – Part 1: Installation in granular materials |
| ISO/TR 10722-1: 1998-03 | Geotextiles and geotextile-related products – Procedure for simulation damage during installation – Part 1: Installation in granular materials |

| Standard | Title |
|---|---|
| EN ISO 11058: 1999-02 | Geotextiles and geotextile-related products – Determination of water permeability characteristics normal to the plane, without load |
| ISO 11058: 1999-02 | Geotextiles and geotextile-related products – Determination of water permeability characteristics normal to the plane, without load |
| E DIN EN 12040: 1995-09 | Geotextiles and geotextile-related products – Determination of water permeability characteristics normal to their plane without load |
| prEN 12040: 1995-07 | Geotextiles and geotextile-related products – Determination of water permeability characteristics normal to their plane without load |
| DIN V ENV 12224: 1996-12 | Geotextiles and geotextile-related products – Determination of the resistance to weathering |
| ENV 12224: 1996-10 | Geotextiles and geotextile-related products – Determination of the resistance to weathering |
| DIN V ENV 12225: 1996-12 | Geotextiles and geotextile-related products – Method for determining the microbiological resistance by a soil burial test |
| ENV 12225: 1996-10 | Geotextiles and geotextile-related products – Method for determining the microbiological resistance by a soil burial test |
| DIN V ENV 12226: 1996-12 | Geotextiles and geotextile-related products – General tests for evaluation following durability testing |
| ENV 12226: 1996-10 | Geotextiles and geotextile-related products – General tests for evaluation following durability testing |
| DIN EN ISO 12236: 1996-04 | Geotextiles and geotextile-related products – Static puncture test (CBR test) |
| EN ISO 12236: 1996-02 | Geotextiles and geotextile-related products – Static puncture test (CBR test) |
| ISO 12236: 1996-10 | Geotextiles and geotextile-related products – Static puncture test (CBR test) |

| Standard | Title |
|---|---|
| DIN V ENV 12447: 1997-11 | Geotextiles and geotextile-related products – Screening test method for determining the resistance to hydrolysis |
| ENV 12447: 1997-10 | Geotextiles and geotextile-related products – Screening test method for determining the resistance to hydrolysis |
| E DIN EN ISO 12956: 1996-02 | Geotextiles and geotextile-related products – Determination of the characteristic opening size |
| EN ISO 12956: 1999-02 | Geotextiles and geotextile-related products – Determination of the characteristic opening size |
| ISO 12956: 1999-02 | Geotextiles and geotextile-related products – Determination of the characteristic opening size |
| DIN EN ISO 12947-2: 1999-04 | Textiles – Determination of the abrasion resistance of fabrics by the Martindale method – Part 2: Determination of specimen breakdown |
| DIN EN ISO 12947-3: 1999-04 | Textiles – Determination of the abrasion resistance of fabrics by the Martindale method – Part 3: Determination of mass loss |
| DIN EN ISO 12947-4: 1999-04 | Textiles – Determination of the abrasion resistance of fabrics by the Martindale method – Part 4: Assessment of appearance change |
| E DIN EN ISO 12957-1: 1998-04 | Geotextiles and geotextile-related products – Determination of friction characteristics – Part 1: Direct shear test |
| prEN ISO 12957-1: 1997-12 | Geotextiles and geotextile-related products – Determination of friction characteristics – Part 1: Direct shear test |
| ISO/DIS 12957-1: 1997-12 | Geotextiles and geotextile-related products – Determination of friction characteristics – Part 1: Direct shear test |
| E DIN EN ISO 12957-2: 1998-04 | Geotextiles and geotextile-related products – Determination of friction characteristics – Part 2: Inclined plane test |

| Standard | Title |
|---|---|
| prEN ISO 12957-2: 1997-12 | Geotextiles and geotextile-related products – Determination of friction characteristics – Part 2: Inclined plane test |
| ISO/DIS 12957-2: 1997-12 | Geotextiles and geotextile-related products – Determination of friction characteristics – Part 2: Inclined plane test |
| E DIN EN ISO 12958: 1996-02 | Geotextiles and geotextile-related products – Determination of water flow capacity in their plane |
| EN ISO 12958: 1999-02 | Geotextiles and geotextile-related products – Determination of water flow capacity in their plane |
| ISO 12958: 1999-02 | Geotextiles and geotextile-related products – Determination of water flow capacity in their plane |
| DIN V ENV ISO 12960: 1999-02 | Geotextiles and geotextile-related products – Screening test method for determining the resistance to liquids |
| ENV ISO 12960: 1998-11 | Geotextiles and geotextile-related products – Screening test method for determining the resistance to liquids |
| ISO/TR 12960: 1998-11 | Geotextiles and geotextile-related products – Screening test method for determining the resistance to liquids |
| DIN EN ISO 13427: 1998-10 | Geotextiles and geotextile-related products – Abrasion damage simulation (Sliding block test) |
| EN ISO 13427: 1998-08 | Geotextiles and geotextile-related products – Abrasion damage simulation (Sliding block test) |
| ISO 13427: 1998-08 | Geotextiles and geotextile-related products – Abrasion damage simulation (Sliding block test) |
| E DIN EN ISO 13431: 1996-01 | Geotextiles and geotextile-related products – Determination of tensile creep and creep rupture behaviour |

| Standard | Title |
|---|---|
| prEN ISO 13431: 1998-02 | Geotextiles and geotextile-related products – Determination of tensile creep and creep rupture behaviour |
| ISO/FDIS 13431: 1998-02 | Geotextiles and geotextile-related products – Determination of tensile creep and creep rupture behaviour |
| ISO/DIS 13433: 1996-03 | Geotextiles and geotextile-related products – Dynamic perforation test (Cone-drop test) |
| DIN EN ISO 13437: 1998-10 | Geotextiles and geotextile-related products – Method for installing and extracting samples in soil, and testing specimens in laboratory |
| EN ISO 13437: 1998-08 | Geotextiles and geotextile-related products – Method for installing and extracting samples in soil, and testing specimens in laboratory |
| ISO 13437: 1998-08 | Geotextiles and geotextile-related products – Method for installing and extracting samples in soil, and testing specimens in laboratory |
| ENV ISO 13438: 1999-02 | Geotextiles and geotextile-related products – Screening test method for determining the resistance to oxidation |
| ISO/TR 13438: 1999-02 | Geotextiles and geotextile-related products – Screening test method for determining the resistance to oxidation |
| E DIN EN ISO 13938-1: 1995-04 | Textiles – Bursting properties of fabrics – Part 1: Hydraulic method for determination of bursting strength and bursting distension |
| E DIN EN ISO 13938-2: 1995-04 | Textiles – Bursting properties of fabrics – Part 2: Pneumatic method for determination of bursting strength and bursting distension |
| E DIN 18200: 1998-12 | Assessment of conformity for construction products – Verification of construction products by certification body |
| DIN 18200: 1986-12 | Inspection of construction materials, structural members and types of construction – General principles |

| Standard | Title |
|---|---|
| DIN EN 20105-C01: 1993-03 | Textiles – Tests for colour fastness – Part C01: Colour fastness to washing – Test 1 |
| DIN EN 20105-C02: 1993-03 | Textiles – Tests for colour fastness – Part C02: Colour fastness to washing – Test 2 |
| DIN EN 20105-C03: 1993-03 | Textiles – Tests for colour fastness – Part C03: Colour fastness to washing – Test 3 |
| DIN EN 20105-C04: 1993-03 | Textiles – Tests for colour fastness – Part C04: colour fastness to washing – Test 4 |
| DIN EN 20105-C05: 1993-03 | Textiles – Tests for colour fastness – Part C05: Colour fastness to washing – Test 5 |
| DIN EN 20105-N01: 1995-03 | Textiles – Tests for colour fastness – Part N01: Colour fastness to bleaching: Hypochlorite |
| DIN EN 20139: 1992-09 | Textiles – Standard atmospheres for conditioning and testing |
| DIN EN 24920: 1992-08 | Textiles – Determination of resistance to surface wetting (spray test) of fabrics |
| DIN EN 25077: 1994-02 | Textiles – Determination of dimensional change in washing and drying |
| DIN EN 26330: 1994-02 | Textiles – Domestic washing and drying procedures for textile testing |
| DIN EN 29073-1: 1992-08 | Textiles – Test method for nonwovens – Part 1: Determination of mass per unit area |
| DIN EN 29073-3: 1992-08 | Textiles – Test method for nonwovens – Part 3: Determination of tensile strength and elongation |
| EN 29073-3: 1992-06 | Textiles – Test method for nonwovens – Part 3: Determination of tensile strength and elongation |
| prEN 30318: 1992-08 | Geotextiles – Vocabulary |
| DIN EN 30320: 1993-07 | Geotextiles – Identification on site |
| EN 30320: 1993-06 | Geotextiles – Identification on site |

| Standard | Title |
|----------|-------|
| prEN 33934-1: 1994-07 | Textiles – Tensile properties of fabrics – Part 1: Determination of maximum force and elongation at maximum force – Strip method |
| DIN 53800: 1979-02 | Testing of textiles – Determination of dry mass by dessication in hot air |
| DIN 53803-1: 1991-03 | Sampling – Statistical basis – One-way layout |
| DIN 53803-2: 1994-03 | Sampling – Practical execution |
| DIN 53808-1: 1982-02 | Testing of textiles – Determination of length of fibres by measuring of individual fibres |
| DIN 53811: 1970-07 | Testing of textiles – Determination of the diameter of fibres from longitudinal view by microscope projection |
| DIN 53814: 1974-10 | Testing of textiles – Determination of water retention power of fibres and yarn cuttings |
| DIN 53835-13: 1983-11 | Testing of textiles – Determination of the elastic behaviour of textile fabrics by a single application of tensile load between constant extension limits |
| DIN 53843-2: 1988-03 | Testing of textiles – Loop tensile test for staple fibres |
| DIN 53855-3: 1979-01 | Testing of textiles – Determination of thickness of textile fabrics – Floor coverings |
| DIN 53857-1: 1979-09 | Testing of textiles – Simple tensile test on strips of textile fabrics, woven fabrics and ribbons |
| DIN 53859-2: 1979-01 | Testing of textiles – Tear growth testing of textile fabrics – Leg tear growth test |
| DIN 53859-4: 1977-02 | Testing of textiles – Tear growth test on textile fabrics, nonwoven textiles |
| DIN 53859-5: 1992-12 | Testing of textiles – Tear growth test on textile fabrics – Trapezoid test |
| DIN 53861-1: 1992-11 | Testing of textiles – Vaulting test and bursting test – Definitions of terms |

| Standard | Title |
|---|---|
| DIN 53861-2: 1978-03 | Testing of textiles – Vaulting test and bursting test, method of test |
| DIN 53861-3: 1970-08 | Testing of textiles – Vaulting test and bursting test, tables for the evaluation of tests |
| DIN 53863-1: 1960-12 | Testing of textiles – Abrasion test methods for textile planar fabrics, principles |
| DIN 53863-2: 1979-02 | Testing of textiles – Abrasion test methods for textile fabrics, rotary abrasion test |
| DIN 53863-4: 1992-11 | Testing of textiles – Abrasion testing of textile fabrics – Martindale abrasion test |
| DIN 53864: 1978-08 | Testing of textiles – Determination of the bending strength, method according to Schlenker |
| DIN 53865: 1998-07 | Testing of textiles – Determination of fibre migration tendency – Tumble-method |
| DIN 53885: 1998-12 | Textiles – Determination of compression of textiles and textile products |
| DIN 53890: 1972-01 | Testing of textiles – Determination of the crease recovery angle of area-measured textiles – Method using an air-dry specimen with horizontal fold and erected free limb |
| DIN 53894-1: 1980-04 | Testing of textiles – Determination of dimensional change of textile fabrics, ironing with a moist ironing cloth on ironing presses |
| DIN 53894-2: 1979-02 | Testing of textiles – Determination of dimensional change of textile fabrics, steaming on ironing machines |
| DIN 53895: 1980-08 | Testing of textiles – Determination of the self-smoothing behaviour of textile fabrics after laundering and drying |
| DIN 53923: 1978-01 | Testing of textiles – Determination of water absorption of textile fabrics |
| DIN 53924: 1997-03 | Testing of textiles – Velocity of soaking water of textile fabrics (Method by determining the rising height) |

| Standard | Title |
|---|---|
| DIN 53933-1: 1992-04 | Testing of textiles – Determination of the resistance of cellulose textiles against micro-organisms (resistance to bacteria and fungi of soil) – Identification of rotting retardant finishing |
| DIN 54003: 1983-08 | Testing of colour fastness of textiles – Determination of colour fastness of dyeings and prints of daylight |
| DIN 54004: 1983-08 | Testing of colour fastness of textiles – Determination of colour fastness to light of dyeings and prints: Xenon arc fading lamp test |
| DIN 54005: 1983-11 | Testing of colour fastness of textiles – Determination of colour fastness of dyeings and prints to water (mild) |
| DIN 54015: 1977-08 | Testing of colour fastness of textiles – Determination of colour fastness of dyeings and prints to washing in presence of peroxide |
| DIN 54016: 1977-08 | Testing of colour fastness of textiles – Determination of colour fastness of dyeings and prints to washing in presence of hypochlorite |
| DIN 54029: 1984-08 | Testing of colour fastness of textiles – Determination of colour fastness of dyeings and prints to brightening |
| DIN 54034: 1984-05 | Testing of colour fastness of textiles – Determination of colour fastness of dyeings and prints to bleaching – hypochlorite (mild) |
| DIN 54046: 1985-06 | Testing of colour fastness of textiles – Determination of colour fastness to chlorination |
| DIN 54200: 1974-06 | Testing of textiles – Quantitative analysis of fibre mixtures by means of solvent method, basis and field of application |
| DIN 54201: 1975-08 | Testing of textiles – Quantitative analysis of fibre mixtures, directions for the work |
| DIN 54204: 1975-08 | Testing of textiles – Quantitative analysis of binary mixtures, wool with other fibres, potassium hydroxide solution method |

| Standard | Title |
|---|---|
| DIN 54205: 1975-12 | Testing of textiles – Quantitative analysis of binary mixtures, natural or regenerated cellulose fibres with polyester fibres, sulphuric acid method |
| DIN 54206: 1975-08 | Testing of textiles – Quantitative analysis of binary mixtures, protein fibres with other fibres, hypochlorite method |
| DIN 54208: 1984-04 | Testing of textiles – Quantitative analysis of binary mixtures – Regenerated cellulose fibres with other fibres, especially cotton – Formic acid/zinc chloride method |
| DIN 54209: 1975-08 | Testing of textiles – Quantitative analysis of binary mixtures, degummed mulberry silk with wool, formic acid/zinc chloride method |
| DIN 54210: 1975-08 | Testing of textiles – Quantitative analysis of binary mixtures, acetate fibres with other fibres, acetone method |
| DIN 54211: 1975-08 | Testing of textiles – Quantitative analysis of binary mixtures, triacetate fibres with other fibres, dichlormethane (methylene chloride) method |
| DIN 54212: 1975-08 | Testing of textiles – Quantitative analysis of binary mixtures, casein fibres with other fibres, trypsin method |
| DIN 54215: 1977-12 | Testing of textiles – Quantitative analysis of binary mixtures, polypropylene fibres with other fibres, xylol method |
| DIN 54216: 1975-08 | Testing of textiles – Quantitative analysis of binary mixtures, polyvinyl chloride fibres with other fibres, carbon disulphide/acetone method |
| DIN 54217: 1975-08 | Testing of textiles – Quantitative analysis of binary mixtures, acrylic, modacrylic and certain polyvinyl chloride fibres with other fibres, dimethyl formamide method |

| Standard | Title |
|---|---|
| DIN 54218: 1975-08 | Testing of textiles – Quantitative analysis of binary mixtures, acetate fibres with polyvinyl chloride fibres, acetic acid method |
| DIN 54220: 1975-08 | Testing of textiles – Quantitative analysis of binary mixtures, polyamide 66 or polyamide 6 fibres with other fibres, formic acid method |
| DIN 54221: 1975-08 | Testing of textiles – Quantitative analysis of binary mixtures, polyamide 66 or polyamide 6 fibres with other fibres, hydrochloric acid method |
| DIN 54270-1: 1976-09 | Testing of textiles – Determination of the limit-viscosity of celluloses, principles |
| DIN 54270-2: 1977-08 | Testing of textiles – Determination of the limit-viscosity of celluloses, Cuen-procedure |
| DIN 54270-3: 1977-08 | Testing of textiles – Determination of the limit-viscosity of celluloses, EWNN-procedure |
| DIN 54270-4: 1977-08 | Testing of textiles – Determination of the limit-viscosity of celluloses, nitrate procedure |
| DIN 54275: 1977-12 | Testing of textiles – Determination of the pH value of fibre materials, method of extrapolation |
| DIN 54278-1: 1995-10 | Testing of textiles – Coatings and attendant materials – Part 1: Determination of materials soluble in organic solvents |
| DIN 54278-2: 1978-02 | Testing of textiles – Coatings and attendant materials – Determination of condensation products which can be stripped by means of hydrochloric acid |
| DIN 54279: 1977-01 | Testing of textiles – Determination of the solubility of wool in urea/bisulphite solution |
| DIN 54280: 1977-04 | Testing of textiles – Determination of acidity of wool |
| DIN 54281: 1971-05 | Testing of textiles – Determination of alkali solubility of wool |

| Standard | Title |
|---|---|
| DIN 54285: 1981-02 | Testing of textiles – Determination of sizing content |
| DIN 54286: 1975-10 | Testing of textiles – Determination of the cysteic acid content in wool hydrolysates |
| DIN 54287: 1977-12 | Testing of textiles – Determination of alkali content of wool |
| DIN 54301: 1977-07 | Testing of textiles – Determination of the needle tearing out resistance of nonwoven textiles |
| DIN 54302: 1977-11 | Testing of textiles – Determination of the recovery of nonwoven textiles |
| DIN 54303: 1991-02 | Testing of textiles – Dry-cleaning of nonwoven fabrics – nonwoven interlinings and nonwoven wadding |
| DIN 54304: 1991-02 | Testing of textiles – Washing of nonwoven fabrics – nonwoven interlinings and nonwoven wadding |
| DIN 54305: 1976-02 | Testing of textiles – Determination of the compression elastic behaviour of fibrous webs and nonwovens |
| DIN 54306: 1979-02 | Testing of textiles – Determination of the drape of textile fabrics |
| DIN 54310: 1980-07 | Testing of textiles – Delamination of fusible interlinings from upper fabrics, mechanical delamination test |
| DIN 54311: 1982-07 | Testing of textiles – Determination of dimensional stability of fusible interlinings |
| DIN 54332: 1975-02 | Testing of textiles – Determination of the burning behaviour of textile floor coverings |
| DIN 54333-1: 1981-12 | Testing of textiles – Determination of burning behaviour – horizontal method – Ignition at the edge of the specimen |
| DIN 54345-1: 1992-02 | Testing of textiles – Electrostatic behaviour – Determination of electrical resistance |

| Standard | Title |
|---|---|
| DIN 54345-2: 1991-09 | Testing of textiles – Electrostatic behaviour – Determination of the charge of textile floor coverings by walking test |
| DIN 54345-4: 1985-07 | Testing of textiles – Electrostatic behaviour – Determination of electrostatic charge of textile fabrics |
| DIN 54345-5: 1985-07 | Testing of textiles – Electrostatic behaviour – Determination of electrical resistance of strips of textile fabrics |
| DIN 54345-6: 1992-02 | Testing of textiles – Electrostatic behaviour – Determination of the electrical resistance of textile floor coverings |
| DIN V 60500-1: 1999-06 | Geotextiles and geotextile related products – Determination of the resistance against damage during installation (Pyramid drop test) |
| E DIN 60500-4: 1997-02 | Testing of geotextiles – Part 4: Determination of permeability of geotextiles normal to their planes under load of constant hydraulic high difference |
| E DIN 60500-8: 1997-03 | Testing of geotextiles – Part 8: Determination of permeability of geotextiles with radial flowthrough within the geotextile plane |
| E DIN 61010-1: 1995-02 | Textiles – Upholstery fabrics for general use in living areas – Part 1: Minimum requirements and tests |
| DIN 66081: 1989-05 | Classification of burning behaviour of textile products – Textile floor coverings |
| DIN 66083: 1997-02 | Classification of burning behaviour of textile products – Textile fabrics for working clothing |
| DIN 75200: 1980-09 | Determination of burning behaviour of interior materials in motor vehicles |
| DIN 75202: 1988-12 | Determination of colour fastness of interior materials in motor vehicles, xenon arc lamp test |
| VDI 3926, Sheet 1: 1994-12 | Testing of filter media for cleanable filters |

# 18
# Testing processes

## 18.1
### Raw materials for nonwovens
M. Mägel, B. Bieber, T. Pfüller

### 18.1.1
### Fibres

The main processes for testing and evaluating the fibre characteristics include analyzing the fibrous materials, testing the fibre linear density, length and crimp, and determining the fibre strength properties.

When deciding on which fibre is most suitable for a particular application, the fibre properties have to be evaluated in terms of their effect on the final quality of the nonwoven fabric. This evaluation must be carried out as a function of the product type, and the threshold limits for the fibre parameters must be specified.

### Analysis of fibrous materials
Various processes and methods have been developed for use in the qualitative evaluation of the properties of fibrous materials.

These include:

– microscopic analysis of the lengthwise view and the fibre cross-sections
– microchemical swelling and dissolving experiments
– determination of the melting point
– staining
– infrared spectroscopy

The following sources may be useful when carrying out the investigations, but this list does not claim to be exhaustive.

| | | Author |
|---|---|---|
| Mikroskopisch-Chemische Bestimmungsschlüssel für organische und anorganische Chemiefaserstoffe [1] | Microscopic-chemical evaluation principles for organic and inorganic chemical fibres | Bobeth, W. |
| Faserstofftabellen [2] | Fibre tables | Koch, P. A. |
| Rezeptbuch für Faserstoff-Laboratorien [3] | Recipe book for fibre testing laboratories | Koch, P. A. |
| Methoden der qualitativen Faseranalyse [4] | Methods for qualitative fibre analysis | Stratmann, M. |
| Mikroskopie der Faserstoffe [5] | Microscopy of fibrous materials | Koch, P. A. |
| Die Identifizierung von Faserstoffen mittels chemischer Reaktionen [6] | The identification of fibrous materials by means of chemical reactions | Stratmann, M. |
| Faserstofflehre [7] | Handbook of fibrous materials | Group of authors |
| Qualitätsbeurteilungen von Textilien [8] | Quality evaluation of textiles | Mahall, K. |
| IR-Spektroskopie [9] | IR spectroscopy | Günzler, H., Heise, H. M. |

**Determination of linear density**

| Method/procedure: | Determination of linear density, Gravimetric method |
|---|---|
| Standard: | DIN EN ISO 1973: 1995-12 |
| Brief description: | Parallelized fibre bundles are cut to a specified length. 5 fibres are removed from each of 10 fibre bundles. The mass of the resulting fibre bundle, which consists of 50 fibres all having the same length, is determined. The quotient of the mass and the cut length, multiplied by the number of fibres in the bundle, gives the fibre linear density value. The mean value of at least 10 bundles, each consisting of 50 fibres, is calculated arithmetically. The linear density is given in dtex. |
| Remarks/limitations: | The process is only suitable for fibres which can be decrimped easily and which lie parallel during the cutting process. The process cannot be used for testing fibres having a tendency to taper. |

The method is time-consuming, and the test can only be carried out by highly experienced personnel.

| | |
|---|---|
| Method/procedure: | **Determination of linear density, Vibroscope method** |
| Standard: | DIN EN ISO 1973: 1995-12 |
| Brief description: | The linear density of individual fibres is determined by the oscillating principle at a constant test length and constant fibre loading. A brief acoustic vibration is applied, which causes the fibre to oscillate transversely. The fibre linear density is calculated from the resonance frequency of the fibre: |

$$Tt = \frac{F_v}{4f^2 l^2}$$

Tt = linear density (dtex)
$F_v$ = pretensioning force
f  = resonance frequency
l  = test length

| | |
|---|---|
| Remarks/limitations: | The interpretation of the results corresponds to the direct gravimetric determination of linear density. The process takes less time and does not lend itself to subjective intervention. |
| | The method cannot be used for testing hollow fibres, fibres with a tape-shaped cross-section, and fibres which taper over the course of the test length. |

Fig. 18-1 shows an example of a testing arrangement that can be used in fibre testing.

**Fig. 18-1**  Vibroscope and Vibrodyn fibre fineness testers from Lenzing AG, Lenzing, Austria

| Method/procedure: | Determination of linear density, Determination of the diameter of fibres from longitudinal view by microscope projection |
|---|---|
| Standard: | DIN 53 811: 1970-07 |
| Brief description: | The fibre diameter can be determined by the microscopic projection technique. A sample is made up of the fibres that have to be measured, by cutting them to a defined length and arranging them in an embedding material. The fibres are magnified 500 times in the microscopic projection device, and the entire sample is laid out in a meandering configuration. |

The number of fibres that have to be measured for a confidence region width of 2% and a 95% interpretation accuracy is n=600.

The result of the measurement shows the length-dependent, mean fibre diameter (µm). The linear density is calculated on the basis of:

$$Tt = \frac{d^2 \rho \pi}{400}$$

Tt = linear density (dtex)
$\rho$ = density of the fibrous material (g/cm$^3$)
d = diameter (µm)

| Remarks/limitations: | The process is time-consuming and the test personnel has to be highly experienced. It can only be used for testing fibres having a circular cross-section. |
|---|---|

**Testing of fibre length**

| Method/procedure: | Determination of length of fibres by measuring of individual fibres |
|---|---|
| Standard: | DIN 53 808-1: 1982-02 |
| Brief description: | A glass plate in a contrasting colour is smeared with a thin layer of paraffin oil or vaseline. The fibres that are to be measured are picked up with a pair of tweezers, laid carefully onto the glass plate, and straightened to remove the crimp (tweezers technique). The fibres should not be stretched at all. |

The uncurled length is measured using a ruler, and the measured length inserted into a grid, together with the corresponding class width.

The number of fibres coming within the length classes is determined. The average fibre length as a function of the number is calculated as follows:

$$L_a = \frac{\sum(n_j x_j)}{\sum(n_j)}$$

$L_a$ = mean fibre length, as a function of the number
$n_j$ = number of fibres falling within class j
$x_j$ = mean length of the class with the figure j

*Cumulative frequency curve (staple diagram)*
The curve is produced when all the fibres, which are arranged end-to-end, are lined up according to length, and their end points joined together. It is the cumulative curve of the continuously added frequencies. It represents the fibre length distribution graphically.

*Histogram (frequency diagram)*
The histogram shows the fibre length distribution, in which the class frequencies are applied as ordinates above the mean classes as an abscissa.

| | |
|---|---|
| Remarks/limitations: | The length of the uncurled fibres is measured. The number of fibres that has to be measured will depend on the prescribed width of the confidence zone, but should not be less than 300 fibres. The test is very time-consuming. The process is not suitable for samples having a high proportion of short fibres. It can only be used for testing fibres that can be decrimped easily. |
| Method/procedure: | **Determination of length of fibres, Almeter method (Zellweger Uster, formerly Peyer)** |
| Standard: | no standard is available |
| Brief description: | A fibre assembly, which has been arranged end-to-end by means of preparatory devices, is scanned capacitively. By changing the capacity, conclusions can be drawn regarding the mean fibre length in terms of number and mass, and also with regard to the proportion of short and long fibres present. |

*Cumulative frequency curve (staple diagram)*
See individual fibre measurement technique.

*Histogram (frequency diagram)*
The percentage frequency of fibre length classes having a 2.5 mm class width is shown in the histogram.

Remarks/limitations:    The length of the tensionless crimped fibres is determined. The result is therefore not directly comparable to that of individual fibre measurement.
Many more fibres are measured than with the individual fibre measurement process. The process can be used during production monitoring, since less time is needed for carrying out the test. Subjective intervention in the process by the test personnel is minimal.

The USTER® FL 100 fibre straightener and the USTER® AL 100 fibre length measuring device can both be used in practice for carrying out the above test (see Fig. 18-2). The fibre straightener is used before the test to arrange the fibres in the sample end-to-end, and the measuring device is used to analyze the fibre length and length distribution on the basis of IWTO Specification TM-17-85.

**Fig. 18-2**  USTER® FL 100 fibre alignment device and USTER® AL 100 fibre length measuring unit from Zellweger Uster, Uster, Switzerland

**Testing of fibre crimp**

Method/procedure:    **Determination of fibre crimp (e.g. the STFI method)**

Standard:    no standard is available

Brief description:    The fibre is clamped in a take-off device with a fixed and a vertically movable clamp under a pretensioning force of 0.05 mN/tex. The clamping length will depend on the nominal length of the fibres. At least 100 fibres are tested. The following parameters can be determined:

*Number of crimp arcs*
The vertical points of the crimp arcs between the clamps are counted.

*Change in decrimped length (mm)*
Once the fibres have been clamped, the lower clamp is moved downwards. The decrimping process is complete as soon as the fibres appear straight. The length change can be specified accurately to within 0.1 mm.

$$\Delta L_K = L_K - L_{RV}$$

*Crip, removal (%)*

$$\delta = \Delta \frac{L_K}{L_{RV}} \cdot 100\%$$

*Crimp stability (%)*
Once the level of decrimping has been determined, the fibres are subjected to a load of 30 mN/tex for 20 s. The load is then removed for a period of 20 s, after which the fibre is decrimped again.

$$B_K = \frac{\delta_2}{\delta_1} \cdot 100\%$$

$\Delta L$ = change in decrimped length (mm)
$L_K$ = fibre length, decrimped (mm)
$L_{RV}$ = starting length (mm)
$\delta$ = decrimping (%)
$B_K$ = crimp stability (%)
$\delta_2$ = decrimping after a load has been applied (%)
$\delta_1$ = decrimping before a load is applied (%)

Remarks/limitations: The test procedure is time-consuming and requires a certain amount of manual dexterity.

**Testing of fibre strength**

Method/procedure: **Determination of breaking force and elongation at break of individual fibres**

Standard: DIN EN ISO 5079: 1996-02
Brief description: The fibre is held under a linear density-dependent pretension of 1 cN/tex in the clamping grippers of a fibre tensile tester, and stretched at a constant strain rate. The gauge length is 20 mm, and may also be 10 mm by arrangement between the client and tester. The strain rate is 10 or 20 mm/min, as a function of the breaking elongation. The fibre is subjected to the load until it breaks, and the tensile strength, the breaking elongation, and the linear density -dependent tensile strength (cN/tex) are determined. Stress/strain curves can also be plotted. At least 50 individual fibres are tested.

Remarks/limitations:      The process is used all over the world, but is time-consuming. At a gauge length of 10 mm, the fibres must be
at least 28 mm long.

**Method/procedure:**      **Determination of breaking tenacity of flat bundles using
Pressley clamps**

Standard:      DIN ISO 3060: 1994-04 (cotton fibres)
ÖTN 076 (flax fibres)

Brief description:      The parallelized fibres are clamped perpendicular in the
form of a flat bundle between two Pressley clamps with
a spacing of 0 or 3.2 mm. The protruding ends are cut
off. The Pressley clamps are pushed into the guide of
the tensile tester and the fibre bundle is subjected to a
load until it breaks. The test is only valid if all the fibres
in the bundle tear. The tensile strength and the mass of
the torn bundle are calculated.

*Calculation according to ÖTN 076:*

$$F_T = \frac{F_B[cN]}{Tt[dtex]}$$

$F_T$=linear density-related tensile strength of the bundle
    (cN/dtex)
$F_B$=tensile strength of the bundle (cN)
$Tt$ =bundle linear density (dtex)

Remarks/limitations:      The process is used all over the world, but is time-consuming

## 18.1.2
## Granulates

In general, granulates consist of particles of synthetic polymer powders which,
like agglomerates, have no uniform, geometrical shape. The granulates used for
producing nonwovens, especially extruded nonwovens, are macromolecular compounds (see also Section 2.2). They can be studied and characterised using polymer analysis techniques [10].

Qualitative analysis, i.e. evaluating the polymers present, involves:
– analyzing the elements present [11]
– microscopic and microchemical processes (see Section 18.1.1) [1–7]
– infrared spectroscopy [12]

The polymer can be characterized by determining the
– molar mass [13]
– molar mass distribution [14]
– physical and chemical properties [10]

The properties of the granulates can be determined using the following standards:

DIN EN 543: 1995-01 — Adhesives – Determination of apparent density of powder and granule adhesives

DIN EN ISO 11 358: 1997-11 — Plastics – Thermogravimetry (TG) of polymers – General principles

DIN ISO 4324: 1983-12 — Surface active agents – Powders and granules – Measurement of the angle of repose

DIN 53 242-4: 1980-01 — Raw materials for paints and varnishes – Sampling, solid materials

DIN 53 492: 1992-11 — Testing of plastics – Determination of pourability of granular plastics

18.1.3
**Binders**

Binders are used for bonding webs and nonwoven fabrics. Adhesive bonding is used if the fibres and binders are not made from the same polymer. Both solid substances (powders, granulates, fibres) as well as dispersions are used as the binders. The solid substances can be analyzed using the methods described in Sections 18.1.1 and 18.1.2.

The binder is usually characterized by its thermal properties. According to [15], the main parameters are: glass transition temperature; transition from a solid to a thermo-elastic state (softening temperature) flow or melt temperature; transition from a thermoelastic state to a liquid (thermoplastic) statedisintegration temperature; beginning of the irreversible breakdown of the material

The following standards and specifications can be used for evaluating the properties of binders:

DIN EN ISO 11 358: 1997-11 — Plastics – Thermogravimetry (TG) of polymers – General principles

DIN 52 007-1: 1980-12 — Testing of bituminous binders – Determination of viscosity; general principles and evaluation

DIN 52 007-2: 1980-12 — Testing of bituminous binders – Determination of viscosity; measurement by drawn-sphere viscometer

DIN 53 177: 1990-11 — Binders for paints and varnishes – Measurement of the dynamic viscosity of liquid resins of isoceles type according to Ubbelohde

VDG P 70: 1989-04 — Testing of binders – Testing of liquid, acid-curable furan resins

VDG P 75: 1989-04 — Testing of binders – Testing of liquid, acid-curable phenolic resins

VDG P 76: 1989-04 Testing of binders – Testing of solid and liquid novolaks

VDG P 77: 1989-04 Testing of binders – Testing of urethane reactants

## 18.2
## Nonwovens
M. Mägel, B. Bieber, U. Bernstein, C. Lewicki, T. Pfüller

### 18.2.1
### Textile-physical tests

The following list shows a selection of the tests that can be used for determining the textile-physical characteristics of nonwovens:

– determination of mass per unit area
– determination of thickness
– determination of tensile strength and elongation and reference strength/reference elongation
– determination of bending length
– determination of bursting strength and bursting distension
– determination of air permeability
– determination of tear resistance
– determination of drape coefficient
– determination of abrasion resistance

### Determination of mass per unit area

| | |
|---|---|
| **Method/procedure:** | **Test method for nonwovens, determination of mass per unit area** |
| Standard: | DIN EN 29 073-1: 1992-08 |
| Brief description: | The area and mass of a test sample are determined, and the quotient of the mass and area is given in $g/m^2$. <br> Size of test sample: $\geq 500$ cm$^2$ <br> Number of test samples: $\geq 3$ |
| Remarks/limitations: | The sampling regime depends on the specific objective of the test and must be agreed on by the client and the tester. |

*Normal case – principle of statistical sampling:*
The test samples are distributed over the laboratory sample in such a way that the same probability of being tested exists at every location.

*Special case – principle of targeted selection of the test samples:*
The anisotropy of the nonwovens, which is produced during manufacture, enables systematic differences to be determined by arranging the test samples appropriately in the lengthwise and crosswise direction of the sample.

The mass per unit area is the reference value for further, specific, textile/physical parameters.

### Determination of thickness

| | |
|---|---|
| Method/procedure: | **Test method for nonwovens, determination of thickness, method A, B or C** |
| Standard: | DIN EN ISO 9073-2: 1997-02 |
| Brief description: | The thickness is measured as the distance between a reference plate on which the nonwoven is lying and a pressure stamp, which is arranged parallel to it under specified conditions. |

*Method A: normal nonwovens with a compressibility < 20%*
Test area: 25 cm$^2$
Testing pressure: 0.5 kPa (5 cN/cm$^2$)
Loading time: 10 s
Arrangement of reference plate/sample/pressure stamp: horizontal

*Method B: bulky nonwovens with a thickness < 20 mm*
Test area: 10 cm$^2$
Testing pressure: 0.02 kPa (0.2 cN/cm$^2$)
Loading time: 10 s
Arrangement of reference plate/sample/pressure stamp: vertical

*Procedure C: bulky nonwovens with a thickness > 20 mm*
Test area: 400 cm$^2$
Testing pressure: 0.02 kPa (0.2 cN/cm$^2$)
Loading time: 10 s
Arrangement of base plate/sample/measuring plate: horizontal

Remarks/limitations: A note must be made in the test record of what process has been used, since the experimental conditions involved in each of the procedures are very different, and therefore the results obtained by the various methods are not comparable. The sampling regime will depend on the specific objective of the test and must be agreed on by the client and the tester (see testing of mass per unit area).

**Determination of tensile strength and elongation and reference strength/reference elongation**

| | |
|---|---|
| Method/procedure: | **Test method for nonwovens, determination of tensile strength and elongation and reference strength/ reference elongation** |
| Standard: | DIN EN 29 073-3: 1992-08 |
| Brief description: | A force is applied to a sample strip of specified dimensions at a constant strain rate. The load is applied until the sample tears, and the tensile force (N) and elongation (%) are determined. |

*Test conditions:*
Width of sample: 50 mm
Gauge length: 200 mm
Take-off rate: 100 mm/min
Pretension: none

The tests are carried out on the nonwoven in the machine and cross directions.

**Fig. 18-3** 'Zwicki' modular testing system from Zwick GmbH & Co, Ulm, Germany

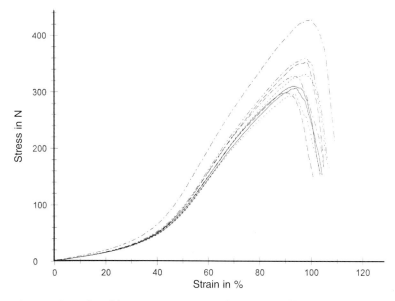

**Fig. 18-4** Examples of the stress/strain curves of a nonwoven fabric

Remarks/limitations:    Plotting stress/strain curves of the individual tests is a useful way of obtaining information on the stress/strain behaviour of nonwovens. The reference strength and elongation can be obtained from these.

In wet tensile testing, the samples are tested after they have been wetted in distilled water for one hour in the presence of 1 g/l nonionic wetting agent.

Tensile testers are available from a number of companies, and may differ in terms of their measurement ranges and applications. Fig. 18-3 shows an example of such a tensile tester.

Fig. 18-4 shows examples of the stress/strain curves plotted for a nonwoven fabric.

## Determination of bending length

Method/procedure:    **Test method for nonwovens, determination of bending length**

Standard:    DIN EN ISO 9073-7: 1998-10

Brief description:    *Principles*
A rectangular nonwoven fabric strip (test sample) is positioned on the horizontal testing table of the testing in-

strument. The strip is pushed forwards so that an increasing proportion hangs over the edge of the testing table and sags under its own weight. The overhanging section of the test sample can move freely, whilst the other end is held securely by the pressure of a steel ruler, which is positioned on top of that piece of the sample which is still lying on the testing table.

*Testing arrangement*
A steel ruler, graduated in millimetres, is placed onto the horizontal testing table of the bending rigidity tester, which is 40 mm wide and 200 mm long. The testing table is fitted with a buffer device so that the steel ruler can be slid along it centrally. A mark is made on the testing table, 10 mm away from the front edge.
Two lines, $L_1$ and $L_2$, inclined at an angle of 40° 30′ to the horizontal, are engraved on the two transparent side faces of the tester, which intersect the upper front edge of the testing surface.

*Test procedure*
Test samples of known size are first of all prepared, and their mass is determined.
To measure the bending rigidity, the sample is positioned between the sliding mechanism and the test surface, so that the three front edges lie exactly on top of each other. The zero point of the steel ruler must be in line with the mark on the testing table.
The sliding mechanism is pushed forward until the sample, which has been carried along with it, sags in such a way that its front edge meets the congruent lines, $L_1$ and $L_2$. After 8 s, the overhanging length at the sliding mechanism is measured on the ruler.
The test must be repeated at the same end of the sample, with the other side of the fabric facing upwards. The test then has to be carried out at the other end of the sample, following the same procedures.

*Calculations*
The bending length is taken as half the overhanging length. The mean bending length for each sample is calculated from the four bending length values of each sample, and the total mean value of the bending length, C, is calculated from the results of all the samples tested in one direction. The bending rigidity, G, is calculated in mN·cm on the basis of:

$$G = m \cdot C^3 \cdot 10^{-3}$$

m = mass per unit area (g/m$^2$)
C = total mean value of the bending length (cm)

The results are given separately for the machine and cross direction.

| | |
|---|---|
| Remarks/limitations: | The process is not suitable for testing unstable, very bulky, or stiff nonwovens or composites which have a tendency to twist. This process is very similar to DIN 53 362, and can be regarded as a tried-and-tested procedure. |

**Determination of bursting strength and bursting distension**

| | |
|---|---|
| Method/procedure: | Test method for nonwovens, determination of bursting strength and bursting distension, pneumatic method |
| Standard: | DIN EN ISO 13938-2: 1999-10 |
| Brief description: | The substrate is placed so that it is flat and firmly gripped on a circular surface that is covered with a membrane. When doing this, care must be taken to ensure that the sample is not distorted while it is being clamped, so that no slippage can occur when the load is applied. Air at a constantly increasing pressure is applied to the side of the membrane facing away from the sample, which causes the membrane and substrate to bulge outwards. The clamping ring must enable highly elastic textiles, which bulge outwards more than half the diameter of the test area, to be tested. |
| | The pressure is increased uniformly until the test sample bursts. The maximum pressure (measured bursting limit) and the resulting bursting height are measured. |
| | Once the experiments have been completed, the membrane is allowed to bulge outwards to the mean bursting height, without the test sample being present. The value obtained for the membrane is used as a correction value for the measured bursting limit. |
| | *Test conditions* |
| | Time taken for the test sample to burst: 20 s |
| | Preferred test area: 50 cm$^2$ |
| | The following are also possible by arrangement: |
| | 100 cm$^2$, 10 cm$^2$, 7.3 cm$^2$, 7.1 cm$^2$ |

Membrane thickness: 2 mm

Deformation property of the membrane: highly elastic

*Calculations*

– The mean value of the measured bursting limit (kPa) is calculated, and the value for the membrane pressure is subtracted from it. The difference is the bursting pressure (kPa)

– Mean value of the bursting height (mm)

For wet testing, the test sample is soaked for 1 hour in water containing 1 g/l wetting agent. The test sample is removed from the water, squeezed briefly between 2 layers of nonwoven fabric, and tested immediately.

Remarks/limitations:       Bursting testing is the most suitable method for determining the strength characteristics of textile substrates which have a marked tendency to pucker at the sides during tensile testing.

The results obtained with different test areas are not comparable!

The process enables the strength and deformation characteristics of a nonwoven fabric to be determined under complex loading at an angle of 360°.

## Determination of air permeability

Method/procedure:       **Test method for nonwovens, determination of air permeability, measurement of air volume at a specified differential pressure**

Standard:       DIN EN ISO 9237: 1995-12

Brief description:       The test sample is clamped so that it lies flat in a circular clamping device of specified size. The airflow is regulated to the required differential pressure via a pressure meter which is connected to the measuring head. The flow rate (mm/s or m/s) is read off after a specified period of time.

*Recommended test conditions:*

Test area: 20 cm$^2$

Differential pressure: 100 Pa (apparel fabrics)
                                      200 Pa (technical textiles)

The values should be read once stable conditions have been established, i.e. after about 1 min.

Remarks/limitations: The volume of air flowing in at the sides can be determined by measuring the sample whilst the test surface is covered with a plastic film. If the volume of lateral inflowing air accounts for >1% of the test result, the amount must be subtracted (corrected air permeability).

The side of the fabric that has been tested (penetration side) must be stated. This is an internationally recognised procedure, which shows good correlation with practical conditions.

Fig. 18-5 shows a tester for use in air permeability testing.

**Fig. 18-5** FX 3300 air permeability tester from TEXTEST AG, Zürich, Switzerland

**Determination of tear resistance**

Method/procedure: **Test method for nonwovens, determination of tear resistance, trapezoid method**

Standard: DIN EN ISO 9073-4: 1997-09

Brief description: A test strip of specified dimensions is marked with a regular trapezoid. A cut is made at right angles in the centre of the short side of the trapezoid.

The test sample is clamped along the non-parallel sides of the trapezoid so that the cut lies in the centre between the gripping clamps.

A force is applied to the test sample at a constant rate, so that it continues to tear at the cut. The resulting stress/strain curve should be plotted.

The tear propagation force (N) is determined as the mean value of a series of significant force peaks within the specified tear propagation path of 64 mm.

Remarks/limitations:

If only one defined force peak occurs in the tear propagation path, this represents the final result.

Test samples which do not tear in the cut should be disregarded.

This process has proved useful for describing the tear propagation characteristics of nonwoven fabrics in the machine and cross directions.

In addition to the trapezoid tear propagation test, other tear propagation tests are also available for testing nonwoven fabrics. Their suitability is usually evaluated by carrying out preliminary experiments first. These processes are documented in:

E DIN EN ISO 13 937-1: 1995-11 Elmendorf method

E DIN EN ISO 13 937-2: 1995-11 Trouser-shaped method

E DIN EN ISO 13 937-4: 1995-11 Tongue-shaped method

**Determination of drape coefficient**

Method/procedure:

**Test method for nonwovens, determination of drape coefficient**

Standard:

DIN EN ISO 9073-9: 1998-10

Brief description:

This process is used to determine the deformation capacity of a circular nonwoven test sample hanging under specified conditions.

The sample is held horizontally between two concentric discs having a diameter of 18 cm, so that the outer ring of the test sample is draped around the lower retaining disc under its own weight, and forms folds. The shadow of the draped sample is cast from below by means of parallel-directed light shining onto a light-permeable paper ring, whose mass had been determined before the test began. The size of the ring corresponds to the proportion of the sample which is lying free.

The outline of the shadow on the paper ring is traced, the paper is cut out along the edges of the shadow, and the mass of the inner section, which represents the shadow, is determined.

*Experimental conditions*

The diameter of the test sample is determined first of all by carrying out a preliminary test.

Time taken to draw around the shadow: 30 s

Number of test samples: 2

Number of measurements/sample: 6

The test is carried out on each sample on the front and reverse sides, and is repeated a further two times.

*Evaluation*

The mean value of draping coefficient, D (%), is calculated separately for the upper and lower sides of the nonwoven fabric.

$$D = \frac{m_{sa}}{m_{pr}} \cdot 100$$

$m_{pr}$ = initial mass of the paper ring (g)

$m_{sa}$ = mass of the part of the paper ring which represents the shadow (g)

Remarks/limitations: The test can only be carried out on nonwovens which drape uniformly around the horizontal disc. If the sample bends along a line on both sides of the retaining disc, this process is not suitable for testing the material being examined.

## Determination of abrasion resistance

Method/procedure: **Determination of abrasion resistance, Martindale method, specimen breakdown**

Standard: DIN EN ISO 12 947-2: 1999-04

Brief description: A circular sample is moved translatory against an abrading means (standard fabric) under a specified load following a Lissajous figure. The sample holder, which holds the abrading fabric, can be rotated easily about its axis in a vertical direction to the plane of the sample. A specified number of inspections are carried out during the abrading procedure, which is calculated on the basis of the expected number of abrading rotations specific to the sample being tested (preliminary test).

The load exerted on the sample is 9 kPa or 12 kPa, and is selected according to the intended end-use of the nonwoven fabric.

The nonwoven sample is destroyed when the diameter of the first hole to be abraded measures 5 mm.

The result of the experiment is taken as the time for the sample to be destroyed.

| | |
|---|---|
| Remarks/limitations: | The standard fabric used is replaced after 50,000 abrading rotations if the sample has not been destroyed. Further tests are then carried out on the sample until it is destroyed.
A magnifying lens or a microscope having 8 times magnification is recommended for monitoring the level of abrasion.
The process is used all over the world. |

| | |
|---|---|
| **Method/procedure:** | **Determination of abrasion resistance, Martindale method, assessment of appearance change** |
| Standard: | DIN EN ISO 12 947-4: 1999-04 |
| Brief description: | See Martindale technique, destruction of the sample.
The abrasion resistance of the textile substrate is determined by evaluating the surface change on an abraded sample, compared to a sample that has not been abraded.

*Abrading process a:*
The sample is abraded for a specified number of rotations and the surface change is evaluated.

*Abrading process b:*
The sample is abraded until a specific change in the surface takes place, and the time taken for the change to occur is calculated. |
| Remarks/limitations: | With process a, the surface change can be evaluated verbally, or by using a graduated comparative standard (e.g. a photograph) to allocate a mark. |

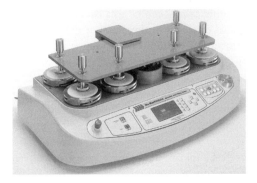

**Fig. 18-6** Nu-Martindale abrasion tester from James H. Heal & Co. Ltd, Halifax, UK

The change in colour tone can also be given a mark in accordance with ISO 105-A02.

The evaluations are subjective, and therefore have to be carried out by experienced test personnel.

The process is used all over the world.

## 18.2.2
### Fastness testing

The colourfastness ratings, which express the resistance of the colour in dyed and printed textiles to various influencing factors, can be divided into processing fastness values and performance-in-use fastness values.

The colourfastness is always tested as a single parameter, e.g. fastness to washing or fastness to rubbing, and not in conjunction with other processes, e.g. washfastness in conjunction with subsequent ironing fastness. This means that the contribution of each process to the change in the test sample can be evaluated. In the future, it is quite likely that it will be possible to carry out combined tests, which will enable several influencing factors to be evaluated.

With the exception of fastness to rubbing, all the colourfastness tests involve evaluating the change in colour taking place on the test sample. An adjacent fabric can be used to test the level of staining which occurs.

Adjacent fabrics are undyed, standard fabrics made from one or more types of fibre, and are used during the test to evaluate staining.

With a few exceptions, the adjacent fabric used in the test is chosen on the basis of the material being tested. If the sample being tested is only made up of one component, it is advisable to use an adjacent fabric made of the same fibre type. The use of a second fabric will depend on the relevant testing standard.

If the test samples are made from blends of fibres, the use of the two single-fibre test fabrics will depend on which fibre accounts for the largest proportion of fibres in the blend.

The properties and production of single-fibre, adjacent fabrics are described in international standard, ISO 105, Part F.

Water of quality type no. 3 is used for preparing the test solutions according to ISO 3696. In order to obtain reproducible test results, these solutions must always be absolutely fresh.

The change in colour of the test sample and staining into the adjacent fabric are evaluated using the relevant grey scales, and the result is given as a fastness rating from 5 to 1. The grey scale for evaluating staining in accordance with DIN EN 20 105 A03 and the grey scale for evaluating the change in colour in accordance with DIN EN 20 105 A02 are used. These grey scales should not be mixed up when carrying out the analysis, since their graduations according to CIELAB values are very different.

The samples can be given marks from 5 to 1 and, if required, intermediate marks can also be awarded (e.g. 4–5 or 3–4). The following marks can be awarded:

Note 5: no change from the original state
Note 4: hardly any change from the original state
Note 3: slight change from the original state
Note 2: marked change from the original state
Note 1: very obvious change from the original state

The test samples should always be evaluated under the same lighting conditions, for which a colour matching chamber operating with daylight D65 is recommended.

If other changes apart from colour changes occur in the test samples, such as surface and dimensional changes, or if the sample becomes shiny, these should also be recorded. When evaluating the level of staining, care must be taken to remove any fibres which may be sticking to the adjacent fabric, since these could give erroneous test results. This often occurs when testing the fastness to rubbing, especially the wet fastness to rubbing.

A selection of the most frequently used testing processes, together with some instructions for carrying out the tests, will now be presented:

| | |
|---|---|
| **Method/procedure:** | **Determination of colour fastness to artificial light: Xenon arc fading lamp test** |
| Standard: | DIN EN ISO 105-B02: 1999-09 |
| Brief description: | The capacity of the colour of dyed or printed materials to resist the effects of D65 standard light is evaluated. The sample is illuminated simultaneously with a light-fastness scale, which consists of a set of 8 standard dyeings carried out on wool fabric using blue dyes. In accordance with the process used, the test sample and light-fastness scale are both covered with special metal templates of different shapes, to enable any differences in colour change to be compared directly. The illuminating conditions are chosen on the basis of the intended end-use of the sample. The settings are controlled by simultaneously illuminating a dyed material as a control (woven cotton fabric dyed red with a combination of naphthol dyes) for determining the moisture content, and also by using the selected standard dyeings of the lightfastness scale. |
| Remarks/limitations: | The results obtained with instrumental lighting cannot be regarded as absolute lightfastness values, since the instrument reproduces uniform lighting conditions, corresponding to the mean annual conditions. They are a measure of the changes which take place under defined conditions, and form an integral part of any technical delivery conditions and arrangements with the client. |

| | |
|---|---|
| **Method/procedure:** | **Determination of colour fastness to washing at washing temperatures of 40 °C, 50 °C and 60 °C** |
| Standards: | DIN EN 20 105 C01: 1993-03 (40 °C)<br>DIN EN 20 105 C02: 1993-03 (50 °C)<br>DIN EN 20 105 C03: 1993-03 (60 °C) |
| Brief description: | The samples are sewn to the relevant adjacent fabrics along the narrow side, and treated in washing solutions at the relevant temperatures. The tests can be carried out using a Linitest instrument. After washing, the samples are rinsed twice in distilled water and then dried. The colour change and amount of staining into the adjacent fabric are evaluated after the samples have been dried. |
| Remarks/limitations: | The processes describe what happens to the colour of a dyed textile material when it is subjected to a gentle wash at the relevant temperature, and does not supply information on the overall effect of the washing process. |
| **Method/procedure:** | **Determination of colour fastness to domestic and commercial laundering, process carried out using different temperatures, additives and levels of mechanical agitation** |
| Standard: | DIN EN ISO 105-C06: 1997-05 |
| Brief description: | The fastness to washing tests carried out in accordance with this standard are user-oriented, in which case, a specially developed test detergent must be used. Sodium perborate can also be added with certain processes. Steel balls can be incorporated during the washing process to simulate an increased level of agitation, but the washing temperature must not exceed 70 °C. These conditions do not correspond to general laundry practices in Europe. |
| Remarks/limitations: | The process describes the fastness of the colour during domestic and commercial laundering. It does not take into account the effects of optical brighteners which are used in commercial detergents. |
| **Method/procedure:** | **Determination of colour fastness to dry cleaning, perchloroethylene process** |
| Standard: | DIN EN ISO 105-D01: 1995-04 |
| Brief description: | A test sample is sewn completely into two cotton twill fabric pieces, together with 12 steel discs. The samples |

are treated in the presence of perchloroethylene for 30 min at a temperature of 30 °C, in a Linitest instrument, for example.

The samples are dried and evaluated for any change in colour using a grey scale. The perchloroethylene used is also compared with the initial solution using the grey scale in order to evaluate staining.

Remarks/limitations:

The testing process describes the colourfastness of textiles to dry cleaning with perchloroethylene in particular. The finishing effects and any treatments carried out during commercial laundering, e.g. spotting treatments or steam ironing, are not considered.

**Method/procedure:** **Determination of colour fastness to water, heavy loading**

Standard: DIN EN ISO 105-E01: 1996-08

Brief description:

The samples are sewn to the relevant adjacent fabrics along the narrow side. Each sample is completely soaked with water of quality type no. 3, and placed in the testing frame between two plates made from acrylic resin or glass, where they are treated for 4 hours at a temperature of 37 °C. Following drying, the adjacent fabric is evaluated for staining, and the change in colour is determined using the relevant grey scales.

Remarks/limitations:

The testing process describes the colourfastness of textiles towards immersion in water.

**Method/procedure:** **Determination of colour fastness to perspiration, acidic and alkaline perspiration**

Standard: DIN EN ISO 105-E04: 1996-08

Brief description:

The samples are sewn to adjacent fabrics along the narrow side and placed into an alkaline solution at a pH value of 8 and an acidic solution having a pH value of 5.5, at a liquor ratio of 1:50, and thoroughly wetted at room temperature. The excess liquid is squeezed out, and the test samples are laid between two plates made from acrylic resin or glass in a testing frame, where they are treated for 4 hours at a temperature of 37 °C. After drying, the amount of staining into the adjacent fabric and the change in colour are evaluated using the relevant grey scales.

| | |
|---|---|
| Remarks/limitations: | The test process describes the colour fastness of textiles in relation to the effects of human perspiration. |

| | |
|---|---|
| **Method/procedure:** | **Determination of colour fastness to spotting: acid** |
| Standard: | DIN EN ISO 105-E05: 1997-05 |
| Brief description: | A sample is prepared for each acid used (acetic, sulphuric and tartaric acid). Acid is applied to the surface until the wetted area has a diameter of approximately 20 mm. No more than 0.5 ml of solution should be applied, even when testing water-repellent fabrics. After the samples have been dried at room temperature, the colour change is evaluated using the grey scale. |
| Remarks/limitations: | The test process describes the colour fastness of textiles to the effects of dilute solutions of organic and inorganic acids. In order to ensure that any changes are not caused solely by the influence of water, the test should also be carried out with water in order to compare the results (DIN EN ISO 105-E07). |

| | |
|---|---|
| **Method/procedure:** | **Determination of colour fastness to spotting: alkali** |
| Standard: | DIN EN ISO 105-E06: 1997-05 |
| Brief description: | Drops of a sodium carbonate solution are applied to the test sample until the wetted area has a diameter of approximately 20 mm. No more than 0.5 ml of solution should be applied, even when testing water-repellent fabrics. The samples are dried at room temperature, and the change in colour is evaluated using the grey scale. |
| Remarks/limitations: | See acid process |

| | |
|---|---|
| **Method/procedure:** | **Determination of colour fastness to spotting: water** |
| Standard: | DIN EN ISO 105-E07: 1997-05 |
| Brief description: | Water droplets of water quality type 3 are applied to the test sample until the wetted area has a diameter of approximately 20 mm. No more than 0.5 ml of solution should be applied, even when testing water-repellent fabrics. The change in colour at the edges of the sample is evaluated after 2 min and after drying, using the grey scale. |
| Remarks/limitations: | The test process describes the colour fastness of textiles to the effects of water droplets. |

| | |
|---|---|
| Method/procedure: | **Determination of colour fastness to rubbing, wet and dry** |
| Standard: | DIN EN ISO 105-X12: 1995-06 |
| Brief description: | The test is carried out using a dry abrading fabric (dry fastness to rubbing) and a wet abrading fabric (wet fastness to rubbing). A pin is moved backwards and forwards over the test path at a force of 9 N, using the appropriate testing instrument. Each sample is tested when wet and dry, in both the machine and cross directions. The staining of the abrading fabric, which is stretched over the abrading pin during the test, is evaluated. Care must be taken to ensure that the abrading fabric is stretched correctly in the device, so that it can rub over the test sample in line with the yarns. |
| Remarks/limitations: | The test process describes the colour fastness of textiles to abrasion and staining from other textiles during use. |

| | |
|---|---|
| Method/procedure: | **Determination of colour fastness of interior materials in motor vehicles, xenon arc lamp test** |
| Standard: | DIN 75 202: 1988-12 |
| Brief description: | The test samples are subjected to the required lighting conditions, together with a set of light fastness scales (see DIN EN ISO 105-B02). The test is carried out using a light exposure instrument fitted with one or more xenon arc lamps and the relevant optical filter system. The values for the spectral distribution on the surface of the sample are calculated on the basis of the spectral distribution of overall radiation in accordance with CIE Recommendation No. 85, and the spectral transmission of window glass having a thickness of 4 mm.
As far as the exposure intensity, the black standard temperature and the humidity are concerned, the values are defined as a function of the process (A or B), and these can be measured and controlled using the relevant testing instruments.
If the fastness to light at high temperatures is to be measured, the samples and the light fastness scales are illuminated simultaneously for one cycle, and the light fastness values compared to the standard dyeings is determined. When evaluating the ageing behaviour, the sample is subjected to the test conditions over several cycles. A cycle corresponds to the level of exposure required for standard dyeing no. 6 to correspond to a |

change of note 3 in accordance with the grey scale. Following on from this, the fastness can be tested using the grey scale to evaluate the colour change. The loss in strength can also be evaluated by carrying out strength tests before and after exposure.

Remarks/limitations:     The test process describes the colour fastness and the ageing behaviour under the influence of artificial light (D65), and also the effect of heat at the same time.

**Method/procedure:**     **Analysis of commodity goods – Testing of coloured children's toys with respect to their resistance to saliva and perspiration, method LMBG § 35, 82.10-1**

Standard:     No standard is currently available

Brief description:     The method is based on the use of two solutions (acid and alkaline) for each sample to be tested. The sample is laid onto an adjacent cotton fabric that has been completely wetted in the test solution and treated in a desiccator for 2 hours at 40 °C. The fastness to saliva and perspiration is determined using the grey scale to evaluate staining.

Remarks/limitations:     The test process describes a method for establishing whether dye from the textile material could pass into the mouth, the mucous membrane or the skin of the wearer.

## 18.2.3
## Testing of burning behaviour

The burning behaviour of nonwoven fabrics is affected by the specific properties of the material, the construction of the nonwoven, its use with other materials in composites, the type of binder, and by a whole number of other influencing factors.

Practical tests carried out to study the burning behaviour of nonwoven fabrics, and consequently the end products made from the textiles, are an important requirement for evaluating their suitability for the intended end-use.

This must be qualified by saying that, when testing the burning behaviour of materials, it is not possible to take into account all the possible external influences that could also have an effect, and therefore the results do not give an unequivocal indication of the potential risk of fire.

It is important for the tests to be carried out under reproducible conditions, in order to ensure that the results can be compared. Even a change in temperature and moisture content can lead to the materials exhibiting a different burning behaviour. The nature of the flame and the way the test materials are arranged can also be a significant factor.

The criteria used in evaluating the behaviour of materials to heat and flames include the ignitability, the flame spread over a predetermined piece of the sample (and the associated time of flame spread) or the time taken for the flame to reach a set of measured marks, the afterglow, the afterburning time of self-extinguishing fabrics, and the evaluation of the secondary effects of burning, such as whether the material melts or forms burning droplets.

The afterburning time is defined as the time taken between removal of the source of ignition to the flame going out on the sample. The duration of afterglow refers to the length of the time in which the material continues to glow after the flame has been extinguished under the stipulated test conditions.

The materials are tested in accordance with testing standards and legal requirements, and in accordance with agreements made with the clients and/or their conditions of delivery.

| | |
|---|---|
| **Method/procedure:** | **Determination of safety of toys – flammability** |
| Standard: | DIN EN 71-2: 1994-01 |
| Brief description: | The test procedures are described in the standard, depending on the type of toy that has to be tested. In every case, the maximum rate of flame spread is an important evaluation criterion. The requirements are such that there is adequate time to remove the child safely if the toy does catch fire. |
| Remarks/limitations: | The test process describes the general requirements and procedures for testing the burning behaviour of all types of toys. |
| **Method/procedure:** | **Protection against heat and flame, method of test for limited flame spread** |
| Standard: | DIN EN 532: 1995-01 |
| Brief description: | The flame spread is tested using a small flame (conventional propane gas) against a vertically arranged textile sample to evaluate burning across the surface. The flame is applied for 10 s, and the characteristics of the flame are specified. After the flame has been removed from the sample, the afterburning and afterglow times (time taken and area affected by the flame) are recorded. Whether the flame has reached the upper and/or sides of the sample is also recorded. Total burning takes place if a hole measuring 5×5 mm is produced by burning, melting or glowing. If the hole does not go completely through the material, this is regarded as incomplete hole formation. In order to avoid |

further damage to the material, the formation of holes in the samples is always evaluated while they are still stretched over the test frame.

The occurrence of other, secondary burning phenomena, such as the formation of burning or melting droplets, is also recorded.

Remarks/limitations: The test process describes the limited flame spread over materials which are used in flame-resistant protective clothing.

The classification criteria are covered in standards DIN EN 531 and DIN EN 533, for example.

Method/procedure: **Assessment of the ignitability of mattresses and uphol-stered bed bases, ignition source: Smouldering cigarette**

Standard: DIN EN 597-1: 1995-01

Brief description: The test arrangement specified in the standard is set up in a suitable fume cupboard, and a representative sample is placed on it. A glowing cigarette is laid on top of the sample, and the process of burning or smouldering is observed. The criteria used in making the evaluation include ignition as a result of continued smouldering, ignition as the result of flame formation, or non-ignition. If the sample does ignite, the test times and the reasons for ignition should be recorded.

Remarks/limitations: The standard describes a test process for determining the ignitability of mattresses, padded bed bases and mattress covers when in contact with a glowing cigarette. The results of these experiments relate solely to the ignitability of the combination of materials used under the test conditions specified. However, the process does not allow conclusions to be drawn relating to every single possible fire hazard, or its consequences.

Method/procedure: **Assessment of the ignitability of mattresses and uphol-stered bed bases, ignition source: Match flame equivalent**

Standard: DIN EN 597-2: 1995-01

Brief description: A defined flame (butane gas producing the same amount of heat as a burning match) is applied for 15 s along a plane surface of the test sample. The burner pipe must be arranged horizontally and it should be in contact with the sample. The evaluation criteria include continued smouldering, and ignition or non-ignition of the sample. If the sample continues to smoulder and then ignite, the sample should be extinguished, and this should be recorded. If the sample does not ignite, the test should be repeated at another location on the sample.

| | |
|---|---|
| Remarks/limitations: | The standard describes a test process for determining the ignitability of mattresses, padded bed bases or mattress covers when subjected to a gas flame comparable to a burning match as the source of ignition. |
| | The tester is also advised to carry out simultaneous tests with a glowing cigarette, since it should not be assumed that, just because a material resists flames, it will automatically resist burning from a smouldering source. |
| **Method/procedure:** | **Assessment of the ignitability of upholstered furniture, ignition source: smouldering cigarette** |
| Standard: | DIN EN 1021-1: 1994-01 |
| Brief description: | An upholstered composite is made up of the relevant materials and stretched over a test frame that is similar to a chair. It should be noted that if it is possible to clean the material, then the relevant pre-treatments should be carried out first. A glowing cigarette is used as the source of ignition and is applied along the join between the horizontal and the vertical sections. The burning process of the cigarette, continued smouldering, and ignition or non-ignition of the test sample should all be observed and noted. The test sample should be observed for up to 1 hour after being in contact with the cigarette. |
| Remarks/limitations: | The standard describes a test process for determining the ignitability of a combination of materials, such as covers and stuffing materials, for upholstered chairs, when a glowing cigarette is used as the source of ignition. This procedure can be used for testing the ignitability of a combination of materials, but not the ignitability of a final piece of furniture. |
| **Method/procedure:** | **Assessment of the ignitability of upholstered furniture, ignition source: match flame equivalent** |
| Standard: | DIN EN 1021-2: 1994-01 |
| Brief description: | The test frame and arrangement of the test sample are identical to those in DIN EN 1021-1. A small flame (bu- |

tane gas), which produces the same amount of heat as a burning match, is brought into contact with the sample. The burner pipe, producing a small flame corresponding to a gas flow of approx. 45 ml/min, is placed along the join between the sitting surface and the back section. The flame is applied for 15 s and then it is carefully removed.

Signs of continued smouldering and/or burning in the padding and/or the cover are noted whilst the burning process is being observed. Any flames, afterglow, smoking or smouldering, which stop within 120 s of the burner pipe being removed, can be ignored. The sample should be extinguished if it ignites, and the reason for the ignition should be recorded. If the sample neither ignites nor continues to smoulder, the test should be repeated at another location on the test sample.

Since continued smouldering cannot always be detected from the outside, the test sample should be taken apart and the interior should also be evaluated.

Remarks/limitations: The standard describes a test process for determining the ignitability of combinations of materials, such as covers and stuffing materials for upholstered chairs, when a small flame is used as the source of ignition. The ignitability of combinations of materials is tested using this process. The result can be used as a point of reference, but is no guarantee of the ignition behaviour of a final piece of furniture.

**Method/procedure:** **Detailed procedure to determine the burning behaviour of fabrics for apparel, vertical plane method**

Standard: DIN EN 1103: 1996-01

Brief description: Before the flame is applied, the samples are subjected to a pre-treatment following the instructions on the care label. If no instructions are available, the process is carried out in accordance with the specifications contained in the standard. The pre-treatment is carried out in order to ensure that the test samples are treated in accordance with normal usage.

The test is carried out in accordance with DIN EN ISO 6941, Section 8.6.1, using a flame exposure time of 10 s. The rate of flame spread between marking threads no. 1 and no. 3 is recorded. If these do not burn through during the course of the test, because the flame did not spread, the afterburning and afterglow times of the sample must be recorded. Burning pieces of the sample, which fall off and ignite a filter paper lying underneath, must be recorded as secondary burning phenomena.

Remarks/limitations: This standard describes a process for testing the burning behaviour of apparel fabrics using the surface ignition test in accordance with DIN EN ISO 6941. Both simple substrates as well as combinations of materials can be tested, depending on the end-use.

Method/procedure: **Respiratory protective devices for self-rescue, self-contained open-circuit compressed air breathing apparatus incorporating a hood – Requirements, testing, marking**

Standard: DIN EN 1146: 1997-05

Brief description: Every article, which is likely to come into contact with a flame during use, must be self-extinguishing. The materials must not burn easily and should burn for no longer than 5 s once the flame has been removed.
Testing in accordance with Section 7.5.3 of this standard describes a procedure in which a complete test sample is passed through a defined flame at a specified speed.

Remarks/limitations: This standard applies to compressed air breathing devices with hoods used in self-rescue applications. Laboratory tests and practical performance trials are described, to ensure that the results correlate with requirements.

Method/procedure: **Fire behaviour of building materials and building components, class B2**

Standard: DIN 4102-1: 1998-05

Brief description: Edge burning tests are carried out on building materials if no protection is used along the edges. The entire surface area is subjected to the flame if there are no free edges. The surface does not have to be ignited if no failure is expected. The relevant test samples are tensioned vertically in the required sample carrier and subjected to a flame in a specified manner. How and when the flame reaches the set of measured marks is recorded, and also whether any burning droplets have been formed.

Remarks/limitations: Building materials are classified in accordance with DIN 4102-1 into Classes A1 and A2 (non-flammable) or

B1, B2 and B3 (not easily flammable, normal flammability or easily flammable). The test for flammability class B2 involves subjecting the sample to a small, defined flame (burning match), whereby the sample should exhibit limited ignitability and flame spread characteristics. For a sample to be allocated to flammability class B2, neither the edges nor the surface of the sample should have burned up to the measured marks after a period of 20 s.

| | |
|---|---|
| Method/procedure: | **Burning behaviour, measurement of flame spread properties of vertically oriented specimens, vertical edge method and vertical plane method** |
| Standard: | DIN EN ISO 6941: 1995-04 |
| Brief description: | The type of ignition selected will depend on the likely end-use of the fabric. Surface burning is usually used in the case of textiles. The samples are subjected to the flame for 5 or 15 s, the burning time between the measured marks is recorded, and the rate of flame spread is calculated from this. If a sample does not burn between the measured marks, the afterburning and afterglow times are recorded, together with the greatest burnt length and width of the sample; a note should also made of whether the flame reached a vertical edge, and whether a hole and/or burning droplets were formed. |
| Remarks/limitations: | The standard describes a test process for measuring the flame spread characteristics of textiles that are going to be used in applications such as clothing, curtains and net curtains. It should be pointed out that the results only apply to evaluating the flame spread under controlled conditions, and do not allow any conclusions to be drawn relating to situations where the conditions may be different. The burning behaviour of textile products, especially textile substrates for work clothing, can be determined in accordance with DIN 66 083. |
| Method/procedure: | **Determination of burning behaviour of interior materials in motor vehicles, horizontal edge method** |
| Standard: | DIN 75 200:1980-09 |
| Brief description: | The test samples are arranged horizontally in a fume cupboard and subjected to a defined flame. The rate of flame spread between three measured marks is recorded. |

Remarks/limitations:      This standard describes a process for testing materials for use in vehicle interiors. A small flame can be used to evaluate the uniformity of production batches as far as their burning behaviour is concerned. This test procedure is not suitable for determining the burning behaviour of materials in the vehicle itself.

## 18.2.4
### Testing of the behaviour during laundering/dry cleaning

In the clothing sector, tests for evaluating the laundering and dry cleaning characteristics are carried out on nonwovens for use as interlinings and padding materials.

Specific standards exist for carrying out washing and dry cleaning treatments in combination with carrier substrates.

| Standard | Content |
|---|---|
| DIN 54 303: 1991-02 | Testing of textiles; dry-cleaning of nonwoven fabrics; nonwoven interlinings and nonwoven wadding |
| DIN 54 304: 1991-02 | Testing of textiles; washing of nonwoven fabrics; nonwoven interlinings and nonwoven wadding |

The properties of the nonwoven fabrics which are influenced by the aftercare treatments, such as the dimensional stability, strength and moisture content, and their resistance to the treatments, are evaluated in accordance with testing standards generally applicable to textiles.

## 18.2.5
### Human ecological tests

The Öko-Tex Standard 100 has become established all over Europe as a private label applicable to the human-ecological testing of textiles, although other labels also exist. The Forschungsinstitut Hohenstein GmbH & Co KG (Hohenstein Research Institute), Bönnigheim, and the Österreichisches Textil-Forschungsinstitut (Austrian Textile Research Institute), Vienna, set up the Öko-Tex Initiative in 1992. Since then, 17 testing institutes in 14 countries have become affiliated to the 'International Federation for Research and Testing in the Field of Textile Ecology' (Öko-Tex). The address of the secretariat is Gotthardstr. 61, Postfach 585, CH-8027 Zürich, Switzerland. In Germany, the Hohenstein Institute is represented on all the relevant committees of this International Federation. In addition to this, the Deutsches Textilforschungszentrum Nord-West e.V. (German Textile Research Centre North-West), Krefeld, the Sächsisches Textilforschungsinstitut e.V. (Saxon Textile Research Institute), Chemnitz, and the Umweltlabor ABC GmbH (ACB Environmental Laboratory) in Münster are co-opted members, and are certified to

**Table 18-1** Öko-Tex product classes and test criteria, situation as at 01/1999

| Test criteria | Product classes/Threshold values | | | |
| --- | --- | --- | --- | --- |
| | I<br>Baby | II<br>In contact<br>with the skin | III<br>Not in<br>contact<br>with the skin | IV<br>Soft<br>furnishings |
| pH-value | 4.0–7.5 | 4.0–7.5 | 4.0–9.0 | 4.0–9.0 |
| Formaldehyde [ppm]: | | | | |
| Law 112 | 20 | 75 | 300 | 300 |
| Emission | 0.1 | | | 0.1 |
| Eluatable heavy metals [ppm]: | | | | |
| Sb | 5.0 | 10.0 | 10.0 | 10.0 |
| As | 0.2 | 1.0 | 1.0 | 1.0 |
| Pb | 0.2 | 1.0 | 1.0 | 1.0 |
| Cd | 0.1 | 0.1 | 0.1 | 0.1 |
| Cr | 1.0 | 2.0 | 2.0 | 2.0 |
| Cr (VI) | n. n. | n. n. | n. n. | n. n. |
| Co | 1.0 | 4.0 | 4.0 | 4.0 |
| Cu | 25.0 | 50.0 | 50.0 | 50.0 |
| Ni | 1.0 | 4.0 | 4.0 | 4.0 |
| Hg | 0.02 | 0.02 | 0.02 | 0.02 |
| Pesticides [ppm]: total | 0.5 | 1.0 | 1.0 | 1.0 |
| Chlorinated phenols [ppm]: | | | | |
| PCP | 0.05 | 0.5 | 0.5 | 0.5 |
| TeCP | 0.05 | 0.5 | 0.5 | 0.5 |
| Dyes: | | | | |
| Splittable arylamines | | not used | | |
| Carcinogenic arylamines | | not used | | |
| Allergenic arylamines | | not used | | |
| Chloro-organic carriers [ppm] | 1.0 | 1.0 | 1.0 | 1.0 |
| Biocidal finishing | none | none | none | |
| Flame-resistant finishing | none | none | none | |
| Colour fastness ratings: | | | | |
| Water fastness | | 3 | 3 | 3 |
| Fastness to perspiration, acidic | | 3–4 | 3–4 | 3–4 |
| Fastness to perspiration, alkaline | | 3–4 | 3–4 | 3–4 |
| Dry rub fastness | 4 | 4 | 4 | 4 |
| Wet rub fastness | 2–3 | 2–3 | 2–3 | 2–3 |
| Fastness to saliva and perspiration | resistant | | | |

**Table 18-1** (continued)

| Test criteria | Product classes/Threshold values | | | |
|---|---|---|---|---|
| | *I*<br>*Baby* | *II*<br>*In contact*<br>*with the skin* | *III*<br>*Not in*<br>*contact*<br>*with the skin* | *IV*<br>*Soft*<br>*furnishings* |
| Emission of volatile components | | | | |
| Toluene | 0.1 | | | 0.1 |
| Styrene | 0.005 | | | 0.005 |
| Vinylcyclohexane | 0.002 | | | 0.002 |
| 4-Phenolcyclohexane | 0.03 | | | 0.03 |
| Butadiene | 0.002 | | | 0.002 |
| Vinylchloride | 0.002 | | | 0.002 |
| Aromatic hydrocarbons | 0.3 | | | 0.3 |
| Volatile organic substances | 0.5 | | | 0.5 |
| Odour testing: | | | | |
| general | | no unusual odour | | |
| SNV 195651 | 4 | | | 4 |
| (for textile floorcoverings only) | | | | |

carry out the relevant test procedures. The German Öko-Tex Certification Centre in Eschborn is responsible for certifying products in Germany.

Four product classes are currently covered by the Öko-Tex criteria. Product Class I covers baby products and all the pre-products and accessories used for producing articles for babies and small children up to the age of two. Product Class II covers all those products that come into contact with the skin, such as blouses, shirts and underwear. Products that do not come into contact with the skin are allocated to Product Class III, and Product Class IV covers soft furnishings, such as tablecloths, textile wall coverings, textile furnishings and curtains, upholstery fabrics, textile floor coverings and mattresses [16].

Table 18-1 shows the requirements relevant to the various applications.

## 18.3
### Test processes relating to end-use
M. MÄGEL, U. BERNSTEIN, C. LEWICKI

### 18.3.1
### Hygiene and medical products

Nonwoven fabrics are used in the medical and hygiene sectors to treat patients both directly and indirectly, and they are also used in work clothing and underwear. A biological evaluation has to be carried out first on textiles that are to be used as medical products. This provides information on the compatibility of materials and products which come into contact with the skin. The functional and general textile-physical and chemical characteristics should also be evaluated as a function of the very different end-uses of the various materials. The main criteria for determining the suitability of textile products are

– barrier efficiency toward bacteria, viruses, liquids and dust particles
– capacity to absorb and store body fluids
– resistance to mechanical influences

The requirements and methods used for analysing the characteristics of products have still not been fully standardised. The following test procedures are used extensively for evaluating medical and hygiene textiles:

| Test parameters/characteristics | Testing standard/specification |
|---|---|
| Barrier efficiency towards bacteria and liquids | SS 8760019 EDANA 190.0-8 |
| | ASTM F 1670-97 |
| | ASTM F 1671-97 |
| | DIN EN 20811 |
| Bacterial loading of products | DIN EN 1174 |
| Particle emissions | EDANA 220.0-96 |
| | EDANA 300.0-84 |
| Liquid absorption capacity | ASTM F 1819-97 |
| Liquid storage capacity | ISO 9073-6 |
| Liquid transfer capacity | Test methods of the leading organisations of German medical insurance companies |
| Textile-physical properties | ISO 9073-3 |
| | ISO 9073-4 |
| | ISO 13938-1 |

Coordinated standards are available for recording the physical and chemical characteristics of nonwovens that are to be used in particular for producing surgical compresses or as packaging materials for products that have to be sterilised.

The testing of nonwovens for medical applications is covered in Chapter 11.

### 18.3.2
### Cleaning cloths and household products

Depending on the intended end-use, the following parameters are particularly important as far as cleaning cloths and household products are concerned:

– moisture absorption
– absorbency
– water retention capacity
– dust pick-up capacity

698 | 18 Testing processes

No DIN standards covering the special requirements of this product sector are currently available. The test criteria are based mainly on existing requirements (e.g. the technical delivery conditions stipulated by the Federal Office for Defence Technology and Acquisition (Bundesamt für Wehrtechnik und Beschaffung) and are based on the test procedures used in the hygiene/medical sectors. Special specifications relating to quality analysis in the testing of polishing materials for optical lenses only are covered by DIN 58 750: 1995-04:

Part 1: Test methods for physical properties
Part 2: Test methods for polishing performance for spectacle lenses of silicate and of synthetic material
Part 3: Test methods for polishing performance and surface condition for connector optics

Chapter 12 has further information on this.

18.3.3
**Household textiles**

Nonwoven fabrics can be used in the following household textile applications:

– floor coverings
– sub-upholstery materials
– webbings

Materials for use in the contract sector have to meet legal specifications regarding their flammability.

Meeting the Öko-Tex criteria is also becoming increasingly important as far as this group of products is concerned.

Special DIN Standards exist only for floor coverings; these are already subject to existing textile-physical testing procedures for use in special applications. These include, for example:

DIN 53 855-3: 1979-01     Testing of textiles – Determination of thickness of textile fabrics – Floor coverings

DIN 54 316: 1983-10     Testing of textiles – Determination of thickness loss of textile floor coverings at static loa

DIN 54 326: 1984-01     Testing of textiles – Determination of appearance retention of textile floor coverings – Tetra pod-walker-test

DIN 54 345-3 1985-07:     Testing of textiles – Electrostatic behaviour – Determination of electrostatic charge of textile floor coverings by machine

As far as other products in the household textile sector are concerned, the requirements will depend on any arrangements and agreements made with the client.

The German Institute for Quality Assurance and Characterization (Deutsches Institut für Gütesicherung und Kennzeichnung e.V.) publishes the following specifications:

RAL 991 A2: 1974-05    Cleaning of textile floor coverings – Definition

RAL-RG 368/2: 1976-08  Polfleece floor coverings

RAL 399 C4: 1956-07    Designation regulations for pad stuffings

## 18.3.4
### Protective clothing

In accordance with an EU directive relating to equipment for personal protection, the procedures for testing this type of product have been standardised extensively in Europe. Unlike the end-uses for other textile products, the requirements for this group have been stipulated quantitatively, and form the basis for classifying clothing in relation to its end-use. In the protective clothing sector, nonwoven fabrics are used in

- products with a limited service life, especially for clothing having a barrier effect
- reusable products, used mainly for thermal insulation

The relevant safety/technical requirements for these sectors are covered in special product standards. These may also include some testing procedures, but special testing standards are also available. Apart from chemical-resistant clothing, nonwoven fabrics are mainly used only as a component of the clothing assembly, so that the tests have to be carried out in combination with other textile substrates present in the clothing assembly.

   The following testing standards relating to product standards and functional characteristics are important for evaluating the protective effect provided by the clothing:

### Heat- and flame-resistant clothing

Nonwovens for thermal insulation and as a component in moisture-barrier assemblies

Product standards

EN 469     Protective clothing for fire fighters – Requirements and test methods for protective clothing for fire fighting

EN 1486    Protective clothing for fire fighters – Test methods and requirements for reflective clothing for specialized fire fighting

EN 470-1   Protective clothing for use in welding and allied processes – Part 1: General requirements

| EN 531 | Protective clothing for industrial workers exposed to heat (excluding fire fighters' and welders' clothing) |
| EN 533 | Protective clothing – Protection against heat and flame – Limited flame spread materials and material assemblies |

Special testing standards

| EN 348 | Protective clothing – Test method: determination of behaviour of materials on impact of small splashes of molten metal |
| EN 366 | Protective clothing – Protection against heat and fire – Method of test: evaluation of materials and material assemblies when exposed to a source of radiant heat |
| EN 367 | Protective clothing – Protection against heat and fire – Method of determining heat transmission on exposure to flame |
| EN 373 | Protective clothing – Assessment of resistance of materials to molten metal splash |
| EN 532 | Protective clothing – Protection against heat and flame – Test method for limited flame spread |
| EN 702 | Protective clothing – Protection against heat and flame – Test method: Determination of the contact heat transmission through protective clothing or its materials |

**Chemical-resistant protective clothing**

Nonwoven fabrics as inserts in suits

Product standards

| prEN 1511 | Protective clothing against liquid chemicals – Performance requirements for limited use chemical protective clothing or suits with liquid-tight connections between different parts of the clothing (type 3 limited use clothing) |
| prEN 1512 | Protective clothing against liquid chemicals – Performance requirements for limited use chemical protective clothing or suits with spray-tight connections between different parts of the clothing (type 4 limited use clothing) |
| prEN 1513 | Protective clothing against liquid and solid chemicals – Performance requirements for limited use chemical protective garments providing chemical protection to parts of the body |

| | |
|---|---|
| prEN 13 034 | Protective clothing against liquid chemicals – Performance requirements for chemical protective suits offering limited protective performance against liquid chemicals (type 6 equipment) |
| prEN 13 982-1 | Protective clothing for use against solid particulate chemicals – Part 1: Performance requirements for chemical protective clothing providing protection to the full body against solid particulate chemicals (type 5 clothing) |

Special testing standards

| | |
|---|---|
| EN 386 | Protective clothing – Protection against liquid chemicals – Test method: resistance of materials to penetration by liquids |
| EN 369 | Protective clothing – Protection against liquid chemicals – Test method: resistance of materials to permeation by liquids |
| EN 463 | Protective clothing for use against liquid chemicals – Test method: Determination of resistance to penetration by a jet of liquid (Jet Test) |
| EN 464 | Protective clothing for use against liquid and gaseous chemicals, including aerosols and solid particles – Test method: Determination of leak-tightness of gas-tight suits (internal pressure test) |
| EN 468 | Protective clothing for use against liquid chemicals – Test method: Determination of resistance to penetration by spray (Spray test) |

**Protective clothing for protecting against humidity, wind and cold**

Nonwovens as a component in moisture-barrier assemblies and for thermal insulation.

There are no special testing standards in this case; the procedures for carrying out the tests are covered in the product standards.

| | |
|---|---|
| ENV 342 | Protective clothing – Ensembles for protection against cold |
| ENV 343 | Protective clothing – Protection against foul weather |

**Clothing for protecting against mechanical influences**

Nonwovens as reinforcing and cut-resistant inserts

Product standards

| | |
|---|---|
| EN 381-5 | Protective clothing for users of hand-held chain saws – Part 5: Requirements for leg protectors |
| EN 381-9 | Protective clothing for users of hand-held chain saws – Part 9: Requirements for chain saw protective gaiters |
| EN 412 | Protective aprons for use with hand knives |

Special testing standards

| | |
|---|---|
| EN 381-1 | Protective clothing for users of hand-held chainsaws – Part 1: test rig for testing resistance to cutting by a chainsaw |
| EN 381-2 | Protective clothing for users of hand-held chain saws – Part 2: Test methods for leg protectors |
| EN 381-8 | Protective clothing for users of hand-held chain saws – Part 8: Test methods for chain saw protective gaiters |
| EN 530 | Abrasion resistance of protective clothing material – Test methods |
| EN 863 | Protective clothing – Mechanical properties – Test method: puncture resistance |

**Special protective clothing**

There are no special testing standards in this case; the testing procedures are covered in the product standards.

| | |
|---|---|
| EN 471 | High-visibility warning clothing |
| prEN 1073 | Protective clothing against radioactive contamination – Requirements and testing |

**General requirements of protective clothing**

| | |
|---|---|
| EN 340 | Protective clothing – General requirements |
| EN 420 | General requirements for gloves |
| EN 510 | Specification for protective clothing for use where there is a risk of entanglement with moving parts |
| prEN 1149 | Protective clothing – Electrostatic properties |

In addition to testing the safety/technical characteristics, the textile-physiological, ergonomic and design aspects of protective clothing are being examined more and more. Following on from this, the testing of complete clothing assemblies by carrying out tests on a mannequin is becoming increasingly important. Examples of this include flammability tests, leakage tests and rain tests. These enable the effects of the combination of materials, the design, and the bonding/joining processes to be studied.

## 18.3.5
### Filter fabrics

Every technical process results in by-products and energy being produced, and these should be avoided as much as possible or else released into the environment with the fewest possible impurities present in the carrier medium. The types of emissions include dust and gaseous and solid impurities, and these can be reduced and even eliminated completely by means of filtration.

One of the most important elements of a filtration unit is the filter medium, which has to separate out the impurities from the carrier medium as efficiently as possible. Filters in which separation takes place inside the filter medium are referred to as depth or storage filters, and these are usually disposed of once the fibrous layer has become fully loaded. On the other hand, if a layer of particles is formed on the surface of the filter, these types of filters are known as purifying or surface filters, and can usually be recycled.

When nonwoven fabrics are to be used as filter media for both wet and dry filtration, e.g. for industrial dust removal or for cleaning atmospheric air, in addition to determining the filtration characteristics relevant to each technical process using suitable testing specifications and standards, the end-use-specific textile-physical and general physical parameters should also be determined:

- mass per unit area
- thickness
- gross density
- percentage of pores
- air permeability
- strength and deformation characteristics
- surface characteristics
- behaviour during further processing
- moisture absorption capacity
- burning behaviour
- electrostatic behaviour

Their resistance to other materials should also be determined:
- chemical resistance
- thermal resistance
- biological resistance, e.g. to bacteria

Determining the fractional separation capacity is an important criterion for characterising filter materials. The separation capacity of the filter is determined as a function of the particle size, and the total separation capacity is calculated from this [17]. Unlike the fractional separation capacity, the total separation capacity depends on the particle size distribution in the raw gas and on the types of volumes being studied.

In addition to selecting a suitable particle measuring unit, e.g. an optical particle measuring unit, which must be able to clearly determine the particle size and concentration over the entire distribution range in the test aerosol being used, the procedures for producing the test aerosols and for sampling the test dusts and aerosols must be specified, since these can have an important influence on the test results [18].

Important instructions for carrying out the tests correctly are covered in VDI 3489 (Selection of the particle size measuring unit), in VDI Specification 3491, Sheets 1 to 16 (Terms, definitions and production of the test aerosols), and in VDI Specification 2066, Sheet 1 (Arrangement of the measuring points for measuring dust in flowing gases).

Filter media are classified in relation to their intended end-use. For example, air filters are classified in accordance with DIN 24 185, airborne particle filters in accordance with DIN EN 1822, purifying filters in accordance with VDI 3926, air filters for internal combustion engines and compressors in accordance with ISO 5011, filters for use in vehicle interiors in accordance with DIN 71 460, and air conditioning filters in accordance with DIN EN 779.

Test rigs for carrying out filtration tests are usually very large and expensive and require a range of accessories. In both the research and industrial sectors, suitable materials often have to be selected quickly and subjected to continuous quality

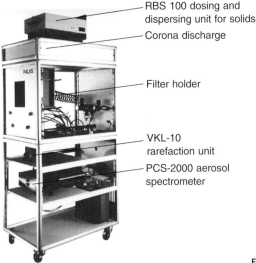

RBS 100 dosing and dispersing unit for solids

Corona discharge

Filter holder

VKL-10 rarefaction unit

PCS-2000 aerosol spectrometer

**Fig. 18-7** MFP-2000 filter testing unit from PALAS® GmbH, Karlsruhe, Germany

control. It is therefore advisable to carry out comparative investigations using a modular system like the one shown in Fig. 18-7, for example.

It is virtually impossible to draw up generally applicable testing methods because of the large number of influencing parameters that exist, and the ways in which they all interact with each other. It is recommended the relevant tests should be carried out to compare the differences in the filtration efficiency of different filter media [19].

| | |
|---|---|
| Method/procedure: | **Particulate air filters for general ventilation – Requirements, testing, marking, testing using atmospheric aerosols and synthetic dust** |
| Standard: | DIN EN 779: 1994-09 |
| Brief description: | Depending on their filtration efficiency, filters can be allocated to a number of groups, which can be further sub-divided into classes: |
| | G1 to G4: Coarse dust filters (starting efficiency, 20%) F5 to F9: Fine dust filters (starting efficiency of between 20 and 98%) |
| | The pressure difference is tested as a starting pressure difference for at least four volume flows (e.g. 50, 75, 100 and 125% of the nominal volume flow). The final pressure difference for classification up to 250 Pa for coarse dust filters and 450 Pa for fine dust filters is also tested. The filtration performance should be tested at a nominal volume flow. This should be set in accordance with the manufacturer's recommendations, or at $0.94 \, \mathrm{m^3/s}$ $(3400 \, \mathrm{m^3/h})$. The efficiency is a measure of the capacity of the test sample to remove atmospheric dust from the test air, and is determined by means of turbidity measurements; the average efficiency is calculated from the starting efficiency. The measure of the capacity of the sample to remove synthetic dust, which has been blown into the test air, is referred to as the separation capacity, and is determined gravimetrically. Coarse dust filters are classified on the basis of their average separation capacity in relation to synthetic dust, and fine dust filters are classified on the basis of their average efficiency in relation to atmospheric dust [20]. |
| Remarks/limitations: | The standard describes the testing process and equipment for determining the filtration performance of particle air filters for use in air conditioning systems, which |

are tested at a volume flow in the range of 0.24 m$^3$/s (850 m$^3$/h) and 1.39 m$^3$/s (5000 m$^3$/h) and whose starting efficiency is less than 98%.

The disadvantage of this testing method is that it provides no information on the efficiency of the filter in relation to specific particle sizes [20].

It should also be noted that the present standard (which was issued in 1993) is currently being revised. The new version of the standard will also cover procedures for determining the fractional separation capacity.

The results of the performance measurements obtained in accordance with this standard cannot be used as a precise indication of the performance of the filter during subsequent use.

| | |
|---|---|
| **Method:** | **High efficiency particulate air filters (HEPA and ULPA)** |
| Standard: | DIN EN 1822: 1998-07, Part 1 to Part 5 |

Part 1: Classification, performance testing, marking
Part 2: Aerosol production, measuring equipment, particle counting statistics
Part 3: Testing flat sheet filter media
Part 4: Determining leakage of filter elements (Scan method)
Part 5: Determining the efficiency of filter elements

Brief description:  EN 1822 covers continuous testing processes using oil droplet aerosols (DEHS or similar) in conjunction with counting of the particles for evaluating the penetration and efficiency. The filter is tested in accordance with the concept of 'Most Penetration Particle Size', i.e. the particle size at which maximum penetration occurs, or at which the filter efficiency is at its lowest. Monodispersed test aerosols, measured using condensation nuclei counters, or polydispersed test aerosols, analysed using high-resolution laser particle spectrometers, are used in the tests [21].

Remarks/limitations:  This standard describes the tests that manufacturers of filter media or filters can carry out in the factory.

| | |
|---|---|
| **Method/procedure:** | **Air filters for motor vehicle passenger compartments – Test procedure for particle filtration** |
| Standard: | Draft E DIN 71 460-1: 1993-06 |
| Brief description: | This draft standard defines the evaluation criteria for selecting and comparing air filters and filter elements under laboratory conditions, and contains the following guidelines relating to: |

– selection of the test dust
– dimensions of the testing channel and position of the sample holder
– introduction of the test dust
– distribution of the test dust in the channel
– sampling of the test specimen and its arrangement in the testing channel
– relative measurement thresholds for the fractional separation capacity
– selective dilution of the fine dust

The tests also provide information on the pressure differences, the separation capacity, the fractional separation capacity, and the dust pick-up capacity of the filter medium [22].

| | |
|---|---|
| Remarks/limitations: | The standard is applicable to air filters used inside motor vehicles for removing dust particles from the internal and external air. Evaluation criteria for use in selecting and comparing air filters and filter elements under laboratory conditions are specified. The filter elements can only be compared directly if they have the same shape and size, and are positioned at the same location in the test channel. |

| | |
|---|---|
| **Method:** | **Testing of filter media for cleanable filters** |
| Standard: | VDI 3926, Sheet 1, issued 1994-12 |
| Brief description: | This specification describes the testing of the filtration behaviour of various filter media on two different testing systems (Type 1 and Type 2), and contains the relevant instructions for carrying out the tests. Comparative tests can be carried out using both testing systems for evaluating the characteristics and quality of cleanable filter media in order to obtain information on their behaviour during operational use. |

Automatic cleaning cycles are carried out under defined operating conditions.

Comparative results obtained by BWF Textil GmbH & Co KG (Type 1) and F.O.S. Filtertechnik GmbH (Type 2) using an improved testing system from the company Palas, gave the following results:

- the pressure difference curves were comparable
- the duration of the cycle with testing system Type 1 was longer in the initial stage. As the number of cycles increased, it was more or less the same for the two testing systems, and the cycle was slightly longer with testing system Type 2 towards the end
- pressure loss in the starting condition (with no dust) Type 1: 0.2 mbar and
  Type 2: 0.5 mbar
- residual pressure loss after cleaning: in the initial stage, it was about 1 mbar lower with testing system Type 1; after 100 cycles, the residual pressure loss was comparable for both testing systems
- the patterns of the pressure difference curves after the first cleaning cycle were different: with Type 1, the curve was S-shaped (after 6 mbar, it was similar to a second order polynomial); with Type 2, it was similar to a second order polynomial as the pressure increased
- the dust emission was higher with testing system Type 2 (especially at the beginning of the measurement sequence)

Similar qualitative test results were obtained with the two testing systems, but they cannot be compared quantitatively [23].

Remarks/limitations:    When publishing and evaluating the data, it should be made clear which of the two testing systems was used in the tests.

## 18.3.6
### Geononwovens

Geosynthetics are divided into water-permeable and water-impermeable products. The water-permeable geosynthetics include geotextiles and geotextile-related products. So woven fabrics, nonwoven fabrics, knitted fabrics and composite materials are all geotextiles.

A whole series of standards and draft standards are either already available, or are in the course of being drafted (Table 18-2). In certain cases, especially where the testing process has not been fully perfected and still requires some discussion, the relevant standards organisations have decided to publish preliminary standards as an interim measure.

**Table 18-2** Published testing processes for geotextiles and related products, situation as at 1999-06

| *Abbreviated title* | | *DIN* | *EN* | *ISO* |
|---|---|---|---|---|
| Geotextiles, terms | | | prEN 30318: 1992-08 | ISO 10318: 1990-11 |
| Identification on the construction site | | DIN EN ISO 10320: 1999-04 | EN ISO 10320: 1999-02 | ISO 10320: 1999-02 |
| Sampling | from the piece | DIN EN 963: 1995-05 | EN 963: 1995-03 | ISO 9862: 1990-08 |
| | from the roll | DIN EN ISO 186: 1996-02 | EN ISO 186: 1996-01 | ISO 186: 1994-11 |
| Mass per unit area | | DIN EN 965: 1995-05 | EN 965: 1995-03 | ISO 9864: 1990-09 |
| Thickness | individual layers | DIN EN 964-1: 1995-05 | prEN 964: 1992-12<br>EN 964-1: 1995-03 | ISO 9863: 1990-10 |
| | multilayer | DIN EN ISO 9863-2: 1996-10 | prEN 964-2: 1995-11<br>EN ISO 9863-2: 1996-08 | ISO 9863-2: 1996-08 |
| Compressibility | | DIN 53885: 1998-12 | | |
| Breaking strength/<br>Breaking elongation | Nonwoven fabrics,<br>50 mm wide | DIN EN 29073-3: 1992-08 | EN 29073-3: 1992-06 | ISO 9073-3: 1989-07 |
| | Woven fabrics, etc.,<br>50 mm wide | DIN EN ISO 13934-1: 1999-04 | EN ISO 13934-1: 1999-02 | ISO 13934-1: 1999-02 |
| | Strips,<br>200 mm wide | DIN EN ISO 10319: 1996-06 | EN ISO 10319: 1996-05 | ISO 10319: 1993-04 |
| | Seams/joins,<br>200 mm | DIN EN ISO 10321: 1996-06 | EN ISO 10321: 1996-05 | ISO 10321: 1992-12 |
| Joins | Geocells | E DIN EN ISO 13426-1: 1998-12 | prEN ISO 13426-1: 1998-08 | ISO/DIS 13426-1: 1998-08 |
| Creep behaviour | Tensile creep | E DIN EN ISO 13431: 1996-01 | prEN ISO 13431: 1998-02 | ISO/FDIS 13431: 1998-02 |
| | Compressive creep | DIN V ENV 1897: 1996-03 | ENV 1897: 1996-01 | |
| Stamping penetration force | | DIN EN ISO 12236: 1996-04 | prEN 776: 1992-05<br>EN ISO 12236: 1996-02 | ISO 12236: 1996-10 |

**Table 18-2** (continued)

| Abbreviated title | | DIN | EN | ISO |
|---|---|---|---|---|
| Falling pyramid test | | DIN V 60500-1: 1999-06 | | |
| Falling ball test | | DIN EN 918: 1996-02 | EN 918: 1995-12 | ISO/DIS 13433: 1996-03 |
| Damage during installation | in granular materials | DIN V ENV ISO 10722-1: 1998-05 | ENV ISO 10722-1: 1998-03 | ISO/TR 10722-1: 1998-03 |
| Control testing of construction site samples | | DIN EN ISO 13437 1998-10 | EN ISO 13437: 1998-08 | ISO 13437: 1998-08 |
| Evaluating the changes in characteristics | | DIN V ENV 12226: 1996-12 | ENV 12226: 1996-10 | |
| Abrasion behaviour | | DIN EN ISO 13427: 1998-10 | EN ISO 13427: 1998-08 | ISO 13427: 1998-08 |
| Resistance to oxidation | | | ENV ISO 13438: 1999-02 | ISO/TR 13438: 1999-02 |
| Resistance to liquid media | | DIN V ENV ISO 12960: 1999-02 | ENV ISO 12960: 1998-11 | ISO/TR 12960: 1998-11 |
| Resistance to hydrolysis | | DIN V ENV 12447: 1997-11 | ENV 12447: 1997-10 | |
| Resistance to microbiological degradation | | DIN V ENV 12225: 1996-12 | ENV 12225: 1996-10 | |
| Weathering resistance | | DIN V ENV 12224: 1996-12 | ENV 12224: 1996-10 | |
| Frictional characteristics | shear box | E DIN EN ISO 12957-1: 1998-04 | prEN ISO 12957-1: 1997-12 | ISO/DIS 12957-1: 1997-12 |
| | slope, plane | E DIN EN ISO 12957-2: 1998-04 | prEN ISO 12957-2: 1997-12 | ISO/DIS 12957-2: 1997-12 |
| Effective aperture width | | E DIN EN ISO 12956: 1996-02 | EN ISO 12956: 1999-02 | ISO 12956: 1999-02 |
| Water permeability $k_V$ | without a load | E DIN EN 12040: 1995-09 | prEN 12040: 1995-07 | |
| | with a load | E DIN 60500-4: 1997-02 | EN ISO 11058: 1999-02 | ISO 11058: 1999-02 |
| Water permeability $k_H$ | parallel | E DIN EN ISO 12958: 1996-02 | EN ISO 12958: 1999-02 | ISO 12958: 1999-02 |
| | radial | E DIN 60500-8: 1997-03 | | |

The textile-physical testing procedures for testing geotextiles are similar to the general testing procedures used for nonwovens. However, the following special cases should be mentioned:

The thickness of the individual layers (DIN EN 964-1) is determined in the same way as described in the normal standards, DIN EN ISO 5084 and DIN EN ISO 9073-2, for testing textiles, but at pressures of 2, 20 and 200 kPa.

For 'normal' tensile tests, i.e. where the width of the sample is 50 mm, the existing standards applicable to textiles should be used.

The tensile test, which uses wide clamps in accordance with DIN EN ISO 10 319 and DIN EN ISO 10 321, i.e. with a sample width of 200 mm, should be used especially for nonwovens and other products having a high cross elongation, and for testing seams.

The stamping penetration force can be used for classifying nonwoven fabrics into so-called 'toughness' classes.

Fig. 18-8 shows an example of a testing unit with the relevant modified tensile tester.

The toughness of the product is characterised by the Geotextile Toughness Class (GTC), and its classification will depend on the results of the strength test and the mass per unit area [24].

Selecting geotextiles with the relevant geotextile toughness class is based on its soil burial behaviour (involving manual or mechanical burial in the ground, compression of the chippings, and walking tests) and the intended end-use (whether it is to be used as a filter or separating layer, taking into account the ground classification of the chipping material).

**Fig. 18-8** Example of a tester for testing the stamping penetration force

DIN EN ISO 13 437 describes a process for burying and digging-up samples, and for testing them subsequently in a laboratory. It contains instructions for burying the control samples, removing them at prescribed time intervals as a function of their intended service life (e.g. every 20 years), and also stipulates which tests should be carried out.

A preliminary standard (DIN V ENV 12 226) describes how to evaluate the changes in characteristics after the resistance tests have been carried out.

A completely new procedure for testing the abrasion behaviour of geotextiles is covered in DIN EN ISO 13 427. In this process, a special abrasion tester is used to test the resistance of woven and nonwoven geotextiles to the effects of abrasion. The decrease in tensile strength is used as a measure of their resistance.

The following points are applicable to the resistance behaviour of geotextiles [24]:

The normal synthetic raw materials used until now are usually resistant to the effects of chemical and micro organisms naturally occurring in the soil and water for the service life of the product. However, proof of the product's resistance must be provided in the following special cases: the resistance of polyethylene and polypropylene to oxidation, and the resistance of polyamide and polyester to hydrolysis.

A preliminary standard (DIN V ENV 12 224) is also available for testing the weathering resistance, and describes a process for carrying out the test using a Global UV testing unit (Fig. 18-9).

The test result (residual strength after weathering) can be divided into 3 categories (high, average and low resistance). Materials with a low resistance have to be covered over or protected after a week, materials with an average resistance have to be covered over/protected after two weeks, and fabrics with a high resistance have to be covered over/protected after two months [24].

**Fig. 18-9** Global UV testing unit from Weiss Umwelttechnik GmbH, Reiskirchen, Germany

The effective aperture width or pore size is used as a measure of the hydraulic filtration efficiency (DIN EN ISO 12 956). This is determined by a wet sieving process using a specific testing surface on the base, and different sizes of sieve. The effective aperture width corresponds to the diameter of the testing surface at which the geotextile holds back 90% of the particles.

In simple cases (low volume of water flow-through and hydrostatic loading), the mechanical filtration efficiency is regarded as being adequate when the effective aperture width lies within the region of $0.06 \, \text{mm} \leq O_{90,w} \leq 0.20 \, \text{mm}$ [24].

The testing process for determining the water permeability under a load specifies that loads of $2 \, \text{kN/m}^2$, $20 \, \text{kN/m}^2$ and $200 \, \text{kN/m}^2$ should be applied. The volume of water flowing through is determined as a function of time. In simple cases, the rule applies that the water permeability of the geotextile filter is adequate if it corresponds at least to 100 times the water permeability value of the soil that has to be drained [24].

Technical Leaflets and Technical Delivery Conditions [24–36] contain information on selecting geosynthetics, as well as details of their requirements' profiles; these are listed in [37].

As far as harmonisation is concerned, i.e. establishing common technical requirements and certification procedures within the EU, attempts are currently being made to introduce the CE Symbol for building products. For example, harmonisation is needed where aspects of safety are concerned [38] and in cases where specific quality requirements have to be met.

With the CE Symbol, the manufacturer is stating that his product meets all the relevant technical requirements relating to the materials, and that the relevant conformity evaluation procedures and other requirements have been met.

In conjunction with the introduction of the CE Symbol, products also have to be classified within different end-use classes in order to provide the user with information on the various relevant, standardised end-use possibilities.

The first draft standards (Tables 18-3 and 18-4) are now available for various end-uses, which have been authorised in accordance with the Construction Products Directive 89/106/EWG [39]; these cover the tests applicable to all the relevant end-uses. Depending on what functions the product has to carry out when used for a particular end-use, the distinction has to be made on the basis of whether the test is

– required in the mandate,
– relevant to all the end-use conditions, but is not prescribed for the classification or specification of products, or
– only relevant to certain end-use conditions.

Specifications for evaluating the test results, for carrying out suitability tests, and for production control in the factory, also allow conclusions to be drawn relating to the conformity of the product. In an appendix to the standards, reference is made to Construction Products Directive 89/106/EWG [39], together with the requirements for its mandate, the system used for certifying conformity, and the requirements for being awarded the CE Symbol.

**Table 18-3** Published product standards for geotextiles and related products, situation as at 1999-06

| Applications | DIN | EN |
|---|---|---|
| Roads, transport surfaces | E DIN EN 13249: 1998-09 | prEN 13249: 1998-06 |
| Railway construction | E DIN EN 13250: 1998-09 | prEN 13250: 1998-06 |
| Earthworks and foundations, support structures | E DIN EN 13251: 1998-09 | prEN 13251: 1998-06 |
| Drainage systems | E DIN EN 13252: 1998-09 | prEN 13252: 1998-06 |
| Erosion protection | E DIN EN 13253: 1998-09 | prEN 13253: 1998-06 |
| Storage basins, dams | E DIN EN 13254: 1998-09 | prEN 13254: 1998-06 |
| Canal construction | E DIN EN 13255: 1998-09 | prEN 13255: 1998-06 |
| Tunnel construction, underground constructions | E DIN EN 13256: 1998-09 | prEN 13256: 1998-06 |
| Removal of solid waste materials | E DIN EN 13257: 1998-09 | prEN 13257: 1998-06 |
| Containment of liquid waste materials | E DIN EN 13265: 1998-09 | prEN 13265: 1998-06 |

**Table 18-4** Published product standards for geomembranes and related products, situation as at 1999-06

| Applications | DIN | EN |
|---|---|---|
| Storage basins, dams | E DIN EN 13361: 1998-12 | prEN 13361: 1998-10 |
| Canal construction | E DIN EN 13362: 1998-12 | prEN 13362: 1998-10 |
| Tunnel construction, underground constructions | E DIN EN 13491: 1999-04 | prEN 13491: 1999-01 |
| Removal of solid waste materials | E DIN EN 13493: 1999-04 | prEN 13492: 1999-01 |
| Containment of liquid waste materials | E DIN EN 13492: 1999-04 | prEN 13493: 1999-01 |

It is also intended that the CE Symbol should act as a quality assurance system. Independent monitoring is recommended for special products, such as reinforcement materials ([24] and DIN 18 200).

The European Union's Directive, 89/106/EWG, which covers construction products (Construction Products Directive) [39], was incorporated into German national legislation under the Building Product Legislation of 10.08.92 [40], and published in the Federal Legislation Leaflet, Part 1 of 14.08.92.

The BauPG-PÜZ Authorisation Directive was set up in order to work with and authorise the various testing, monitoring and certification centres [41]. This should be used whenever harmonised standards for the relevant building products/end-uses are available.

# References to Chapter 18

[1] Bobeth W (1965) Mikroskopisch-chemische Bestimmungsschlüssel für organische und anorganische Chemiefaserstoffe, Deutsche Textiltechnik 15: 1

[2] Koch PA (1952–1993) Faserstofftabellen, Zeitschr fd ges Textilind

[3] Koch PA (1960) Rezeptbuch für Faserstoff-Laboratorien, Springer-Verlag Berlin – Heidelberg – New York

[4] Stratmann M (1969) Methoden der qualitativen Faseranalyse, Zeitschr fd ges Textilind

[5] Koch PA (1951) Mikroskopie der Faserstoffe, Dr. Spohr Verlag

[6] Stratmann M (1973) Die Identifizierung von Faserstoffen mittels chemischer Reaktionen, Handb f Textilingenieure, Dr. Spohr Verlag

[7] Autorenkollektiv (1967) Faserstofflehre, Fachbuchverlag Leipzig

[8] Mahall K (1989) Qualitätsbeurteilungen von Textilien, Verlag Schiele & Schön

[9] Günzler H, Heise HM (1996) IR-Spektroskopie – eine Einführung, VCH-Verlagsges Weinheim

[10] Schmiedel H (1977) Prüfung hochpolymerer Werkstoffe, DeutschER Verlag f Grundstoffindustrie Leipzig

[11] Schröder E, Franz J, Hagen E (1976) Ausgewählte Methoden zur Plastanalytik, Akademie-Verlag Berlin

[12] Hummel O, Scholl F (1978) Atlas der Polymer- und Kunststoffanalyse, 2nd edition, Verlag Chemie, Weinheim, New York

[13] Schröder E, Müller G, Arndt KF (1982) Leitfaden der Polymercharakterisierung, Akademie-Verlag Berlin

[14] Glöckner G (1981) Polymercharakterisierung durch Flüssigkeitschromatographie, Deutscher Verlag d Wissensch Berlin

[15] Rouette HK (1995) Lexikon für Textilveredlung, Laumann-Verlag Dülmen

[16] Pressemitteilung, Öko-Tex Standard 100: Neue Struktur und erweiterte Kriterien, Forschungsinst Hohenstein (1997)

[17] Mölter L (1999) Voraussetzungen zur Reproduzierbarkeit von Meßergebnissen bei der Charakterisierung von Filtern und Abscheidern, Kolloquium Tiefenfilter 1999

[18] Helsper Ch (1995) Probleme der Staubprobenahme bei der Filterprüfung, F & S Filtrieren und Separieren 9, 1: 5

[19] Böttcher P (1995) Textile Filtermedien – Beschaffenheit und Eigenschaften, F & S Filtrieren und Separieren 9, 1: 41

[20] Förster B, Ein neues Zellenmodell zur Bestimmung von Abscheidegrad und Druckverlust der in der Klimatechnik verwendeten Filtermedien, Dissertation Univ GHS Essen, Fachber 12

[21] Horn HG (1999) Leckprüfung und Abscheidegradmessung an hocheffizienten Schwebstofffiltern nach der neuen Prüfnorm EN 1822, Kolloquium Tiefenfilter 1999

[22] Luftfilter für Kraftfahrzeuginnenraum, F & S Filtrieren und Separieren 9, 1: 41 (1995)

[23] Praxisrelevanter Test von abreinigbaren Filtermedien – Technisch verbesserter Prüfstand nach VDI 3926, Typ 2, 11; Palas ATS-Seminar 97

[24] Wilmers W (ed) (1994) Merkblatt für die Anwendung von Geotextilien und Geogittern im Erdbau des Straßenbaus, Forschungsgesellsch f Straßen- und Verkehrswesen, Arbeitsgr Erd- und Grundbau

[25] DBAG-TL 918039, Technische Lieferbedingungen Geokunststoffe, Ausg 1997-04

[26] Empfehlungen für Bewehrungen aus Geokunststoffen – EBGEO, Deutsche Gesellsch f Geotechnik eV (DGGT) (ed),Verlag Ernst & Sohn, Berlin, 1st edition (1997)

[27] Technische Lieferbedingungen für Geotextilien und Geogitter für den Erdbau im Straßenbau (TL-Geotex E-StB), Forschungsges f Straßen- und Verkehrswesen eV, Arbeitsgr Erd- und Grundbau, Köln (1995)

[28] Zusätzliche Technische Vertragsbedingungen und Richtlinien für Erdarbeiten im Straßenbau (ZTVE-StB 94), Bundesministerium für Verkehr, Abt Straßenbau (1994)

[29] Empfehlungen der Arbeitskreise zur „Geotechnik der Deponien und Altlasten: GDA", Deutsche Gesellsch f Geotechnik eV (DGGT) (ed), Verlag Ernst & Sohn, Berlin, 3rd edition (1997)

[30] Merkblatt – Anwendung von geotextilen Filtern an Wasserstraßen (MAG), Verkehrsblatt, Amtlicher Teil (1994) 2: 130

[31] Richtlinien für bautechnische Maßnahmen an Straßen in Wassergewinnungsgebieten (RiStWag), Forschungsgesellsch f Straßen- und Verkehrswesen/Deutscher Verein des Gas- und Wasserfachs/Länderarbeitsgemeinschaft Wasser (1982)

[32] Zitscher FF (ed) (1989) Anwendung und Prüfung von Kunststoffen im Erd- und Wasserbau; Empfehlung des Arbeitskreises 14 der Deutschen Gesellschaft für Erd- und Grundbau eV (DGEG), DVWK-Schrift 76/89, Verlag Paul Parey Hamburg/Berlin, 2nd edition

[33] Zitscher FF (ed) (1992) Anwendung von Geotextilien im Wasserbau, DVWK-Merkblatt 221/92, Verlag Paul Parey Hamburg/Berlin

[34] Zitscher FF (ed) (1992) Anwendung von Kunststoffdichtungsbahnen im Wasserbau und für den Grundwasserschutz, DVWK-Merkblatt 225/92, Verlag Paul Parey Hamburg/Berlin

[35] Zusätzliche Technische Vertragsbedingungen und Richtlinien für den Bau von Straßentunneln, Part 1 – Geschlossene Bauweise (Spritzbetonbauweise), ZTV-Tunnel, Verkehrsblatt – Dokument – No. B 5330, Verkehrsblatt-Verlag Dortmund (1995)

[36] Druckwasserhaltende Abdichtung von Verkehrstunnelbauwerken und anderen Bauwerken mit Doppeldichtungssystemen aus Kunststoffdichtungsbahnen (EDT), Deutsche Gesellsch f Geotechnik eV (DGGT) (ed), Verlag Ernst & Sohn Berlin, 1st edition (1997)

[37] Mägel M (1994) Normen und Richtlinien für Geotextilien und geotextilverwandte Produkte, Techn Textilien 37, 2: T50–T55 (Beilage in Chemiefasern Text Ind 44/96 (1994) 5).

[38] Berghaus H (1991) Europäische Entwicklungen im Bereich der Prüfstellen und Zertifizierung in: Akkreditierung von Prüflaboratorien und Zertifizierungsstellen, Tagungsbericht BDI-Tagung am 21.09.90 in Bonn, editor TGA – Trägergemeinschaft für Akkreditierung GmbH i. G. unter Mitwirkung der Bundesanstalt für Materialforschung u -prüfung (BAM).

[39] Richtlinie des Rates v 21. Dezember 1988 zur Angleichung der Rechts- und Verwaltungsvorschriften der Mitgliedsstaaten über Bauprodukte (89/106/EWG), geändert durch die Richtlinie 93/68/EWG des Rates vom 22. Juli 1993

[40] Gesetz über das Inverkehrbringen von und den freien Warenverkehr mit Bauprodukten zur Umsetzung der Richtlinie 89/106/EWG des Rates v 21. Dezember 1988 zur Angleichung der Rechts- und Verwaltungsvorschriften der Mitgliedsstaaten über Bauprodukte (Bauproduktengesetz – BauPG) vom 10.08.92, Bundesgesetzblatt, Part I 14.08.92 (1992)

[41] Kiehne H (ed) Bauproduktengesetz – Materialsammlung, Beuth Verlag GmbH, Berlin, Wien, Zürich

# 19

# Quality surveillance systems and quality assurance systems

N. RITTER, R. GEBHARDT

Quality-conforming development and production of nonwovens with defined properties requires the quality characteristics to be observed while at the same time assuring and controlling the production parameters. This is called quality assurance and surveillance. By quality it's understood that all the features and characteristic values of a product or a service fulfill fixed and expected requirements with regard to their suitability [1]. It follows from this definition that production should not be as good as possible, but as good as necessary. Customers demand from nonwovens producers to ensure product properties of a necessary and steady quality. The state of quality assurance in the textile industry is described in detail in [2].

In production of nonwovens so far samples were taken during the individual production stages and properties were determined in the laboratory in conformity with prescribed standards (see Chapter 18). The test results are then available at a later date than when production takes place. When there are any deviations, it generally is no longer possible to influence the running production. Conclusions as to necessary modifications of the production parameters can only be drawn by comparison and coordination with the records of the line operators and gathered from the shift reports. These records and offline quality inspections will also be required in the future for documentation of production. Producers have various rules as to how this should be handled in detail, but the method should be clearly laid down in the quality management manual [1].

To an increasing extent, however, quality-conforming production requires direct monitoring of the production process and influencing of the process stages.

Apart from the examination of properties with testing methods that are separated from the production process, online production monitoring makes ever more complex demands on the measuring and testing technology.

Besides the use of new sensors for measuring of characteristics, there are also higher demands made on the testing and evaluation processes. While in the past it were the classic statistical parameters [3] such as mean value, variance of distribution, coefficient of variation, confidence intervals etc., for which documentation was done by means of control charts, statistical process control (SPC) [1] is employed today. The objective is to find out systematically the decisive parameters from among all the process parameters so that the process can be positively influenced by these parameters and the appropriate measures. SPC comprises machine capability testing (MCT), process capability testing (PCT) and process monitoring with quality control charts (QCC).

Accelerated testing of machinery and plants with regard to the fulfillment of specific quality requirements for the given manufacturing task is expressed by cm (capability machine) and machine capability index $cm_k$.

By means of long-term testing all influencing factors of the manufacturing process can be examined with regard to their suitability. The characteristic value cp (capability process) as a ratio of tolerance to process variation allows to determine whether the task can be fulfilled with regard to given quality requirements. The process capability index $cp_k$ makes allowance for the position of the mean value as compared to the given tolerance limits.

Table 19-1 provides a survey of process capability index evaluation.

The application of quality control charts is useful for monitoring and control of a process. Fig. 19-1 shows a quality control chart for the mean value where the warning and intervention limits and the limiting values result from the requested safety values as e.g. confidence interval $2\,\sigma$ or $3\,\sigma$.

So far only important production parameters such as revolutions per minute speeds, strokes etc. were recorded. New acquisition systems, however, are required to provide statements regarding performance of the process values throughout the production period. This is normally combined with the respective visualization of data. In this area measuring and control techniques offer many possibilities.

Online process monitoring allows the acquired data to be immediately evaluated so that the production parameters can be influenced in order to assure the quality specified by the customer at a high productivity and minimized raw material losses.

The following parameters are important for characterization of nonwovens [4]:

– mass per surface unit
– nonwoven thickness
– elasticity and air permeability

**Table 19-1** Survey of process capability evaluation

| cp | Process evaluation | |
| | $cp_K$ | Process |
| --- | --- | --- |
| ≥1.33 | ≥1.33 | capable<br>controlled |
| ≥1.33 | ≥1.33<br>>1.00 | capable<br>partially controlled |
| ≥1.33 | <1.00 | capable<br>not controlled |
| <1.33 | <1.00 | not capable<br>not controlled |

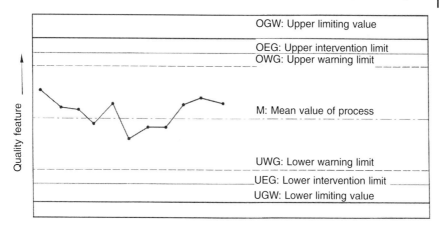

**Fig. 19-1**  Quality control chart for mean value

The following are therefore the relevant parameters for the nonwoven production process:

– variations of mass and/or thickness in relation to the surface
– speeds
– drafts
– parameters for characterization of nonwovens structure
  (air permeability [5], cloudiness etc.)

Any production process involves errors and tolerances. It cannot and will not be the objective of quality assurance to avoid making mistakes or completely eliminate mistakes. Basic knowledge of statistics and calculation of errors, however, can help to reduce the tolerances to the necessary size. Subdivision of nonwovens production processes into sub-processes and their control has positive effects on the entire process. With a possible reduction of the variance of distribution, e.g. during fibre preparation, a reduction of the total variance of distribution can be achieved. Therefore the production of a nonwoven from several web layers allows the variance of the nonwoven to be reduced and the uniformity to be increased. The theoretical considerations are substantiated by practical examinations shown in Fig. 19-2. It is shown that the coefficient of variation and consequently the variance of distribution is reduced with an increasing number of webs n by the factor $1\sqrt{n}$.

According to Böttcher/Kittelmann [4] the production of nonwovens is realized in four main process stages:

– fibre preparation
– web formation
– web bonding
– nonwoven enhancement/finishing

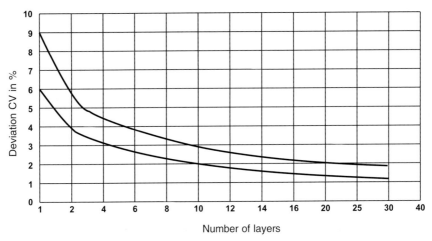

Fig. 19-2   Influence of layer number on coefficient of variation
(from records of Spinnbau GmbH Bremen)

In most cases, the first three stages are combined into a continuous process. Set-point entries and comparison with actual values of process data during the individual process stages and for the entire process form the basis for control of non-wovens production. The individual stages are connected by a process control and monitoring system with a software allowing simple and safe operation.

The following figures show how process control systems work. Figs. 19-3 to 19.5 illustrate graphic menus of Protec GmbH for a double doffer card, a drawing section and a needling machine with two needle boards.

As shown in the schematic survey of Fig. 19-3, the drives of intake, worker/ stripper, drum, doffer, condensing rollers and roller draw-off units as well as draw-off belt can be adjusted separately. In this way, a good distribution of the material can be ensured. Speed setting for the individual roller duos or trios in the drawing section (Fig. 19-4) allows drafting of the respective material depending on the material properties. Apart from the adjustable draft between intake and draw-off unit of the needling machine (Fig. 19-5), the respective needling depth as well as the gap can be automatically adjusted, also during operation. The gap width is maintained constant during adjustment, unless this value shall be changed also.

Fig. 19-6 shows a schematic survey of the water circulation system in the Fleissner AQUAJET spunlacing line.

The water supply pressure to each jet head and the required suction capacity is entered specifically for each product with combined operation of high-pressure pumps being possible.

A coloured status indication of all units, filling levels and operation statuses allows the operator to quickly view the machine status.

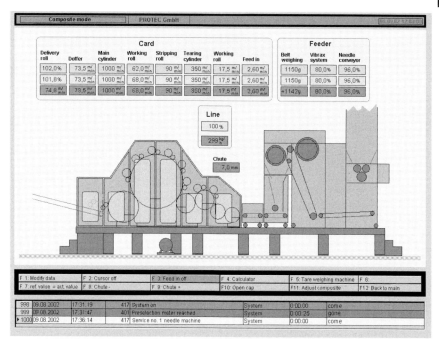

**Fig. 19-3** Schematic survey of card

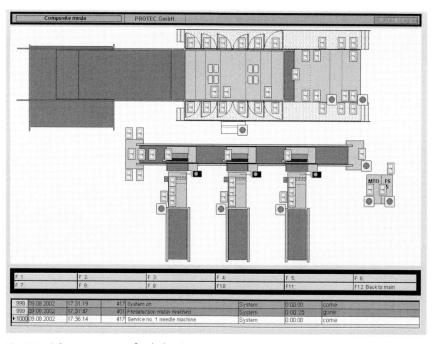

**Fig. 19-4** Schematic survey of web drawing section

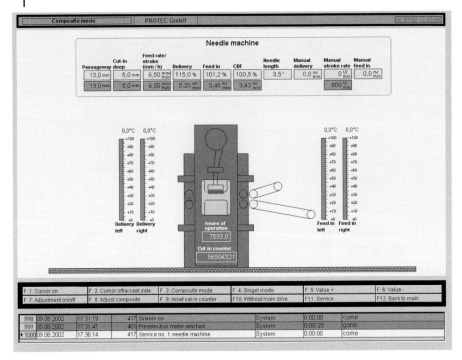

**Fig. 19-5** Schematic survey of a needling machine

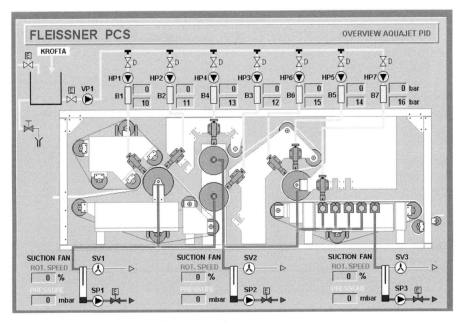

**Fig. 19-6** Schematic survey of a spunlacing line water circuit

Any deviations between setpoint and actual value and any malfunctions in the water circuit are immediately indicated in this schematic survey by a change of colour of the respective units and are at the same time recorded in a protocol.

These extensive data quantities can no longer be efficiently entered or processed by conventional control techniques and discrete operating and indicating elements.

Process control systems offer the following functions:

– full-graphics visualization of machine and its components
– process operation and parameter assignment
– fault indication and logging
– recipe management
– data acquisition and logging
– representation of process values in trend form
– logging of all parameter changes
– storage of fault messages and parameter changes over long periods

Further conditions for efficient project planning, line control and visualization of the characteristic values are modular software systems for the programmable logic controls (PLC).

All current process data as well as status and fault messages are provided through interfaces. These interfaces offer the possibility to Quality Assurance to prepare statistics with regard to failure frequency of certain machine components and consequently work out plans for preventive maintenance. At the same time, production protocols can be prepared for the individual products which serve to prove that the production parameters have been met.

Fig. 19-7 shows the survey menu of the process control system of a Fleissner fibre line. This survey menu shows all essential actual values of the process data such as speeds of draw stands in m/min, loading of drives in amperes (A), temperatures in °C, drawing ratios. By selecting the desired line component, as e.g. crimper unit (Fig. 19-8), a detailed representation of the selected machine can be displayed.

As shown by this example, all operating statuses of the individual machine components can be presented down to the last detail in the detail representation. The motor statuses, for example, are indicated by a change of colour (green=drive running, red=drive at fault, grey=drive deselected, yellow=drive ready for start). The pipes of the heating/cooling circuits change colour when the respective valves are opened. By fading in and out of graphic symbols various operating modes can be displayed in an easily surveyed representation. When malfunctions occur in the line, notes for maintenance and service work can be displayed at the respective points. The process control system used also allows to schedule and request preventive maintenance work and monitor execution of this work.

Recipe management is usually included when using process control systems. Product-specific recipes, which can be selected by clear product names, are used to store all relevant setting data for the process. When changing products, the complete data set of all setpoint parameters can be transmitted to the PLC by simply selecting the respective recipe name. In this way, short change-over times for product changes can be

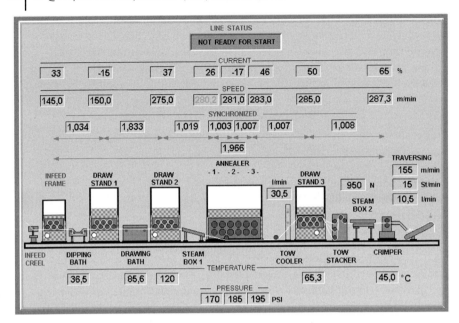

**Fig. 19-7** Schematic survey of a fiber drawing line

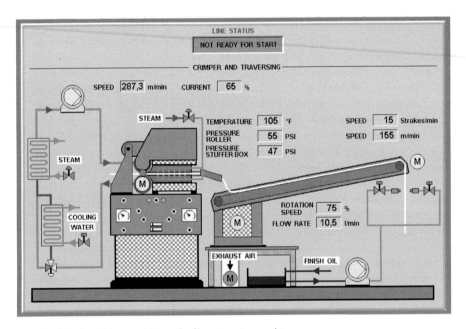

**Fig. 19-8** Detail representation of a fiber crimping machine

achieved and a high reproducibility of technological data for the products is ensured. Any influence of faulty process parameters entered by the operator is largely excluded. Fig. 19-8 is a detail view of a Fleissner fibre crimping machine.

Complete digital processing of all data from operator entry or recipe selection to speed acquisition of converters allows the entire technological process to be monitored and optimally controlled. Any deviations of process values from the setpoints as well as any malfunctions are detected and reported. Any process parameter change by the operator is logged.

The development of process control systems also offered the possibility to determine certain web properties as, for example, surface mass, thickness, air permeability and "internal tension" during production in the nonwovens line by means of suitable sensors. When knowing the relations between these product properties, which can be measured on-line, and the process variables, quality surveillance and quality control during production can be ensured.

With regard to the measuring methods, problems occur in that widths of more than three meters are produced. Suitable on-line measuring systems therefore have to provide reliable measuring values not only along the fabric length, but also across the entire width. This makes increased demands on measuring and control technology.

Simultaneous measurement at selected points distributed across the width is difficult and expensive. Therefore measuring methods are used which move the measuring head across the fabric width. In this process, different characteristics are measured by various methods [7]. Consequently there are also various suppliers of on-line measuring systems. Radiometric measurements for surface mass and dynamic thickness measurements used in production lines are described in [8]. Measuring the air permeability [5] allows to draw conclusions as to surface mass, thickness and tension.

For evaluation of process data, the offered systems, apart from visualization of process data, also comprise mechanisms for process control and influencing.

One of the major criteria for quality of a nonwoven is the surface mass. For delivery of nonwovens, the roll length and thus observance of nonwovens mass is therefore specified apart from quality requirements.

The machine settings are usually influenced when the measured surface mass deviates from the given value by 10% upwards or downwards. If this is not precise enough, the value can be lowered to 5%.

When a value measured on-line exceeds the given limit, a decision must be made:

1. Stopping of process and changing of machine parameters to counteract the change in surface mass. In this case, it has to be noted that exceeding of the limit may be accidental.
2. Monitoring of surface mass mean value and variance over the last period. In this way, accidental deviations can be excluded with a high degree of certainty. In order to avoid measurements over the entire period, the use of a sliding mean value presents itself. This is done by using values of a defined period for formation of a mean value. As a result, trends are more easily detected.

## References to Chapter 19

[1] Hering E, Triemel J, Blank H-P (1993) Qualitätssicherung für Ingenieure, VDI-Verlag, Düsseldorf
[2] Wulfhorst B (1996) Qualitätssicherung in der Textilindustrie: Methoden und Strategien, Hanser-Verlag München
[3] Klemm L, Riehl H-J, Siegel H, Troll W (1974) Statistische Kontrollmethoden in der Textilindustrie, Fachbuchverlag Leipzig
[4] Böttcher P, Kittelmann W (1998) Trends der mechanischen Vliesverfestigung, ITB Vliesstoffe – Techn Textilien 4: 8–16
[5] Eschweiler W (1997) Zuverlässige Online-Messung von technischen Textilien und Vliesstoffen, ITB Vliesstoffe – Techn Textilien 1: 36–38
[6] Schmitt P, Schmitt K (1997) Quality control on line, Edana's 1997 Nordic nonwovens symposium VIII: 1–12
[7] Müller P (1997) Carding plus airlaid, Edana's 1997 Nordic nonwovens symposium, VII: 1–6
[8] Kittelmann W, Brodtka M: On-line-Messung der Qualität in der Nadelvliesproduktion, ITB Vliesstoffe – Techn Textilien 2: 42–46
[9] Leifeld F (1989) Neueste Entwicklungen für die Herstellung von Krempelvliesen, ITB Vliesstoffe – Techn Textilien 1: 26–31
[10] Holliday T (1998) Testing Properties of Nonwovens, Nonwovens Industry 04: 28–30

## 20
# Coming development in nonwovens industry

In the last decades, the manufacture of nonwovens has developed from an industry utilizing all kinds of textile waste to an independent textile branch which complements the classical textile production.

At the same time, the rapid growth of nonwovens indicates products have found a market and shows how wide the range of goods may be. It has turned out that single fibre characteristics constitute the basis for specialization. It is important to see that fibre characteristics are predominant in product and quantity development. The selection of basic substances, fibre fineness and external features of fibres such as crimping, fibre staple, fibre cross-section are essential in the design of nonwovens. That means in practice, the nonwovens industry will encourage the producers of man-made fibres to make fibres with characteristics not called for in other textile branches. The figures available in the different chapters of this book show the permanently growing capacity of nonwoven production. Thus, it should not be a problem to ever better relate the characteristics of single fibre to the final product, up to the manufacture of biodegradable products.

The present division of nonwoven products into disposables and durables is well-proven. In both groups of products further quantity increase can be prognosticated. The quantity increase with regard to the disposables can be explained with the increasing application of hygiene products and with their use in a wide range of countries. As for durable products, further development will focus on groups of products already well-known. There is little doubt such products will be completed by a number of technical nonwovens, e.g. in electronics, in the field of building textiles (inclusive of geotextiles), in interior design, in medicine and in the sector of protective clothing. They will also be used to reinforce matrices in the widest sense of the word.

It is most likely nonwovens in all sectors will lose in weight in the future. On the one hand that means a smaller amount of fibres will be used, on the other hand, fibre fineness will be reduced in most cases. This entails fibres of fineness will be processed which, so far, have rarely been applied and whose processing on industrial scale needs to be solved. This applies to both spunbonded nonwovens and nonwovens manufactured from fibres. To meet all requirements, hetero fibres of a wide variety of compositions and designs will be used. A cross-section, too, as fares hollow profiles will play a growing roll with regard to fibre characteristics, e.g. stiffness. In the context of fibre bonding, the points of bonding between the fibres will become more and more important for nonwoven characteristics. This can be helped by means of selecting the proper polymers, fibre fineness as well as by means of the design of suitable hetero fibres. However, one question which will take further research is how best to bond the fibres creating the nonwoven so as to achieve a better utilization of strength-strain behaviour.

The development of machines to manufacture nonwovens, in particular of machines to form webs, is generally based on the well-known processes of fibre preparation and carding as found in the manufacture of yarn. It is here that novel technologies will come up that will, due to special purpose fibres being used, surpass the current state-of-the-art. Questions of how best to orientate the fibres in the web and how most optimally to route the air flow in the machine will be very important. It may even be possible electric fields help to solve these questions.

While with batch production it will be essential to use highly efficient machines (velocity, working width, integrated steps, process control), it will also be necessary to develop equipment able to make special purpose products. In this context, it makes good sense to think of the manufacture of three-dimensional rather than two-dimensional nonwovens. Novel processes provide ways to make nonwovens other than in roll goods, such as fibre-blowing (fibre injection moulding), spunbonding and above all, the manufacture of nonwovens from films. The authors of this book feel sure there will be a number of developments concerning bonding processes. Hydroentangling now successfully complementing needle-punching technologies, it is of great interest to find out how air or other gaseous media can be effectively used. In the medical sector, chemical substances used to bond nonwovens can only be applied within certain limits. For the medical reasons, cellulose will play a more important role, which goes well together with the trend to allow product degradability. The quantity of such nonwovens continuing to grow in the future, it will be most important to answer the question of how to dispose them.

Nonwoven composites will also grow in importance. Requirements can be met by a wide variety of combinations of the well-known bonding processes. Very particular effects can be achieved if the single elements of different fibres (polymer, fineness, cross-sections) are used together with new technologies to bond both the single layers and the composites.

The choice of final products will also grow. This draws particular attention to practice-related finishing. This concerns chemical means to finish the nonwovens (as far as odour control and auxiliaries) as well as measures to enlarge the surface, e.g. by means of fluffy short fibres. There is still considerable potential to develop finishing processes. In addition to all this, nonwovens designed for the purpose may be excellent carrier materials for a wide range of substances, as far as catalysts.

The above-set is based in previous developments and in the requirements of possible fields of application know today. Moreover, it will be of growing interest to utilize ideas that allow to reduce mass in order to achieve certain characteristics. Research focuses on reinforcement so as to improve properties. Polymer characteristics and large fibre surface will help to achieve this effect.

The nonwoven industry is open to considerable further development. To utilize the potential it is necessary, more than in other branches of the textile industry, that all concerned with the manufacture of nonwovens work together so as to produce articles of optimal quality.

# Index of names, companies, organizations, and products

# Subject index